Algebra and Galois Theories

Régine Douady · Adrien Douady

Algebra and Galois Theories

 Springer

Régine Douady
Université Paris Denis-Diderot
Paris, France

Adrien Douady (1935–2006)
Université Paris-Sud Orsay
Paris, France

Translated by
Urmie Ray
Sceaux, France

ISBN 978-3-030-32798-9   ISBN 978-3-030-32796-5   (eBook)
https://doi.org/10.1007/978-3-030-32796-5

Mathematics Subject Classification (2010): 15-01, 18-01, 55-01, 12F10, 14E20, 14F35, 14H, 14H30

Translation from the French language edition: *Algèbre et théories galoisiennes* by Régine and Adrien Douady, © Published by Cassini 2004. All Rights Reserved.

This Springer imprint is published by the registered company Springer Nature Switzerland AG
The registered company address is: Gewerbestrasse 11, 6330 Cham, Switzerland

Introduction

Similarities between Galois theory and the theory of covering spaces are so striking that algebraists use geometric language to talk of field extensions whereas topologists talk of Galois covers. Here we have tried to develop these theories in parallel, beginning with that of coverings. The reader will thereby be better able to visualize. This similarity can sometimes be found in specific formulations:

(4.5.5) *Proof of the proposition. Let* φ : $S(X) \to S(Y)$ *be a morphism in* G-$\mathcal{S}ets$. *There is a morphism* $\varphi_* : E \times S(X) \to E \times S(Y)$ *defined by* $\varphi_*(t, s) = (t, \varphi(s))$ *corresponding to the morphism* φ. *The morphism* φ_* *is compatible with the G-operation* $\perp$ ...

(5.7.4) *Proof of the proposition. Let* φ : $S(B) \to S(A)$ *be a morphism in* G-$\mathcal{S}ets$. *There is a morphism of* L-*algebras* φ^* : $L^{S(A)} \to L^{S(B)}$ *defined by* $\varphi^*(h) = h \circ \varphi$ *corresponding to* φ. *The homomorphism* φ^* *is compatible with the G-operation* $\perp$ ...

The presentation chosen highlights this resemblance, the intention being to transfer to Galois theory any eventual geometric intuition gained in the framework of coverings. This presentation is unlikely to be the best suited for each individual context. The reader will recognize the influence of Grothendieck... and of Bourbaki. The similarity is not a perfect correspondence: in the algebraic context, profinite groups occur whereas discrete groups do in coverings. In the context of Galois theory, there is nothing comparable to points of a space, nor to loops; the choice of an algebraic closure of a field can only very indirectly be considered as similar to the choice of a base point. In the study of Riemann surfaces in Chapter 6, this similarity takes the shape of an equivalence of categories. This enables us in Section 6.4 to obtain through transcendental means a purely algebraic result without any known purely algebraic proof (Proposition 6.5.6). The end of that chapter deals with the study of automorphisms of Riemann surfaces, and some Fuchsian groups are considered. In each of the Chapters 4, 5, 6, the central theorem is expressed as an equivalence of categories (4.5.3, 4.6.8, 5.7.3, 6.2.4). We have therefore included a preliminary chapter on categories: Chapter 2. To avoid difficulties stemming from the impossibility of considering the set of all sets, our approach is that of universes; so we only consider "small categories", namely, categories whose underlying set is indeed a set. This in no way impacts on our goal as seen in "lemmas about cardinalities" (4.4.7, 5.2.5). Infinite Galois theory forces us to consider sets and profinite groups. Tychonoff's theorem (1.7.7), based on the axiom of choice and Zorn's theorem, is essential in the theory of profinite groups. These theorems are proved in an initial chapter (Chapter 1). In 1.5.3, we mention the places where Zorn's theorem occurs in mathematics. Excepting Sections 2.5.14, 2.5.15, and 2.7.7 and paragraphs 2.8 and 2.9 (profinite spaces and groups), written in view of infinite Galois theory, Chapter 2 is independent from Chapter 1. We have inserted a chapter on rings, modules, and tensor products (Chapter 3). This chapter is broader than strictly needed for the development of Galois theory.

In this new edition, we have added a chapter on Grothendieck's theory of "dessins d'enfants". Through topology (theory of ramified coverings), it reduces problems about the classification of number fields to combinatorics questions.

We would like to thank the readers of the manuscript for their comments and for the improvements they suggested, in particular, Antoine Chambert-Loir and Christian Houzel.

Contents

About the Authors

Régine Douady was born in 1934. After a Ph.D. in the didactics of mathematics, she became a Lecturer at the University Paris Denis-Diderot, as well as the head of the IREM (Institut de Recherche sur l' Enseignement des Mathematiques). She was made a chevalier dans l'ordre des palmes academiques. Now retired, she has ceaselessly endeavored to introduce into teaching the necessity to address concepts from different standpoints—the motivating idea behind this book.

Adrien Douady was born in 1935. After receiving his Ph.D. in mathematics in the field of complex analytic geometry, he later joined the University Paris-Sud (Orsay). A recipient of the Ampere prize, he was a member of the French Academy of Sciences, as well as of the highly influential informal Bourbaki group. Throughout his life, he remained interested in several areas. Yet his odyssey always brought him back to complex numbers. He passed away in 2006 and is notably survived by his wife, Regine Douady.

Chapter 1
Zorn's Lemma

Introduction

The purpose we have in mind is infinite Galois theory. The notion of inverse limits of finite groups will be needed for this. By Tychonoff's theorem, these are compact groups. The example in 6.4.6 shows that countable products are not sufficient. Tychonoff's theorem is required in full, and hence so is Zorn's lemma, and thus the axiom of choice.

This chapter introduces the axiom of choice as well as the two results enabling its application: Zorn's Lemma and Zermelo's theorem. Both play more or less the same role, and either can usually be chosen. In our opinion, Zorn's lemma is more powerful, while Zermelo's theorem sometimes casts a more significant light.

Well-ordered sets form the framework for Zermelo's theorem. After outlining their theory, we proceed with the proof of Zorn's lemma. The quickest way to prove Tychonoff's theorem uses ultrafilters, and this is the method we follow.

1.1 Choice Functions

1.1.1

Definition Let E be a set. Set $\mathfrak{P}^*(E)$ be the set of its non-empty subsets. The **choice function** over E is a map $\tau : \mathfrak{P}^*(E) \to E$ such that $\forall X \in \mathfrak{P}^*(E), \tau(X) \in X$.

© Springer Nature Switzerland AG 2020
R. Douady and A. Douady, *Algebra and Galois Theories*,
https://doi.org/10.1007/978-3-030-32796-5_1

1.1.2

AXIOM OF CHOICE. For any set E, there exists a choice function defined on E.

1.1.3   Commentary

The axiom of choice is now accepted by most mathematicians. This was not always the case: in the early XX-th century, it remained highly controversial. It is only slowly that it was realized that all the axiom does is give a weaker meaning to the expression "there exists" (and to the quantifier ∃), at least weaker than "it can be explicitly constructed".

As Zorn's lemma is equivalent to it, the axiom of choice is a powerful tool which we do not wish to forgo. But given the monsters whose existence it ensures (see for example the Banach-Tarski paradox 1.1.5), the emergence of mathematical contradictions could be feared. Only in 1938 did Gödel prove that there is no such danger: he showed that if (ZF) set theory together with the axiom of choice is inconsistent, then it remains inconsistent without the axiom of choice. For example, let x and y be two real numbers explicitly computable with arbitrary precision. If a large inequality $x \leqslant y$ can be proved with the axiom of choice, it must then be true. Indeed, if their computation showed that $x > y$, this would contradict the axiom of choice. Gödel's theorem would then ensure that a contradiction can be obtained without the axiom of choice.

In 1963, Cohen showed, in the event that the axiom of choice is discarded, the axiom stating that there is no choice function over $\mathbb{R}$ can be added to set theory without making it inconsistent unless it already is so. This addition amounts to giving the expression "there exists" a strong meaning. Cohen's result tells us that an explicit algorithm giving a choice function over $\mathbb{R}$ would result in inconsistent mathematics. Even the stronger statement (see 1.1.4) that every subset of $\mathbb{R}$ is Lebesgue measurable can be accepted.

1.1.4   Example 1: Existence of Non-measurable Sets

A non-measurable set can be constructed by means of a choice function τ over $\mathbb{T} = \mathbb{R}/\mathbb{Z}$:

Let $\alpha \in \mathbb{T}$ be an irrational element. Define $\varpi : \mathbb{T} \to \mathbb{T}$ by $\varpi(x) = \tau(\{x + n\alpha\}_{n \in \mathbb{Z}})$. Set X to be the image of ϖ and R_α the map $t \mapsto t + \alpha$. Then $\mathbb{T} = \bigcup_{n \in \mathbb{Z}} R_\alpha{}^n(X)$, and these sets are disjoint. Hence X is not measurable for the Lebesgue measure μ. Indeed, $\mu(X) = 0$ would imply $\mu(\mathbb{T}) = 0$, and $\mu(X) > 0$ would imply $\mu(\mathbb{T}) = \infty$.

1.1.5 Example 2: The Banach-Tarski Paradox

Theorem (Banach-Tarski) *There exist a finite partition* $(A_i)_{i \in I}$ *of the unit sphere* $S^2 \subset \mathbb{R}^3$, *isometries* $(F_i)_{i \in I}$ *of* S^2 *and a partition of the index set* I *into two subsets* I′ *and* I″, *such that* $F_i(A_i)$ *for* $i \in I'$ *form a partition of* S^2, *as well as for* I″. *The same holds for unit balls.*

See 1.1, Exercise 5, for a proof.

Commentary. This shows that there can be no "finitely additive measure", preserved by rotations, on every subset of S^2, with respect to which the latter has measure 1. Indeed, the union X′ of A_i for $i \in I'$ should have measure 1, similarly for I″, which would yield measure 2 for S^2. By contrast, there are such "measures" on the circle (1.1, Exercise 4), and contradictions arise only if countable additivity is insisted upon. It is now known that for a coherent theory of integration, countable additivity is of prime importance; this was not so obvious at the time (1926).

1.1.6

Proposition *Every surjection* $f : X \to Y$ *between two sets* X *and* Y *has a section, i.e. a map* $\sigma : Y \to X$ *such that* $f \circ \sigma = I_Y$.

Proof Let τ be a choice function on X. The map defined by $\sigma(y) = \tau(f^{-1}(y))$ is a right inverse. $\qquad\square$

Remark The axiom of choice also follows from this proposition. Indeed, let E be a set, and $\Gamma = \{(x, X) \in E \times \mathfrak{P}^*(E) | x \in X\}$. The second projection $\pi : \Gamma \to \mathfrak{P}^*(E)$ is surjective. By the proposition, it has a section of the form $X \mapsto (\tau(X), X)$, and the function τ defined thereby is a choice function. This proposition is therefore equivalent to the axiom of choice.

Exercises 1.1. (Choice Functions)

1.—(a) Construct a choice function on $\mathbb{Q}$, and on $\mathbb{Q}^2$.

(b) Let X and Y be two sets, each equipped with a choice function. Construct a choice function on $X \times Y$.

2.—Construct a choice function on the set of non-empty open subsets of $\mathbb{R}^2$ (i.e, a function τ which associates to all non-empty open subsets U of $\mathbb{R}^2$ a point of U). Do the same for closed subsets. Is it possible to repeat this for the locally closed subsets? for the Borel subsets? (see 1.1.4).

3.—Let E denote the set of $\mathbb{R}$-valued continuous functions on [0, 1], and F the quotient of E by the following equivalence relation:

$$f \sim g \iff f = g \text{ or } f = -g \, .$$

Define a section σ of the canonical quotient map $\chi : E \to F$, i.e. a map $\sigma : F \to E$ such that $\chi \circ \sigma = 1_F$.

Can this be repeated for the set of all functions $[0, 1] \to \mathbb{R}$?

4. (*Finitely additive measure on* $\mathbb{T}$.)—Set $\mathbb{T} = \mathbb{R}/\mathbb{Z}$. For $\alpha \in \mathbb{T}$, let R_α denote the map $x \mapsto x + \alpha$, and R_α^* the map $f \mapsto f \circ R_\alpha$ from the space E of all bounded functions $\mathbb{T} \to \mathbb{R}$ into itself.

(a) Show that the constant function 1 cannot be written as follows:

$$(*) \qquad \left(\sum_1^n f_i - R_{\alpha_i}^* f_i \right) - g \quad \text{with } f_i \in E \text{ and } g \geqslant 0 ;$$

(for such an h, there is an inequality of the form:

$$\sum_{0 \leqslant k_i \leqslant N} h\left(\sum k_i \alpha_i \right) \leqslant C.N^{n-1}).$$

(b) Show that there is a linear form μ on E such that $\mu(h) \leqslant 0$, where h can be expressed as $(*)$, and $\mu(1) = 1$.

(c) For $x \in \mathbb{T}$, write $\mu(X)$ pour $\mu(\chi_X)$, where $\chi_X(x) = 1$ if $x \in X$ and 0 if $x \notin X$. Then μ is a finitely additive measure on $\mathbb{T}$, invariant under all R_α.

5. (*Banach-Tarski Paradox.*)—Let S^2 be the unit sphere of $\mathbb{R}^3$, and G the group SO_3 of direct isometries of $\mathbb{R}^3$, and e its identity element. Define the following equivalence relation on the set of subsets of S^2: $X \sim Y$ if and only if there exist finite partitions $(A_i)_{i \in I}$ and $(B_i)_{i \in I}$ of X and Y respectively (with the same index set I, and a family $(f_i)_{i \in I}$ of elements of G such that for all $i \in I$, $f_i(A_i) = B_i$.

The aim is to show, using the axiom of choice, that there exists a partition of S^2 into two subsets X, Y such that $X \sim Y \sim S^2$.

A. (a) Let $f \in G$ be an irrational rotation (counting the angle in turns), and $E \subset S^2$ be a set intersecting each orbit at most at one point, and only containing fixed points of f. Show that $S^2 - E \sim E$.

(b) Show that for each countable subset $E \subset S^2$, $S^2 - E \sim S^2$.

B. (a) Find a subset Γ in G isomorphic to the "free product" $\mathbb{Z}/2 * \mathbb{Z}/3$ (a sum in the category of groups (2.6.2, Example 3)), i.e. two rotations a and b satisfying relations $a^2 = b^3 = 1$, and not satisfying any relation deducible from these.

(b) Show that $G = A \sqcup B \sqcup C$ so that

$$aA = B \cup C, \quad bA = B, \quad b^2 A = C .$$

The elements of Γ can be written as words using the letters a, b, b^2 using the following rule contingent on the parity of the length n of the word and on its first letter: $e \in A$, and

$$\text{parity de } n \begin{cases} 0 \quad \text{first letter} \begin{cases} a & A \\ b & A \\ b^2 & B \end{cases} \\ 1 \quad \text{first letter} \begin{cases} a & C \\ b & B \\ b^2 & C \end{cases} \end{cases}$$

C. Let E be the set of fixed points of Γ, and set $\Omega = S^2 - E$.

(a) Show that Γ acts freely on Ω.

(b) Using the axiom of choice, show that there is a partition of Ω into three sets $\mathscr{A}, \mathscr{B}, \mathscr{C}$ satisfying $\mathscr{A} \sim \mathscr{B} \sim \mathscr{C} \sim \mathscr{B} \cup \mathscr{C}$.

(c) Show that the sets S^2, Ω, $\mathscr{A}$, $\mathscr{B}$, $\mathscr{C}$ are all equivalent.

D. Prove the Banach-Tarski theorem (1.1.5).

1.2 Well-Ordered Sets

1.2.1 Ordered Sets

An **ordered set** is a set E equipped with an **order relation**, i.e. a reflexive, transitive and antisymmetric binary relation, denoted by $x \leqslant y$, in other words, satisfying:

(O$_1$) (reflexivity) $\forall x \in E,\ x \leqslant x$;

(O$_2$) (transitivity) $\forall x \in E, \forall y \in E, \forall z \in E,\ (x \leqslant y \text{ and } y \leqslant z) \Rightarrow x \leqslant z$;

(O$_3$) (antisymmetry) $\forall x \in E, \forall y \in E, (x \leqslant y \text{ and } y \leqslant x) \Rightarrow x = y$.

An ordered set is said to be *totally ordered* if its order relation satisfies:

(O$_4$) $\forall x \in E,\ \forall y \in E,\ x \leqslant y \text{ or } y \leqslant x$.

A *chain* in an ordered set E is a totally ordered subset X, endowed with the induced order.

1.2.2 Upper and Lower Bounds

Let E be an ordered set, X a subset of E and $a \in E$. The element a is said to be an *upper bound* of X if, $\forall x \in X, a \geqslant x$. The element a is said to be *the greatest element* (resp. a *strict upper bound*) of X if a is an upper bound of X and $a \in X$ (resp. $a \notin X$). *Lower bounds, least element* and *strict lower bounds* are defined likewise. The element a is said to be the *supremum* (resp. *infimum*) of X and is written $a = \sup X$ (resp. $a = \inf X$) if a is the least upper bound (resp. the greatest lower bound) of X. The subset X is an *upper set* of E if

$$\forall x \in X, \ \forall y \in E, \quad y \leqslant x \Rightarrow y \in X.$$

Lower sets are defined likewise.

A subset Y of E is said to be *cofinal* if every element of E is bounded above by an element of Y.

1.2.3

Definition A **well-ordered** set E is an ordered set all of whose non-empty subsets have a least element.

A well-order is a total order. Indeed, for $x, y \in E$, the set $\{x, y\}$ must have a least element.

1.2.4 *Examples*

(1) A totally ordered set is well-ordered.

(2) $\mathbb{N}$ is well-ordered.

(3) $\overline{\mathbb{N}} = \mathbb{N} \cup \{\infty\}$ is well-ordered. More generally, if E is a well-ordered set, then so is the set $\overline{E}$ obtained by adjoining an element $\omega \notin E$ to E such that $\forall x \in E, \omega > x$.

(4) Let A and B be two well-ordered sets with non-trivial intersection. Equip $E = A \cup B$ with the order inducing the given orders on A and B, and such that $\forall x \in A, \forall y \in B, y > x$. The set E is well-ordered. Denote it by $A + B$. Note that $A + B$ is not necessarily isomorphic to $B + A$.

(5) Every subset of a well-ordered set is well-ordered by the induced order.

(6) The image in an ordered set of a well-ordered set under an increasing map is well-ordered.

(7) The *lexicographic* order on $\mathbb{N} \times \mathbb{N}$ is defined by

$$(x, y) \leqslant (x', y') \Longleftrightarrow x < x' \text{ or } (x = x' \text{ and } y \leqslant y').$$

Equipped with this order, $\mathbb{N} \times \mathbb{N}$ is well-ordered. More generally, the lexicographic order can be defined on the product of two arbitrary ordered sets. The product of two ordered sets, equipped with the lexicographic order, is well-ordered.

1.2.5

Proposition *Let* E *be a well-ordered set,* A *and* B *two upper sets of* E *and* f *an isomorphism from* A *onto* B. *Then* $A = B$ *and* $f = 1_A$.

Proof Suppose this is false. Let x be the least element of A such that $f(x) \neq x$. Write S_x for the set of strict lower bounds of x. Then x is the least element of $E - S_x$, and so is $f(x)$, and so $f(x) = x$; a contradiction. $\qquad\square$

1.2.6

Proposition *Let* E *and* F *be two well-ordered sets. Then at least one of the following statements holds:*

(i) E *is isomorphic to an upper set of* F*;*
(ii) F *is isomorphic to an upper set of* E*.*

Proof Let Γ be the set of isomorphisms from an upper set of E onto an upper set of F. Its order by extension (i.e. inclusion order on graphs) is a total order. Indeed, E being totally ordered, if A and A$'$ are two upper sets, then $A \subset A'$ or $A' \subset A$. Let $f : A \to B$ and $f' : A' \to B'$ be two elements of Γ, where $A \subset A'$ and $f'|_A \neq f$. Writing x_0 for the least $x \in A$ such that $f'(x) \neq f(x)$,

$$f(x_0) = \inf(F - f[0, x_0[) = \inf(F - f'[0, x_0[) = f'(x_0),$$

a contradiction.

The union of graphs of elements $g : A_g \to B_g$ of Γ is again the graph of an element $\bar{g} \in \Gamma$. Indeed, setting $\overline{A} = \bigcup A_g$ and $\overline{B} = \bigcup B_g$, for all $x \in \overline{A}$ there exists $g \in \Gamma$ such that $x \in A_g$, and the element $\bar{g}(x) = g(x)$ does not depend on the choice of g; The map $\bar{g}$ thus defined is clearly strictly increasing and its image is $\overline{B}$.

Let us prove by contradiction that $\overline{A} = E$ or $\overline{B} = F$. Otherwise, let $x = \inf(E - \overline{A})$ and $y = \inf(F - \overline{B})$. The map $\bar{\bar{g}} : \overline{A} \cup \{x\} \to \overline{B} \cup \{y\}$ which extends $\bar{g}$ by $\bar{\bar{g}}(x) = y$ is an element of Γ strictly extending $\bar{g}$, a contradiction. $\qquad\square$

Exercises 1.2. (Well-ordered Sets)

1. (*Farey Isomorphism.*)—Given two rationals $x_1 = p_1/q_1$ and $x_2 = p_2/q_2$, write $M(x_1, x_2)$ *the Farey mean* $(p_1 + p_2)/(q_1 + q_2)$. Let $\mathbf{D}_2$ denote the set of dyadic numbers, i.e. of the form $m/2^k$, and I the interval $[0, 1]$. Define $\varphi(x)$ for $x = m/2^k \in \mathbf{D}_2 \cap I$ by recursion on k by setting:

$$\varphi(0) = 0, \quad \varphi(1) = 1, \quad \varphi((2m + 1)/2^{k+1}) = M(\varphi(m/2^k), \varphi((m + 1)/2^k)).$$

Show that this gives an isomorphism of well-ordered sets from $\mathbf{D}_2 \cap I$ onto $\mathbb{Q} \cap I$.

Show that φ can be extended to a homeomorphism Φ of I. Is the function Φ Hölderian (i.e. does it satisfy an equality of the form $|\Phi(y) - \Phi(x)| \leqslant c.|y - x|^\beta$)?

2.—Let I be an ordered set and $(E_i)_{i \in I}$ a family of ordered sets. Let $E = \bigsqcup_{i \in I} E_i$ be the set of pairs (i, x) such that $i \in I$ and $x \in E_i$ (summation set, or disjoint union).

Define an order on E by $(i, x) \leqslant (j, y) \Longleftrightarrow i < j$ or $(i = j)$ and $x \leqslant y$. Show that if I and all E_i are well ordered, then so is E.

3.—Fix and integer k, and let A_k the set of numbers of type

$$a_0 - \cfrac{1}{a_1 - \cfrac{1}{a_2 - \cfrac{1}{\cdots - \cfrac{1}{a_k}}}}$$

where the a_i are integers $\geqslant 2$. Show that A_k is a well-ordered set (with respect to the order induced by $\mathbb{R}$), isomorphic to lexicographic $\mathbb{N}^{k+1}$. Is the set $\bigcup_k A_k$ well-ordered?

4.—(a) Show that a well-ordered set extends to $\mathbb{R}$ (with the induced order) if and only if it is countable. In this case, can it always be embedded into $\mathbb{Q}$?

(b) Let A be a well-ordered subset of $\mathbb{R}$, show that its closure $\overline{A}$ with respect to the topology of $\mathbb{R}$ is well-ordered.

5.—Let E be a well-ordered set. For every $x \in E$, denote by S_x the set of strict lower bounds of x in E.

(a) *The principle of transfinite induction*: Let F be a subset of E. Suppose that

$$(\forall x \in E) \quad (S_x \subset F \Rightarrow x \in F).$$

Show that $F = E$.

(b) *Construction by transfinite induction*: Let U be a set, Γ the set of maps from the upper sets of E to U and φ a map from Γ to U.

Show that there exists a unique map $f : E \to U$ such that for all $x \in E$, $f(x) = \varphi(f|_{S_x})$. (The proof given should be formally correct).

(c) *Construction by transfinite induction with choice and stopping*: let Φ be a map from Γ to $\mathfrak{P}(U)$. Show that there is an upper set J of E and a map f from J to U such that the following hold simultaneously

(i) $(\forall j \in J) \quad f(j) \in \Phi(f|_{S_j})$;
(ii) $J = E$ or $\Phi(f) = \varnothing$.

6.—Let U be a set. Denote by Ω the set of the isomorphism classes of well-ordered sets whose underlying set is contained in U. Equip Ω with the order defined by: $\alpha \leqslant \beta$ if α has as representative the upper set of a representative of β. Show that this gives a well-ordering of Ω.

7.—Let X be an uncountable set, denote by E the set of finite subsets of X. Show that:

(a) every totally ordered subset of E is well-ordered;
(b) there are no totally ordered cofinal subsets.

8.—Let I be a well-ordered set and $(E_i)_{i \in I}$ a family of totally ordered sets. The **lexicographic product** of the sets E_i is the set $E = \prod_{i \in I} E_i$ equipped with the following order relation: "$(x_i)_{i \in I} \leqslant (y_i)_{i \in I}$" if $(x_i)_{i \in I} = (y_i)_{i \in I}$ or $(x_i) \neq (y_i)$ and $x_{i_0} < y_{i_0}$, where $i_0 = \inf\{i \mid x_i \neq y_i\}$. Show that:

(a) E is totally ordered.

(b) If the sets E_i are well-ordered, then E is not necessarily well-ordered.

(c) Suppose that I is countable and that each E_i satisfies the following property: every well-ordered subset of E_i is countable. Then so does E.

(d) As an ordered set, the lexicographic product $\mathbb{N}^{\mathbb{N}}$ is isomorphic to $\mathbb{R}_+$.

9.—Show that the lexicographic product $\mathbb{R} \times \mathbb{R}$ is not embedded in $\mathbb{R}$. For this, show that $\mathbb{R}$ satisfies the following property: if Λ is a totally ordered set, $(x_\lambda)_{\lambda \in \Lambda}$ and $(y_\lambda)_{\lambda \in \Lambda}$ increasing families of real numbers such that $(\forall \lambda \in \Lambda)\ x_\lambda < y_\lambda$ and $(\forall \lambda \in \Lambda)\ (\forall \mu \in \Lambda)\ \lambda < \mu \Rightarrow y_\lambda < x_\mu$, then Λ is countable.

10.—Let K be a compact space and $f : K \to \mathbb{R}$ a continuous map. Suppose that f has a local maximum at some point of K. Show that $f(K)$ is well-ordered (with respect to the order induced by that of $\mathbb{R}$). Deduce that it is countable.

1.3 τ-Chains

1.3.1

Definition Let E be a well-ordered set, τ a choice function on E and A a subset of E. The set A is said to be a τ-**chain** if, for any upper set $C(\neq A)$ of A, the least element of the set $A - C$ is $\tau(M_C)$, where M_C is the set of strict upper bounds of C in E.

Remark If a τ-chain A is not empty, then its least element is $\tau(E)$. If moreover A does not reduce to $\tau(E)$, then the least element of $A - \tau(E)$ is $\tau\left(M_{\{\tau(E)\}}\right)$.

The remark shows that τ-chains need to satisfy strong conditions. This will be addressed in greater detail in 1.3.4.

1.3.2 Example

Let τ be a choice function defined on the ordered set $\mathbb{R}$. There is a unique τ-chain $(a_0, a_1, \ldots, a_n, \ldots)$ of $\mathbb{R}$ isomorphic to $\mathbb{N}$: it is defined by

$$a_0 = \tau(\mathbb{R}), \ a_1 = \tau(]a_0, +\infty[), \ldots, \ a_n = \tau(]a_{n-1}, +\infty[), \ldots$$

1.3.3

Proposition *Let* E *be an ordered set and* τ *a choice function on* E. *Then every* τ-*chain of* E *is well-ordered.*

Proof Let A be a τ-chain of E and X a non-empty subset of A. We show that X has a least element. The set C of strict lower bounds of X in A is an upper set of A, and C is distinct from A since A $-$ C contains the non-empty subset X. Since A is a τ-chain, $x_0 = \tau(M_C)$ is the least element of A $-$ C, and so x_0 is a lower bound of X, but not a strict one. Consequently, x_0 is the least element of X and A is well-ordered. $\square$

1.3.4

Proposition *Let* E *be an ordered set,* τ *a choice function,* A *and* A' *two* τ-*chains. Then, either* A *is an upper set of* A', *or* A' *is an upper set of* A.

Proof Denote by C the union of upper sets common to A and A': it is the largest among them. Suppose that C $\neq$ A and C $\neq$ A'. Then $\tau(M_C) = \inf(A - C) = \inf(A' - C)$ necessarily holds. But then C $\cup \{\tau(M_C)\}$ is an upper set common to A and A' strictly containing C, contradicting the definition of C, and so C $=$ A or C $=$ A', $\square$

1.3.5

Proposition *Let* E *be an ordered set and* τ *a choice function on* E.
   (a) *The union* $\overline{A}$ *of all* τ-*chains of* E *is a* τ-*chain.*
   (b) *The* τ-*chain* $\overline{A}$ *has no strict upper bound.*

Proof (a) Let C be an upper set of $\overline{A}$ distinct from $\overline{A}$, and $x \in \overline{A} - C$. There is a τ-chain A containing x. We show that C is an upper set of A. It suffices to show that C is contained in A. Let $y \in C$ and B a τ-chain containing y. By 1.3.4, either B is an upper set of A and $y \in A$, or A is an upper set of B. The elements x and y belong to the totally ordered set B. Since $y \in C$ and $x \notin C$, it necessarily follows that $y < x$, and if $x \in A$ then $y \in A$.

   (b) Suppose that the set X of strict upper bounds of $\overline{A}$ is not empty. Adjoining $\tau(X)$ to $\overline{A}$ gives a τ-chain of E strictly containing $\overline{A}$, which is impossible. Hence X is empty. $\square$

Exercises 1.3. (τ-chains)
1.—Let E be a set and τ a choice function on E. There is a well-order on E such that, for any non-empty subset X of E, $\tau(X)$ is the least element of X if and only if for all subsets X and Y of E with Y $\subset$ X and $\tau(X) \in$ Y, $\tau(Y) = \tau(X)$.

2.—Show that every totally ordered set has a well-ordered cofinal subset.

3. (*Direct proof of Zermelo's theorem.*)—Let E be a set, $\tau : \mathfrak{P}(E) - \{\varnothing\} \to E$ a choice function on E. In this exercise, by a τ-chain, we will mean a pair (F, ω) where F is a subset of E and ω an order on F such that for any upper set $A \neq F$ of F, $\tau(E - A)$ is the least element of $F - A$. (This notion is not the same as in the text since E is not ordered.) Show that:

(a) if (F, ω) is a τ-chain, F is well-ordered by ω;

(b) every upper set of a τ-chain equipped with the induced order is a τ-chain;

(c) if A and A' are two τ-chains, then one of the them is the upper set of the other with respect to the induced order;

(d) let (F_i, ω_i) be the family of all τ-chains. There is a unique order ω on $F = \bigcup F_i$ inducing ω_i on each F_i. The pair (F, ω) is a τ-chain and $F = E$.

Deduce that E is well-ordered.

1.4   Inductive Sets and Zorn's Lemma

1.4.1

Definition An ordered set is **inductive** if every ordered subset is bounded above.

1.4.2   Examples

(0) The empty set is not inductive: the empty chain is not bounded above.

(1) The set of subsets of a set ordered by inclusion is inductive.

(2) Let E be an inductive set. The set of upper bounds of any $x \in E$ is inductive.

(3) Let X be a set, E a non-empty subset of the set of its subsets. The set E is said to be **of finite character** if for any subset A of X the following conditions are equivalent:

(i) $A \in E$;

(ii) For any $B \subseteq A$, $B \in E$.

Examples of inductive sets are given by the next result:

Every set E of finite character ordered by inclusion is inductive.

Indeed, let $(A_i)_{i \in I}$ be a chain of E, set $A = \bigcup_{i \in I} A_i$. Any finite subset B of A is contained in some A_i, hence in E.

Example of sets of finite character: the set $\Gamma(X, Y)$ of graphs of maps from subsets of a set X to a set Y is a subset of $\mathfrak{P}(X \times Y)$ of finite character. Indeed, for any $A \subseteq X \times Y$, the following properties are equivalent:

(i) $A \in \Gamma(X, Y)$;

(ii) the projection $\pi_A : A \to X$ is injective;

(iii) For any finite $B \subseteq A$, the projection $\pi_B : B \to X$ is injective.

1.4.3

> **Theorem** (ZORN'S LEMMA) *Every inductive ordered set has a maximal element.*

Proof Let E be an inductive ordered set, τ a choice function on E. The union $\overline{A}$ of all τ-chains of E has an upper bound m. The element m is maximal in E. Indeed, if m was strictly bounded above by m', so would $\overline{A}$, which is impossible by 1.3.5 (b). $\square$

1.4.4

Corollary *Let* E *an inductive ordered set. Every element of* E *is bounded above by a maximal element of* E.

Indeed, the set of upper bounds of a point x of E is inductive, and so has a maximal element which is maximal in E.

Exercises 1.4. (Inductive sets and Zorn's lemma)

1. (*Path- and arc-connectedness.*)—Let I denote the interval $[0, 1]$. Let X be a Hausdorff topological space, a and b points of X. Suppose there is a path from a to b in X, i.e. a continuous map $\gamma : I \to X$ such that $\gamma(0) = a$ and $\gamma(1) = b$. The aim is to show that there is an injective path $\eta : I \to X$ from a to b.

(a) Consider the set Ω of open subsets U of I not containing 0 or 1 and such that, for any connected component $J =]s, s'[$ of U, $\gamma(s) = \gamma(s')$. Show that Ω ordered by inclusion is inductive.

(b) Show that, if the open subset W of I is a maximal element of Ω, the only isolated points of the closed subset $I - W$ are 0 or 1.

(c) Let U be an open subset of I for which 0 or 1 are the only isolated points of $I - U$. Show that there is an increasing continuous function $h : I \to I$, constant on each connected component of U, satisfying the following condition: if $h(t) = h(t')$ with $t \neq t'$, then there is a connected component J of U such that t and t' belong to the closure of J.

(d) Conclude.

1.5 Applications of Zorn's Lemma

1.5.1

Theorem *Let* X *and* Y *be two sets. There is an injection from* X *into* Y *or an injection from* Y *into* X.

Proof Let E be the set of graphs of bijections from subsets of X onto subsets of Y ordered by inclusion. It is also the set of subsets G of $X \times Y$ such that the projections $\pi_1 : G \to X$ and $\pi_2 : G \to Y$ are injective. The set E is of finite character and so is inductive. By Zorn's lemma, E has as maximal element the graph H of a bijection from a subset A of X onto a subset B of Y. Then $A = X$ or $B = Y$. Otherwise $H \cup \{x, y\}$, where $x \in X - A$ and $y \in Y - B$, would be a strict upper bound of H in E. $\square$

1.5.2

Theorem (ZERMELO'S THEOREM) *Every set can be well-ordered.*

Proof Let E be a set, and Ω denote the set of pairs (F, ω), where $F \subseteq E$ is well-ordered by ω. The set Ω can be ordered as follows: "$(F, \omega) \leqslant (F', \omega')$" if F is an upper set of F' with respect to ω' and if $\omega = \omega'|_F$. With this order, Ω is inductive. Indeed, let $(F_i, \omega_i)_{i \in I}$ be a totally ordered family of elements of Ω. There is an order ω on $F = \bigcup F_i$ such that, for all $i \in I$, $\omega|_{F_i} = \omega_i$. Equipped with ω, the set F is well-ordered. Let $(\overline{F}, \bar{\omega})$ be a maximal element of Ω. We show by contradiction that $\overline{F} = E$. Let $a \in E - \overline{F}$. Let ω' be the order on $F' = \overline{F} \cup \{a\}$ inducing $\bar{\omega}$ on $\overline{F}$ and such that $\forall x \in \overline{F}, a > x$. Then (F', ω') is a strict upper bound of $(\overline{F}, \bar{\omega})$, contradicting the maximality of $(\overline{F}, \bar{\omega})$. $\square$

1.5.3

Zorn's lemma is also applied to prove the following results:

1. Every commutative ring $A \neq 0$ contains an ideal $\mathfrak{m}$ such that $A/\mathfrak{m}$ is a field. (Krull's Theorem, 3.1.5.)

2. Let A be a commutative ring. The intersection of prime ideals of A is the set of nilpotent elements of A (3.1.6).

3. Every vector space has a basis (3.4.2).

4. Let A be a principal ring. Every submodule of a free A-module is free (3.5.1). Every divisible A-module is injective (3.10.11, Corollary, b).

5. Every product of compact spaces is compact. (Tychonoff's Theorem, 1.7.7.)

6. Every open cover of a paracompact space has a shrinking (N. Bourbaki, *General Topology, Chap. 5 §10* [1], Chap. 9, §4, cor. 1).

7. Let E be a normed vector space over $\mathbb{R}$ or $\mathbb{C}$ and $x \in E$ with norm 1. There is a linear form ξ on E such that $\|\xi\| = 1$ and $\xi(x) = 1$. (Hahn-Banach Theorem.)

Exercises 1.5 (Applications of Zorn's lemma)
1. (*Transfinite line.*)—(a) Show that there is a well-ordered set Ω, unique up to isomorphism, such that Ω is not countable but every upper set other than Ω is countable.

(b) The *transfinite line* (or rather transfinite half-line) is the set $L = \Omega \times [0, 1[$ ordered by the lexicographic product. Endow L with the order topology: a fundamental system of neighbourhoods of x is given by the intervals $]x', x''[$ where $x' < x < x''$. Set 0 to be the least element $(0, 0)$ and $L^* = L - \{0\}$. Show that, for all $x \in L$, the interval $[0, x]$ is homeomorphic to $[0, 1]$, but that L is not homeomorphic to $\mathbb{R}_+$. Show that L is neither a metric space, nor the countable union of compact spaces. Show that any increasing function $f : L \to \mathbb{R}$ is stationary (i.e. $(\exists a)\ (\forall x > a)\ f(x) = f(a)$).

(c) Show that every increasing sequence in L converges. Show that L is not compact, but that every sequence in L has a convergent subsequence.

Show that, if A and B are two disjoint closed subsets of L, then one of them is bounded (suppose this does not hold and construct an increasing sequence (x_n) such that $x_n \in A$ for odd n and $x_n \in B$ for even n). Deduce that L is normal (two disjoint closed subsets have disjoint neighbourhoods). Show that every continuous function $f : L \to \mathbb{R}$ is stationary.

(d) Show that there is a continuous function $f : L \to L$ such that $(\forall x)\ f(x) \geqslant x$ and such that the set of x satisfying $f(x) > x$ is unbounded, but that there is no such function f with $f(x) > x$ for all x. Show that every continuous map $f : L \to L$ such that $(\forall x)\ f(x) < x$ is stationary.

(e) Set $M = \{(x, y) \in L^2 \mid 0 < x < y\}$. Show that M, endowed with the projection $(x, y) \to x$, is a non-trivial fiber bundle (4.1.7) with fiber $\mathbb{R}$ over L^*, but that its restriction to any compact subspace of L^* is a trivial bundle.

Show that L is not contractible.

(f) Show that L^* has a structure of a real-analytic manifold (assume that every real-analytic manifold homeomorphic to $\mathbb{R}$ is isomorphic to $\mathbb{R}$).

1.6   Noetherian Ordered Sets

1.6.1

Proposition and Definition *Let* E *be an ordered set. The following properties are equivalent:*

(i) every increasing sequence of elements of E *is stationary;*
(ii) Every non-empty subset X *of* E *has a maximal element in* X.

If these properties hold, the ordered set E *is said to be* **Noetherian.**

Proof (ii) $\Rightarrow$ (i): if (x_n) is an increasing sequence of elements of E, then the set of these elements has a maximal element x_{n_0}; for $n \geqslant n_0$, $x_n \geqslant x_{n_0}$ and so $x_n = x_{n_0}$.

(i) $\Rightarrow$ (ii): Suppose that the nonempty subset X of E has no maximal element and let τ be a choice function on X. Define a sequence of elements $(x_n) \in X$ inductively by setting $x_0 = \tau(X)$ and $x_{n+1} = \tau(M(x_n))$, where $M(x_n)$ is the set of strict upper bounds for x_n in X. The sequence (x_n) is strictly increasing, a contradiction. $\qquad\square$

An ordered set will be said to be *co-Noetherian* if it is Noetherian with respect to the dual order.

Example The ordered set $\mathbb{N}$ is co-Noetherian. More generally, every well-ordered set is co-Noetherian.

1.6.2

Proposition *Let* E *be an ordered set. If there is a strictly increasing map* $\varphi : E \to \mathbb{N}$ *(i.e. such that $x < y \Rightarrow \varphi(x) < \varphi(y)$), then* E *is co-Noetherian.*

Proof If there were a strictly decreasing sequence $(x_n)_{n\in\mathbb{N}}$ in E, then the sequence $(\varphi(x_n))$ would be strictly decreasing in $\mathbb{N}$, which is impossible. $\qquad\square$

Examples (1) Let X be a set. The set of finite subsets of X is co-Noetherian with respect to inclusion.

(2) Let E be a vector space. The set of finite dimensional vector subspaces of E is co-Noetherian with respect to inclusion.

1.6.3

Proposition (Principle of Noetherian Induction.) *Let* E *be a Noetherian ordered set, and for all* $x \in E$, $M(x)$ *the set of strict upper bounds for* x *in* E. *Let* F *be a subset of* E *such that* $(\forall x \in E)\ M(x) \subset F \Rightarrow x \in F$. *Then* $F = E$.

Proof Set $X = E - F$. If $X \neq \varnothing$, let x be a maximal element of X. Then $M(x) \subset F$ and $x \notin F$, a contradiction.   □

(This proposition will be needed in 3.6.4 and 3.6.7.)

1.7   Tychonoff's Theorem

1.7.1   Filters

Let X be a topological space. A **filter** on X is a subset $\mathscr{F}$ of $\mathfrak{P}(X)$ satisfying the following conditions:

(F_1) $\mathscr{F}$ is closed under finite intersection.
(F_2) $F \in \mathscr{F}$ and $G \supset F \Rightarrow G \in \mathscr{F}$.
(F_3) $\varnothing \notin \mathscr{F}$.

Example   For all $x \in X$, the set $\mathscr{V}_x$ of neighbourhoods of x is a filter on X.

Given two filters $\mathscr{F}$ and $\mathscr{G}$, $\mathscr{F}$ is said to be *finer* than $\mathscr{G}$ if $\mathscr{F} \supset \mathscr{G}$. If $\mathscr{H}$ is finer than both $\mathscr{F}$ and $\mathscr{G}$, then $\mathscr{F}$ and $\mathscr{G}$ are said to be *compatible*. Equivalently, $\forall F \in \mathscr{F}, \forall G \in \mathscr{G}, F \cap G \neq \varnothing$. The supremum of $\mathscr{F}$ and $\mathscr{G}$ is then called the *intersection filter* and is written filter $\mathscr{F} \wedge \mathscr{G} = \{F \cap G\}_{F \in \mathscr{F}, G \in \mathscr{G}}$. The filter $\mathscr{F}$ is *convergent* to $x \in X$ (resp. has x as *cluster point*) if $\mathscr{F}$ is finer than (resp. compatible with) $\mathscr{V}_x$. The filter $\mathscr{F}$ has x as cluster point if and only if for all F of $\mathscr{F}$, x belongs to the closure $\overline{F}$ of F.

Commentary. A property that may or may not hold for a point of X is described by using a subset of X: the set of x satisfying this property. A filter is meant to describe a more subtle condition that may be variably satisfied, each set of the filter (or of a a base of this filter) satisfying it to some degree.

1.7.2   Ultrafilters

An **ultrafilter** is a maximal filter, namely one for which there is no finer one.

Example   For all $x \in X$, the set $\mathscr{U}_x$ of $F \subset X$ such that $x \in F$ is an ultrafilter.

If x is a cluster point of an ultrafilter $\mathscr{U}$, then $\mathscr{U}$ converges to x. Indeed $\mathscr{U} \wedge \mathscr{V}_x$ is finer than $\mathscr{U}$, and so equal to $\mathscr{U}$, and $\mathscr{U}$ is finer than $\mathscr{V}_x$.

Proposition   *For every filter $\mathscr{F}$ on X, there is an ultrafilter finer than $\mathscr{F}$.*

Proof It readily follows that the set of of filters, ordered by the fineness relation, is inductive. The proposition is then a consequence of 1.4.4.   □

1.7.3 Characterization of Compact Sets

Proposition *Let* X *be a Hausdorff topological space. The following conditions are equivalent:*

   (i) X *is compact;*
  (ii) *every filter on* X *has a cluster point;*
 (iii) *every ultrafilter on* X *is convergent.*

Proof (i) $\Leftrightarrow$ (ii). By definition, (i) means that every open cover of X has a finite subcover. Considering their complements in X, (i) is seen to be equivalent to

(i') *if* $(F_i)_{i \in I}$ *is a closed family of* X *such that* $\bigcap_{j \in J} F_j \neq \varnothing$ *for all finite* $J \subset I$,
      then $\bigcap_{i \in I} F_i \neq \varnothing$.

   (ii) $\Rightarrow$ (i') follows by taking the filter generated by the subsets F_i, and (i') $\Rightarrow$ (ii) by taking the family $(\overline{F})_{F \in \mathscr{F}}$.
   (ii) $\Leftrightarrow$ (iii). Since an ultrafilter having a cluster point is convergent, (ii) $\Rightarrow$ (iii). Suppose (iii) holds and let $\mathscr{F}$ be a filter. By Proposition 1.7.2 there is an ultrafilter $\mathscr{U}$ finer than $\mathscr{F}$, and if $\mathscr{U}$ converges to x, then x is a cluster point of the filter $\mathscr{F}$. $\square$

1.7.4 Image Filter

Let $f : X \to Y$ be a map and $\mathscr{F}$ a filter on X. The set $f_* \mathscr{F}$ of subsets $G \subset Y$ such that $f^{-1}(G) \in \mathscr{F}$ is a filter on Y, called the *image filter* of $\mathscr{F}$ by f.

   If X and Y are topological spaces, if f is continuous and if $\mathscr{F}$ converges to a point $x \in X$, the image filter $f_* \mathscr{F}$ converges to $f(x)$.

1.7.5

Proposition *Let* $(X_i)_{i \in I}$ *be a family of topological spaces. Set* $X = \prod_{i \in I} X_i$ *and* p_i *the projection* $X \to X_i$. *A filter* $\mathscr{F}$ *on* X *is convergent if and only if for all* $i \in I$, *so is* $(p_i)_* \mathscr{F}$.

Proof The condition is obviously necessary. Let us show it is sufficient. For all i, let $x_i \in X_i$ be such that $(p_i)_* \mathscr{F}$ is convergent to x_i, and set $x = (x_i)$.
   If J is a finite subset of I and $(V_j)_{j \in J}$ a family where V_j is a neighbourhood of x_j, the set $\bigcap_{j \in J} p_j^{-1} V_j$ belongs to $\mathscr{F}$. Since such sets form a fundamental system of neighbourhoods of x, the filter $\mathscr{F}$ is convergent to x. $\square$

Remark Even when, for all i, the filter $(p_i)_* \mathscr{F}$ has a cluster point, this does not necessarily imply that this is also the case for $\mathscr{F}$.

1.7.6

Proposition *The image filter of an ultrafilter is an ultrafilter.*

Proof A filter $\mathscr{F}$ on X is an ultrafilter if and only if, for every $A \subseteq X$, $A \in \mathscr{F}$ or $X - A \in \mathscr{F}$ (consider $\{F \cap A\}_{F \in \mathscr{F}}$ and $\{F \cap (X - A)\}_{F \in \mathscr{F}}$). If $\mathscr{U}$ is an ultrafilter on X, then, for all $B \subset Y$,

$$f^{-1}(B) \in \mathscr{U} \quad \text{or} \quad f^{-1}(Y - B) = X - f^{-1}(B) \in \mathscr{U}. \qquad \square$$

1.7.7

Theorem (TYCHONOFF'S THEOREM) *An arbitrary product of compact spaces is compact.*

Proof Let $(X_i)_{i \in I}$ be a family of compact spaces and set $X = \prod_{i \in I} X_i$. Let $\mathscr{U}$ be an ultrafilter on X. For all i, the filter $p_{i*}\mathscr{U}$ is an ultrafilter 1.7.6, and so is convergent 1.7.3.

Hence, $\mathscr{U}$ is convergent 1.7.5. Since this holds for all ultrafilters on X, the space X is compact 1.7.3, $\qquad \square$

Exercises 1.7 (Filters and ultrafilters, Tychonoff's theorem)
1. (*Moore-Smith convergence.*)—An ordered set is said to be *directed* if all finite subsets are bounded above, in other words, if it is nonempty and every pair of elements has an upper bound. Any totally ordered set is directed, in particular $\mathbb{N}$. The following generalizes the notion of sequence.

Given a set X, a *Moore-Smith sequence* in X is a family $(x_i)_{i \in I}$ where I is a directed set. If X is a topological space and $a \in X$, then (x_i) is said to converge to a if for all neighbourhoods V of a, $\exists k \in I, \forall i \geqslant k, x_i \in V$.

Given two sets X and Y, a map $f : X \to Y$ and a filter $\mathscr{F}$ on X, the subsets V of Y such that $f^{-1}(V) \in \mathscr{F}$ form a filter on Y written $f_*\mathscr{F}$ and called *image filter*. If X is a topological space, for $a \in X$, denote by $\mathscr{V}_{X,a}$ the neighbourhood filter of a in X.

(a) Let X and Y be two topological spaces, $f : X \to Y$ a map, let a be a point of X and $b = f(a)$. Show that the following conditions are equivalents:

  (i) f is continuous at a;
  (ii) $f_*\mathscr{V}_{X,a} \supset \mathscr{V}_{Y,b}$;
  (iii) For every Moore-Smith sequence (x_i) convergent to a in X, the (Moore-Smith) sequence $(f(x_i))$ is convergent to b in Y.

(b) Let (iii$_\mathbb{N}$) be the condition analogous to (iii) for ordinary sequences. Give an example showing that (iii$_\mathbb{N}$) does not imply (iii). Show that Lebesgue's dominated convergence theorem provides such an example.

(The convergence of Moore-Smith sequences provides a language which, like that of filters, enables us to tackle non-metric topological spaces.)

2.—Let X be an infinitely countable set. Set A to be the ring $\mathbb{R}^X$. For $f \in A$, denote by $V(f)$ the set of $x \in X$ such that $f(x) = 0$.

(a) For $f, g \in A$, show that f divides g if and only if $V(f) \subset V(g)$.

(b) Let $I(\neq A)$ be an ideal of A. Show that $\mathscr{F}_I = \{V(f)\}_{f \in I}$ is a filter on X, and that $I = \{f; V(f) \in \mathscr{F}_I\}$.

(c) Show that, if $\mathscr{F}$ is a filter on X, the set $I_{\mathscr{F}}$ of $f \in A$ such that $V(f) \in \mathscr{F}$ is a strict ideal of A.

Show that $I \mapsto \mathscr{F}_I$ and $\mathscr{F}_I \mapsto I_{\mathscr{F}}$ are mutually inverse bijections, from the set of filters on X onto the set of strict ideals of A.

(d) Let $\mathscr{F}$ be a filter on X and $I = I_{\mathscr{F}}$. Show that the following three conditions are equivalent:

(i) I is a maximal ideal;
(ii) I est a prime ideal;
(iii) $\mathscr{F}$ is an ultrafilter.

3. (*Hyper-reals.*)—We keep the notation of the previous exercise. Let $\mathscr{U}$ be an ultrafilter on X and $\mathbb{R}_{\mathscr{U}}$ the field $A/I_{\mathscr{U}}$. The aim is to prove properties stating that $\mathbb{R}_{\mathscr{U}}$ is a model of the real field in non-standard analysis.

(a) $\mathbb{R}_{\mathscr{U}}$ is a totally ordered field and there is a natural embedding of $\mathbb{R}$ in $\mathbb{R}_{\mathscr{U}}$. For any subset E of $\mathbb{R}$, set $E_{\mathscr{U}}$ to be the image of E^X in $\mathbb{R}_{\mathscr{U}}$. For a function $\varphi : \mathbb{R}^n \to \mathbb{R}$, let $\varphi_{\mathscr{U}}$ be the map $\mathbb{R}_{\mathscr{U}}^n \to \mathbb{R}_{\mathscr{U}}$ induced by $(f_1, ..., f_n) \to \varphi \circ (f_1, ..., f_n)$ from A^n to A. If φ is $x \to |x|$, $(x, y) \to x + y$, etc., write $x \to |x|$, $(x, y) \to x + y$, etc. the map $\varphi_{\mathscr{U}}$.

There are elements $\omega \in \mathbb{R}_{\mathscr{U}}$ such that $\forall r \in \mathbb{R}$, $\omega > r$. They are called *infinitely large*, while $\varepsilon \in \mathbb{R}_{\mathscr{U}}$ is *infinitely small* if $(\forall r \in \mathbb{R}, r > 0) |\varepsilon| < r$. If $\xi - x$ is infinitely small, $\xi \in \mathbb{R}_{\mathscr{U}}$ is said to have *shadow* $x \in \mathbb{R}$.

(b)   For $E, E' \subset \mathbb{R}$, $(E \cap E')_{\mathscr{U}} = E_{\mathscr{U}} \cap E'_{\mathscr{U}}$, $(E \cup E')_{\mathscr{U}} = E_{\mathscr{U}} \cup E'_{\mathscr{U}}$. For $f : \mathbb{R} \to \mathbb{R}$, $f_{\mathscr{U}}(E_{\mathscr{U}}) = (f(E))_{\mathscr{U}}$.

(c) Let $E \subset \mathbb{R}$. The set $E_{\mathscr{U}}$ contains all infinitely large elements if and only if $\exists r \in \mathbb{R}, E \supset [r, \infty[$.

(d) Let $\varphi : \mathbb{R} \to \mathbb{R}$ be a function. φ is continuous if and only if $\forall x \in \mathbb{R}, \forall \xi \in \mathbb{R}_{\mathscr{U}}$, $(\xi$ has shadow $x) \Rightarrow (\varphi_{\mathscr{U}}(\xi)$ has shadow $\varphi(x))$.

Reference

1. N. Bourbaki, *Topologie générale*, ch. 5 à 10. (Hermann, Paris, 1974)

Chapter 2
Categories and Functors

Introduction

Some mathematical constructions are "natural" because they do not involve any arbitrary choice. These constructions can be transferred from one model to another representing the same situation. Category theory has been elaborated so as give this vague but powerful idea a precise meaning which can be used in mathematical proofs.

A logical difficulty arises. It is well-known that "the set of all sets" is an ill-defined concept as it leads to a contradiction: For any property $\mathscr{P}$, one should be able to define the set of sets satisfying $\mathscr{P}$, in particular $Y = \{X \mid X \notin X\}$. But then $Y \in Y$ implies $Y \notin Y$, and $Y \notin Y$ implies $Y \in Y$ (Russell's paradox). Likewise, it is not possible to talk of the set of all groups, etc.

This difficulty can be overcome in two ways. One involves the notion of "classes" that are not sets. This however is not altogether satisfactory. The categories constructed from these classes are not veritably objects of the theory, and can only be manipulated with great caution. The use of quantifiers cannot be justified, statements do not really amount to theorems that can be formally applied by substituting to an object a letter governed by the quantifier "$\forall$". No proof of Theorem 2.3.5 is known in this framework (the proof we give uses the axiom of choice).

Another way to avoid the problem, and which we have adopted here, amounts to considering "small" categories, i.e. categories whose set of objects is truly a set. The category of groups is not a well-defined concept, but that of groups whose underlying set belongs to a predetermined set U is. Although "statements" are not as general as one would wish, they are truly statements on which formal reasoning can be built with total peace of mind. Drawbacks are minor, and they completely disappear in practice if the "rag-bag" U is a universe (2.1.3). Admitting the universe axiom, whose role is not critical, greatly simplifies the language.

© Springer Nature Switzerland AG 2020
R. Douady and A. Douady, *Algebra and Galois Theories*,
https://doi.org/10.1007/978-3-030-32796-5_2

2.1 Categories

2.1.1

Definition A **category** consists of a set $\mathscr{C}$ whose elements are called objects,

(a) for all objects $X, Y \in \mathscr{C}$, of a set $\mathrm{Hom}_\mathscr{C}(X, Y)$ or simply $\mathrm{Hom}(X, Y)$, called the set of **morphisms** of $\mathscr{C}$ from X to Y;

(b) and, for any three objects $X, Y, Z \in \mathscr{C}$ $(f, g) \mapsto g \circ f$ from $\mathrm{Hom}(X, Y) \times \mathrm{Hom}(Y, Z)$ to $\mathrm{Hom}(X, Z)$ called *composition*;
satisfying the following conditions:

(C_1) $(\forall X, Y, Z, T \in \mathscr{C})$ $(\forall f \in \mathrm{Hom}(X, Y))$
$\qquad (\forall g \in \mathrm{Hom}(Y, Z))$ $(\forall h \in \mathrm{Hom}(Z, T))$ $(h \circ g) \circ f = h \circ (g \circ f)$.
(C_2) $(\forall X \in \mathscr{C})$ $(\exists e \in \mathrm{Hom}(X, X))$ $(\forall Y \in \mathscr{C})$
$\qquad ((\forall f \in \mathrm{Hom}(X, Y))$ $f \circ e = f$ and $(\forall g \in \mathrm{Hom}(Y, X))$ $e \circ g = g)$.

The element e whose existence is predicated in (C_2) is unique. It is called the *identity* of X and is written 1_X.

A morphism $f \in \mathrm{Hom}(X, Y)$ is often written $f : X \to Y$.

An *endomorphism* of X is a morphism from X to X and $\mathrm{End}(X) = \mathrm{Hom}(X, X)$.

A morphism $f : X \to Y$ from X to Y is said to be an *isomorphism* from X onto Y if there is a morphism $g : Y \to X$ such that $g \circ f = 1_X$ and $f \circ g = 1_Y$. Then g is unique and is called the inverse of f. If f is an isomorphism from X onto X, then f is said to be an *automorphism* of X.

Two objects X and Y de $\mathscr{C}$ are *isomorphic* if there is an isomorphism from X onto Y. We then write $X \approx Y$.

2.1.2 Examples

Let U be a set.

(1) The set $\mathscr{C}$ of all sets belonging to U with $\mathrm{Hom}(X, Y)$ the set of maps from X to Y, for $X, Y \in \mathscr{C}$, composition being map composition, is a category called the category of sets of U and of maps, written $\mathscr{E}\!n\!s_U$.

(2) The set $\mathscr{C}$ of groups (resp. commutative unitary rings, resp. modules over a given ring A, resp. topological spaces) whose underlying set belongs to U, with the morphisms being group (resp. unitary ring, resp. A-linear, resp. continuous) homomorphisms or maps, is a category $\mathscr{G}r_U$ called the category of groups and group homomorphisms of U (resp. the category $\mathscr{A}nn_U$ of commutative unitary rings and unitary ring homomorphisms, resp. the category A-$\mathscr{M}od_U$ of A-modules and A-linear maps, resp. the category $\mathscr{T}op_U$ of topological spaces and continuous maps).

(3) Let G be a group. A G-set is a set X endowed with a left action of G on X, i.e. a map $(g, x) \mapsto g \cdot x$ from $G \times X$ to X such that $\forall x \in X \; e \cdot x = x$ where e is the identity element of G and $\forall g \in G \; \forall h \in G \; \forall x \in X \; g \cdot (h \cdot x) = (g \cdot h) \cdot x$.

This amounts to defining a homomorphism ρ_X from G to the group $\mathfrak{S}(X)$ of permutations of X. If X and Y are two G-sets, a G-*morphism* from X to Y is a map $f : X \to Y$ such that $\forall g \in G \; \forall x \in X \; f(g \cdot x) = g \cdot f(x)$. Equivalently, $\forall g \in G$, the diagram

$$
\begin{array}{ccc}
X & \xrightarrow{\;f\;} & Y \\
\rho_X(g) \downarrow & & \downarrow \rho_Y(g) \\
X & \xrightarrow{\;f\;} & Y
\end{array}
$$

is commutative.

A category G-$\mathscr{Ens}_U$ is defined by taking as its objects the G-sets whose underlying sets belong to U and as its morphisms the G-morphisms.

(4) Let X and Y be two topological spaces, f_0 and f_1 two continuous maps from X to Y. The maps f_0 and f_1 are said to be homotopic if there is a continuous map h from $[0, 1] \times X \to Y$ such that, for all $x \in X$, $h(0, x) = f_0(x)$ and $h(1, x) = f_1(x)$. The homotopy relation is an equivalence relation on the set of continuous maps from X to Y.

Let $\mathscr{C}$ be the set of topological spaces whose underlying sets belong to U and with the homotopy classes of continuous maps as morphisms, composition being induced by map composition on passing to the quotient. This gives a category called the *category of topological spaces and homotopy classes of continuous maps*.

(5) Let I be a preordered set (i.e. endowed with a reflexive and transitive binary relation, see 1.2.1). The set I becomes a category by setting $\mathrm{Hom}_I(i, j)$ to be a singleton if $i \leq j$ and empty otherwise, composition being the only possible one.

(6) Let M be a monoid (i.e. a set endowed with an associative composition law with an identity element). A category $\mathscr{M}$ can be defined by taking a singleton with the element x and setting $\mathrm{Hom}(x, x) = M$, composition being that of M.

2.1.3 Universe

In Examples 1–4 of 2.1.2 and other similar ones, if U is taken to be the set to which all the sets in question belong, then, for the sake of simplicity, any mention of U is omitted, and the category is said to be that of sets, groups, etc., and is written $\mathscr{Ens}$, $\mathscr{Gr}$, $\mathscr{Top}$, etc.

The notion of universe provides a framework where the drawbacks of this simplification are alleviated. The set U is said to be a **universe** if it satisfies the following conditions:

(U_1) $\varnothing \in U$.

(U_2) $\forall X \in U, \{X\} \in U$.

(U_3) $\forall X \in U, \forall Y \in X, Y \in U$.

(U_4) For all sets X and Y belonging to U, the product set $X \times Y$ belongs to U.

(U_5) For any set X belonging to U, $\mathfrak{P}(X) \in U$.

(U_6) If $(X_i)_{i \in I}$ is a family of sets

$$(I \in U \text{ and } (\forall i \in I) \, X_i \in U) \Rightarrow \bigcup_{i \in I} X_i \in U.$$

The following axiom can be added to the axioms of the theory of sets:

> Axiom of universes: $(\forall X)(\exists U \text{ universe}) \, X \in U$.

It follows that, if U is a universe, the objects constructed from an object $X \in U$ using usual mathematical methods belong to U. For example, if E is a uniform space belonging to U, its completion $\widehat{E}$ belongs to U. Indeed, each element of $\widehat{E}$ is a set of filters on E, and so $\widehat{E} \in \mathfrak{P}(\mathfrak{P}(\mathfrak{P}(\mathfrak{P}(E))))$. Likewise, let A be a commutative ring, E and F two A-modules. If A, E and F belong to U, so does $E \otimes_A F$, which as a quotient of $A^{(E \times F)}$ is an element of $\mathfrak{P}(\mathfrak{P}(\mathfrak{P}(E \times F \times A)))$. However, beware that U, $\mathcal{E}ns_U$, $\mathcal{G}r_U$, etc. are not elements of U.

As far as we know, as a result of adding the axiom of universes, the theory of sets may become inconsistent (assuming it is not already the case) since the converse has not been proved.

2.1.4

Definition Let $\mathscr{C}$ and $\mathscr{C}'$ be two categories, $\mathscr{C}'$ is said to be a **subcategory** of $\mathscr{C}$ if

  1. $\mathscr{C}' \subset \mathscr{C}$;

  2. $(\forall X, Y \in \mathscr{C}') \, \mathrm{Hom}_{\mathscr{C}'}(X, Y) \subset \mathrm{Hom}_{\mathscr{C}}(X, Y)$;

  3. Composition in $\mathscr{C}'$ is inherited from composition in $\mathscr{C}$;

  4. For any object X of $\mathscr{C}'$, the identity element of X in $\mathscr{C}'$ is the identity element of X in $\mathscr{C}$.

$\mathscr{C}'$ is said to be *a full subcategory* of $\mathscr{C}$ if, for all $X, Y \in \mathscr{C}'$, $\mathrm{Hom}_{\mathscr{C}'}(X, Y) = \mathrm{Hom}_{\mathscr{C}}(X, Y)$.

2.1.5 Examples

(1) The category of metric spaces and continuous maps is a full subcategory of $\mathcal{T}\!op$.

(2) The category of local rings and local homomorphisms is a (not full) subcategory of $\mathcal{A}nn$. The category $\mathcal{A}nn$ is a subcategory of the category of (not necessarily unitary or commutative) rings and homomorphisms.

(3) The category of complex Banach spaces and $\mathbb{C}$-linear continuous maps is a full subcategory of the category of topological vector spaces over $\mathbb{C}$ and $\mathbb{C}$-linear continuous maps.

(4) Let $\mathscr{C}$ be a category. A subcategory of $\mathscr{C}$ can be defined by taking its objects to be those of $\mathscr{C}$ and for its morphisms the isomorphisms of $\mathscr{C}$.

2.1.6 Definition

Let $\mathscr{C}'$ and $\mathscr{C}''$ be two categories. The *product category* is the category $\mathscr{C}$ defined by $\mathscr{C} = \mathscr{C}' \times \mathscr{C}''$, $\operatorname{Hom}_{\mathscr{C}}((X' \times X''), (Y' \times Y'')) = \operatorname{Hom}_{\mathscr{C}'}(X', Y') \times \operatorname{Hom}_{\mathscr{C}''}(X'', Y'')$ and $(g', g'') \circ (f', f'') = ((g' \circ f'), (g'' \circ f''))$.

2.1.7 Definition

For any category $\mathscr{C}$, another category $\mathscr{C}^\circ$, called the *opposite category*, can be defined by $\mathscr{C}^\circ = \mathscr{C}$, $\operatorname{Hom}_{\mathscr{C}^\circ}(X, Y) = \operatorname{Hom}_{\mathscr{C}}(Y, X)$ and $g \circ_{\mathscr{C}^\circ} f = f \circ_{\mathscr{C}} g$.

Exercises 2.1. (Categories)

1.—Show that a category $\mathscr{C}$ can be defined by taking rings for its objects, isomorphism classes of (A, B)-bimodules for morphisms from A to B and $(M, N) \mapsto M \otimes_B N$ for composition, where M is a (A, B)-bimodule and N a (B, C)-bimodule.

2.—Let $\mathscr{C}$ be a category. An object $A \in \mathscr{C}$ is *initial* (resp. *terminal*) if $\forall X \in \mathscr{C}$, $\operatorname{Hom}_{\mathscr{C}}(A, X)$ (resp. $\operatorname{Hom}_{\mathscr{C}}(X, A)$) is a singleton.

Give an initial object and a terminal one in each of the following categories: $\mathscr{E}ns$, $\mathscr{G}r$, $\mathscr{A}nn$, $\mathscr{T}op$, G-$\mathscr{E}ns$, A-$\mathcal{M}od$, and the category of Banach algebras over $\mathbb{C}$. Which of these categories contains an object which is both initial and terminal?

3.—Let α be an initial object and ω a terminal object in a category $\mathscr{C}$. Show that, if there is a morphism from ω to α, it is unique and is an isomorphism.

2.2 Functors

2.2.1 Definition

Let $\mathscr{C}$ and $\mathscr{C}'$ be two categories. A **covariant functor** (or simply functor) from $\mathscr{C}$ to $\mathscr{C}'$ is a map F from $\mathscr{C}$ to $\mathscr{C}'$ such that, for all objects X, Y $\in \mathscr{C}$, there is a map $F_{X,Y} :$ Hom$_{\mathscr{C}}$(X, Y) $\to$ Hom$_{\mathscr{C}'}$(F(X), F(Y)) satisfying the following conditions:

(F$_1$) ($\forall$X $\in \mathscr{C}$) $F_{X,X}(1_X) = 1_{F(X)}$.
(F$_2$) ($\forall$X, Y, Z $\in \mathscr{C}$) ($\forall f : X \to Y$) ($\forall g : Y \to Z$)
$$F_{X,Z}(g \circ f) = F_{Y,Z}(g) \circ F_{X,Y}(f).$$

A **contravariant functor** from $\mathscr{C}$ to $\mathscr{C}'$ is a map F from $\mathscr{C}$ to $\mathscr{C}'$ such that, for all X, Y $\in \mathscr{C}$, there is a map $F_{X,Y} :$ Hom$_{\mathscr{C}}$(X, Y) $\to$ Hom$_{\mathscr{C}'}$(F(Y), F(X)) satisfying condition (F$_1$) and

(F$_2^*$) ($\forall$X, Y, Z $\in \mathscr{C}$) ($\forall f : X \to Y$) ($\forall g : Y \to Z$)
$$F_{X,Z}(g \circ f) = F_{X,Y}(f) \circ F_{Y,Z}(g).$$

In other words, a contravariant functor from $\mathscr{C}$ to $\mathscr{C}'$ is a covariant functor from $\mathscr{C}^o$ to $\mathscr{C}'$, or, a covariant functor from $\mathscr{C}$ to $\mathscr{C}'^o$. The map $F_{X,Y}$ is often written $f \mapsto f_*$ (resp. $f \mapsto f^*$) when F is a covariant (resp. contravariant) functor.

When defining a functor, often only F is given, leaving it to the reader to specify all the $F_{X,Y}$.

2.2.2 Examples

(1) **Forgetful functor.** Consider the categories $\mathscr{Ann}$ and $\mathscr{Gr}$. Assigning to every ring, the underlying additive group, and to every ring homomorphism f, f considered as a group homomorphism, defines a functor called the forgetful functor from $\mathscr{Ann}$ to $\mathscr{Gr}$.

Let $\mathscr{C}$ be the category of topological vector spaces and $\mathbb{R}$-linear continuous maps, the category $\mathscr{Vect}_{\mathbb{R}}$ of $\mathbb{R}$- vector spaces and linear maps. Define a functor from $\mathscr{C}$ to $\mathscr{Vect}_{\mathbb{R}}$ by assigning to every object of $\mathscr{C}$ its underlying vector space.

Functors obtained in this manner are called **forgetful functors**. Thus there are forgetful functors $\mathscr{Gr} \to \mathscr{Ens}$, $\mathscr{Top} \to \mathscr{Ens}$, etc.

(2) **Inclusion functor.** Let $\mathscr{C}$ be a category, $\mathscr{C}'$ a subcategory of $\mathscr{C}$. The canonical inclusions $\iota : \mathscr{C}' \to \mathscr{C}$ and $\iota_{X,Y} :$ Hom$_{\mathscr{C}'}$(X, Y) $\to$ Hom$_{\mathscr{C}}$(X, Y) define a functor from $\mathscr{C}'$ to $\mathscr{C}$.

(3) For any topological space X write $\mathscr{C}(X, \mathbb{R})$ for the ring of continuous functions defined on X with values in $\mathbb{R}$. For any continuous map $f : X \to Y$ write f^* for the map from $\mathscr{C}(Y, \mathbb{R})$ to $\mathscr{C}(X, \mathbb{R})$ defined by $f^*(h) = h \circ f$.

This defines a contravariant functor X $\to \mathscr{C}(X, \mathbb{R})$ from $\mathscr{Top}$ to $\mathscr{Ann}$.

(4) **Duality functor.** Let A be a commutative ring. Define a contravariant functor from the category A-$\mathcal{M}od$ to itself by assigning to every A-module X, its dual $X^\top = \mathrm{Hom}(X, A)$ and to every A-linear map its transpose.

(5) **Bidual functor.** Define a covariant functor from the category A-$\mathcal{M}od$ to itself by assigning to every A-module its bidual and to every A-linear map its bitranspose.

(6) Let $\mathscr{C}$ be a category, X an object of $\mathscr{C}$. Define a covariant functor $\check{X} : \mathscr{C} \to \mathscr{E}ns$ by setting

$$(\forall\, Y \in \mathscr{C}) \quad \check{X}(Y) = \mathrm{Hom}(X, Y)$$
$$(\forall\, Y, Z \in \mathscr{C})\,(\forall f \in \mathrm{Hom}(Y, Z))\,(\forall h \in \mathrm{Hom}(X, Y)) \quad f_*(h) = f \circ h\,.$$

(7) Let $\mathscr{C}$ be a category, X an object of $\mathscr{C}$. Define a contravariant functor $\widehat{X} : \mathscr{C} \to \mathscr{E}ns$ by setting

$$(\forall\, Y \in \mathscr{C}) \quad \widehat{X}(Y) = \mathrm{Hom}(Y, X)$$
$$(\forall\, Y, Z \in \mathscr{C})\,(\forall f \in \mathrm{Hom}(Y, Z))\,(\forall h \in \mathrm{Hom}(Z, X)) \quad f^*(h) = h \circ f\,.$$

2.2.3 Definition

Let $\mathscr{C}$, $\mathscr{C}'$, $\mathscr{C}''$ be categories, $F : \mathscr{C} \to \mathscr{C}'$ and $G : \mathscr{C}' \to \mathscr{C}''$ functors. Define the *composite functor* $G \circ F : \mathscr{C} \to \mathscr{C}''$ as follows:

$$(\forall\, X \in \mathscr{C}) \quad G \circ F(X) = G(F(X))\,;$$
$$(\forall\, X, Y \in \mathscr{C})(\forall f \in \mathrm{Hom}_\mathscr{C}(X, Y)) \quad (G \circ F)_{X,Y}(f) = G_{F(X),F(Y)}(F_{X,Y}(f)).$$

The composite of two functors, one covariant and the other contravariant, as well as of two contravariant functors can be likewise defined. The former is contravariant, while the latter is covariant.

2.2.4 Examples

(1) The composite of the contravariant duality functor from A-$\mathcal{M}od$ to itself (2.2.2, Example 4) and of the forgetful functor from A-$\mathcal{M}od$ to $\mathscr{E}ns$ (2.2.2, Example 1) is a contravariant functor $\widehat{A} : $ A-$\mathcal{M}od \to \mathscr{E}ns$ (2.2.2, Example 7).

(2) The contravariant biduality functor is the composition of the duality functor with itself.

(3) The composite of the contravariant functor $X \mapsto \mathscr{C}(X, \mathbb{R})$ from the category of compact spaces and continuous maps to the category $\mathscr{B}$ of Banach spaces over $\mathbb{R}$ and of the contravariant functor $E \mapsto E^\top$ from $\mathscr{B}$ to itself is a covariant functor

$X \to \mathcal{M}(X)$. The space $\mathcal{M}(X)$ is the space of *measures* on X endowed with the norm topology, and for $f : X \to Y$ and $\mu \in \mathcal{M}(X)$, $f_*(\mu) \in \mathcal{M}(Y)$ is the direct image under f of the measure μ.

2.2.5 Category of Categories

Let U be a universe. Denote by $\mathscr{C}at_U$ the category whose objects are categories $\mathscr{C}$ with the following property: the set $\mathscr{C}$ belongs to U and for all objects X, Y $\in \mathscr{C}$, $\mathrm{Hom}_{\mathscr{C}}(X, Y) \in U$. Morphisms from $\mathscr{C}at_U$ are covariant functors.

Note that $\mathscr{E}ns_U$ does not belong to $\mathscr{C}at_U$; however if U' is a universe such that $U \in U'$, then $\mathscr{E}ns_U \in \mathscr{C}at_{U'}$.

2.2.6 Another Example of a Functor

Let $f : G \to H$ be a group homomorphism and X a H-set.

Denote by f^*X the G-set obtained by endowing the set X with the action $(g, x) \mapsto f(g) \cdot x$. This defines a functor f^* from H-$\mathscr{E}ns$ to G-$\mathscr{E}ns$. More precisely, if U is a universe, then f^* is a functor from H-$\mathscr{E}ns_U$ to G-$\mathscr{E}ns_U$.

Let U and U' be two universes such that $U \in U'$, $G \to G$-$\mathscr{E}ns_U$ from $\mathscr{G}r_U$ to $\mathscr{C}at_{U'}$ is a contravariant functor.

2.2.7

Proposition *Let G and H be two groups, $\omega_G : G\text{-}\mathscr{E}ns \to \mathscr{E}ns$ and $\omega_H : H\text{-}\mathscr{E}ns \to \mathscr{E}ns$ the forgetful functors. Let $\Phi : H\text{-}\mathscr{E}ns \to G\text{-}\mathscr{E}ns$ be a functor such that the diagram*

$$
\begin{array}{ccc}
H\text{-}\mathscr{E}ns & \xrightarrow{\Phi} & G\text{-}\mathscr{E}ns \\
 & \searrow{\omega_H} \quad \swarrow{\omega_G} & \\
 & \mathscr{E}ns &
\end{array}
$$

commutes. Then there is a unique homomorphism $f : G \to H$ such that $\Phi = f^*$.

Proof For any H-set X, its image $\Phi(X)$ is the set X endowed with an action $(g, x) \mapsto g \perp x$ from $G \times X$ to X. If $\Phi = f^*$, then $g \perp x = f(g) \cdot x$. In particular for X = H endowed with left translations, $f(g) = g \perp e$, hence uniqueness.

Define $f : G \to H$ by $f(g) = g \perp e$. Let X be a H-set, then for all $g \in G$ and all $x \in X$, $g \perp x = f(g)x$. Indeed, the map $\delta_x : h \mapsto h \cdot x$ from H to X is a H-morphism, its image $\Phi(\delta_x)$ is a G-morphism, but the maps $\Phi(\delta_x)$ and δ_x agree, and so $g \perp x = g \perp \delta_x(e) = \delta_x(g \perp e) = \delta_x(f(g)) = f(g) \cdot x$. In particular, for X = H and $x = f(g')$,

$$f(g) \cdot f(g') = g \perp f(g') = g \perp (g' \perp e) = gg' \perp e = f(gg'),$$

and hence f is a homomorphism, and $\Phi = f^*$. $\qquad\square$

2.3 Morphisms of Functors

2.3.1 Definition

Let $\mathscr{C}$ and $\mathscr{C}'$ be two categories, F and G two functors from $\mathscr{C}$ to $\mathscr{C}'$. A *functorial morphism* or **morphism of functors** Φ from F to G, denoted by $\Phi : F \to G$, is defined by assigning to each object X of $\mathscr{C}$ a morphism $\Phi_X : F(X) \to G(X)$ such that, for any morphism $f : X \to Y$ of $\mathscr{C}$ the following diagram commutes:

$$
\begin{array}{ccc}
F(X) & \xrightarrow{F(f)} & F(Y) \\
\Phi_X \downarrow & & \downarrow \Phi_Y \\
G(X) & \xrightarrow{G(f)} & G(Y)
\end{array}
$$

If F and G are contravariant functors from $\mathscr{C}$ to $\mathscr{C}'$, define the morphisms from F to G by considering F and G as covariant functors from $\mathscr{C}^o$ to $\mathscr{C}'$ (not from $\mathscr{C}$ to $\mathscr{C}'^o$).

2.3.2 Examples

(1) **Biduality morphism.** Let $\mathscr{C} = \mathscr{C}' = A\text{-}\mathcal{M}od$. Write $1_{\mathscr{C}}$ for the identity functor of $\mathscr{C}$ and G for the biduality functor. Let X be an A-module, $x \in X$ and $\delta_x : X^\top \to A$ the linear form $h \mapsto h(x)$. The map $\delta_X : x \mapsto \delta_x$ from X to $X^{\top\top}$ is A-linear, and so is a morphism of A-$\mathcal{M}od$. Define a functorial morphism from $1_{\mathscr{C}}$ to G by assigning to each A-module X the morphism δ_X.

(2) **Kronecker morphism.** Let $\mathscr{C}$ be the category $(A\text{-}\mathcal{M}od)^o \times A\text{-}\mathcal{M}od$. For every object (X, Y) of $\mathscr{C}$, set $F(X, Y) = X^\top \otimes Y$ and $G(X, Y) = \text{Hom}(X, Y)$. This defines two functors F and G from $\mathscr{C}$ to A-$\mathcal{M}od$. For every object (X, Y) of $\mathscr{C}$, the Kronecker morphism $K_{X,Y}$ from $X^\top \otimes Y$ to $\text{Hom}(X, Y)$ defined by $K(\alpha \otimes y)(x) = \alpha(x) \cdot y$ is a morphism from F(X, Y) to G(X, Y). The family $(K_{X,Y})_{(X,Y)\in\mathscr{C}}$ is a functorial morphism from F to G. This will be checked in (3.9).

2.3.3 Functor Category

Let $\mathscr{C}$ and $\mathscr{C}'$ be two categories. Construct a category by taking as its objects the functors from $\mathscr{C}$ to $\mathscr{C}'$ and for its morphisms the morphisms of functors. If $\Phi : F \to G$ and $\Psi : G \to H$ are morphisms of functors, define the composition $\Psi \circ \Phi : F \to H$ by $(\Psi \circ \Phi)_X = \Psi_X \circ \Phi_X$ for any object X of $\mathscr{C}$.

A morphism of functors $\Phi : F \to G$ is an *isomorphism* if there is a morphism $\Psi : G \to F$ such that $\Psi \circ \Phi$ is the identity morphism 1_F of F and $\Phi \circ \Psi = 1_G$.

A morphism of functors Φ is an isomorphism if and only if for any object $X \in \mathscr{C}$, the morphism $\Phi_X : F(X) \to G(X)$ is an isomorphism. Indeed the morphisms Φ_X^{-1} then define a morphism from G to F.

Remark F(X) may be isomorphic to G(X) for any X even when the functors F and G are not (2.3, Exercise 1).

2.3.4 Definitions

Let $\mathscr{C}$ and $\mathscr{C}'$ be two categories, $F : \mathscr{C} \to \mathscr{C}'$ a functor.

1. The functor F is *essentially surjective* if for any $X' \in \mathscr{C}'$, there exists $X \in \mathscr{C}$ such that X' is isomorphic to $F(X)$; in other words if the isomorphism induced by F from the set of isomorphism classes of objects of $\mathscr{C}$ to the set of isomorphism classes of objects of $\mathscr{C}'$ is surjective.

2. The functor F is said to be *faithful* (resp. *fully faithful*) if, for all $X, Y \in \mathscr{C}$, the map $F_{X,Y} : \mathrm{Hom}_{\mathscr{C}}(X, Y) \to \mathrm{Hom}_{\mathscr{C}'}(F(X), F(Y))$ is injective (resp. bijective).

2.3.5 Equivalence of Categories

Theorem and Definition *Let $\mathscr{C}$ and $\mathscr{C}'$ be two categories, $F : \mathscr{C} \to \mathscr{C}'$ a functor. The following conditions are equivalent:*

(i) F is fully faithful and essentially surjective;
(ii) there is a functor $G : \mathscr{C}' \to \mathscr{C}$ such that $G \circ F$ is isomorphic to $1_{\mathscr{C}}$ and $F \circ G$ is isomorphic to $1_{\mathscr{C}'}$.

The functor F is said to be an **equivalence of categories** *if it satisfies one of the two conditions.*

A functor G satisfying (ii) *is called a* **quasi-inverse** *of* F.

Proof (i) $\Rightarrow$ (ii). Since F is essentially surjective, by the axiom of choice, there is a map $G : \mathscr{C}' \to \mathscr{C}$ and a family $(\alpha_{X'})_{X' \in \mathscr{C}'}$ such that, for all X', the morphism $\alpha_{X'}$ is an isomorphism from $F(G(X'))$ onto X'. For any morphism $f : X' \to Y'$ of $\mathscr{C}'$, set $G_{X',Y'}(f) = F_{X,Y}^{-1}(\alpha_{Y'}^{-1} \circ f \circ \alpha_{X'})$, where $X = G(X')$ and $Y = G(Y')$, which is possible since F is fully faithful. The functor G together with $(G_{X',Y'})$ can be checked to be a functor from $\mathscr{C}'$ to $\mathscr{C}$ satisfying the conditions of (ii).

(ii) $\Rightarrow$ (i). Since F $\circ$ G is isomorphic to $1_{\mathscr{C}'}$, every object X′ of $\mathscr{C}'$ is isomorphic to F(G(X′)). Hence F is essentially surjective.

We show that F is fully faithful. Let $\alpha = (\alpha_X)_{X \in \mathscr{C}}$ be an isomorphism from G $\circ$ F onto $1_{\mathscr{C}}$ and $\beta = (\beta_{X'})_{X' \in \mathscr{C}'}$ an isomorphism from F $\circ$ G onto $1_{\mathscr{C}'}$. Let X, Y $\in \mathscr{C}$; set X′ = F(X), Y′ = F(Y), X″ = G(X′), Y″ = G(Y′), $\varphi_{X,Y}(f) = \alpha_Y^{-1} \circ f \circ \alpha_X$ for $f \in \mathrm{Hom}_{\mathscr{C}}(X, Y)$ and $\psi_{X',Y'}(g) = \beta_{Y'}^{-1} \circ g \circ \beta_{X'}$ for $g \in \mathrm{Hom}_{\mathscr{C}'}(X'Y')$.

The following diagram commutes:

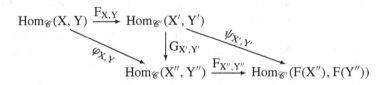

The maps $\varphi_{X,Y}$ and $\psi_{X',Y'}$ are bijective. Hence $G_{X',Y'}$ is surjective and injective and consequently, $F_{X,Y} = G_{X',Y'}^{-1} \circ \varphi_{X,Y}$ is itself bijective. $\qquad\square$

An *anti-equivalence of categories* is a contravariant factor F : $\mathscr{C} \to \mathscr{C}'$ which is an equivalence from $\mathscr{C}^\circ$ to $\mathscr{C}'$.

2.3.6

Proposition *Let* $f : G \to H$ *be a group homomorphism. The functor* f^* : H-$\mathit{8ns}$ $\to$ G-$\mathit{8ns}$ *is an equivalence of categories if and only if f is an isomorphism.*

Proof (a) *Sufficiency.* $(f^{-1})^*$ is an inverse of f^*.

(b) f^* *is essentially surjective* $\Rightarrow$ *f injective.* If $f(g) = e$, then g acts trivially on any G-set f^*Y, hence on any G-set, in particular on G itself by left translations, and so $g = e$.

(c) f^* *fully faithful* $\Rightarrow$ *f surjective.* Let X $\in$ H-$\mathit{8ns}$ be a singleton with H acting trivially. For all Y $\in$ H-$\mathit{8ns}$, $\mathrm{Hom}_H(X, Y)$ can be identified with the set of fixed points of Y under the action of H. Take Y to be the set $H/f(G)$ together with the action of H induced by left translations. Then $\mathrm{Hom}_H(X, Y)$ is empty unless f is surjective, but $\mathrm{Hom}_G(f^*X, f^*Y)$ is not empty, since the image of the identity element in Y is kept fixed by G and so f is necessarily surjective. $\qquad\square$

2.3.7

Let U and U' be two sets and $U \subset U'$. Suppose U is hereditary, i.e. for any set $X \in U$, $\mathfrak{P}(X) \subset U$ (a far weaker condition than the existence of universes, which supposes $\mathfrak{P}(X) \in U$).

Let $\mathscr{C}$ be a category whose set of objects is contained in U'. Suppose that $\mathscr{C}$ is a *category with underlying sets and transfer of structure* (in U'). This means there is a functor (the forgetful functor) $\Omega : \mathscr{C} \to \mathscr{Ens}_U$, and that for any object X of $\mathscr{C}$ and any bijection $f : \Omega(X) \to F$ with $F \in U'$, there is an object Y of $\mathscr{C}$ and a morphism $\varphi : X \to Y$ such that $\Omega(Y) = F$ and $\Omega(\varphi) = f$ (this is the case for $\mathscr{Ens}_{U'}$, $\mathscr{Gr}_{U'}$, $\mathscr{Top}_{U'}$, A-$\mathscr{Mod}_{U'}$, etc.).

Denote by $\mathscr{C}_U$ the full subcategory of $\mathscr{C}$ consisting of objects whose underlying sets belong to U. Suppose there is a set $E \in U$ such that, for any object X of $\mathscr{C}$, $\text{Card } \Omega(X) \leq \text{Card } E$.

Then the inclusion morphism $\mathscr{C}_U \to \mathscr{C}$ is an equivalence of categories.

Indeed it is fully faithful since it is the inclusion morphism of a full subcategory, and essentially surjective since any object of $\mathscr{C}$ is isomorphic to an object whose underlying set is a subset of E.

Thanks to the lemmas about cardinalities (4.4.7, 5.2.5), this is why, up to equivalence, the category of coverings of a given space, or that of algebraic extensions of a given field, do not depend on the choice of the universe.

Exercises 2.3. (Morphisms of functors)

1.—Let $\mathscr{C}$ be the category of finite dimensional vector spaces over a field k and of isomorphisms. Let F be the identity functor of $\mathscr{C}$ and $G : \mathscr{C} \to \mathscr{C}$ the functor defined by $G(X) = X$ and $G(f) = (f^\top)^{-1}$. Show that, for any object X of $\mathscr{C}$, the objects $F(X)$ and $G(X)$ are isomorphic but that the functors F are G not.

2.—Show that the set of endomorphisms of the identity functor of the category of abelian groups can be identified with $\mathbb{Z}$.

3.—Is there a map $f : \mathbb{N} \to \mathbb{N}$ such that

$$\forall x \in \mathbb{N}, \forall y \in \mathbb{N} \quad f(y)^{f(x)} = x^y \ ?$$

Is the category $\mathscr{Setf}$ of finite set its opposite category?

What about the category $\mathscr{Ens}$.

4.—Let $\mathscr{C}$ be the category of finite commutative groups. The aim is to show that $\mathscr{C}$ is equivalent to its opposite category. For this, assume that every object of $\mathscr{C}$ is a product of cyclic groups (3.5.8, Corollary 3).

(a) Show that for any group $X \in \mathscr{C}$, the group $X^\vee = \text{Hom}_{\mathscr{Gr}}(X, \mathbb{Q}/\mathbb{Z})$ is finite and isomorphic to X. (First consider the case where X is a cyclic group.)

(b) Show that the natural map $X \mapsto X^{\vee\vee}$ is an isomorphism. Conclude.

(c) Let $\mathscr{C}'$ be the category of objects of $\mathscr{C}$ and of isomorphisms. Show that the identity functor $1_{\mathscr{C}'}$ and the functor $X \mapsto X^{\vee}$ from $\mathscr{C}'$ to itself are not isomorphic (apply Exercise 1 to $k = \mathbb{Z}/(p)$).

5.—The category of locally compact commutative groups can be shown to be equivalent to its opposite category (Pontryagin duality, the natural framework of the Fourier transform). The aim is to show that there is a subcategory for which this is the case.

Let $\mathscr{C}$ be the category of commutative topological groups G satisfying the following properties:

(i) the connected component G_0 of the identity element of G is isomorphic to the quotient of a real finite dimensional vector space by a discrete subgroup;
(ii) G_0 is open and the quotient G/G_0 is a finitely generated $\mathbb{Z}$-module.

(a) Show that every object of $\mathscr{C}$ is a finite product of groups of the following type: $\mathbb{R}, \mathbb{R}/\mathbb{Z}, \mathbb{Z}, \mathbb{Z}/(n)$.

(b) Let $G \in \mathscr{C}$. Show that $G^{\vee} = \text{Hom}_{\mathscr{C}}(G, \mathbb{R}/\mathbb{Z})$ equipped with the topology of compact convergence (2.4.9, Example 2) is an object of $\mathscr{C}$.

Determine $G^{\vee}$ where G is successively $\mathbb{R}, \mathbb{R}/\mathbb{Z}, \mathbb{Z}$ and $\mathbb{Z}/(n)$. Show that $G^{\vee\vee} = G$. Conclude.

6.—Let U and U' be two universes with $U \in U'$. Show that the categories $\mathscr{E}ns_U$ and $\mathscr{E}ns_{U'}$ are not equivalent.

Denote by $\mathbb{R}$-vectfin$_U$ the category of finite dimensional vector spaces over $\mathbb{R}$ belonging to U. Show that $\mathbb{R}$-vectfin$_U$ and $\mathbb{R}$-vectfin$_{U'}$ are equivalent.

2.4 Representable Functors

2.4.1 Notation

For each object X of a category $\mathscr{C}$, set $\widehat{X}$ to be the contravariant functor from $\mathscr{C}$ to $\mathscr{E}ns$ defined by $\widehat{X}(Y) = \text{Hom}(Y, X)$.

2.4.2 Definition

Let $\mathscr{C}$ be a category and F a contravariant functor from $\mathscr{C}$ to $\mathscr{E}ns$. Then F is said to be representable if there is an object X of $\mathscr{C}$ such that F is isomorphic to $\widehat{X}$.

2.4.3

The previous definition implicitly involves an isomorphism from $\widehat{X}$ onto F. We next give a description of the morphisms from $\widehat{X}$ to F.

Scholium. Let $\mathscr{C}$ be a category, F a contravariant functor from $\mathscr{C}$ to $\mathscr{E}ns$ and X an object of $\mathscr{C}$. The morphisms from $\widehat{X}$ to F are in bijective correspondence with the elements of F(X).

More precisely, for $\xi \in F(X)$, $Y \in \mathscr{C}$, and $f \in \mathrm{Hom}(Y, X)$, write $\hat{\xi}_Y(f)$ for the element $f^*(\xi)$ of F(Y). This defines a map $\hat{\xi}_Y$ from $\widehat{X}(Y)$ to F(Y) and $\hat{\xi} = (\hat{\xi}_Y)_{Y \in \mathscr{C}}$ is a morphism of functors from $\widehat{X}$ to F.

Proposition (Yoneda Lemma) *The map $\xi \mapsto \hat{\xi}$ is a bijection from* F(X) *onto the set of morphisms from* $\widehat{X}$ *to* F.

Proof We first show that $\hat{\xi}$ is a morphism of functors. Consider the following diagram:

$$
\begin{array}{ccc}
\widehat{X}(Z) & \xrightarrow{\widehat{X}(g)} & \widehat{X}(Y) \\
{\scriptstyle \hat{\xi}_Z} \downarrow & & \downarrow {\scriptstyle \hat{\xi}_Y} \\
F(Z) & \xrightarrow{F(g)} & F(Y)
\end{array}
$$

where $Y, Z \in \mathscr{C}$ and $g \in \mathrm{Hom}(Y, Z)$. Then, $(\hat{\xi}_Y \circ \widehat{X}(g))(\alpha) = F(\alpha \circ g)(\xi)$ and $F(g)(\hat{\xi}_Z(\alpha)) = F(g) \circ F(\alpha)(\xi) = F(\alpha \circ g)(\xi)$. So the diagram commutes.

For any morphism Φ from $\widehat{X}$ to F, set $\eta(\Phi) = \Phi_X(1_X) \in F(X)$.

We show that the maps $\xi \mapsto \hat{\xi}$ and $\Phi \mapsto \eta(\Phi)$ are mutual inverses. $\eta(\hat{\xi}) = \hat{\xi}_X(1_X) = \xi$ and, for all $Y \in \mathscr{C}$ and all $f \in \mathrm{Hom}(Y, X)$

$$
\widehat{\eta(\Phi)}_Y(f) = F(f)(\eta(\Phi)) = F(f)(\Phi_X(1_X)) = \Phi_Y \circ \widehat{X}(f)(1_X) = \Phi_Y(f).
$$

$\square$

2.4.4 Definition

Keeping the notation of 2.4.3, the functor F is said to be *represented by* (X, ξ) if $\hat{\xi}$ is an isomorphism from $\widehat{X}$ onto F. In other words, F is represented by (X, ξ) if and only if

$$
(\forall Y \in \mathscr{C})(\forall \eta \in F(Y)) \quad (\exists| f : Y \to X) \; f^*(\xi) = \eta.
$$

2.4.5 Comments. Universal Problems

Choosing a pair (X, ξ) representing F amounts to identifying F(Y) with $\widehat{X}(Y)$. Hence for simplicity's sake, for any object Y of $\mathscr{C}$, one writes $\mathrm{Hom}(Y, X) = F(Y)$. Such a relation is called a *universal property of* X.

The search for a pair (X, ξ) representing F is called a **universal problem** defined by the functor F.

2.4.6 Definition

A covariant functor F from a category $\mathscr{C}$ to $\mathscr{Ens}$ is *representable* if F is a representable functor from $\mathscr{C}^o$ to $\mathscr{Ens}$.

2.4.7 Example of a Universal Problem

Tensor product. Let A be a commutative ring, E and F A-modules. We want to find an A-module T such that, for any A-module G, the set of bilinear maps from $E \times F$ to G can be identified with the set of linear maps from T to G.

More precisely, we want to find an A-module T and a bilinear map $\theta : E \times F \to T$ such that, for any A-module G and any bilinear map $f : E \times F \to G$, there is a unique linear map $g : T \to G$ such that $f = g \circ \theta$.

Put differently, let B be the covariant functor from A-$\mathscr{Mod}$ to $\mathscr{Ens}$ which assigns to a module G the set of bilinear maps from $E \times F$ to G. We want to find a pair (T, θ) where T is an A-module and $\theta \in B(T)$ such that, for any A-module G the map $g \mapsto g^*(\theta) = g \circ \theta$ from $\mathrm{Hom}(T, G)$ to $B(G)$ is bijective. In other words, we are trying to represent the covariant functor B.

This universal problem always has a solution: the *tensor product* $E \otimes F$ of the A-modules E and F (3.8.1).

2.4.8 Other Examples of Representable Covariant Functors

(1) **Quotient.** Let X be a set, R an equivalence relation on X. We want to find a set Y and a map $\chi : X \to Y$ such that

(i) $(\forall x, x' \in X) \quad x \, R \, x' \Rightarrow \chi(x) = \chi(x')$;
(ii) for any set Z and any map $f : X \to Z$ such that $x \, R \, x' \Rightarrow f(x) = f(x')$, there is a unique map $g : Y \to Z$ satisfying $f = g \circ \chi$.

This universal problem has a solution: the quotient set $Y = X/R$ endowed with the canonical quotient map $\chi : X \to X/R$.

In other words, the quotient represents the covariant functor which assigns to a set Z the set of maps $f : X \to Z$ for which $x \, R \, x' \Rightarrow f(x) = f(x')$.

(2) **Quotient modules.** Let A be a commutative ring, E an A-module, and $(x_i)_{i \in I}$ a family of elements of E. Denote by $\Phi(G)$ the set of linear maps from E to an A-module

G which vanish at x_i for all $i \in I$. The functor $G \mapsto \Phi(G)$ from A-$\mathcal{M}od$ to $\mathcal{E}ns$ is representable. The pair (F, χ), where F is the quotient module of E by the submodule generated by all x_i and $\chi : E \to F$ the canonical quotient map, represents Φ.

(3) **Free module.** Let I be a set. We want to find an A-module L such that for any A-module F,

$$\mathrm{Hom}(L, F) = F^I$$

where F^I is the set of maps from I to F.

This universal problem has a solution: the free A-module $A^{(I)}$, constructed over I (3.4.6).

(4) **Exterior products.** Let E be an A-module, p a positive integer. We want to find an A-module T such that, for any A-module F, the $\mathrm{Hom}(T, F)$ can be identified to the set of p-linear alternating maps from E^p to F.

This universal problem has a solution: the pth exterior power[1] $\bigwedge^p(E)$.

(5) **Completion.** Let E be a normed vector space. We want to find a Banach space $\widehat{E}$ and an embedding $\iota : E \to \widehat{E}$ such that, for any Banach space F, every continuous linear map from E to F has a unique continuous extension from $\widehat{E}$ to F.

In other words, we want to represent the functor Φ defined on the category of Banach spaces and continuous linear maps with values in $\mathcal{E}ns$ by $\Phi(F) = \{f : E \to F \ f \text{ linear continuous}\}$.

The completion of E, the space $\widehat{E}$, endowed with the canonical embedding from E into $\widehat{E}$ is the desired solution.

(6) **Action of a group on itself.** Let G be a group. The forgetful functor G-$\mathcal{E}ns \to \mathcal{E}ns$ is represented by $(^{\cdot}G, e)$, where $^{\cdot}G$ is the set G on which the group G acts by left translations, and e is the identity element of G.

2.4.9 Examples of Representable Contravariant Functors

(1) **Induced topology.** Let X be a topological space, Y a subset of X. We want to find a topological space Y' such that, for any topological spaces T, the set of continuous maps from T to Y' can be identified with the set of continuous maps from T to X whose images are contained in Y. In other words, we want to represent the functor F from $\mathcal{T}op$ to $\mathcal{E}ns$ defined by

$$F(T) = \{f \in \mathrm{Hom}(T, X) \mid f(T) \subset Y\}.$$

This universal problem has a solution: the set Y equipped with the topology induced by that of X.

---

[1] See N. Bourbaki, *Algebra, ch. 1 to 3* (1970), ch. 3, §7, n° 4, (A III.80).

(2) **Topology of compact convergence.** Let X and Y be two topological spaces. We want to find a topological space F such that, for any topological space T,

$$\operatorname{Hom}_{\mathcal{T}op}(T \times X, Y) = \operatorname{Hom}_{\mathcal{T}op}(T, F).$$

If X is locally compact, the functor $T \mapsto \operatorname{Hom}_{\mathcal{T}op}(T \times X, Y)$ is represented by $F = \operatorname{Hom}_{\mathcal{T}op}(X, Y)$ endowed with a topology called the topology of compact convergence, more precisely by (F, φ) where, for all $f \in F$ and all $x \in X$, $\varphi(f, x) = f(x)$.

2.4.10

Let $\mathcal{C}$ be a category, F a contravariant functor from $\mathcal{C}$ to $\mathcal{S}ns$, (X, ξ) and (X', ξ') two pairs representing F. There is a unique morphism $h : X \to X'$ such that $h^*\xi' = \xi$.

Definition The morphism h is the **canonical morphism** from X onto X'.

2.4.11 Uniqueness of the Solution of the Universal Problem

Proposition *keeping the above notation,*
(a) the canonical morphism h is an isomorphism;
(b) let (X, ξ), (X', ξ') (X'', ξ'') be three pairs representing F, and $h : X \to X'$, $h' : X' \to X''$, $h'' : X \to X''$ the canonical morphisms.
Then $h'' = h' \circ h$.

Proof (b) follows from the uniqueness of canonical morphisms. Apply (b) to the case $(X'', \xi'') = (X, \xi)$. Then $h' \circ h = 1_X$ and $h \circ h' = 1_{X'}$, which proves (a).

In practice, the two solutions of a same universal problem are identified, i.e. the two representatives of a same functor, and for the sake of simplicity, any of them is referred to as *the* solution to the universal problem.

2.4.12

Let F and G be representable contravariant functors, (X, ξ) and (Y, η) respective representatives. The functorial morphisms from F to G are in bijective correspondence with morphisms from X to Y. More precisely, for any morphism $\Phi : F \to G$, there is a unique morphism $\varphi : X \to Y$ such that $\Phi_X(\xi) = \varphi^*(\eta)$.

This result is an immediate consequence of 2.4.3. It can also be stated as follows:

Proposition *Let $\mathcal{C}$ be a category, and $\widehat{\mathcal{C}}$ the category of contravariant functors from $\mathcal{C}$ to $\mathcal{S}ns$. The functor $X \to \widehat{X}$ from $\mathcal{C}$ to $\widehat{\mathcal{C}}$ is fully faithful.*

Exercices 2.4. (Representable functors)

1.—Let X be a compact metric space. Denote by F the set of closed subsets of X. For closed subsets A and B of X, set $\sigma(A, B) = \sup_{x \in A} d(x, B)$ and $d(A, B) = \sup(\sigma(A, B), \sigma(B, A))$. Show that

(a) d is a distance on F;

(b) F is compact with respect to the topology defined by this distance;

(c) F represents the functor $\Phi : \mathcal{T}\!op \to \mathcal{E}\!ns$ which assigns to each topological space S the set of closed subsets of $S \times X$ whose projection onto S is open.

Is the functor $\psi : \mathcal{T}\!op \to \mathcal{E}\!ns$ which assigns to each space S the set of closed subsets of $S \times X$ representable? If yes, is it represented by a Hausdorff space?

2.—Let the locally compact metric space X be the union of countable compact spaces. Define two functors from $\mathcal{T}\!op$ to $\mathcal{E}\!ns$ by assigning respectively to each topological space S, the set of closed subsets of $S \times X$ whose projection onto S is open, and the set of closed subsets of $S \times X$ whose projection onto S is both proper (4.1.5) and open. Are these functors representable? Can the assumption that X is metrizable be removed?

3.—Let E and F be normed vector spaces, B_E and B_F the unit balls of E and F respectively, and $B_E \otimes B_F$ the set of xy for $x \in B_E$ and $y \in B_F$.

(a) Show that the convex envelope from $B_E \otimes B_F$ to $E \otimes F$ is the unit ball open with respect to the semi-norm $\pi(E, F)$ on $E \otimes F$.

(b) Give an expression for $\pi(E, F)$.

(c) Formulate the universal problem having $(E \otimes F, \pi(E, F))$ as solution.

(d) Let E_1 be a subspace of E equipped with the induced semi-norm (resp. a quotient of E by a closed subspace equipped with the quotient semi-norm). Is the semi-norm $\pi(E_1, F)$ on $E_1 \otimes F$ induced by (resp. quotient of) the semi-norm $\pi(E, F)$?

(e) Is the semi-norm $\pi(E, F)$ a norm?

2.5 Products and Inverse Limits

2.5.1 Definition

Let $\mathscr{C}$ be a category, X and Y two objects of $\mathscr{C}$. Suppose that the functor $T \mapsto \mathrm{Hom}(T, X) \times \mathrm{Hom}(T, Y)$ from $\mathscr{C}$ to $\mathcal{E}\!ns$ is representable and let $(P, (p, q))$ be a representative. Equipped with (p, q), P is said to be a product of X and Y in $\mathscr{C}$ and, for the sake of simplicity, we write $P = X \times Y$, $p = \mathrm{pr}_1$, $q = \mathrm{pr}_2$.

$\mathscr{C}$ is said to be a *product category* if there is a product in $\mathscr{C}$ for any two arbitrary objects of $\mathscr{C}$.

Let $\mathscr{C}$ and $\mathscr{C}'$ be two product categories and F a functor from $\mathscr{C}$ to $\mathscr{C}'$. F is said to *commute with the products of two objects* if

$$(\forall\, X, Y \in \mathscr{C})\quad F(X \times Y) = F(X) \times F(Y)\,,$$

more precisely, if $F(X \times Y)$ equipped with $(F(\mathrm{pr}_1), F(\mathrm{pr}_2))$ is a product of $F(X)$ and $F(Y)$ in $\mathscr{C}'$.

2.5.2 Examples

Let $\mathscr{C}$ be one of the following categories: $\mathscr{Gr}$, $\mathscr{Ann}$, A-$\mathscr{Mod}$, $\mathscr{Top}$. Then $\mathscr{C}$ is a product category and the forgetful functor from $\mathscr{C}$ to $\mathscr{Ens}$ commutes with products.

In other words, this gives the product $X \times Y$ in $\mathscr{C}$ of two objects of $\mathscr{C}$ by equipping the product set with the underlying sets of a particular structure.

2.5.3

Let $\mathscr{C}$ be a category, $(X_i)_{i\in I}$ a family of objects of $\mathscr{C}$. Suppose the functor $T \mapsto \prod_i \mathrm{Hom}(T, X_i)$ from $\mathscr{C}$ to $\mathscr{Ens}$ is representable and let $(P, (p_i)_{i\in I})$ be a representative. Together with the family $(p_i)_{i\in I}$, P is said to be a *product of the family* $(X_i)_{i\in I}$. We write $P = \prod_{i\in I} X_i$ and $\mathrm{pr}_i = p_i$ is called the *projection of index i*.

Let U be a universe and $\mathscr{C}$ be one of the U-categories $\mathscr{Gr}$, $\mathscr{Ann}$, A-$\mathscr{Mod}$, $\mathscr{Top}$. Then any family of objects indexed by a set belonging to U has a product in $\mathscr{C}$, and the forgetful functor $\mathscr{C} \to \mathscr{Ens}$ commutes with products.

In $\mathscr{Top}$, the product $P = \prod_{i\in I} X_i$ is the product set endowed with the topology generated by the sets $\mathrm{pr}_i^{-1}(U)$, for $i \in I$ and U open in X_i. This topology is called the product topology. A base for this topology is given by $\prod_{i\in I} U_i$ where, for all i, (U_i) is a family such that U_i is open in X_i and $U_i = X_i$ except for finitely many i. Let $x = (x_i) \in P$, and for all i, let $(V_{i,\alpha})_{\alpha\in A}$ be a fundamental system of neighbourhoods of x_i in X_i. The sets $V_{i_1,\alpha_1} \times \cdots \times V_{i_k,\alpha_k} \times \prod_{j\in I-\{i_1,\ldots,i_k\}} X_j$ form a fundamental system of neighbourhoods of x.

Let $\mathscr{C}$ and $\mathscr{C}'$ be two categories. A functor F from $\mathscr{C}$ to $\mathscr{C}'$ is said to *commute with products when these exist* if, for any family (X_i) of objects of $\mathscr{C}$ having a product P in $\mathscr{C}$, the object $F(P)$ is a product of $F(X_i)$ in $\mathscr{C}'$.

2.5.4 Definition

Let X, Y, S be sets, $f : X \to S$ and $g : Y \to S$ maps. The **fibered product** of X and Y over S with respect to f and g is the set $X \times_S Y = \{(x, y) \mid (x, y) \in$

$X \times Y$ and $f(x) = g(y)$}. Next let $\mathscr{C}$ be a category, X, Y, S objects of $\mathscr{C}$, $f : X \to S$ and $g : Y \to S$ morphisms. Consider the contravariant functor $T \mapsto \widehat{X}(T) \times_{\widehat{S}(T)} \widehat{Y}(T)$ from $\mathscr{C}$ to $\mathbf{Ens}$. If this functor is representable and if $(P, (p, q))$ is a representative, then P together with (p, q) is said to be a *fibered product* of X and Y over S with respect to *à* f and g, and is written $P = X \times_S Y$.

2.5.5 *Definition*

Let X and Y be sets, f and g maps from X to Y. The *kernel of a double arrow* (f, g) is the set

$$\mathrm{Ker}(f, g) = \{x \mid x \in X \text{ and } f(x) = g(x)\}.$$

Let $\mathscr{C}$ be a category, $X, Y \in \mathscr{C}$, f and g morphisms from X to Y. Consider the functor $F : T \mapsto \mathrm{Ker}(f_*, g_*)$, where $f_* : \widehat{X}(T) \to \widehat{Y}(T)$ is defined by $f_*(\alpha) = f \circ \alpha$ and g_* is defined likewise. If this functor is representable, then a representative of this functor is called a **kernel of a double arrow** (f, g).

2.5.6 *Definition*

An ordered set I is said to be *right directed* or simply directed if

$$(\forall i, j \in I) \; (\exists k \in I) \quad k \geq i \text{ and } k \geq j \,.$$

Let I be a directed set and $\mathscr{C}$ a category. An *inverse system in $\mathscr{C}$ indexed by* I consists of a family $(X_i)_{i \in I}$ of objects of $\mathscr{C}$ and, for any pair $(i, j) \in I^2$ such that $i \leq j$, of a morphism $f_i^j : X_j \to X_i$ satisfying the following conditions:

(i) $(\forall i \in I) \; f_i^i = 1_{X_i}$;
(ii) $(\forall i, j, k \in I \mid i \leq j \leq k) \; f_i^j \circ f_j^k = f_i^k$.

In other words, an inverse system in $\mathscr{C}$ indexed by I is a contravariant functor from $\mathscr{J}$ to $\mathscr{C}$, where $\mathscr{J}$ is the category associated to the ordered set I (2.1.2, Example 5).

2.5.7 *Definition*

Let I be a directed set and $((X_i), (f_i^j))$ an inverse system in $\mathbf{Ens}$ indexed by I. The **inverse limit** of the inverse system of the given sets is the set

$$\left\{ x = (x_i)_{i \in I} \in \prod_{i \in I} X_i \; \middle| \; (\forall i \in I) \; (\forall j \in I)_{i \leq j} \; f_i^j(x_j) = x_i \right\}.$$

It is written $\varprojlim_{i \in I} X_i$, or simply $\varprojlim X_i$.

2.5.8 *Definition*

Let $\mathscr{C}$ be a category, I a directed set and $((X_i)_{i\in I}, (f_i^j))$ an inverse system in $\mathscr{C}$ indexed by I. If the functor $T \mapsto \varprojlim \mathrm{Hom}(T, X_i)$ is representable, a representative $(X, (p_i))$, i.e. a pair $(X, (p_i))$ where X is an object of $\mathscr{C}$ and $(p_i : X \to X_i)_{i\in I}$ a family of morphisms such that, for any pair (i, j) $i \le j$, $p_i = f_i^j \circ p_j$ is called the *inverse limit* of the inverse system $((X_i), (f_i^j))$.

In other words, the inverse limit $\varprojlim X_i$ equipped with morphisms $p_i : \varprojlim X_i \to X_i$ is characterized by the following universal property:

$$(\forall T \in \mathscr{C})\ (\forall (g_i)_{i\in I} \in \varprojlim \mathrm{Hom}(T, X_i))$$

$$(\exists|\ g \in \mathrm{Hom}(T, \varprojlim X_i))\ (\forall i \in I)\quad p_i \circ g = g_i\,.$$

2.5.9 *Example*

Let E be a set, I a non-empty directed set and $(X_i)_{i\in I}$ a decreasing sequence of subsets of E. For $i \le j$, let f_i^j be the canonical injection from X_j to X_i. This defines an inverse system whose inverse limit can be identified with the intersection of the subsets X_i.

2.5.10

Proposition *Let $\mathscr{C}$ be a category and $\mathfrak{w}$ an infinite cardinal. If every family $(X_i)_{i\in I}$ with* Card I $\le \mathfrak{w}$ *has a product in $\mathscr{C}$ and if every double arrow $X \rightrightarrows Y$ has a kernel, then every inverse system indexed by an ordered set I of cardinality* Card I $\le \mathfrak{w}$ *has an inverse limit.*

Proof Let $((X_i), (f_i^j))$ be an inverse system indexed by an ordered set é I such that Card I $\le \mathfrak{w}$. For every pair of indices (i, j) with $i \le j$, set $Y_{ij} = X_i, \alpha_{ij} = \mathrm{pr}_i$ where ù $\mathrm{pr}_i : \prod_{i\in I} X_i \to X_i$ is the projection of index i and $\beta_{ij} = f_i^j \circ p_j$. This defines two morphisms $\alpha = (\alpha_{ij})$ and $\beta = (\beta_{ij})$ from $\prod_{i\in I} X_i$ to $\prod_{\substack{(i,j)\in I^2 \\ i\le j}} Y_{ij}$.

The kernel for the double arrow (α, β) is the inverse limit of the inverse system $((X_i), (f_i^j))$. $\square$

2.5.11 Examples

Let $\mathscr{C}$ be one of the following U-categories $\mathscr{Gr}$, $\mathscr{Ann}$, A-$\mathscr{Mod}$, $\mathscr{Top}$. Then every inverse system (indexed by a set belonging to U) has an inverse limit and the forgetful functor from $\mathscr{C}$ to $\mathscr{Ens}$ commutes with inverse limits.

Remark In $\mathscr{Top}$, let $((X_i), (f_i^j))$ be an inverse system, and setting $X = \varprojlim X_i$, write p_i for the canonical map $X \to X_i$. Then the topology of X is induced by the product topology on $\prod_{i \in I} X_i$. Indeed, let $x = (x_i)$ be a point of X and, for all i, let $(V_{i,\alpha})_{\alpha \in A_i}$ be a fundamental system of neighbourhoods of x_i in X. Set $W_{i,\alpha} = p_i^{-1}(V_{i,\alpha})$. The subsets $W_{i_1,\alpha_1} \cap \cdots \cap W_{i_k,\alpha_k}$ clearly form a fundamental system of neighbourhoods of x. Let $i \in I$ be a common upper bound of $i_1, \ldots, i_k$. Let $(f_{i_1}^i)^{-1}(V_{i_1,\alpha_1}) \cap \cdots \cap (f_{i_k}^i)^{-1}(V_{i_k,\alpha_k})$ be a neighbourhood of x_i in X_i; it thus contains some $V_{i,\alpha}$, with $\alpha \in A_i$, and $W_{i,\alpha} \subset W_{i_1,\alpha_1} \cap \cdots \cap W_{i_k,\alpha_k}$.

2.5.12 Morphisms of Inverse Systems

Let $\mathscr{C}$ be a category, I a directed set, $\mathscr{X} = ((X_i), (f_i^j))$ and $\mathscr{Y} = ((Y_i), (g_i^j))$ inverse systems in $\mathscr{C}$ indexed by I. A morphism from $\mathscr{X}$ to $\mathscr{Y}$ is a family (φ_i) of morphisms such that, for $i \leq j$, the following diagram is commutative

$$
\begin{array}{ccc}
X_j & \xrightarrow{\; f_i^j \;} & X_i \\
\downarrow{\scriptstyle \varphi_j} & & \downarrow{\scriptstyle \varphi_i} \\
Y_j & \xrightarrow[\; g_i^j \;]{} & Y_i
\end{array}
$$

If $\mathscr{X}$ and $\mathscr{Y}$ have inverse limits X_∞ and Y_∞, then for any morphism (φ_i) from $\mathscr{X}$ to $\mathscr{Y}$, there is a unique morphism $\varphi_\infty : X_\infty \to Y_\infty$ such that $\varphi_i \circ p_i = q_i \circ \varphi_\infty$ for all i, where $p_i : X_\infty \to X_i$ and $q_i : Y_\infty \to Y_i$ are canonical morphisms. The morphism φ_∞ is said to be *induced by the morphisms φ_i on passing to the inverse limit*.

2.5.13 Cofinal Sets

Let I be a directed set and J a subset of I. We remind the reader (1.2.2) that J is said to be *cofinal* in I if $(\forall i \in I)(\exists j \in J)\, i \leq j$. If J is cofinal in I, then J is a directed set under the induced order.

Let $((X_i), (f_i^j))$ be an inverse system of sets. If J is a cofinal subset of I, the canonical map $\varprojlim_{i \in I} X_i \to \varprojlim_{i \in J} X_i$ is bijective.

It follows that, if $((X_i)_{i \in I}, (f_i^j))$ is an inverse system in a category $\mathscr{C}$ and if J is a cofinal subset of I, the inverse system $((X_i)_{i \in J}, (f_i^j))$ has an inverse limit if and only if so does $((X_i)_{i \in I}, (f_i^j))$, and these limits are identical if they exist.

A *cofinal sequence* in I is an increasing sequence (i_n) such that the set of i_n is cofinal in I.

Proposition *Every countable non-empty directed set has a cofinal sequence.*

Proof Let $n \mapsto x_n$ be a surjective map from $\mathbb{N}$ over I. Define i_n inductively by setting $i_0 = x_0$ and choosing for i_{n+1} a common upper bound of i_n and x_{n+1}. The sequence (i_n) is cofinal.

2.5.14 Inverse Limits of Compact Spaces

Theorem *In $\mathscr{T}\!op$, an inverse limit of non-empty compact is non-empty and compact.*

Proof Let $((X_i)_{i \in I}, (f_i^j))$ be non-empty compact projective spaces. By Tychonoff's Theorem (1.7.7), the space $P = \prod_{i \in I} X_i$ is compact. For $k \in I$, set L_k to be the subspace of P of elements $x = (x_i)$ such that $x_i = f_i^k(x_k)$ for $i \leq k$; it is closed in P. If τ is a choice function (1.1.2) on $\bigcup X_i$, then the element x defined by $x_k = \tau(X_k)$, $x_i = f_i^k(x_k)$ if $i \leq k$ and $x_i = \tau(X_i)$ otherwise, belongs to L_k. Then $(L_k)_{k \in I}$ are non-empty directed closed sets in P, and so $\varprojlim X_i = \bigcap_{k \in I} L_k$ is non-empty and compact. $\square$

2.5.15

Corollary *In $\mathscr{T}\!op$, let $((X_i), (f_i^j))$ be an inverse system of compact spaces, and set $X_\infty = \varprojlim X_i$. If every f_i^j is surjective, then so are $f_i^\infty : X_\infty \to X_i$.*

Proof Let $x \in X_i$, and for $j \geq i$ set $Y_j = (f_i^j)^{-1}(x_i)$. By the theorem, $\varprojlim Y_j \neq \varnothing$. $\square$

Exercises 2.5. (Products and inverse limits)
1.—Show that in the category of locally compact spaces and proper maps there are fibered products but no products.

2.—Let $\mathscr{C}$ be a category of finite products, containing a terminal object E. Let $X \in \mathscr{C}$. A *group law on X in $\mathscr{C}$* is a morphism $\mu : X \times X \to X$ satisfying the following properties:

(i)

$$
\begin{array}{ccc}
X \times X \times X & \xrightarrow{\ \mu \times 1_X\ } & X \times X \\
{\scriptstyle 1_X \times \mu} \downarrow & & \downarrow {\scriptstyle \mu} \\
X \times X & \xrightarrow{\ \ \mu\ \ } & X
\end{array}
$$

commutes.

(ii) There is a map $\varepsilon : E \to X$ such that the diagrams

commute, $e : X \to E$ being the unique morphism from X to E.

(iii) There is a morphism $\sigma : X \to X$ such that the diagrams

$$
\begin{array}{ccc}
X & \xrightarrow{\ 1_X, \sigma\ } & X \times X \\
{\scriptstyle e} \downarrow & & \downarrow {\scriptstyle \mu} \\
E & \xrightarrow{\ \ \varepsilon\ \ } & X
\end{array}
\qquad \text{and} \qquad
\begin{array}{ccc}
X & \xrightarrow{\ \sigma, 1_X\ } & X \times X \\
{\scriptstyle e} \downarrow & & \downarrow {\scriptstyle \mu} \\
E & \xrightarrow{\ \ \varepsilon\ \ } & X
\end{array}
$$

commutate.

(a) Show that, if μ is a group law, then the morphisms ε and σ are unique.

(b) Let $\mu : X \times X \to X$ be a morphism. It is a group law if and only if for any object T of $\mathscr{C}$, the composition law defined by μ on $\mathrm{Hom}_{\mathscr{C}}(T, X)$ is a group law.

(c) Show that defining a group law on X amounts to defining a functor $\Phi : \mathscr{C} \to \mathscr{G}r$ such that $\omega \circ \Phi = \widehat{X}$ where $\omega : \mathscr{G}r \to \mathscr{E}ns$ is the forgetful functor.

(d) Which groups have a group law in the category $\mathscr{G}r$?

(e) Let X be an object of $\mathscr{C}$ and $F : \mathscr{C} \to \mathscr{G}r$ a functor commuting with products. Show that if there is a group law μ on X in $\mathscr{C}$, then $F(X)$ is commutative and $F(\mu)$ agrees with the group law of $F(X)$.

3.—Let E_0 be a vector space $\mathbb{R}[X]$, and for all n, let E_n denote the subspace $X^n \mathbb{R}[X]$. The map $f : E_0 \to \mathbb{R}$ defined by $f(P) = P(1)$ induces a surjective map $f_n : E_n \to \mathbb{R}$ for each n. Show that the induced map on passing to the inverse limit is not surjective.

4.—Let $((X_i)_{i \in I}, (f_i^j)_{j \geq i})$ be an inverse system of sets such that all f_i^j are surjective. Set $X_\infty = \varprojlim X_i$.

(a) Show that if I is countable, $f_i^\infty : X_\infty \to X_i$ is surjective for all i.

(b) The aim is to show that the countability assumption can be omitted: let E be an uncountable infinite set and F a countable infinite set. For any countable subset A of E, let X_A be the set of injective maps from A to F for which the complement of the image is infinite; for $B \subset A$, denote by p_B^A the restriction $X_A \to X_B$. Show that the maps p_B^A are surjective, but that the inverse limit of the sets X_A is empty.

5.—The aim is to prove the following result:

Theorem (Mittag-Leffler) *Let* $(0 \to E_i' \xrightarrow{u_i} E_i \xrightarrow{v_i} E_i'' \to 0, (f'^j_i, f^j_i, f''^j_i))$ *be an inverse system of exact sequences of Fréchet spaces (i.e. complete and metrizable vector spaces), indexed by a countable directed set* I. *Suppose that, for any pair* (i, j) *such that* $j \geq i$, *the image of the map* $f'^j_i : E_j' \to E_i'$ *is dense. Then, passage to the inverse limit gives an exact sequence* $0 \to E_\infty' \to E_\infty \to E_\infty'' \to 0.$

(a) Show that $0 \to E_\infty' \to E_\infty \to E_\infty''$ is an exact sequence.

(b) Show that the question reduces to the case $I = \mathbb{N}$, where E_n is equipped with a distance $(x, y) \mapsto p_n(x - y)$, E_n' being a closed subspace of E_n equipped with the induced distance and all f_n^{n+1} are contraction maps (i.e. $p_n(f_n^{n+1}(x)) \leq p_{n+1}(x)$ for $x \in E_{n+1}$).

(c) Let $x'' = (x_n'')$ be an element of E''. Construct inductively a sequence (y_n) such that $y_n \in E_n$, $v_n(y_n) = x_n''$ and $p_n(f_n^{n+1}(y_{n+1}) - y_n) \leq \frac{1}{2^n}$.

(d) Set $x_{n,k} = f_n^{n+k}(y_{n+k})$. Show that, for fixed n, $(x_{n,k})_{k \in \mathbb{N}}$ is a Cauchy sequence in E_n. Conclude.

(e) Show that condition "the images of f'^j_i are dense" can be replaced by the following weaker one:

$$(\forall i)(\exists j \geq i)(\forall k \geq j)\ \mathrm{Im}\, f'^k_i \text{ dense in } \mathrm{Im}\, f'^j_i.$$

6.—Let X be a set with an anti-reflexive symmetric relation $\mathcal{R}$. A *k-colouring* of X is a map f from X to a set with k elements such that $\mathcal{R}(x, y) \Rightarrow f(x) \neq f(y)$. (For instance, the set X of the regions of a geographical map and $\mathcal{R}(x, y) \Leftrightarrow x$ and y are adjacent.)

Show that, if every finite subset of X has a k-colouring (for fixed k), so has X.

7.—Let A be a group (resp. a ring) and (I_n) a decreasing sequence of normal subgroups (resp. ideals) With $I_0 = A$. Let (ε_n) be a sequence approaching 0. A function $h : A \to \mathbb{R}_+$ can be defined by $h(x) = \varepsilon_n$ if $x \in I_n - I_{n+1}$ and $h(x) = 0$ if $x \in \cap_n I_n$. Let d be the mapping $A \times A$ to $\mathbb{R}_+$ defined by $d(x, y) = h(xy^{-1})$ (resp. $h(x - y)$). Show that

(a) d is a distance on A, the composition law (resp. laws) on A is continuous with respect to the topology defined by d.

(b) The Hausdorff completion of A with respect to this distance can be identified with the inverse limit $\varprojlim A/I_n$.

(c) When $A = \mathbb{Z}$ and I_n is the ideal generated by p^n where $p \in \mathbb{Z}$:

- for prime p, $\widehat{\mathbb{Z}}_p = \varprojlim \mathbb{Z}/p^n\mathbb{Z}$ is an integer ring;
- for $p = p_1 \times p_2$, with g.c.d.$(p_1, p_2) = 1$, $\widehat{\mathbb{Z}}_p = \widehat{\mathbb{Z}}_{p_1} \times \widehat{\mathbb{Z}}_{p_2}$;
- for $p = p_1^r$, $\widehat{\mathbb{Z}}_p = \widehat{\mathbb{Z}}_{p_1}$.

(d) The elements of $\widehat{\mathbb{Z}}_{10}$ can be represented by positive non terminating decimal expansions to the left.

What is the expansion of -1?

Give an algorithm for the decimal expansions of the idempotents of $\widehat{\mathbb{Z}}_{10}$.

(e) Show that the profinite completion 2.9.5 $\widehat{\mathbb{Z}}$ of $\mathbb{Z}$ can be identified with the product of $\widehat{\mathbb{Z}}_p$ over primes p.

2.6 Sums and Direct Limits

2.6.1 Definition

Let $\mathscr{C}$ be a category, $(X_i)_{i \in I}$ a family of objects of $\mathscr{C}$. The **sum** of X_i in $\mathscr{C}$, written $\bigsqcup_{i \in I} X_i$, is the product, if it exists, of all X_i in the opposite category $\mathscr{C}^o$, i.e. an object S of $\mathscr{C}$ equipped with a family of morphisms $(s_i : X_i \to S)_{i \in I}$ such that, for any object T of $\mathscr{C}$ and for any family $(f_i : X_i \to T_i)_{i \in I}$ of morphisms, there is a unique morphism $f : S \to T$ for all $i \in I$ satisfying $f \circ s_i = f_i$.

Let $X, Y, S \in \mathscr{C}$, $f : S \to X$ and $g : S \to Y$ morphisms. An *amalgamated sum* of X and Y over S with respect to f and g, written $X \sqcup_S Y$, is the fibered product (if it exists) in the opposite category $\mathscr{C}^o$ (2.5.4 and 2.1.7).

Let $X, Y \in \mathscr{C}$, f and g morphisms from X to Y. The *cokernel of a double arrow* (f, g) is the kernel of the double arrow (f, g) (if it exist) in the opposite category $\mathscr{C}^o$ (2.5.5).

2.6.2 Examples

The categories considered in the following examples are U-categories, and the index sets are elements of U.

(1) Let $(X_i)_{i \in I}$ be a family of sets. If the sets X_i are disjoint, then the set $\bigcup_{i \in I} X_i$ equipped with canonical injections is the sum of all X_i in $\mathscr{E}ns$. In the general case, $\bigsqcup_{i \in I} X_i = \bigcup_{i \in I} \{i\} \times X_i$ equipped with maps s_i defined by $s_i(x) = (i, x)$ is the sum of all X_i in $\mathscr{E}ns$.

(2) In $\mathscr{T}op$ every family of topological spaces has a sum and the forgetful functor from $\mathscr{T}op$ to $\mathscr{E}ns$ commutes with the sum.

Let $(X_i)_{i \in I}$ be a family of topological spaces. The topological space $\bigsqcup_{i \in I} X_i$ is given by the summation set $\bigsqcup_{i \in I} X_i$ equipped with the topology whose open subsets are $\bigsqcup_{i \in I} U_i$, where, for each i, U_i is open in X_i.

In the following examples, the sums exist but the forgetful functors do not commute with them. To avoid confusion, the sums will be written using a symbol other than $\sqcup$.

(3) In $\mathscr{G}r$ any family $(G_i)_{i\in I}$ of groups has a sum also called a **free product** of the groups G_i. It is written $*_{i\in I}G_i$, where $G_1 * G_2 * \cdots * G_k$ if $I = \{1, ..., k\}$.

The free product of a family $(G_i)_{i\in I}$ of groups is obtained as follows: let S be the summation set of the underlying sets of the groups G_i, $s_i : G_i \to S$ the canonical injections, and M the set $\bigsqcup_{n\in\mathbb{N}} S^n$. Define a composition law on M by assigning to the pair $((x_1, ..., x_p), (y_1, ..., y_q))$ the element $(x_1, ..., x_p, y_1, ..., y_q) \in S^{p+q}$. This law is associative and has an identity element: the empty set, which is the unique element of S^0. For any group H and any family $f = (f_i : G_i \to H)_{i\in I}$ of morphisms, define a map $\alpha_f : S \to H$ by $\alpha_f \circ s_i = f_i$ for all $i \in I$ and a map $\beta_f : M \to H$ par $\beta_f(x_1, ..., x_p) = \alpha_f(x_1) \cdots \alpha_f(x_p)$. The map β is a homomorphism. In M, write $x \sim y$ if, for any group H and any family of morphisms $f = (f_i : G_i \to H)_{i\in I}$, $\beta_f(x) = \beta_f(y)$. This is an equivalence relation compatible with the composition law of M, and hence there is an associative quotient law with an identity element on $L = M/\sim$; moreover, setting χ for the canonical map $M \to L$, the element $\chi(x_1, \ldots, x_p)$ of L has as inverse $\chi(x_p^{-1}, \ldots, x_1^{-1})$. Hence L is a group. Identify S with S^1 and write λ_i for the morphism $\chi \circ \iota_1 \circ s_i$ where ι_1 is the canonical injection from S^1 to $\bigsqcup S^n$. The group L equipped with the morphisms λ_i is the desired free product.

(4) In the category of Abelian groups, any family $(G_i)_{i\in I}$ of objects has a sum, called the **direct sum** of the groups G_i and written $\bigoplus_{i\in I} G_i$. If I is finite, the direct sum can be identified with the product $\prod_{i\in I} G_i$. In the general case, $\bigoplus_{i\in I} G_i$ can be identified with the subgroup of $\prod_{i\in I} G_i$ consisting of elements having finitely many non-trivial coordinates.

In general, the free product $*G_i$ is not commutative (for example $\mathbb{Z} * \mathbb{Z}$); the direct sum of the groups G_i can be identified with a quotient of the free product.

(5) Let A be a commutative ring. In the category of unital associative commutative A-algebras, the sum of two algebras E and F is the tensor product algebra $E \otimes_A F$ (3.8.15).

2.6.3 Definition

Let $\mathscr{C}$ be a category, and I a directed set.

A **direct system** in $\mathscr{C}$ indexed by I is an inverse system in $\mathscr{C}^\circ$ indexed by I, i.e. for every $i \in I$, consists of an object X_i of $\mathscr{C}$ and, for every pair (i, j) such that $i \leq j$, of a morphism $f_i^j : X_i \to X_j$ such that, for all i,

$$f_i^i = 1_{X_i},$$

and for all $i \leq j \leq k$,

$$f_j^k \circ f_i^j = f_i^k.$$

2.6.4 Definition

Let $\mathscr{C}$ be a category, I a directed set, and $((X_i), (f_i^j))$ a direct system in $\mathscr{C}$. If $((X_i), (f_i^j))$ has a direct limit $(X_\infty, (\sigma_i))$ in $\mathscr{C}^0$, then $(X_\infty, (\sigma_i))$ is said to be the **direct limit** in $\mathscr{C}$ of the given direct system. It is written $X_\infty = \varinjlim X_i$.

In other words, the direct limit X_∞ equipped with morphisms $\overrightarrow{\sigma_i} : X_i \to X_\infty$ is characterized by the following universal property: for all $T \in \mathscr{C}$, for all families of morphisms $(f_i : X_i \to T)_{i \in I}$ such that $f_i = f_j \circ f_i^j$, for all pairs (i, j) with $i \leq j$, there is a unique morphism $g : X_\infty \to T$ such that, for all i, $g \circ \sigma_i = f_i$.

2.6.5 Direct Limits of Sets

Let $((X_i), (f_i^j))$ be a direct system of sets indexed by a directed set I. On the summation set $S = \bigsqcup_{i \in I} X_i$ define an equivalence relation by setting the class of (i, x) to consist of all (j, y) for which there is an upper bound k of i and j such that $f_i^k(x) = f_j^k(y)$. The quotient set $X_\infty = S/\sim$, together with the maps $\sigma_i = \chi \circ s_i$, where χ is the canonical map $S \to X_\infty$ and s_i the canonical map $X_i \to S$, is the direct limit of the given system.

The set X_∞ together with the maps σ_i is characterized by the following properties:

(i) $\sigma_i = \sigma_j \circ f_i^j$ pour tout (i, j);
(ii) $X_\infty = \bigcup_{i \in I} \sigma_i(X_i)$;
(iii) $(\forall i \in I) (\forall x, y \in X_i) \, \sigma_i(x) = \sigma_i(y) \Rightarrow (\exists j \geq i) \, f_i^j(x) = f_i^j(y)$.

Let $((X_i), (f_i^j))$ and $((Y_i), (g_i^j))$ be two direct systems of sets indexed by the same set, and $(X_\infty, (\sigma_i))$ and $(Y_\infty, (\tau_i))$ their direct limits. Then, $\varinjlim(X_i \times Y_i) = X_\infty \times Y_\infty$. Indeed, $X_\infty \times Y_\infty$, equipped with the maps $\sigma_i \times \tau_i : X_i \times Y_i \to X_\infty \times Y_\infty$, satisfy above conditions (i), (ii) and (iii).

2.6.6

In the categories $\mathscr{G}r$, $\mathscr{A}nn$, A-$\mathscr{M}od$, $\mathscr{T}op$, any direct system has a direct limit, and the forgetful functors commute with the direct limits.

In $\mathscr{T}op$, the direct limit $\varinjlim X_i$ is obtained by equipping the direct limit X_∞ with the finest topology for which the canonical maps $\sigma_i : X_i \to X_\infty$ remain continuous. A subset U of X_∞ is open with respect to this topology if and only if, for any i, $\sigma_i^{-1}(U)$ is open in X_i.

Note that the space X_∞ need not be Hausdorff even if the spaces X_i are.

2.6.7

Let $((X_i), (f_i^j))$ and $((Y_i), (g_i^j))$ be two direct systems of sets indexed by the same set I, and $(\varphi_i : X_i \to Y_i)_{i \in I}$ a morphism of direct systems (i.e. a family of maps commuting with all f_i^j and g_i^j). Write φ_∞ for the map from X_∞ to Y_∞ induced by the morphisms the φ_i on passing to the direct limit. If the φ_i are injective (resp. surjective), φ_∞ is injective (resp. surjective); in general,

$$\varphi_\infty(X_\infty) = \varinjlim \varphi_i(X_i).$$

Remark If $((X_i), (f_i^j))$ and $((Y_i), (g_i^j))$ are inverse systems of sets, and if $(\varphi_i)_{i \in I}$ is a morphism of inverse systems such that, for all i, φ_i is injective, then φ_∞ is also injective; but if, for all i, φ_i is surjective, φ_∞ need not be surjective (2.5, Exercise 4).

2.6.8 Direct Limit of Categories

If $\mathscr{C}$ is a category, write $\mathscr{F}\!\ell(\mathscr{C})$ for the set of morphisms of $\mathscr{C}$, namely

$$\mathscr{F}\!\ell(\mathscr{C}) = \bigsqcup_{(X,Y) \in \mathscr{C} \times \mathscr{C}} \mathrm{Hom}(X, Y),$$

and define the maps α and β from $\mathscr{F}\!\ell(\mathscr{C})$ to $\mathscr{C}$ by $\alpha(f) = X$ and $\beta(f) = Y$ if $f \in \mathrm{Hom}(X, Y)$. The composition law is a map $(\mathscr{F}\!\ell(\mathscr{C}), \beta) \times_{\mathscr{C}} (\mathscr{F}\!\ell(\mathscr{C}), \alpha) \to \mathscr{F}\!\ell(\mathscr{C})$.

Let $((\mathscr{C}_i), (F_i^j))$ be a direct system in $\mathscr{C}\!\mathit{at}$: all $\mathscr{C}_i$ are categories and F_i^j functors. Define a category $\mathscr{C}_\infty$ by setting its objects to be the direct limits $\mathscr{C}_i$ and by setting $\mathscr{F}\!\ell(\mathscr{C}_\infty) = \varinjlim \mathscr{F}\!\ell(\mathscr{C}_i)$, the maps α and β and the composition law being induced by the corresponding maps in $\mathscr{C}_i$ on passing to the direct limit.

The category $\mathscr{C}_\infty$ obtained is the direct limit of the given system in $\mathscr{C}\!\mathit{at}$.

Let X and Y be objects of $\mathscr{C}_\infty$, and let (i, A) and (j, B) be representatives of X and Y respectively. Then

$$\mathrm{Hom}_{\mathscr{C}_\infty}(X, Y) = \varinjlim_{k \geq i, k \geq j} \mathrm{Hom}_{\mathscr{C}_k}(F_i^k(X), F_j^k(Y)).$$

Exercises 2.6. (Sums and direct limits)

1.—Let $\mathscr{C}$ be a category with sum. Let $X \in \mathscr{C}$, and i_1 and i_2 the canonical morphisms from X to $X \sqcup X$. The *codiagonal* morphism of X is the unique morphism $\nabla : X \sqcup X \to X$ such that $\nabla \circ i_1 = \nabla \circ i_2 = 1_X$. Describe the codiagonal morphism in the following categories: $\mathscr{E}\!\mathit{ns}$, the category $\mathscr{T}\!\mathit{op}$ of point spaces (topological spaces with what is known as a basepoint, the morphisms from X to Y being the continuous maps from X to Y sending the basepoint of X to the basepoint of Y), A-$\mathcal{M}\!od$, and the category $\mathcal{G}\!r$ of unital associative commutative algebras over a ring A.

2.—Let $\mathcal{C}$ be a product category with sums, having an object O which is both initial and terminal. For X, Y $\in \mathcal{C}$, let $O_{X,Y}$ be the unique morphism from X to Y which can be factorized through O, and define $\varphi_{X,Y} : X \sqcup Y \to X \times Y$ by $\varphi_{X,Y} \circ i_1 = (1_X, O_{X,Y})$ and $\varphi_{X,Y} \circ i_2 = (O_{Y,X}, 1_Y)$.

(a) Suppose that, for all X and Y, the morphism $\varphi_{X,Y}$ is an isomorphism. For every object X, define $\alpha_X : X \times X \to X$ by $\alpha_X = \nabla_X \circ \varphi_{X,X}^{-1}$, where ∇_X is the codiagonal morphism of X. Show that α_X is a commutative and associative law in $\mathcal{C}$. For all X, Y, the morphism α_Y gives a composition law $\alpha_{X,Y}$ on Hom(X, Y). This law is associative, commutative and has $O_{X,Y}$ as identity element. Let X, Y, Z $\in \mathcal{C}$. For fixed $f \in$ Hom(X, Y) (resp. for $g \in$ Hom(Y, Z)), the map $g \mapsto g \circ f$ (resp. $f \mapsto g \circ f$) is a homomorphism from Hom(Y, Z) (resp. Hom(X, Y)) to Hom(X, Z).

(b) Assume that, for all X, Y $\in \mathcal{C}$, there is a composition law $\mu_{X,Y}$ on Hom(X, Y) with $O_{X,Y}$ the identity element. Suppose that, as above, the maps $g \mapsto g \circ f$ and $f \mapsto g \circ f$ are homomorphisms. Show that $\varphi_{X,Y}$ is an isomorphism for all X and Y, and that $\mu_{X,Y} = \alpha_{X,Y}$. In particular $\mu_{X,Y}$ is associative and commutative.

3.—Let X be a Hausdorff topological space.

(a) Show that the following conditions are equivalent:

(i) X is the direct limit of compact subspaces;
(ii) A subset A of X is closed if and only if, for any compact subspace K of X, A ∩ K is closed.

(b) Show that every metrizable space is the direct limit of compact subspaces and that the same holds for every locally compact space.

(c) Show that in the category $\mathcal{C}$ of Hausdorff direct limits of compact subspaces, there are products and that the product X × Y in $\mathcal{C}$ of X, Y $\in \mathcal{C}$ is the set X × Y equipped with a topology finer than the product topology. Give examples where these two topologies agree.

(d) In $\mathbb{R}^{(\mathbb{N})}$, write X_n for the union $\bigcup_{k \le n}[0, e_k]$ (where the elements e_n form the canonical basis of $\mathbb{R}^{(\mathbb{N})}$) equipped with the topology induced by that of $\mathbb{R}^{\mathbb{N}}$. Set $X = \varinjlim X_n$ (i.e. $\bigcup X_n$) equipped with the direct limit topology. For any sequence $\varepsilon = (\varepsilon_n)$ of strictly positive numbers, set $V_\varepsilon = \bigcup [0, \varepsilon_n e_n[$.

Show that the sets V_ε form a fundamental system of neighbourhoods in X, and that any compact subset of X is contained in some X_n.

(e) Let Y be the unit ball of ℓ^2. Write $(f_p)_{p \in \mathbb{N}^*}$ for the canonical Hilbert basis of Y. In X × Y, for any $n \in \mathbb{N}^*$, consider the set $A_n = \{(\frac{1}{p}e_n, \frac{1}{n}f_p)\}$. Show that A_n is closed in X × Y. Let $A = \bigcup A_n$. Show that the intersection of A and of a compact subset of X × Y is closed, but that every neighbourhood of 0 meets A.

Deduce that the topology of the product $X \times Y$ in the category $\mathscr{C}$ is strictly finer than the product topology.

4. *(Alternative proof of the existence of the free product of a family of groups.)* Let $(G_i)_{i \in I}$ be a family of groups.

(a) (Lemma about cardinality.) Let H be a group and $(f_i : G_i \to H)_{i \in I}$ a family of homomorphisms. Show that the cardinality of the subgroup of H generated by all $f_i(G_i)$ is bounded above by the supremum $\mathfrak{w}$ of the cardinalities of I, G_i and $\aleph_0$.

(b) Let X be a set with cardinality $\mathfrak{w}$, and Λ the set of pairs (H, f), where H is a group whose underlying set is contained in X and $f = (f_i)_{i \in I}$ a family of homomorphisms $f_i : G_i \to H$.

Set $P = \prod_{(H,f) \in \Lambda} H$ and write φ_i for the homomorphism $(f_i)_{(H,f) \in \Lambda}$ from G_i to P. Show that the subgroup L of P generated by all $\varphi_i(G_i)$, equipped with the family of homomorphisms $(\varphi_i)_{i \in I}$ is a free product of the groups G_i.

5.—With the notation of 2.6.2, Example 3, an element of S^n is called a *word* of length n. A word $(x_1, ..., x_n)$ is said to be *reduced* if no x_k is the identity element for $k \in \{1, ..., n\}$, and if x_k and x_{k+1} belong to different groups for $k \in \{1, ..., n-1\}$.

(a) Show that every element of the free product L can be represented by a reduced word.

(b) Give an algorithm associating to any two reduced words $(x_1, ..., x_p)$ and $(y_1, ..., y_q)$ a reduced word of length $\grave{a} \leq p + q$ representing the same element of L as $(x_1, ..., x_p, y_1, ..., y_q)$. Show that the law thereby defined in the set of reduced words is associative.

(c) Deduce that every element of L can be uniquely represented by a reduced word.

2.7 Adjoint Functors

2.7.1 Definition

Let $\mathscr{C}$ and $\mathscr{C}'$ be categories, $F : \mathscr{C} \to \mathscr{C}'$ and $G : \mathscr{C}' \to \mathscr{C}$ functors. G is said to be a **right adjoint** to F, or F a **left adjoint** to G, if the functors

$$(X, Y) \mapsto \mathrm{Hom}_{\mathscr{C}'}(F(X), Y) \text{ and } (X, Y) \mapsto \mathrm{Hom}_{\mathscr{C}}(X, G(Y))$$

from $\mathscr{C}^o \times \mathscr{C}'$ to $\mathbf{Sns}$ are isomorphic.

When this is the case, for the sake of simplicity, one writes:

$$\mathrm{Hom}_{\mathscr{C}'}(F(X), Y) = \mathrm{Hom}_{\mathscr{C}}(X, G(Y)).$$

2.7.2

Proposition *Let $\mathscr{C}$ and $\mathscr{C}'$ be categories. A functor* $\mathrm{F} : \mathscr{C} \to \mathscr{C}'$ *has a right adjoint if and only if, for all* $\mathrm{Y} \in \mathscr{C}'$, *the contravariant functor* $\mathrm{X} \mapsto \mathrm{Hom}(\mathrm{F}(\mathrm{X}), \mathrm{Y})$ *from* $\mathscr{C}$ *to* $\mathscr{E}ns$ *is representable.*

Proof The condition is obviously necessary. We show that it is sufficient. Let $\widehat{\mathscr{C}}_0$ be the category of representable contravariant functors from $\mathscr{C}$ to $\mathscr{E}ns$. By definition, the functor $\Lambda : \mathrm{X} \mapsto \widehat{\mathrm{X}}$ from $\mathscr{C}$ to $\widehat{\mathscr{C}}_0$ is fully faithful (2.4.12, proposition) and essentially surjective. Hence it is an equivalence of categories (2.3.5). Let $\mathrm{M} : \widehat{\mathscr{C}}_0 \to \mathscr{C}$ be a functor such that $\Lambda \circ \mathrm{M}$ is isomorphic to the identity functor of $\widehat{\mathscr{C}}_0$. Let $\mathrm{H} : \mathscr{C}' \to \widehat{\mathscr{C}}_0$ be the functor defined by $\mathrm{H}(\mathrm{Y})(\mathrm{X}) = \mathrm{Hom}(\mathrm{F}(\mathrm{X}), \mathrm{Y})$. Set $\mathrm{G} = \mathrm{M} \circ \mathrm{H}$. Then $\Lambda \circ \mathrm{G}(\mathrm{Y})(\mathrm{X}) = \mathrm{Hom}(\mathrm{X}, \mathrm{G}(\mathrm{Y}))$, and $\Lambda \circ \mathrm{G} = \Lambda \circ \mathrm{M} \circ \mathrm{H} \approx \mathrm{H}$. So F and G are adjoint functors. $\qquad\square$

2.7.3

Proposition *Let $\mathscr{C}$ and $\mathscr{C}'$ be two categories and* G *a functor from $\mathscr{C}'$ to $\mathscr{C}$. If* G *admits a left adjoint, then* G *commutes with products, kernels of double arrows, fibered products and inverse limits when these exist.*

Proof We only give the proof for inverse limits, the other cases being similar. Let F be a left adjoint to G. Let (Y_i) be an inverse system of objects of $\mathscr{C}'$ having an inverse limit. Then, for all $\mathrm{X} \in \mathscr{C}$

$$\mathrm{Hom}(\mathrm{X}, \mathrm{G}(\varprojlim \mathrm{Y}_i)) = \mathrm{Hom}(\mathrm{F}(\mathrm{X}), \varprojlim \mathrm{Y}_i) = \varprojlim \mathrm{Hom}(\mathrm{F}(\mathrm{X}), \mathrm{Y}_i)$$
$$= \varprojlim \mathrm{Hom}(\mathrm{X}, \mathrm{G}(\mathrm{Y}_i)) = \mathrm{Hom}(\mathrm{X}, \varprojlim \mathrm{G}(\mathrm{Y}_i))$$

and so $\mathrm{G}(\varprojlim \mathrm{Y}_i)$ is an inverse limit of $(\mathrm{G}(\mathrm{Y}_i))$. $\qquad\square$

Corollary *Let* F *be a functor from $\mathscr{C}$ to $\mathscr{C}'$. If* F *admits a right adjoint, then* F *commutes with sums, amalgamated sums, cokernels of double arrows and direct limits when these exist.*

Proof Apply the proposition to opposite categories.

2.7.4 Examples

(1) Let $\mathscr{C}$ be the category of normed vector spaces over $\mathbb{R}$ and $\mathscr{C}'$ the category of Banach spaces. The functor from $\mathscr{C}$ to $\mathscr{C}'$ which assigns to every normed space its completion is a left adjoint to the inclusion functor from $\mathscr{C}'$ to $\mathscr{C}$. Indeed, for any normed space E and any Banach space F, $\mathrm{L}(\widehat{\mathrm{E}}, \mathrm{F}) = \mathrm{L}(\mathrm{E}, \mathrm{F})$.

(2) Let A be a ring. The functor from $\mathscr{E}ns$ to A-$\mathscr{M}od$ which assigns to every set X the free module $A^{(X)}$ constructed on X is a left adjoint to the forgetful functor A-$\mathscr{M}od \to \mathscr{E}ns$. Indeed, for any set X and any A-module E, $\mathrm{Hom}_{A\text{-}\mathscr{M}od}(A^{(X)}, E) = \mathrm{Hom}_{\mathscr{E}ns}(X, E)$ (3.4.6).

Similarly, the functor from $\mathscr{E}ns$ to the category A-$\mathscr{A}lg$ of unital associative commutative algebras which assigns to each set X the polynomial algebra A[X] is a left adjoint to the forgetful functor from A-$\mathscr{A}lg$ to $\mathscr{E}ns$.

(3) Let A and B be two rings and $A \to B$ a ring homomorphism. The extension of scalars functor $E \mapsto B \otimes_A E$ from A-$\mathscr{M}od$ to B-$\mathscr{M}od$ is a left adjoint to the restriction of scalars functor B-$\mathscr{M}od \to$ A-$\mathscr{M}od$. Indeed, for any A-module E and any B-module F, $\mathrm{Hom}_B(B \otimes_A E, F) = \mathrm{Hom}_A(E, F)$.

(4) Let A be a commutative ring and F an A-module. The functor $E \mapsto E \otimes_A F$ from the category A-$\mathscr{M}od$ to itself is a left adjoint to the functor $G \mapsto \mathrm{Hom}_A(F, G)$. Indeed, if E and G are A-modules, then $\mathrm{Hom}_A(E \otimes_A F, G) = \mathrm{Hom}_A(E, \mathrm{Hom}_A(F, G))$.

(5) Let Y be a locally compact space. The functor $X \mapsto X \times Y$ from the category $\mathscr{T}op$ to itself has for right adjoint $\grave{a}$ the functor $Z \mapsto C(Y, Z)$, where $C(Y, Z)$ is the set of continuous functions from Y to Z equipped with the topology of compact convergence.

(6) Let $\mathscr{C}$ be the category of point spaces, i.e. of pairs (X, x_0) where X is a topological space and $x_0 \in X$, with

$$\mathrm{Hom}_{\mathscr{C}}((X, x_0), (Y, y_0)) = \{f : X \to Y \mid f \text{ continuous and } f(x_0) = y_0\}.$$

For $(X, x_0) \in \mathscr{C}$, let $\Omega(X, x_0)$ be the loop space of (X, x_0), i.e. the set of continuous maps $\gamma : [0, 1] \to X$ such that $\gamma(0) = \gamma(1) = x_0$, equipped with the topology of compact convergence and basepoint $\bar{x}_0 : t \mapsto x_0$. This defines a functor Ω from the category $\mathscr{C}$ to itself. This functor admits a left adjoint S defined as follows: $S(X, x_0)$ is the quotient space of $X \times [0, 1]$ by the equivalence relation contracting the set

$$(X \times \{0\}) \cup (X \times \{1\}) \cup (\{x_0\} \times [0, 1])$$

to a point, equipped with this point as basepoint.

In other words, for X, Y $\in \mathscr{C}$, $\mathrm{Hom}(S(X), Y) = \mathrm{Hom}(X, \Omega(Y))$. The space S(X) is called the *reduced suspension* of X.

2.7.5

Let $\mathscr{C}$ and $\mathscr{C}'$ be categories, $F : \mathscr{C} \to \mathscr{C}'$ and $G : \mathscr{C}' \to \mathscr{C}$ functors such that G is the right adjoint to F. Write FX for F(X), GF for G ∘ F, etc. Define a functorial morphism $\alpha : 1_{\mathscr{C}} \to GF$ by setting α_X to be the element of $\mathrm{Hom}_{\mathscr{C}}(X, GFX)$ corresponding to $1_{FX} \in \mathrm{Hom}_{\mathscr{C}'}(FX, FX)$. If $u : X' \to X$ is a morphism of $\mathscr{C}$, the morphism $F(u) \in \mathrm{Hom}_{\mathscr{C}'}(FX', FX)$ is the one that can be identified with $\alpha_X \circ u \in \mathrm{Hom}_{\mathscr{C}}(X', GFX)$.

Likewise define $\beta_Y \in \text{Hom}_{\mathscr{C}'}(\text{FGY}, Y)$ corresponding to 1_{GY}, and for $v \in \text{Hom}_{\mathscr{C}}(Y, Y')$ the morphism $G(v) \in \text{Hom}_{\mathscr{C}}(\text{GY}, \text{GY}')$ can be identified with $v \circ \beta_Y \in \text{Hom}_{\mathscr{C}'}(\text{FGY}, Y')$.

Proposition *With this notation,* F *is fully faithful if and only if α is an isomorphism.*

Proof (a) Suppose α is an isomorphism. For X and X' in $\mathscr{C}$, the map $u \mapsto \alpha_X \circ u$ is a bijection of $\text{Hom}_{\mathscr{C}}(X', X)$ onto $\text{Hom}_{\mathscr{C}}(X', \text{GFX}) = \text{Hom}_{\mathscr{C}'}(\text{FX}', \text{FX})$, which must be $F_{X',X}$.

(b) Conversely, suppose F is fully faithful. For $X \in \mathscr{C}$, the map $u \mapsto \alpha_X \circ u = F(u)$ from $\text{Hom}_{\mathscr{C}}(\text{GFX}, X)$ to $\text{Hom}_{\mathscr{C}}(\text{GFX}, \text{GFX}) = \text{Hom}_{\mathscr{C}'}(\text{FGFX}, \text{FX})$ is a bijection. Hence there is a unique $w \in \text{Hom}_{\mathscr{C}}(\text{GFX}, X)$ such that $\alpha_X \circ w = 1_{\text{GFX}}$. Then $\alpha_X \circ w \circ \alpha_X = \alpha_X$, namely $F(w \circ \alpha_X) = F(1_X)$, and so $w \circ \alpha_X = 1_X$, and α_X is an isomorphism with $\alpha_X^{-1} = w$, $\square$

Corollary G *is fully faithful if and only if β is an isomorphism.*

Apply the proposition to opposite categories.

Comments. Hence if F is fully faithful, the functor GF is isomorphic to $1_{\mathscr{C}}$. In other words, for a fully faithful functor, a right adjoint functor is a left inverse up to isomorphism (the side change is only due to conventions regarding composition).

Saying that F is fully faithful means that F is an equivalence between $\mathscr{C}$ and the full subcategory of $\mathscr{C}'$. A right or left adjoint then gives a "retraction up to isomorphism" from $\mathscr{C}'$ onto $\mathscr{C}$. Note that, if F admits a right and a left adjoint, they need not necessarily be isomorphic (2.7, Exercise 3).

For examples, see 2.7.4, Example 1 for the corollary, and 2.7, Exercises 1 and 2 for the proposition.

2.7.6 Topology and Algebra

For any compact space X, let C(X) denote the algebra of continuous functions $X \to \mathbb{C}$. This defines a contravariant functor C from the category $\mathscr{K}$ of compact spaces and continuous maps to the category $\mathscr{A}$ of unital associative commutative Banach algebras over $\mathbb{C}$ and unital continuous algebra homomorphisms. To complete the definition of C as a functor, it remains to specify that $f^*h = h \circ f$ for $f : X \to X'$ continuous and $h \in C(X')$.

We will show that the contravariant functor C is fully faithful 2.7.8, and admits an adjoint S (2.7.7, where we will indicate in what sense).

Comments. The functor C thus establishes an analogy between the category $\mathscr{K}$ and the opposite category $\mathscr{A}^\circ$. This is our first encounter with a very general and fruitful situation, where topological or geometric objects are made to correspond to algebraic ones (in Banach algebras, the norm does not really play any role, see 2.7.7, Lemma 2). Thus, the algebra of C^∞-functions from M to $\mathbb{R}$ corresponds to the C^∞-manifold

M, and the algebra of functions induced on X by polynomials, etc. to an algebraic
variety.

Concepts or viewpoints can thereby be transferred from a topological or geometric framework to an algebraic one, where their domain of validity is extended.
In turn, in a given algebraic situation, thanks to versions of the functor S, topological objects can be constructed, enabling us to apply geometric reasoning. This is
what Grothendieck's "theory of schemes" consists in (an important part of the initial
algebraic data is retained in the structure to avoid any loss of information). More
generally, this method underscores every important recent progress in algebra and
arithmetic.

As the correspondence is not perfect, the geometric framework is enhanced by
"forcing" it (algebraic geometry with nilpotent elements, and more recently "non-
commutative geometry").

Chapter 6 addresses situations where this correspondence is reflected by an equivalence of categories (6.2.4 and 6.3.5).

2.7.7 Spectra and Gelfand Transforms

We keep the notation of 2.7.6. Let A be a unital associative commutative Banach
algebra over $\mathbb{C}$, and S(A) the set of all unital algebra homomorphisms A $\rightarrow$ $\mathbb{C}$,
equipped with the weak topology. Then S(A) is a compact space (Lemma 3 below)
called the **spectrum** of A. For $a \in A$, define $\hat{a} : S(A) \rightarrow \mathbb{C}$ by $\hat{a}(\xi) = \xi(a)$. The
function $\hat{a}$ is continuous, and $a \mapsto \hat{a}$ is a homomorphism $\Lambda_A : A \rightarrow C(S(A))$, called
the **Gelfand transform.**

Proposition *The compact space* S(A), *together with* Λ_A, *satisfies the following
universal property:*
For all compact spaces X *and unital algebra homomorphisms* $\Phi : A \rightarrow C(X)$,
there is a unique continuous map $\varphi : X \rightarrow S(A)$ *such that* $\Phi = \varphi^* \circ \Lambda_A$.

This can be restated in terms of categories as follows:

Corollary *For* $X \in \mathcal{K}$ *and* $A \in \mathcal{A}$, *the map* $\Psi_{X,A} : \varphi \mapsto \varphi^* \circ \Lambda_A$ *from* $\mathrm{Hom}_{\mathcal{K}}(X, S(A))$
to $\mathrm{Hom}_{\mathcal{A}}(A, C(X))$ *is a bijection. The family of maps* $\Psi_{X,A}$ *is an isomorphism of
functors from* $\mathcal{K}^o \times \mathcal{A}^o$ *to* $\mathcal{E}ns$. *Thus* $S : \mathcal{A}^o \rightarrow \mathcal{K}$ *is a right adjoint to* $C : \mathcal{K} \rightarrow \mathcal{A}^o$.

Comments. Since C and S are contravariant functors, to use the given definition of
adjoint functors, they must be made covariant by replacing one of the two categories
$\mathcal{K}$ and $\mathcal{A}$ by its opposite. We have chosen to do so for $\mathcal{A}$. Were $\mathcal{K}$ chosen, S would
become the left adjoint of C.

The corollary follows from the proposition by a direct application of Proposition
2.7.2. The proof of the proposition uses the following lemmas:

Lemma 2.1 *Let* A $\in \mathcal{A}$. *Every algebra homomorphism* $\varphi : A \rightarrow \mathbb{C}$ *is continuous
and has norm* ≤ 1.

Proof Since $\varphi(1)^2 = \varphi(1)$, $\varphi(1) = 1$ or 0. If $\varphi(1) = 0$, then $\varphi = 0$; we may assume that $\varphi(1) = 1$.

Were φ not continuous and its norm $\nleq 1$, then there would exist $u \in A$ such that $|\varphi(u)| > \|u\|$. We may assume that $\varphi(u) = 1$. As $\|u\| < 1$, $1 - u$ is invertible and $\varphi(1 - u) = 0$, we get a contradiction. $\qquad\square$

Lemma 2.2 *Let* $A \in \mathscr{A}$ *and* $X \in \mathscr{K}$. *Every algebra homomorphism* $\varphi : A \to C(X)$ *is continuous and has norm* ≤ 1.

Proof By Lemma 1, for all $x \in X$, the map $a \mapsto \varphi(a)(x)$ is continuous and has norm ≤ 1. So $|\varphi(a)(x)| \leq \|a\|$, and

$$(\forall a \in A) \quad \|\varphi(a)\| = \sup_{x \in X} |\varphi(a)(x)| \leq \|a\|,$$

$\qquad\square$

Lemma 2.3 *Let* $A \in \mathscr{A}$. *The set* $S(A) = \mathrm{Hom}_{\mathscr{A}}(A, \mathbb{C})$, *equipped with the weak topology, is compact.*

Proof Let D_r to be the closed disk of radius $r(\in \mathbb{R}_+)$ in $\mathbb{C}$. By Lemma 1, the space $S(A)$ is a subspace of $\prod_{a \in A} D_{\|a\|}$ which, by Tychonoff's theorem, is compact. The set $S(A)$ being defined by the equalities

$$\varphi(a + b) = \varphi(a) + \varphi(b),$$

$$\varphi(ab) = \varphi(a) \cdot \varphi(b), \quad \varphi(1) = 1,$$

is closed in $\prod_{a \in A} D_{\|a\|}$, and so is compact. $\qquad\square$

Proof of the Proposition The set $\mathrm{Hom}_{\mathscr{K}}(X, S(A))$ of all continuous maps from X to $S(A)$ equipped with the weak topology, can be identified with the set of functions $f : X \times A \to \mathbb{C}$ satisfying:

(i) for $a \in A$, the map $x \mapsto f(x, a)$ is continuous;
(ii) for $x \in X$, $a \mapsto f(x, a)$ is a (necessarily continuous) unital algebra homomorphism.

This set can be identified to the set of (necessarily continuous) unital algebra homomorphisms $A \to C(X)$. Hence

$$\mathrm{Hom}_{\mathscr{K}}(X, S(A)) = \mathrm{Hom}_{\mathscr{A}}(A, C(X)),$$

and, for $X = S(A)$, the homomorphism corresponding to the identity element is readily seen to be the Gelfand transform. $\qquad\square$

2.7.8

Proposition *The contravariant functor* $C : \mathcal{K} \to \mathcal{A}$ *is fully faithful.*

Proof Thanks to Proposition 2.7.7, all we need to prove is that, for a compact space X, the natural map $X \to S(C(X))$ is a homeomorphism. This map is $x \mapsto \delta_x$, where $\delta_x(f) = f(x)$. It is clearly continuous. Since X is compact, it is sufficient to show that it is bijective, i.e. that every homomorphism $\xi : C(X) \to \mathbb{C}$ is of type δ_x for some unique $x \in X$.

Hence let $\xi : C(X) \to \mathbb{C}$ be a homomorphism, $\mathfrak{m}$ its kernel and $V(\mathfrak{m})$ the intersection of all $h^{-1}(0)$ for $h \in \mathfrak{m}$. Were the set $V(\mathfrak{m})$ empty, there would be a finite family $h_1, \ldots, h_k$ in $\mathfrak{m}$ without any common roots; then $h_1 h_1 + \cdots + h_k h_k$ would be a function everywhere > 0, hence invertible, and so $1 \in \mathfrak{m}$, an absurdity.

Let $x \in V(\mathfrak{m})$. Then $\mathrm{Ker}\xi \subset \mathrm{Ker}\delta_x$, but both are hyperplanes, and so $\mathrm{Ker}\xi = \mathrm{Ker}\delta_x$. Since $\xi(1) = \delta_x(1) = 1$, $\xi = \delta_x$. Hence the map $x \mapsto \delta_x$ is surjective.

Injectivity follows from Urysohn's theorem: if x and y are two distinct points of X, then there is a continuous function $h : X \to \mathbb{C}$ for which $h(x) \neq h(y)$. So $\delta_x(h) \neq \delta_y(h)$, which implies that $\delta_x \neq \delta_y$. □

Exercises 2.7. (Adjoint functors)
1.—Let G be a group. The *commutator group* of G is the subgroup G' of G generated by all elements $xyx^{-1}y^{-1}$, where $x, y \in G$.

(a) Show that G' is normal and that $A(G) = G/G'$ is commutative.

(b) Show that the functor $G \to A(G)$ from $\mathcal{Gr}$ to the category $\mathcal{Ab}$ of Abelian groups is the adjoint to the inclusion functor $\mathcal{Ab}$ in $\mathcal{Gr}$; determine whether it is a left or right one; state this property in elementary terms.

(c) Show that there is no adjoint functor on the other side.

2.—Let X be a topological space, I the segment [0, 1] and E the set of continuous maps from X to I. Define a map $\delta : X \to I^E$ by $\delta(x) = (f(x))_{f \in E}$. Equip I^E with the product topology. Then δ is continuous. The *Cech compactification* of X, written $\check{X}$, is the closure of $\delta(X)$ in I^E.

(a) The space X is said to be *completely regular* if, for all all neighbourhoods U of all elements $x \in X$, there is a continuous function $f : X \to I$ such that $f(x) = 1$ and $f|_{\complement U} = 0$. Show that:

• if X is completely regular, then δ induces a homeomorphism from X onto a subset of $\check{X}$;
• if X is compact, then $\check{X} = X$;
• if X is locally compact, then X is open in $\check{X}$.

(b) Let K be a compact space, and $f : X \to K$ a continuous map. Show that there is a continuous map $\check{f} : \check{X} \to K$ such that $f = \check{f} \circ \delta$. State this property in terms of adjoint functors.

(c) Is there an adjoint functor on the other side?

3.—Let $\mathscr{T}\text{-}op\ell c$ be the category of locally connected topological spaces, and identify $\mathscr{E}ns$ with the full subcategory of discrete spaces. Show that the inclusion functor $\iota : \mathscr{E}ns \to \mathscr{T}\text{-}op\ell c$ has both a left and a right adjoint, and that these are not equal.

4. *(Fourier series.)*—Consider the Banach space $\ell^1(\mathbb{Z})$ consisting of complex sequences $\mathbf{a} = (a_n)_{n \in \mathbb{Z}}$ such that $\sum |a_n| < \infty$, equipped with the norm defined by $\|\mathbf{a}\| = \sum_{n \in \mathbb{Z}} |a_n|$. Set $a(n) = a_n$, and define the basis element $\mathbf{e_n}$ by $\mathbf{e_n}(p) = 1$ if $p = n$ and 0 otherwise.

(a) Show that there is a unique continuous bilinear multiplication law $*$ on $\ell^1(\mathbb{Z})$ such that $\mathbf{e_p} * \mathbf{e_q} = \mathbf{e_{p+q}}$ for $p, q \in \mathbb{Z}$. This law is called *convolution*. Let A be the algebra obtained by equipping $\ell^1(\mathbb{Z})$ with it. Write c_n in terms of the elements a_p and b_q when $\mathbf{c} = \mathbf{a} * \mathbf{b}$.

(b) Show that if $\xi : A \to \mathbb{C}$ is a unital homomorphism, then $\|\xi(\mathbf{e_1})\| = 1$. Show that S(A) can be identified with the unit circle of $\mathbb{C}$, i.e. with $\mathbb{T} = \mathbb{R}/\mathbb{Z}$.

Show that, for $\mathbf{a} \in A$, the Gelfand transform $\widehat{\mathbf{a}} : \mathbb{T} \to \mathbb{C}$ is given by the formula for Fourier series $\widehat{\mathbf{a}}(t) = \sum_{n \in \mathbb{Z}} a_n e^{2i\pi nt}$.

(c) Show that the elements a_n can be retrieved from $\widehat{\mathbf{a}}$ by the inverse Fourier formula $a_n = \int_{\mathbb{T}} \widehat{\mathbf{a}}(t) e^{-2i\pi nt} \, dt$. Deduce that the Gelfand transform of $\Lambda : A \to C(\mathbb{T})$ is injective.

For $g \in L^1(\mathbb{T}, \mathbb{C})$, i.e. integrable over $\mathbb{T}$ with values in $\mathbb{C}$, set $\hat{g}(n) = \int_{\mathbb{T}} g(t) e^{-2i\pi nt} \, dt$ (nth Fourier coefficient of g).

(d) The aim is to show that Λ is not surjective. We remind the reader of the

Banach Theorem *The inverse of a continuous bijective linear map between Banach spaces is continuous.*

Let $g : \mathbb{T} \to \mathbb{C}$ be the function taking value 1 on $[0, 1/2]$ and -1 on $]1/2, 1]$. Show that $\sum |\hat{g}(n)| = \infty$.

For $k > 4$, let f_k be the continuous function, affine on $[-1/k, 1/k]$ and on $[1/2 - 1/k, 1/2 + 1/k]$, and agreeing with g elsewhere. Show that $\sum_n \widehat{f_k}(n)$ tends to $\hat{g}(n)$ as $k \to \infty$ for fixed n. Deduce that the upper bounds of $\sum_n |\widehat{f_k}(n)|$ depend on k. Conclude.

(e) Give an explicit example of a continuous function on $\mathbb{T}$ which is not the Gelfand (here Fourier) transform of $\ell^1(\mathbb{Z})$.

2.8 Profinite Spaces

In this section and the next one, we study profinite spaces and groups with infinite Galois theory in mind.

2.8.1 Pro-objects of a Category

Let $\mathscr{C}$ be a category. Define a category $\overleftarrow{\mathscr{C}}$ as follows: an object of $\overleftarrow{\mathscr{C}}$ is an inverse system[2] $((X_i)_{i \in I}, (f_i^j)_{i \leq j})$ in $\mathscr{C}$. If $X = (X_i)_{i \in I}$, and $Y = (Y_j)_{j \in J}$ are two objects of $\overleftarrow{\mathscr{C}}$, set $\mathrm{Hom}(X, Y) = \varprojlim_j \varinjlim_i \mathrm{Hom}(X_i, Y_j)$. Let $X = (X_i)_{i \in I}$, $Y = (Y_j)_{j \in J}$ and $Z = (Z_k)_{k \in K}$ be inverse systems, $\varphi = (\varphi_j) \in \mathrm{Hom}(X, Y)$, where $\varphi_j \in \varinjlim \mathrm{Hom}(X_i, Y_j)$, and $\psi = (\psi_k) \in \mathrm{Hom}(Y, Z)$. Define $\psi \circ \varphi$ as follows: for all $k \in K$, let (j, ν), where $j \in J$ and $\nu \in \mathrm{Hom}(Y_j, Z_k)$, be a representative of ψ_k, and define $\omega_k = \nu \circ \varphi_l \in \varinjlim \mathrm{Hom}(X_l, Z_k)$. The element ω_k is independent of the choice of (j, ν).

Indeed, if (j', ν') is another representative of ψ_k, then there exists j'' greater than j and j' such that $\nu \circ g_j^{j''} = \nu' \circ g_{j'}^{j''}$, and so $\nu \circ \varphi_j = \nu \circ g_j^{j''} \circ \varphi_{j''} = \nu' \circ g_{j'}^{j''} \circ \varphi_{j''} = \nu' \circ \varphi_{j'}$. Then $\omega_k = h_k^{k'} \circ \omega_{k'}$. Thus the family $(\omega_k)_{k \in K}$ defines an element $\psi \circ \varphi = \omega \in \mathrm{Hom}(X, Z)$. Checking that $\overleftarrow{\mathscr{C}}$ is a category is tedious but not difficult.

The objects of $\overleftarrow{\mathscr{C}}$ are called **pro-objects** de $\mathscr{C}$. Considering an object of $\mathscr{C}$ as an inverse system indexed by a singleton identifies $\mathscr{C}$ with a full subcategory of $\overleftarrow{\mathscr{C}}$. A functor $F : \overleftarrow{\mathscr{C}} \to \overleftarrow{\mathscr{C}}'$ corresponds to a functor $F : \mathscr{C} \to \mathscr{C}'$. If F is an equivalence of categories, so is $\overleftarrow{F}$.

2.8.2 Totally Disconnected Spaces

A topological space X is said to be *totally disconnected* if for any two distinct elements x and y of X, there is a clopen (both open and closed) subset U of X such that $x \in U$ and $y \notin U$.

Examples The discrete spaces, the field $\mathbb{Q}$ of rationals, the Cantor set $\{0\} \cup \{1/n\}_{n \in \mathbb{N}}$, are totally disconnected spaces.

Every subspace of a totally disconnected space is totally disconnected. Every product $E = \prod E_i$ of totally disconnected spaces is totally disconnected. Indeed, let x and y be distinct elements of E. Their projections on at least some E_i must be different. Let i be such that $x_i \neq y_i$ and U be a clopen subset in E_i containing x but not y. Then the inverse image of U in E is clopen in E, contains x but not y.

Every inverse limit of totally disconnected spaces is totally disconnected.

---

[2]More precisely, assume there is an universe U (2.1.3) such that $\mathscr{C} \subset U$. For the definition of $\overleftarrow{\mathscr{C}}$, we only consider inverse systems $(X_i)_{i \in I}$ such that $I \in U$. Then $\overleftarrow{\mathscr{C}} \subset U$.

2.8.3

Proposition and Definition *Let* X *be a topological space. The following conditions are equivalent:*

  (i) X *is compact and totally disconnected;*
  (ii) X *is homeomorphic to the inverse limit of an inverse system of finite discrete sets.*

If these conditions are satisfied, then X *is said to be a* **profinite space.**

Proof (ii) $\Rightarrow$ (i). Let (X_i) be an inverse system of finite discrete sets such that $\varprojlim X_i = X$. Then X is closed in the product $\prod X_i$, which is compact by Tychonoff's theorem (1.7.7) and totally disconnected since so are all X_i. Hence X is compact and totally disconnected.

(i) $\Rightarrow$ (ii). Let Λ be the set of equivalence relations λ on X such that the quotient X_λ is finite and discrete, and for all $x \in X$ denote by x_λ the class of x for the relation λ. Write $\lambda \leq \mu$ if μ is finer than λ. The set Λ is a directed set. Indeed, $\forall \lambda \in \Lambda$ $\forall \lambda' \in \Lambda$ the map $x \mapsto (x_\lambda, x_{\lambda'})$ from X to $X_\lambda \times X_{\lambda'}$ defines an equivalence relation μ on X finer than λ and λ'.

Let X_λ form an inverse system. For all $x \in X$, $(x_\lambda)_{\lambda \in \Lambda} \in \varprojlim X_\lambda$. We show that the map $\theta : x \mapsto (x_\lambda)_{\lambda \in \Lambda}$ from X to $\varprojlim X_\lambda$ is a homeomorphism. It is obviously continuous. Since X is compact, it is image-closed and if θ is injective, θ is a homeomorphism onto its image. Hence it is sufficient to show that θ is injective and image-dense. Let $x \neq y \in X$ and let (U, V) be a partition of X into two open subsets such that $x \in U$ and $y \in V$. For the relation λ associated to this partition, $x_\lambda \neq y_\lambda$, and so the images of x and y in $\varprojlim X_\lambda$ are distinct. Consequently θ is injective.

The homomorphism θ is image-dense because of the following lemma:

Lemma *Let* X *be a set,* (Y_λ) *an inverse system of topological spaces,* $f : X \to \varprojlim Y_\lambda$ *a map. If, for all* λ, *the map* $f_\lambda : X \to Y_\lambda$ *induced by* f *is surjective, then* f *is image-dense.*

Let $a = (a_\lambda)_{\lambda \in \Lambda} \in \varprojlim Y_\lambda$. For every finite family $\lambda_1, ..., \lambda_k \in \Lambda$ the set $V_{\lambda_1,...,\lambda_k}$ $(a) = \{b = (b_\lambda)_{\lambda \in \Lambda}, (\forall i = 1, ..., k)\ b_{\lambda_i} = a_{\lambda_i}\}$ is a neighbourhood of a and the sets $V_{\lambda_1,...,\lambda_k}(a)$ form a fundamental system of neighbourhoods of a. Let $L = \{\lambda_1, ..., \lambda_k\}$ be a finite subset of Λ and μ an upper bound of L in Λ. Let $x \in X$ be such that $f_\mu(x) = a_\mu$; for all $\lambda_i \in L$, $f_{\lambda_i}(x) = a_{\lambda_i}$ and $\text{Im} f \cap V_{\lambda_1,...,\lambda_k}(a)$ is non-empty. Hence f is image-dense. $\square$

2.8.4

Let $\mathscr{E}$ be the category of profinite spaces and continuous maps and Setf that of finite sets. The category $\mathscr{E}$ is a full subcategory of $\mathscr{Top}$.

Theorem *The functor* $F : (X_i) \rightarrow \varprojlim X_i$ *from* Setf *to* $\mathscr{E}$ *is an equivalence of categories.*

Definition 2.8.3 states that F is essentially surjective. We show (2.8.6) that F is fully faithful. For this the following lemma is needed:

2.8.5

Lemma *Let* (X_i, f_i^j) *be an inverse system of finite sets. Denote by* (X_∞, f_i^∞) *its inverse limit. Then*

$$(\forall i)\ (\exists j \geq i)\ \operatorname{Im} f_i^\infty = \operatorname{Im} f_i^j \,.$$

Proof Let $x \in X_i$. For all $j \geq i$, set $Y_j = (f_i^j)^{-1}(x)$. These form an inverse system of finite sets and $\varprojlim Y_j = (f_i^\infty)^{-1}(x)$. Since

$$x \in \cap \operatorname{Im} f_i^j \iff \forall j \geq i\ Y_j \neq \varnothing \iff \varprojlim Y_j \neq \varnothing \iff x \in \operatorname{Im} f_i^\infty$$

by 2.5.14, $\bigcap_{j \geq i} \operatorname{Im} f_i^j = \operatorname{Im} f_i^\infty$. The family $(\operatorname{Im} f_i^j)_{j \geq i}$ is a decreasing family of finite sets, whose intersection is equal to one of them. $\square$

2.8.6

Proposition *Let* (X_i, f_i^j) *be an inverse system of finite sets and* Y *a finite set. Then,*

$$\operatorname{Hom}_{\mathscr{Top}}(\varprojlim X_i; Y) = \varinjlim \operatorname{Hom}_{\mathscr{Ens}}(X_i; Y) \,.$$

Proof Let $\varphi \in \varinjlim \operatorname{Hom}_{\mathscr{Ens}}(X_i; Y)$. Represent φ by $\varphi_i : X_i \rightarrow Y$. The map $x = (x_k) \mapsto \varphi_i(x_i)$ from $\varprojlim X_i$ to Y does not depend on the choice of the representative φ_i of φ. Denote this map by $\alpha(\varphi)$. We show that α is injective and surjective.

Let φ and ψ be such that $\alpha(\varphi) = \alpha(\psi)$. Then φ_i and ψ_i are equal on $\operatorname{Im} f_i^\infty$, hence, for some j, also on $\operatorname{Im} f_i^j$ by Lemma 2.8.5. So $\varphi = \psi$.

Let $\varphi : X_\infty \rightarrow Y$ be a continuous map, Γ_φ (resp. Γ_i) the graph of the equivalence relation defined by φ (resp. f_i^∞) i.e. the inverse image of the diagonal of $Y \times Y$ under the map $(\varphi, \varphi) : X_\infty \times X_\infty \rightarrow Y \times Y$ (resp. of the diagonal of $X_i \times X_i$ under (f_i^∞, f_i^∞)). These graphs are all clopen in $X_\infty \times X_\infty$ and, for all i and $j \geq i$, $\Gamma_j \subset \Gamma_i$. Let Δ be the diagonal of $X_\infty \times X_\infty$. Then, $\Delta \subset \Gamma_\varphi$, and $\Delta = \cap_i \Gamma_i$ is closed. Therefore

$$\varnothing = \left(\bigcap_i \Gamma_i\right) \cap \complement\Gamma_\varphi = \bigcap_i (\Gamma_i \cap \complement\Gamma_\varphi) \,.$$

The intersection of the decreasing closed sets $\Gamma_i \cap \complement \Gamma_\varphi$ is empty, hence so is one of them. Thus, there is an index j such that $\Gamma_j \subset \Gamma_\varphi$. Thus $f_j^\infty(x) = f_j^\infty(y) \Rightarrow \varphi(x) = \varphi(y)$.

Therefore there is a map from $\operatorname{Im} f_j^\infty$ to Y making the diagram

$$
\begin{array}{ccc}
X_\infty & \xrightarrow{f_j^\infty} & \operatorname{Im} f_j^\infty \\
 & \searrow{\scriptstyle\varphi} & \downarrow \\
 & & Y
\end{array}
$$

commutative.

By Lemma 2.8.5 there exists $k \geq j$ such that $\operatorname{Im} f_j^\infty = \operatorname{Im} f_j^k$. Composing with f_j^k then gives a map φ^k from X_k to Y such that the diagram

$$
\begin{array}{ccc}
X_\infty & \xrightarrow{f_k^\infty} & X_k \\
 & \searrow{\scriptstyle\varphi} & \downarrow{\scriptstyle\varphi^k} \\
 & & Y
\end{array}
$$

commutes.   □

Corollary *The functor* F *of theorem 2.8.4 is fully faithful.*

Indeed, if (X_i) and (Y_j) are two inverse systems of finite sets, then

$$
\operatorname{Hom}_{\mathcal{Top}}(\varprojlim X_i; \varprojlim Y_j) = \varprojlim_j \operatorname{Hom}_{\mathcal{Top}}(\varprojlim X_i; Y_j)
$$
$$
= \varprojlim_j \varinjlim_i \operatorname{Hom}_{\mathcal{Ens}}(X_i; Y_j) .
$$

Exercises 2.8. (Profinite spaces)
1.—Let X be a compact space.

(a) Let Y be closed in X and let $(V_i)_{i \in I}$ be a directed family of closed sets of X such that $\cap V_i = Y$. Show that every neighbourhood of Y contains some V_i.

(b) Let $a \in X$. Show that the intersection of the clopen sets of X containing a is the connected component of a in X.

(c) A compact space is totally discontinuous as defined in (2.8.2) if and only if every connected component is reduced to a point.

(d) For the equivalence given in (c), can the assumption of compactness of X be omitted?

2.   *(Cantor set.)*—Define a sequence $(A_n)_{n \in \mathbb{N}}$ of subsets of $[0, 1]$ as follows:

$$
\begin{aligned}
A_0 &= [0, 1], \\
A_1 &= [0, 1] -]1/3, 2/3[, \\
A_2 &= A_1 - (]1/9, 2/9[\cup]7/9, 8/9[).
\end{aligned}
$$

Each A_n is a finite union of closed disjoint intervals and A_{n+1} is the third of the open middle of A_n. The set $K = \bigcap_{n \in \mathbb{N}} A_n$ is called the *triadic Cantor set*.

(a) Show that K is a profinite space.

(b) Show that K is the set of elements x of $[0, 1]$ having a base 3 expansion without the number 1.

(c) Show that K is homeomorphic to $\{0, 1\}^{\mathbb{N}}$.

(d) Write $[0, 1] - K$ as the disjoint union of intervals $]\alpha_i, \beta_i[_{i \in I}$, where I is a countable set. Show that the three sets $\{\alpha_i\}$, $\{\beta_i\}$ and $K - (\{\alpha_i\} \cup \{\beta_i\})$ are dense in K and that the interior of K in $[0, 1]$ is empty.

(e) Show that the quotient of K by the equivalence relation which equates α_i and β_i is homeomorphic to $[0, 1]$.

(f) Let $(X_n)_{n \in \mathbb{N}}$ be an inverse system of finite sets. Suppose that for all n, f_n^{n+1} : $X_{n+1} \to X_n$ is such that, $\forall x \in X_n$, $(f_n^{n+1})^{-1}(x)$ contains at least two point. Show that K is homeomorphic to $\varprojlim X_i$.

(g) Show that every metrizable profinite space without isolated points is homeomorphic to K.

3.—Let K be the Cantor set and $\gamma \in K - (\{\alpha_i\} \cup \{\beta_i\})$ (see Exercise 2 for notation). Set $X = ((K \cap [0, \gamma]) \times \{0\}) \cup ((K \cap [\gamma, 1]) \times \{1\})$. Show that the projection $(k, i) \mapsto k$ from X to K has no continuous section.

4.—Let $\mathscr{C}$ be a category having inverse limits.

(a) Define a functor $\lambda : (X_i)_{i \in I} \mapsto \varprojlim X_i$ from $\overleftarrow{\mathscr{C}}$ to $\mathscr{C}$. Let ι be the inclusion functor $\mathscr{C} \to \overleftarrow{\mathscr{C}}$. Show that $\lambda \circ \iota \approx I_{\mathscr{C}}$, but that in general λ and ι are not equivalences of categories.

(b) Let $\mathscr{C}'$ be another category. Show that for every functor $F : \mathscr{C}' \to \mathscr{C}$, the functor $\overleftarrow{F}$ from $\overleftarrow{\mathscr{C}'}$ to $\overleftarrow{\mathscr{C}}$ commuting with inverse limits of objects of $\mathscr{C}$ is its unique extension. Suppose that $\mathscr{C}'$ has inverse limits: does the diagram

$$
\begin{array}{ccc}
\overleftarrow{\mathscr{C}'} & \xrightarrow{\ \overleftarrow{F}\ } & \overleftarrow{\mathscr{C}} \\
{\scriptstyle \lambda'}\downarrow & & \downarrow{\scriptstyle \lambda} \\
\mathscr{C}' & \xrightarrow{\ F\ } & \mathscr{C}
\end{array}
$$

necessarily commute?

(c) Show that, if $\mathscr{C} = \mathscr{S}\!et\!f$, the inclusion $\overleftarrow{\mathscr{C}} \to \overleftarrow{\overleftarrow{\mathscr{C}}}$ is not an equivalence of categories.

2.9 Profinite Groups

2.9.1

Proposition and Definition *Let* G *be a topological group. The following conditions are equivalent:*

  (i) G *is the inverse limit of finite discrete groups;*
 (ii) *the underlying topological of* G *is profinite.*

   If these conditions are satisfied, G *is said to be a* **profinite group.**

Proof (i) $\Rightarrow$ (ii) is obvious. We show that (ii) $\Rightarrow$ (i).

For any open normal subgroup H of G, the quotient G/H is discrete (because the inverse image of each point under the canonical projection is open) and compact (because it is the continuous image of the compact space G in the Hausdorff space G/H), and so is finite. The canonical homomorphism $\gamma : G \to \varprojlim_{H} G/H$ (inverse limit of the set of open normal subsets, relative to the inclusion order) is image-dense by Lemma 2.8.3; as G is compact, $\gamma(G)$ is closed and equal to its closure, so that γ is surjective. We show it is injective. Suppose that $x \in G$ is not the identity element e. There is an open normal subgroup H not containing x. Indeed, as G is totally discontinuous, there is an open and closed subset U in G such that $e \in U$ and $x \notin U$. Set $V = \{g \in G, gU = U\}$, $K = \{g \in G, gU \not\subset U\}$ and $K' = \{g \in G, U \not\subset gU\}$. Then $V = G - (K \cup K')$ and $V \subset U$. As the images of $U \times (G - U)$ under the respective continuous maps $(x, y) \mapsto yx^{-1}$ and xy^{-1}, the subspaces K and K' of G are compact. Hence V is open in G. The subgroup $H = \cap_{g \in G} gVg^{-1}$ is normal, its complement $G - H$ is the image of $G \times (G - V)$ under $(g, x) \mapsto gxg^{-1}$, i.e. the image of a compact set under a continuous map. So $G - H$ is compact and H is open. Since G is compact, γ is a homeomorphism. $\qquad\square$

2.9.2

Denote by $\mathscr{G}r\!f$ the category of finite groups and by $\mathscr{G}r\mathscr{T}\!op$ that of topological groups.

Proposition *The functor* $G : (G_i) \to \varprojlim G_i$ *is an equivalence from* $\mathscr{G}r\!f$ *onto the category of profinite groups (full subcategory of* $\mathscr{G}r\mathscr{T}\!op$ *).*

Proof By 2.9.1, this functor is essentially surjective. We show that it is fully faithful, i.e. that for $G = \varprojlim G_i$ and $H = \varprojlim H_i$, $\mathrm{Hom}_{\mathscr{G}r\mathscr{T}\!op}(G, H) = \varprojlim \varinjlim \mathrm{Hom}(G_i, H_j)$. Passing to the inverse limit, this reduces to the following lemma:

Lemma *Let* (G_i) *be an inverse system of finite groups,* $G = \varprojlim G_i$ *and* H *a finite group. Then,* $\mathrm{Hom}_{\mathscr{G}r\mathscr{T}\!op}(G, H) = \varinjlim \mathrm{Hom}_{\mathscr{G}r}(G_i, H)$.

Proof of the Lemma The canonical map $\varinjlim \operatorname{Hom}(G_i, H) \to \operatorname{Hom}_{\mathscr{G}r\mathscr{T}op}(G, H)$ is injective since it is induced by the canonical bijection $\varinjlim \operatorname{Hom}_{\mathscr{E}ns}(G_i, H) \to \operatorname{Hom}_{\mathscr{T}op}(G, H)$ (2.8.6). We show that it is surjective, i.e. that for any continuous homomorphism $\varphi : G \to H$, there is an index j and a continuous homomorphism $\varphi_j : G_j \to H$ such that $\varphi = \varphi_j \circ f_j^\infty$. By 2.8.6 there is an index i and a continuous map $\varphi_i : G_i \to H$ such that $\varphi = \varphi_i \circ f_i^\infty$; the map φ_i is not necessarily a group homomorphism, but its restriction to $f_i^\infty(G)$ is. By 2.8.5, $f_i^j(G_j) = f_i^\infty(G)$ for some $j \geq i$. So $\varphi_j = \varphi_i \circ f_i^j : G_j \to H$ is a homomorphism and $\varphi = \varphi_j \circ f_j^\infty$, $\square$

2.9.3

Let G be a topological group and X a topological space. A *continuous operation* from G onto X is a continuous map $(g, x) \mapsto g \cdot x$ from $G \times X$ to X which is an action from G onto X (2.1.2, Example 3).

Remark If X is finite and discrete, an action from G onto X is continuous if and only if, for all $x \in X$, the stabilizer of x is an open subgroup of G or equivalently that the homomorphism $g \mapsto (x \mapsto g \cdot x)$ from G to the group $\mathfrak{S}(X)$ of permutations of X equipped with the discrete topology is continuous.

2.9.4

Let G be a profinite group. Denote by G-$\mathscr{S}et\!f$ the category of finite sets on which G acts continuously. If G is finite, the category G-$\mathscr{S}et\!f$ is a full subcategory of the category G-$\mathscr{E}ns$ (2.1.2, Example 3). Let $((G_i), (f_i^j))$ be an inverse system of finite groups. There is a direct system $(G_i$-$\mathscr{S}et\!f)$ of categories corresponding to it (2.2.6). Set $G = \varprojlim G_i$.

Proposition *The category* G-$\mathscr{S}et\!f$ *is the direct limit of* G_i-$\mathscr{S}et\!f$.

Proof An object of G-$\mathscr{S}et\!f$ is a pair (X, ρ), where $X \in \mathscr{S}et\!f$ and $\rho \in \operatorname{Hom}_{\mathscr{G}r\mathscr{T}op}(G, \mathfrak{S}(X))$. Applying Lemma 2.9.2 to $H = \mathfrak{S}(X)$ gives $\operatorname{Hom}_{\mathscr{G}r\mathscr{T}op}(G, \mathfrak{S}(X)) = \varinjlim_i \operatorname{Hom}_{\mathscr{G}r}(G_i, \mathfrak{S}(X))$. Therefore,

$$\bigsqcup_{X \in \mathscr{S}et\!f} \operatorname{Hom}_{\mathscr{G}r\mathscr{T}op}(G, \mathfrak{S}(X)) = \bigsqcup_X \varinjlim_i \operatorname{Hom}(G_i, \mathfrak{S}(X))$$

$$= \varinjlim_i \bigsqcup_X \operatorname{Hom}(G_i, \mathfrak{S}(X)),$$

i.e. the objects of G-$\mathscr{S}et\!f$ and of $\varinjlim G_i$-$\mathscr{S}et\!f$ are the same.

We show that, for all i and G_i-sets X and Y, $\mathrm{Hom}_{G\text{-}\mathit{Setf}}(X, Y) = \varinjlim_{j \geq i} \mathrm{Hom}_{\mathit{GrSetf}}$

(X, Y), X and Y being G-sets by f_i^∞ and G_j-sets by f_i^j. Note that, for all j, $\mathrm{Hom}_{\mathit{GrSetf}}(X, Y)$ is a subset of $\mathrm{Hom}_{\mathit{Gns}}(X, Y)$, and so the direct limit of $\mathrm{Hom}_{\mathit{GrSetf}}(X, Y)$ amounts to a union. Let us show that a G-morphism $\varphi : X \to Y$ is a G_j-morphism for some j. Now, if φ is a G-morphism, then $(\forall g \in G) \, (\forall x \in X)$ $\varphi(g \cdot x) = g \cdot \varphi(x)$ or $(\forall g \in \mathrm{Im} f_i^\infty) \, \varphi(g \cdot x) = g \cdot \varphi(x)$. Choose an index j for which $\mathrm{Im} f_i^\infty = \mathrm{Im} f_i^j$ (2.8.5). Then, φ is a G_j-morphism. $\square$

2.9.5

Let G be a group, which can be endowed with the discrete topology. A **profinite completion** of G is the inverse limit of the groups G/N, where N runs over all normal subgroups of finite index in G ordered by reverse inclusion.

Proposition *Let* G *be a discrete group. Denote its profinite completion by* $\widehat{G}$*. Then the categories* G-Setf *and* $\widehat{G}$-Setf *are equivalent.*

Proof Let X be a finite set. The kernel of a homomorphism from G to $\mathfrak{S}(X)$ is a normal subgroup of finite index, and so $\mathrm{Hom}_{\mathit{Gr}}(G, \mathfrak{S}(X)) = \varinjlim \mathrm{Hom}(G/N, \mathfrak{S}(X)) = \mathrm{Hom}_{\mathit{GrTop}}(\widehat{G}, \mathfrak{S}(X))$. If X and Y are finite G/N-sets, then

$$\mathrm{Hom}_G(X, Y) = \mathrm{Hom}_{G/N}(X, Y) = \varinjlim_{N' \subset N} \mathrm{Hom}_{G/N'}(X, Y) \, .$$

Therefore, G-Setf $= \varinjlim(G/N\text{-}\mathit{Setf}) = \widehat{G}$-$\mathit{Setf}$. $\square$

2.9.6

Let G be a profinite group and (X_i) an inverse system of finite G-sets. Then $X = \varprojlim X_i$ is a profinite space on which G acts continuously. Indeed, $G \times X = \varprojlim(G \times X_i)$, and the action $G \times X \to X$ is induced by the actions $G \times X_i \to X_i$.
Let G-Prof denote the category of profinite spaces on which G acts continuously.

Proposition *Let* G *be a profinite group. The category* G-Prof *is equivalent to* G-$\overleftarrow{\mathit{Setf}}$*.*
More precisely, the functor G-$\overleftarrow{\mathit{Setf}}$ $\to$ G-Prof *which assigns to an inverse system its inverse limit is an equivalence.*

Proof (a) *It is essentially surjective.* Let X be a profinite space on which G acts continuously. We know (2.8.3, pf. (i) $\Rightarrow$ (ii)) that X is the inverse limit of the inverse system $(X_\lambda)_{\lambda \in \Lambda}$, where Λ is the set of equivalence relations on X with open classes and X_λ the quotient X/λ. In general, X_λ is not a G-set. We show that the subset Λ' of Λ consisting of equivalence relations compatible with the action of G on X is cofinal in Λ, i.e.

$$(\forall \lambda \in \Lambda) \; (\exists \mu \geq \lambda) \quad x \sim_\mu y \Rightarrow (\forall g \in G) \quad gx \sim_\mu gy .$$

Let $\lambda \in \Lambda$, and define μ by $x \sim_\mu y \Longleftrightarrow (\forall g \in G) \; gx \sim_\lambda gy$.

The relation μ is finer than λ and is compatible with the action of G. We show that its classes are open. It suffices to show that the graph Γ_μ of μ is open in $X \times X$. Consider the maps $X \times X \times G \underset{q}{\overset{p}{\rightrightarrows}} X \times X$ defined by $p(x, y, g) = (x, y)$ and $q(x, y, g) = (gx, gy)$.

Then $(X \times X) - \Gamma_\mu = p(q^{-1}((X \times X) - \Gamma_\lambda))$, where ú Γ_λ is the graph of λ. Since the classes of λ are open, Γ_λ is open; since X and G are compact, and p and q continuous, $(X \times X) - \Gamma_\mu$ is closed. So Γ_μ is open.

Hence $X = \varprojlim_{\lambda \in \Lambda} X_\lambda = \varprojlim_{\mu \in \Lambda'} X_\mu$, where all X_μ are finite G-sets and the identification is compatible with the actions of G.

(b) *It is fully faithful.* Let (X_i) be an inverse system of finite G-sets, and Y a finite G-set. We show that the natural map $\alpha : \varinjlim \mathrm{Hom}_{G\text{-}\mathcal{S}etf}(X_i, Y) \to \mathrm{Hom}_{G\text{-}\mathcal{P}rof}(\varprojlim X_i, Y)$ is a bijection.

In the following commutative diagram:

$$\begin{array}{ccc}
\varinjlim \mathrm{Hom}_{G\text{-}\mathcal{S}etf}(X_i, Y) & \hookrightarrow & \varinjlim \mathrm{Hom}_{\mathcal{S}etf}(X_i, Y) \\
\alpha \downarrow & & \downarrow \\
\mathrm{Hom}_{G\text{-}\mathcal{P}rof}(\varprojlim X_i, Y) & \hookrightarrow & \mathrm{Hom}_{\mathcal{T}op}(\varprojlim X_i, Y)
\end{array}$$

the arrow on the right is a bijection by 2.8.6, and so α is injective.

We show that it is surjective. Let $\varphi : \varprojlim X_i \to Y$ be a morphism. There is an index i and a map $\varphi_i : X_i \to Y$ making the diagram

$$\begin{array}{ccc}
& \varprojlim X_i & \\
f_i^\infty \downarrow & & \searrow^{\varphi} \\
X_i & \xrightarrow[\varphi_i]{} & Y
\end{array}$$

commutative. The map φ_i is compatible with the action of G on $\mathrm{Im} f_i^\infty$. Choose j such that $\mathrm{Im} f_i^j = \mathrm{Im} f_i^\infty$. Then $\varphi_i \circ f_i^j$ is a G-morphism.

Let (Y_j) be an inverse system of finite G-sets. Then,

$$\varprojlim_j \varinjlim_i \mathrm{Hom}_G(X_i, Y_j) \approx \varprojlim_j \mathrm{Hom}_{G\text{-}\mathcal{P}rof}(\varprojlim X_i, Y_j)$$

$$\approx \mathrm{Hom}_{G\text{-}\mathcal{P}rof}(\varprojlim X_i, \varprojlim Y_j).$$

$\square$

2.9.7

Let G be a profinite group, e its identity element, and for $g \in G$, ρ_g the right translation $x \mapsto x \cdot g$. Let ${}^{\cdot}G$ be the profinite G-set obtained by making G act on itself by left translations. The maps ρ_g are the automorphisms of ${}^{\cdot}G$.

The pair $({}^{\cdot}G, e)$ represents the forgetful functor G-$\mathcal{Prof} \to \mathcal{Ens}$: for any profinite G-set X, the map $x \mapsto \delta_x$, where $\delta_x(g) = g \cdot x$ is a bijection from the set X onto Hom$({}^{\cdot}G; X)$, and its inverse is $\phi \mapsto \phi(e)$. Then $\delta_{g \cdot x} = \delta_x \circ \rho_g$.

If X is finite, Hom$({}^{\cdot}G; X)$ is finite; in the general case X can be written $\varprojlim X_i$ with X_i finite. Then Hom$({}^{\cdot}G; X) = \varprojlim \text{Hom}({}^{\cdot}G; X_i)$ is the inverse limit of finite sets and can be considered a profinite set.

The map $(g, \phi) \mapsto \phi \circ \rho_g$ defines a *left*-action of G on Hom$({}^{\cdot}G; X)$ which can be identified with the action of G on X defining the G-set structure of X.

2.9.8

Let G and G′ be two profinite groups and $\phi : G' \to G$ a morphism (i.e. a continuous homomorphism). Define a functor $\phi^* : $ G-$\mathcal{Prof} \to $ G′-$\mathcal{Prof}$ as follows: if X is a profinite G-set, $\phi^* X$ is the space X on which G-acts by $(g', x) \mapsto \phi(g') \cdot x$; if $f : X \to Y$ is a morphism of G-sets, then $\phi^*(f) = f$. This functor induces a functor G-$\mathcal{Ens} \to $ G′-$\mathcal{Ens}$ also written ϕ^*.

If ϕ is an isomorphism, the functor ϕ^* is an equivalence of categories with $(\phi^{-1})^*$ as inverse.

Conversely:

Proposition *Let G and G′ be two profinite groups. If the categories G-$\mathcal{Prof}$ and G′-$\mathcal{Prof}$ are equivalent, the profinite groups G and G′ are isomorphic.*

Every equivalence of categories $\Phi : $ G-$\mathcal{Prof} \to $ G′-$\mathcal{Prof}$ *is isomorphic to a functor* ϕ^*, *where* $\phi : G' \to G$ *is an isomorphism.*

Proof Let $\Phi : $ G-$\mathcal{Prof} \to $ G′-$\mathcal{Prof}$ be an equivalence of categories. The object $O = \varnothing$ of G-$\mathcal{Prof}$ is characterized by Hom$(X; O) = \varnothing$ if $X \not\approx O$. Likewise for G′. Hence $\Phi(\varnothing) = \varnothing$.

The group G acts transitively on an object X of G-$\mathcal{Prof}$ if and only if there is no object $Y \sqcup Z$ with Y and Z non-empty isomorphic to X. Therefore, if G acts transitively on X, the group G′ acts transitively on $\Phi(X)$.

If X is an object of G-$\mathcal{Prof}$ on which G acts transitively, then $X \approx {}^{\cdot}G$ if and only if $X \neq \varnothing$ and Hom$(X, Y) \neq \varnothing$ for all $Y \neq \varnothing$. Likewise for G′. Therefore $\Phi({}^{\cdot}G) \approx {}^{\cdot}G'$.

Let $\alpha : {}^{\cdot}G' \to \Phi({}^{\cdot}G)$ be an isomorphism in G-$\mathcal{Prof}$. Then $\alpha_! : f \mapsto \alpha \circ f \circ \alpha^{-1}$ is a group isomorphism from Aut$({}^{\cdot}G')$ onto Aut$(\Phi({}^{\cdot}G))$. The map $\Phi_{\cdot G} : $ Aut$({}^{\cdot}G) \to $ Aut$(\Phi({}^{\cdot}G))$ is also an isomorphism, and so is $\Phi_{\cdot G}^{-1} \circ \alpha_! : $ Aut$({}^{\cdot}G') \to$

Aut($^{\cdot}$G). The latter is of the form $\rho_{g'} \mapsto \rho_{\phi(g')}$, where $\phi : G' \to G$ is an isomorphism of profinite groups.

If X is a profinite G-set, define a bijection $\beta = \beta_X : X \to \Phi(X)$ by the commutativity of the diagram:

$$
\begin{array}{ccc}
X & \xrightarrow{\quad\quad \beta \quad\quad} & \Phi(X) \\
{\scriptstyle x \mapsto \delta_x} \downarrow & & \downarrow {\scriptstyle x' \mapsto \delta_{x'}} \\
\mathrm{Hom}(^{\cdot}G; X) & \xrightarrow{\;\Phi\;} \mathrm{Hom}(\Phi(^{\cdot}G); \Phi(X)) \xrightarrow{\;\alpha^*\;} & \mathrm{Hom}(^{\cdot}G'; \Phi(X))
\end{array}
$$

This bijection is a homeomorphism, as can be seen by writing X as an inverse limit of finite G-sets. We show that β_X is a ϕ^{-1}-morphism:

Let $g \in G$ and $g' \in G'$ be such that $g = \phi(g')$, and $x \in X$. We need to show that $\beta(g \cdot x) = g' \cdot \beta(x)$, i.e. that $\delta_{\beta(g \cdot x)} = \delta_{g' \cdot \beta(x)} = \delta_{\beta(x)} \circ \rho_{g'}$. By definition of β_X, $\delta_{\beta(x)} = \Phi(\delta_x) \circ \alpha$. Hence it amounts to showing that $\Phi(\delta_{g \cdot x}) \circ \alpha = \Phi(\delta_x) \circ \alpha \circ \rho_{g'}$, i.e. that $\Phi(\delta_{g \cdot x}) = \Phi(\delta_x) \circ \alpha_!(\rho_{g'})$. However, $\alpha_!(\rho_{g'}) = \Phi(\rho_g)$ by definition of ϕ since $g = \phi(g')$, and Φ being a functor, $\Phi(\delta_{g \cdot x}) = \Phi(\delta_x \circ \rho_g) = \Phi(\delta_x) \circ \Phi(\rho_g)$.

Indeed, $\beta(g \cdot x) = \beta(\phi(g') \cdot x) = g' \cdot \beta(x)$. Hence $\beta = \beta_X$ can be considered an isomorphism in G'-$\mathscr{Prof}$ from $\phi^*(X)$ onto $\Phi(X)$. The family (β_X) where X runs through the objects of G-$\mathscr{Prof}$ is a functorial isomorphism from ϕ^* onto Φ. □

Corollary *Let G and G' be two profinite groups. If the categories* G-$\mathscr{Setf}$ *and* G'-$\mathscr{Setf}$ *are equivalent, then the profinite groups* G *and* G' *are isomorphic.*

Every equivalence of categories Φ : G-$\mathscr{Setf}$ $\to$ G'-$\mathscr{Setf}$ *is isomorphic to a functor* ϕ^*, *where* $\phi : G' \to G$ *is an isomorphism.*

Indeed every equivalence G-$\mathscr{Setf}$ $\to$ G'-$\mathscr{Setf}$ can be extended to an equivalence from G-$\mathscr{Prof}$ onto G'-$\mathscr{Prof}$ by passing to the inverse limit.

Exercises 2.9. (Profinite Groups)

1.—Let G be a compact group and H an open subgroup of G. Show that H contains an open normal subgroup.

2.—Show that the category of profinite Abelian groups is anti-equivalent to the category of torsion Abelian groups (i.e. where every element has finite order) (use 2.3, Exercise 4).

3.—Let Σ be the set of prime numbers. $\mathbb{N}^*$ can be identified with $\Sigma^{(\mathbb{N})}$. Set $\widetilde{\mathbb{N}}^* = \Sigma^{\widetilde{\mathbb{N}}}$ o ú $\widetilde{\mathbb{N}} = \mathbb{N} \cup \{\infty\}$. The elements of $\widetilde{\mathbb{N}}^*$ are called *supernatural numbers*. They are formal products $n = \prod_{p \in \Sigma} p^{r_p}$, where $r_p \in \widetilde{\mathbb{N}}$. The product $n_1 n_2$ is defined in the obvious manner. The *support* of n is the set of p such that $r_p \neq 0$. n_1 and n_2 are said to be relatively prime if they have disjoint supports. Let G be a profinite group. Define the order #G of G to be the supernatural number $\sup \mathrm{Card}\,(G/N)$, where N is subgroup of G of finite index. If H is a closed subgroup of G, define the index $(G : H)$ of H to be the supernatural number $\sup \mathrm{Card}\,(G/H')$ where H' is of finite index and contains H.

(a) Show that #G = #H · (G : H).

Deduce that there is no non-zero homomorphism from G to G_1 when the orders of G and G_1 are relatively prime.

(b) If H_1 and H_2 are two subgroups of G, let (H_1, H_2) denote the closed subgroup generated by all $aba^{-1}b^{-1}$ with $a \in H_1$ and $b \in H_2$. Define the *lower central series* (C_i) by $C_1 = G$, $C_{n+1} = (G, C_n)$. For $p \in \Sigma$, G is said to be a *p-group* if #G is a (finite or infinite) power of p.

Show that if G is a profinite p-group, then $\cap C_i = \{e\}$. (Assume this result holds for finite groups.)

(c) Let G be a profinite group, $p \in \Sigma$. A *Sylow p-subgroup* of G is a closed subgroup which is also a p-group whose index is relatively prime to p. Extend the following two Sylow theorems to profinite groups:

Theorem 1 *Every finite group has a Sylow p-subgroup.*

Theorem 2 *All Sylow p-subgroups are conjugate.*

(The reader may write G as $\varprojlim G_i$, where all $G_j \to G_i$ are surjective, and that for each i, and consider the set X_i of Sylow p-subgroups of G_i.)

4.—Let G be a compact group, S_α a descending filtration of closed subgroups of G. Set $S = \cap S_\alpha$. Show that $G/S = \varprojlim G/S_\alpha$.

5.—Let G be a profinite group, F a closed subgroup of G. The aim is to show the existence of a continuous section from G/F to G (which may not be a group homomorphism even when F is normal).

(a) Show that, if S is a closed subgroup of F of finite index in F, then there is a continuous section $G/F \to G/S$.

(b) Let $\mathscr{E}$ be the set of pairs (S, σ), where S is a closed subgroup of F and $\sigma : G/F \to G/S$ is a continuous section. Define an order with respect to which $\mathscr{E}$ is direct.

(c) Using (a), show that $S = \{e\}$ for any maximal element of $\mathscr{E}$. Conclude. Compare with 2.8, Exercise 3.

Reference

1. N. Bourbaki, *Algèèbre, ch. 1 à 3*, Hermann, 1970; *èAlgèbre, ch. 10, algèbre homologique*, Masson, 1980

Chapter 3
Linear Algebra

Unless stated otherwise, all rings considered in this chapter are assumed to be commutative, to have an identity element, and all homomorphisms to be unital (i.e. to send the identity onto the identity).

3.1 Integral and Principal Ideal Domains, Reduced Rings

3.1.1 Integral Domains

Let A be a ring. An element $a \in$ A is said to be a *zero divisor* in A if the map $x \mapsto ax$ from A to A is not injective.

The ring A is an **integral domain** if A $\neq 0$ and 0 is its only zero divisor.

3.1.2 Examples of Integral Domains

(1) All subrings of an integral domain.

(2) The ring $\mathbb{Z}$ of all integers.

(3) The ring $K[X_1, ..., X_n]$ of polynomials in n variables, and the ring $K[[X_1, ..., X_n]]$ of formal series over an integral domain K.[1]

(4) The ring $\mathbb{C}\{X_1, ..., X_n\}$ of convergent power series in n variables.

(5) The ring $\mathcal{O}(U)$ of all holomorphic functions on any open and connected subset U of $\mathbb{C}$.[2]

---

[1] Queysanne [1], §186, Lelong-Ferrand and Arnaudiès [2], cor. of th. 4.7.2, p. 151.

[2] Lelong-Ferrand and Arnaudiès [3], th. 4.11.2, p. 387.

© Springer Nature Switzerland AG 2020
R. Douady and A. Douady, *Algebra and Galois Theories*,
https://doi.org/10.1007/978-3-030-32796-5_3

Counterexamples

The following rings are not integral domains.

(1) The ring $C(X, \mathbb{R})$ of continuous functions on a non-zero metric space X having more than one point.

(2) The ring $\mathscr{E}(I)$ of C^∞-functions on a non-empty interval I of $\mathbb{R}$ with more than one point (3.1, Exercise 2).

(3) The product ring of two non-zero rings A and B. Indeed $(1, 0) \cdot (0, 1) = (0, 0)$.

3.1.3 Field of Fractions

Let A be an integral domain. Set $A^* = A - \{0\}$. For (a, b) and $(c, d) \in A \times A^*$, write $(a, b) \sim (c, d)$ if $ad = bc$. This defines an equivalence relation on $A \times A^*$; let K denote the quotient of $A \times A^*$ by this relation, and a/b be the class of (a, b). For any two elements x and y of K, there are representatives (a, b) and (c, d) of x and y such that $b = d$ (reduction to the same denominator). Addition in K is defined by $x + y = (a + c)/b$, which is an element only dependent on x and y. Multiplication in $A \times A^*$ is compatible with the equivalence relation and defines a multiplication on K. Together with these two operations, K is a field called the *field of fractions* of A. The map $a \mapsto a/1$ from A to K is an injective homomorphism embedding A in K.

Examples (1) $\mathbb{Q}$ is the field of fractions of $\mathbb{Z}$.

(2) If K is a field, the field of fractions of the polynomial ring $K[X_1, ..., X_n]$ is the *field of rational fractions* $K(X_1, ..., X_n)$.

(3) If U is open and connected in $\mathbb{C}$, the field of fractions of the ring $\mathcal{O}(U)$ of holomorphic functions on U is the field $\mathscr{M}(U)$ of meromorphic functions on U (3.1, Exercise 17).

3.1.4 Prime Ideals and Maximal Ideals

Let A be a ring and I an ideal of A. I is said to be **prime** if the quotient A/I is integral. For the sake of simplicity, I is called **maximal** if I is maximal among all strict ideals of A, i.e. if $I \neq A$ and there is no ideal J such that $I \subsetneq J \subsetneq A$.

Proposition *Let* A *be a ring and* I *an ideal of* A. *Then* A/I *is a field if and only if* I *is maximal.*

Proof The ring $B = A/I$ is a field if and only if $B \neq 0$ and its only ideals are 0 and B. As there is a bijection between the ideals of A containing I, and the ideals of A/I, the result follows.

Corollary *Every maximal ideal is prime.*

Examples (1) The prime ideals in the ring $\mathbb{Z}$ are 0 and $\mathbb{Z}p$, where p is a prime number. The non-zero prime ideals are all maximal. This will be seen to always be the case in a principal ideal domain.

(2) Let A and B be two rings and $\varphi : A \to B$ a homomorphism. If B is an integral domain, then the ideal $\text{Ker } \varphi = \varphi^{-1}(0)$ in A is prime.

(3) Let K be a field. In the ring of polynomials $K[X, Y]$, the ideal (Y) generated by Y is prime. Indeed, it is the kernel of the homomorphism $\rho : K[X, Y] \to K[X]$ defined by $\rho(X) = X$, $\rho(Y) = 0$ (see 3.7.2); in terms of polynomial functions, $\rho(f)$ is the function $x \mapsto f(x, 0)$. This ideal is not maximal: it is strictly contained in the ideal (X, Y).

(4) Let X be a metric space and A the ring $C(X, \mathbb{R})$ or $C(X, \mathbb{C})$. For $x \in X$, the ideal $\mathfrak{m}_x$ of functions vanishing at x is maximal: it is the kernel of the homomorphism δ_x from A onto $\mathbb{R}$ or $\mathbb{C}$, where $\delta_x(f) = f(x)$. If X is not discrete, then A contains non-maximal ideals (3.1, Exercise 18).

3.1.5

Proposition (Krull's Theorem) *A nonzero ring* A *contains at least one maximal ideal.*

Proof The set E of strict ideals of A is direct. Indeed, if (I_λ) is a totally ordered family of strict ideals, then $I = \bigcup I_\lambda$ is a strict ideal of A since $(\forall \lambda)\ 1 \notin I_\lambda$, and so $1 \notin I$. By Zorn's theorem (1.4.3), E has a maximal element. $\quad\square$

Corollary *Let* A *be a ring. Every strict ideal of* A *is contained in a maximal ideal.*

Indeed, the ideals of A containing I correspond to the ideals of A/I.

3.1.6 Nilpotent Elements, Nilradical

Let A be a ring. An element x of A is said to be **nilpotent** if there exists $n \in \mathbb{N}$ such that $x^n = 0$.

Examples (1) In an integral domain, only 0 is nilpotent.

(2) Let X be a topological space. In $C(X, \mathbb{R})$, only 0 is nilpotent.

(3) In the ring A of C^∞-functions $f : \mathbb{R} \to \mathbb{R}$, the functions f such that $f(0) = f'(0) = 0$ form and ideal I. The ring $B = A/I$ is isomorphic to $\mathbb{R}[X]/(X^2)$. In this ring, X is nilpotent.

(4) In the ring $\mathbb{Z}/(1000)$, the class of 150 is nilpotent. More generally, in $\mathbb{Z}/(n)$, the class of m is a nilpotent element if and only if every prime divisor of n divides m.

Theorem and Definition *Let* A *be a ring. The set* $\mathfrak{N}$ *of nilpotent elements of* A *is the intersection of prime ideals of* A *and is called the* **nilradical** *of* A.

Proof If $x \in A$ is a nilpotent element then, for every prime ideal $\mathfrak{p}$, $\chi(x) = 0$, where $\chi : A \to A/\mathfrak{p}$ is the quotient map, since $\chi(x)$ is nilpotent in the integral domain $A/\mathfrak{p}$. So $x \in \mathfrak{N}$.

Conversely, let $x \in A$ be a non-nilpotent element, and $S = \{x^n\}_{n \in \mathbb{N}}$. The set E of ideals I de A such that $I \cap S = \varnothing$ is directed, and so, by Zorn's lemma, has a maximal element $\mathfrak{p}$. We show that $\mathfrak{p}$ is prime. Let $A' = A/\mathfrak{p}$ and x' be the image of x in A'. The element x' is not nilpotent in A' since $\mathfrak{p} \cap S = \varnothing$. For any nonzero $y \in A'$, there exists $n \in \mathbb{N}$ such that $x'^n \in A'y$; otherwise, the inverse image of $A'y$ in A would be an ideal J strictly containing $\mathfrak{p}$ and such that $J \cap S = \varnothing$. If $y, z \in A'$ are nonzero, then there exists n and p such that $x'^n \in A'y$ and $x'^p \in A'z$; so $x'^{n+p} \in A'yz$, and thus $yz \neq 0$. Therefore $\mathfrak{p}$ is prime. As $x \notin \mathfrak{p}$, $x \notin \mathfrak{N}$. □

3.1.7 Reduced Rings

Proposition and Definition *Let* A *be a ring. The following conditions are equivalent:*

 (i) A *does not have any nonzero nilpotent element;*
 (ii) A *is isomorphic to a subring of a product of fields.*

If these conditions hold, then A *is said to be* **reduced.**

Proof Let Δ be the homomorphism $x \mapsto (\chi_\mathfrak{p}(x))$ from A to $\prod_{\mathfrak{p} \in S} K_\mathfrak{p}$, where S is the set of prime ideals of A and where, for $\mathfrak{p} \in S$, $K_\mathfrak{p}$ is the field of fractions of the integral domain $A/\mathfrak{p}$ and $\chi_\mathfrak{p} : A \to A/\mathfrak{p}$ is the quotient map. The kernel of Δ is the nilradical of A. If (i) holds, Δ is injective, implying (ii). The converse is immediate. □

3.1.8

Let A be a ring. The **spectrum** of A is the set S of all prime ideals of A. For all $\mathfrak{p} \in S$, let $K_\mathfrak{p}$ denote the field of fractions of the ring $A/\mathfrak{p}$. If $x \in A$, define the function $\hat{x}$ on S by $\hat{x}(\mathfrak{p}) = \chi_\mathfrak{p}(x) \in K_\mathfrak{p}$ (its values are in varying fields). Then $\widehat{x + y} = \hat{x} + \hat{y}$, $\widehat{xy} = \hat{x}\hat{y}$.

If A is reduced, the map $x \mapsto \hat{x}$ is injective. Assume A is integral and let K be a field of fractions of A. For any $x = \frac{a}{b} \in K$, let $\hat{x}$ be the function defined by $\hat{x} = \frac{\hat{a}}{\hat{b}}$ on the elements q of S for which $\hat{b}(q) \neq 0$. It is said to have a zero of order $\geqslant k$ in $\mathfrak{p}$ if x of the form $\frac{a}{b}$ with $a \in \mathfrak{p}^k$ and $b \notin \mathfrak{p}$. The functions $\hat{x}$ and $\hat{y}$ are said to have the same polar part in $\mathfrak{p}$ if $\widehat{x - y}$ is defined in $\mathfrak{p}$.

Considering functions $\hat{x}$ and using the above language enable us to draw on methods of analysis to solve algebra problems (3.1, Exercises 6 and 16); compare also the proofs of the Chinese theorem (5.1.5) and of Lagrange's interpolation formula.[3]

Let $\varphi : A \rightarrow B$ be a ring homomorphism. For every prime ideal q of B, $\varphi^{-1}(q)$ is a prime ideal of A. This gives a map φ^* from the spectrum of B to that of A, and thus defines a contravariant functor from $\mathscr{A}nn$ to $\mathscr{E}ns$. This functor factorizes naturally through the functor Spec from $\mathscr{A}nn$ to the category $\mathscr{T}op$ of (not necessarily Hausdorff) topological spaces (see 3.1, Exercise 7).

Comments As in 2.7.6, thanks to the functor Spec : $\mathscr{A}nn \rightarrow \mathscr{T}op$, the opposite category of rings and $\mathscr{T}op$ bear some resemblance.

3.1.9 Principal Ideal Domains

An ideal I in a ring A is said to be *principal* if $I = Ax$ for some $x \in A$. An integral domain in which every ideal is principal is a **principal ideal domain** (PID).

As will be seen, in a principal ideal domain, every nonzero prime ideal is maximal (3.2.10, Lemma).

3.1.10 Euclidean Rings

As integral domain A is said to be *Euclidian* if there is a map w from $A^* = A - \{0\}$ to $\mathbb{N}$ such that

(S$_1$) $(\forall x, y \in A^*)$ $w(xy) \geqslant w(y)$;
(S$_2$) $(\forall a \in A)$ $(\forall b \in A^*)$ $(\exists\, q, r \in A)$ $a = bq + r$ and $(r = 0$ or $w(r) < w(b))$.

Proposition *Every Euclidean ring is principal.*

Proof Let A be a Euclidean ring, $w : A^* \rightarrow \mathbb{N}$ satisfying (S$_1$) and (S$_2$) and I a nonzero ideal of A. Choose an element $b \neq 0$ in I such that $w(b)$ is minimum. For all $a \in I$, there exists q and r in A satisfying the conditions of (S$_2$); then $r = a - bq \in I$, and soù $r = 0$ since $w(b)$ is minimum. Hence I is generated by the b. $\square$

3.1.11 Examples

(1) $\mathbb{Z}$ is a Euclidian ring: take $w(x) = |x|$. So it is principal. In fact, every additive subgroup of $\mathbb{Z}$ is a principal ideal domain.

---

[3]Queysanne [1], §192, Exercise 5, p. 413, Lelong-Ferrand and Arnaudiès [2], chap. 4, Exercise 17, p. 476.

(2) If K is a ring, the ring K[X] is Euclidean (for $w(f)$ take the degree of f).

For other examples see 3.1, Exercises 13 and 14. For an example of a non-Euclidean principal ideal domain, see 3.7, Exercise 9.

Exercises 3.1. (Integral and principal ideal domains, reduced rings)

1.—(a) Let A be a ring. Show that A has idempotents $u \neq 0, 1$ (i.e. $u^2 = u$) if and only if $A = B \times C$, for some nonzero rings B and C.

(b) Let X be a non-empty topological space. Show that $C(X, \mathbb{R})$ has idempotents different from 0 and 1 if and only if X is not connected.

2.—(a) Show that the function $f : \mathbb{R} \to \mathbb{R}$ defined by $f(x) = e^{-1/x}$ for $x > 0$ and $f(x) = 0$ for $x \leqslant 0$ is infinitely differentiable. Show that the ring $C^\infty(\mathbb{R}, \mathbb{R})$ of infinitely differentiable functions on $\mathbb{R}$ is not an integral domain.

(b) Show that the convolution ring of distributions with compact support on $\mathbb{R}$ is an integral domain. (Use the fact that the Fourier transform of a distribution with compact support is analytic, or the existence of the Laplace transform.)

3.—Let φ be an analytic map from $\mathbb{R}$ to $\mathbb{R}^2$, $V = \varphi(\mathbb{R})$, and I be the ideal of $\mathbb{R}[X, Y]$ consisting of polynomials f such that $f(x, y) = 0$ for all $(x, y) \in V$.

(a) Show that I is prime.

(b) Give a generating family for I in the following cases:

$$\varphi(t) = (\cos t, \sin t); \quad (\mathrm{ch}\, t, \mathrm{sh}\, t); \quad (\cos 2t, \cos 3t);$$
$$(\cos t, \sin 2t); \quad (\cos 2t, \sin t); \quad (t, e^t);$$
$$\left(\frac{2t}{1+t^2}, \frac{1-t^2}{1+t^2}\right); \quad (t^2 - 1, t^3 - t^2).$$

(c) In each case describe the set W of (x, y) such that $f(x, y) = 0$ for all $f \in I$. When does $W = V$?

(d) Give an example showing that, if φ is only assumed to be C^∞, the ideal I need not be prime.

4.—Show that a direct (resp. inverse) limit of integral domains is an integral domain. Can the field of fractions of the limit be identified with the limit of the field of fractions? Investigate the case where the homomorphisms of the system are injective.

5.—Let A be the quotient ring $\mathbb{R}[X, Y]/I$, where $I = (X^2 + X^3 - Y^2)$.

(a) Construct the curve $\Gamma = \{(x, y) \in \mathbb{R}^2 \mid x^2 + x^3 - y^2 = 0\}$.

(b) Let $f \in \mathbb{R}[X, Y]$. Show that $f \in I$ if and only if $f|_\Gamma = 0$. Deduce that A can be identified with the ring of functions on Γ induced by the polynomials.

(c) Show that A is an integral domain (see Exercise 3).

(d) Let $\mathfrak{m}$ be the ideal of functions of A such that $f(0) = 0$. The sequence $(\mathfrak{m}^k)$, $k \in \mathbb{N}$ gives a fundamental system of neighbourhoods 0 with respect to the topology on A called the $\mathfrak{m}$-adic topology. Show that the completion $\widehat{A}$ of A with respect to this topology is not an integral domain.

6.—(a) Prove that any rational fraction $f \in \mathbb{C}(X)$ can be uniquely written as $f(z) = \sum\limits_{i \in I} \frac{c_i}{(z-a_i)^{r_i}} + P(z)$, where I is a finite set and P a polynomial. (Expand in the neighbourhood of each pole.)

(b) Show that the rational number $x \in \mathbb{Q}$ can be uniquely written as $\sum\limits_{i \in I} \frac{a_i}{p_i^{r_i}} + n$, where I is finite, p_i primes, $r_i \in \mathbb{N}$, $0 < a_i < p_i$, $n \in \mathbb{Z}$.

7.—Let A be a ring and S its spectrum. If I is an ideal of A, denote by V(I) the set of $\mathfrak{p} \in S$ such that $\hat{x}(\mathfrak{p}) = 0$ for all $x \in I$.

(a) Show that $V(\sum I_\lambda) = \bigcap V(I_\lambda)$ if (I_λ) is a family of ideals of A.

(b) Show that $V(I \cap J) = V(I \cdot J) = V(I) \cup V(J)$ if I and J are ideals of A.

(c) Deduce that the sets of type V(I) are the closed sets of a topology on S. This topology is called the *Zariski topology*. Show that, if $f : A \to B$ is a ring homomorphism, then the map f^* from the spectrum of B to that of A is continuous when both spectra are equipped with the Zariski topology.

(d) Describe the Zariski topology on $\text{Spec}(\mathbb{Z})$. Show that it is not Hausdorff.

(e) Show that, for any ring A, the space $\text{Spec}(A)$ is quasi-compact.

(f) Let I be an ideal of A. Show that the set of $x \in A$ such that $\hat{x}(\mathfrak{p}) = 0$ for all $\mathfrak{p} \in V(I)$ is the set of x such that $(\exists r)\ x^r \in I$.

8.—For every open subset U of $\mathbb{C}$, let $\mathcal{O}(U)$ be the ring of holomorphic functions on U, and for every compact subset K of $\mathbb{C}$ and open neighbourhood U of K, $\mathcal{O}(K) = \varinjlim\limits_{U \supset K} \mathcal{O}(U)$.

Let K be a compact connected set; the aim is to show that $\mathcal{O}(K)$ is a principal ideal domain.

(a) Show that every nonzero $f \in \mathcal{O}(K)$ can be uniquely written as the product of a polynomial all of whose roots are in K and of an invertible element.

(b) Suppose all the roots of $P \in \mathbb{C}[Z]$ are in K. For all $h \in \mathcal{O}(K)$, show that there is a unique pair $(g, R) \in \mathcal{O}(K) \times \mathbb{C}[Z]$ such that $h = gP + R$ with $d^\circ R < d^\circ P$.

(c) Deduce that $\mathcal{O}(K)$ is Euclidian. Conclude.

(d) What are the ideals of $\mathcal{O}(K)$?

(e) Let $f_1, \ldots, f_n$ be elements of $\mathcal{O}(K)$ represented by functions without common roots. Show that there exist $u_1, \ldots, u_n \in \mathcal{O}(K)$ such that $\sum_{i=1}^{n} u_i f_i = 1$.

9.—Let A be a ring, but not a field. Prove that A[X] is not a principal ideal domain. For this, the ideal of A[X] consisting of the polynomials whose constant terms are in a non-zero ideal $I \neq 0$, A in A may be shown not to be principal.

10. *(Discrete valuation ring)*—(a) Let K be a field. Prove that the ring K[[X]] of formal sequences is a principal ideal domain.

(b) More generally, let A be a ring and $w : A \to \overline{\mathbb{N}}$ a map satisfying the following properties:

(i) $w(xy) = w(x) + w(y)$ for x and y in A;

(ii) $w(x + y) \geqslant \inf(w(x), w(y))$;

(iii) $w(x) < \infty$ for $x \neq 0$;

(iv) $w(x) > 0$ for non-invertible x .

Prove that A is a principal ideal domain with a unique maximal ideal.

(c) Conversely, let A be a principal ideal domain with a unique maximal ideal $\mathfrak{m}$. For all x, let $w(x)$ be the largest integer k such that $x \in \mathfrak{m}^k$. Show that w satisfies conditions (i) to (iv) (to check (iii), let x be a generator of $\mathfrak{m}$ and y a generator of $\bigcap \mathfrak{m}^k$ and consider $w(y/x)$). (Rings satisfying these properties are called discrete valuation rings)

(d) Prove that the ring E of germs at 0 of C^∞-functions on $\mathbb{R}$ has a unique maximal ideal $\mathfrak{m}$, and that this ideal is principal. However E is not a discrete valuation ring: $\bigcap \mathfrak{m}^k \neq 0$.

11.—Show that every nonzero prime ideal in a principal ideal domain is maximal.

12.—Let A be the ring of analytic functions $f : \mathbb{R} \to \mathbb{R}$ satisfying $(\forall x) f(x + 2\pi) = f(x)$.

(a) Show that A is an integral domain.

(b) Show that every maximal ideal of A is of the form $\mathfrak{m}_x = \{f \mid f(x) = 0\}$ with $x \in \mathbb{R}$.

(c) Show that every nonzero ideal is of the form $\mathfrak{m}_{x_1}^{k_1} \cdots \mathfrak{m}_{x_r}^{k_r}$. In particular every nonzero prime ideal is maximal.

(d) Let $x \in \mathbb{R}$ and $f \in A$ a function having x as a simple root. Show that, for sufficiently small $\varepsilon > 0$, f has a root between $x + \varepsilon$ and $x + 2\pi - \varepsilon$. Deduce that the ideal $\mathfrak{m}_x$ is not principal.

(e) The ideal $\mathfrak{m}_{x_1}^{k_1} \cdots \mathfrak{m}_{x_r}^{k_r}$ is principal if and only if $k_1 + \cdots + k_r$ is even.

(f) Let B be the ring of $\mathbb{R}$-analytic functions $f : \mathbb{R} \to \mathbb{C}$ such that $(\forall x)$ $f(x + 2\pi) = f(x)$. Show that B is principal (note that $t \mapsto e^{it} - e^{ix}$ has a simple root at x). Is the ring B Euclidian?

13.—Show that the subrings $\mathbb{Z}[i]$ and $\mathbb{Z}\left[\frac{1+i\sqrt{3}}{2}\right]$ of $\mathbb{C}$ are principal (they are Euclidian for $z \mapsto |z|^2$).

14.—Let A be an integral domain. Show that if A is Euclidean, then there is a non-invertible element $x \in A$ such that every class of $A/(x)$ contains an invertible invertible or zero (reduce to the case where $w(A^*)$ is an upper set of $\mathbb{N}$ and take x such that $w(x) = 1$, defining w as in 3.1.10). Show that the ring $\mathbb{Z}\left[\frac{1+i\sqrt{d}}{2}\right]$ is not Euclidian if $d \geqslant 5$. ($\mathbb{Z}\left[\frac{1+i\sqrt{19}}{2}\right]$ may be shown to be a principal ideal domain. For another example of a non-Euclidean principal ideal domain, see 3.7, Exercise 9.)

15.—Let A be a principal ideal domain. Show that, for any multiplicatively stable subset $S \subset A^*$, the set $S^{-1}A$ of fractions with numerator in A and denominator in S is a principal ideal domain.

16.—(a) Give a degree 2 expansion of $\sqrt{1 + t}$ in the neighbourhood of $t = 0$.

(b) Let A be a commutative ring in which 2 is not invertible, and $q \in A$ a multiple of $p \in A$. Show that $\left(1 + \frac{q}{2} - \frac{q^2}{8}\right) \equiv 1 + q \pmod{p^3}$.

(c) Find $u \in \mathbb{Z}$ such that $u^2 \equiv 11 \pmod{125}$.

17.—The aim is to prove some algebraic properties of the ring $\mathcal{O}(U)$ of holomorphic functions on U, where U is a connected open subset of $\mathbb{C}$. The reader interested in the case $U = D_R$ (or more generally, given Riemann's uniformization theorem, U simply connected) can omit part B.

A. (a) For $n, k \in \mathbb{N}$, let $\eta_{n,k}$ be the primitive of $((n+1)...(n+k+1)/k!)\, z^n (1-z)^k$ vanishing at 0. Show that $1 - \eta_{n,k}$ has a root of order $\geqslant k$ at 1, and that, as n tends to infinity for fixed k, $\eta_{n,k}$ converges uniformly to 0 on every disc of radius < 1 centered at 0.

(b) Let $a \in \mathbb{C}^*$, $r < |a|$, h a holomorphic function on $D_{r'}$ with $r' > r$, $\varepsilon > 0$, g a holomorphic function in the neighbourhood of a and $k \in \mathbb{N}$. Show that there is a polynomial f on D_r such that $|f - h| \leqslant \varepsilon$ and that $f - g$ has a root of order $\geqslant k$ at a.

(c) Let $R \in]0, \infty]$, and set $U = D_R$ (hence $U = \mathbb{C}$ if $R = \infty$). Let (a_n) be a set of distinct points of U such that $|a_n|$ tends to R; for all n, let g_n be a holomorphic function in the neighbourhood of a_n and $k_n \in \mathbb{N}$. Show that there is a holomorphic function $f \in \mathcal{O}(U)$ such that, for all n, the function $f - g_n$ has a root of order $\geqslant k_n$ at a_n.

B. (a) Let K be a compact subset of the Riemann sphere $\widehat{\mathbb{C}} = \mathbb{C} \cup \{\infty\}$, a_1 and a_2 two points in the same connected component of $\widehat{\mathbb{C}} - K$, $f_1 : \widehat{\mathbb{C}} \to \widehat{\mathbb{C}}$ a rational function such that $f_1^{-1}(\infty) = a_1$ and $\varepsilon > 0$. Show that there is a rational function f_2 such that $f_2^{-1}(\infty) = a_2$ and $|f_2 - f_1| < \varepsilon$ on K (first consider the case where there is a disk in $\widehat{\mathbb{C}} - K$ containing a_1 and a_2, with a_2 mapped to ∞ by a homography).

(b) Let U be an open subset of $\mathbb{C}$. A compact subset $K \subset U$ is said to be *complete* (in U) if $U - K$ does not have any relatively compact connected component in U. Prove the existence of a sequence (K_n) of complete compact subsets in U such that any compact subset of U is contained in some K_n.

(c) Let $a = (a_n)$ be a sequence of distinct points of U tending to infinity in U (i.e. such that compact subsets of U only contain finitely many a_n); for all n, let g_n be a holomorphic function in the neighbourhood of a_n and $k_n \in \mathbb{N}$. Show that there is a holomorphic function $f \in \mathcal{O}(U)$ such that, for all n, the function $f - g_n$ has a root of order $\geqslant k_n$ at a_n.

C. Let U be a connected open subset of $\mathbb{C}$.

(a) Let f be a meromorphic function on U (see 6.1.3). Prove that $f = g/h$ for some $g, h \in \mathcal{O}(U)$ with $h \neq 0$. This identifies the field $\mathcal{M}(U)$ of meromorphic functions on U with the field of fractions of $\mathcal{O}(U)$.

(b) Let $a = (a_n)$ be a sequence of distinct points of U tending to infinity in U. Let $\mathfrak{U}$ be an ultrafilter on $\mathbb{N}$, and set

$$I(a, \mathfrak{U}) = \{f \in \mathcal{O}(U) \mid (\exists A \in \mathfrak{U})\, (\forall n \in A)\, f(a_n) = 0\}.$$

Show that $I(a, \mathfrak{U})$ is a maximal ideal of $\mathcal{O}(U)$ whose codimension as a vector subspace over $\mathbb{C}$ is not 1 (1.7, Exercise 2). Show that any ideal of $\mathcal{O}(U)$ that cannot be written as $\operatorname{Ker} \delta_x$ (where $\delta_x(f) = f(x)$) is of the form $I(a, \mathfrak{U})$.

(c) Let $J(a, \mathfrak{U})$ be the ideal of $\mathcal{O}(U)$ consisting of f such that, $\lim_{\mathfrak{U}} k_n = \infty$, where k_n denotes the order of vanishing of f at a_n. Show that $J(a, \mathfrak{U})$ is a prime ideal but not a maximal one. Construct a strictly decreasing sequence of prime ideals in $\mathcal{O}(U)$.

18.—Let X be a metric space and A the ring $C(X, \mathbb{R})$ or $C(X, \mathbb{C})$. Let $(x_n)_{n \in \mathbb{N}}$ be a sequence of distinct points of X converging to a point $a \in X$, with $x_n \neq a$ for all n. For a filter $\mathscr{F}$ on $\mathbb{N}$ tending to ∞, set $I_{\mathscr{F}}$ to be the set of $f \in A$ such that $\{n \in \mathbb{N} \mid f(x_n) = 0\} \in \mathscr{F}$.

Show that $I_{\mathscr{F}}$ is an ideal of A strictly contained in the maximal ideal $\mathfrak{m}_a$, and that it is a prime ideal if and only if $\mathscr{F}$ is an ultrafilter.

3.2   Unique Factorization Domains

3.2.1   Monoids

A **monoid** is a set with an associative composition law that has an identity element. The law is written either additively or multiplicatively, the additive notation being usually used for commutative monoids. The morphisms are the maps preserving the composition law and the identity element.

If M is a monoid, the set G of invertible elements of M is a group for the induced law. If M is commutative, the relation

$$(\exists g \in G) \quad y = gx \,,$$

between elements x and y of M is an equivalence relation compatible with the law of composition.

A monoid M is said to be *regular* if, for all $x, y, z \in M$,

$$(xz = yz) \Rightarrow (x = y) \text{ and } (zx = zy) \Rightarrow (x = y) \,.$$

Let M be a monoid, e its identity element. A subset M′ of M is said to be a *submonoid* of M if M′ is stable under the composition law and $e \in M'$. If M′ is stable under the law of M and has an identity element $e' \neq e$ for the induced law, then M′ is not a submonoid of M.

Let M be a monoid and $(x_i)_{i \in I}$ a family of elements of M. A submonoid M generated by the family (x_i) is the smallest submoid M′ of M such that $(\forall i \in I)$ $x_i \in M'$. If M is commutative and written additively, the submonoid generated by the family (x_i) is the set of finite linear combinations of x_i with coefficients in $\mathbb{N}$.

3.2.2   Divisibility

Let M be a commutative monoid. For $x, y \in M$, x is said to *divide* y, in which case it is written[4] $x \langle y$ if $(\exists z \in M)$ $y = zx$, and to *strictly divide* y if $x \langle y$ and $y \nmid x$. If M

---

[4]It is usually written $x|y$. The disadvantage of the symbol | is that crossing it is problematic and it cannot be reversed.

is regular and e its only invertible element, then the divisibility relation is an order on M. The identity element is the smallest element for this order. An element of M is *irreducible* if it is minimal in $M - \{e\}$, i.e. if it cannot be written xy with $x \neq e$, $y \neq e$.

3.2.3 Support of a Family

Let $x = (x_i)_{i \in I}$ be a family of elements in a monoid M with identity element e. The *support* of x is the set of indices $i \in I$ such that $x_i \neq e$.

If M is commutative and written additively, and if $(x_i)_{i \in I}$ is a family of elements of M with finite support, then set

$$\sum_{i \in I} x_i = \sum_{i \in J} x_i$$

where J is the support of the family (x_i), or an arbitrary finite family of I containing this support.

3.2.4 Free Commutative Monoids on a Set

Let I be a set, and $\mathbb{N}^{(I)}$ the additive monoid of families $(n_i)_{i \in I}$ of elements of $\mathbb{N}$ with finite support; $\mathbb{N}^{(I)}$ is said to be the **free commutative monoid on** I. For $i \in I$, let e_i be the element $(\delta_{ij})_{j \in I}$ of $\mathbb{N}^{(I)}$ defined by $\delta_{ii} = 1$ and $\delta_{ij} = 0$ if $i \neq j$. The (e_i) are the irreducible elements of $\mathbb{N}^{(I)}$.

UNIVERSAL PROPERTY. *Let* M *be a commutative monoid and* $(x_i)_{i \in I}$ *a family of elements of* M. *There is a unique homomorphism* f *from* $\mathbb{N}^{(I)}$ *to* M *such that* $f(e_i) = x_i$ *for all* $i \in I$.

Proof Write M additively. If f satisfies the stated property, then for all $n = (n_i)_{i \in I} \in \mathbb{N}^{(I)}$,

$$f(n) = f\left(\sum n_i e_i\right) = \sum n_i x_i \,,$$

and uniqueness follows. The map f defined by

$$f(n) = \sum n_i x_i$$

for $n = (n_i)_{i \in I} \in \mathbb{N}^{(I)}$ is the desired one.   □

The map f is surjective if and only if the family (x_i) generates M. If f is an isomorphism, the family (x_i) is said to be a *basis* of M.

3.2.5   Free Commutative Monoids

A monoid M is said to be *free* if it has a basis. Unlike other structures, a commutative monoid has a unique basis, up to permutation: the basis elements are necessarily irreducible.

A free commutative monoid M is a lattice for the divisibility order: every finite subset has an upper bound called l.c.m., and, if it is not empty, a lower bound called g.c.d.. Writing M multiplicatively, for x, $y \in$ M, $xy = $ g.c.d.$(x, y) \cdot$ l.c.m.(x, y).

3.2.6

Proposition *Let* M *be a regular commutative monoid and* e *its only invertible element. Then* M *is free if and only if it satisfies the following two conditions:*

(i) *any decreasing sequence of elements of* M *for the divisibility order is stationary;*
(ii) *for any irreducible* $m \in$ M *and* x, $y \in$ M, $m \langle xy \Rightarrow m \langle x$ *or* $m \langle y$ *(writing* M *multiplicatively).*

Proof To show that conditions (i) and (ii) are necessary, we may assume that M $= \mathbb{N}^{(I)}$. We use additive notation, and define $\psi :$ M $\to \mathbb{N}$ by $\psi(x) = \sum x_i$. The homomorphism ψ is increasing and

$$(x \leqslant y \text{ and } \psi(x) = \psi(y)) \Rightarrow x = y.$$

So (i) follows. If $m = e_i$ divides $x + y$, then $x_i + y_i > 0$, and so $x_i > 0$ or $y_i > 0$. So (ii) holds.

Shifting back multiplicative notation, we show by contradiction that condition (i) implies that M is generated by the irreducible elements. Let M$'$ be the submonoid of M generated by the irreducible elements, and let $x \in$ M $-$ M$'$. We inductively define a strictly decreasing sequence (x_n) of elements of M $-$ M$'$ such that $x_0 = x$. Suppose that x_n has been defined. As x_n is not irreducible, $x = yz$ for some $y \neq e$ and $z \neq e$, and we may assume that $y \notin$ M$'$. Set $x_{n+1} = y$; then x_{n+1} strictly divides x_n since $z \neq e$. This contradicts condition (i).

Suppose condition (ii) holds. We show by induction on r that if $m_1 \ldots m_r = m_1' \ldots m_r'$, where all m_i and m_i' are irreducible, then $r = r'$ and there is a permutation $\sigma \in \mathfrak{S}_r$ such that $m_i = m_{\sigma(i)}'$ for $i \in \{1, ..., r\}$. This readily follows for $r = 0$ or 1. Suppose the property holds for $r - 1$. Then m_r divides some m_i', and hence is equal to it. We may assume this to be $m_{r'}'$. Then $m_1 \ldots m_{r-1} = m_1' \ldots m_{r'-1}'$ since M is regular, and hence by induction $r - 1 = r' - 1$ and there is a permutation $\sigma' \in \mathfrak{S}_{r-1}$ such that $m_i = m_{\sigma'(i)}'$ for $i \leqslant r - 1$. Extending σ' to $\sigma \in \mathfrak{S}_r$ by $\sigma(r) = r$, the result follows.

Hence, if M satisfies (i) and (ii), it is free. $\square$

3.2.7 Monoids Associated to Integral Domains

Let A be an integral domain. Write Mon(A) for the multiplicative quotient monoid of A by the equivalence relation identifying x and ux if u is invertible. Let χ be the canonical map A $\rightarrow$ Mon(A) and Mon*(A) the image of $A^* = A - \{0\}$ in Mon(A). Then Mon*(A) is called the *monoid associated* to A.

For $x, y \in A^*$, $\chi(x) = \chi(y)$ if and only if x and y generate the same ideal. An element $x \in A^*$ is said to be *irreducible* if $\chi(x)$ is irreducible in Mon*(A), i. e. if x is not invertible and if, for all pairs (u, v) such that $x = uv$, u or v is invertible.

3.2.8 Unique Factorization Domains

An integral domain A is said to be a **unique factorization domain** (UFD) if Mon*(A) is a free commutative monoid. This amounts to saying that every element $x \in A^*$ can be written as a product $u \cdot p_1 \dots p_r$ of irreducible elements p_i and some invertible u, and this representation is unique up to permutation of the elements p_i and multiplication of u and p_i by invertible elements.

Thanks to Proposition 3.2.6, A is a UFD if and only if it satisfies the following two conditions:

  (i) *every increasing sequence of principal ideals is stationary;*
 (ii) *for any irreducible element $m \in A^*$, the ideal Am is prime.*

If A is a UFD, every finite subset of Mon*(A) has a lower bound (g.c.d.) and an upper bound (l.c.m.)[5] for divisibility. If (x_i) is a family of finite elements of A, an element x of A is a g.c.d. (resp. a l.c.m.) of x_i if $\chi(x)$ is the g.c.d. (resp. the l.c.m.) of $\chi(x_i)$. If x is a l.c.m. of x_i, then $Ax = \bigcap Ax_i$.

3.2.9

Examples The polynomial ring $K[X_1, \dots, X_n]$ for a field K will be seen to be a UFD. Likewise for the ring of formal series $K[[X_1, \dots, X_n]]$. The ring $\mathbb{C}\{X_1, \dots, X_n\}$ of convergent series can be shown to be a UFD; the ring of germs at 0 of analytic functions on a Banach space is a UFD.

Counterexample Let K be a field. The subring of K[X, Y] consisting of polynomials containing only even powers is not a UFD. Indeed, the element XY of A is irreducible because it has minimal degree. It divides neither X^2, nor Y^2, but does divide X^2Y^2, and so A does not satisfy condition (ii) of 3.2.8.

---

[5]Abbreviations of "greatest common divisor" and "least common multiple".

3.2.10

Theorem *Every PID is a UFD.*

Proof Let A be a PID. We show that it satisfies conditions (i) and (ii) of 3.2.8.

(i) Let (I_n) be an increasing sequence of ideals and set $I = \bigcup I_n$. There exists x such that $I = Ax$. This element x is in I_n; then $Ax \subset I_n \subset I = Ax$. So $I_n = I$, and $I_{n'} = I_n$ for $n' \geqslant n$.

(ii) follows from the following lemma:

Lemma *Let A be a PID and* $I = Ax$ *a nonzero ideal of A. The following conditions are equivalent:*

(i) the element x is irreducible;
(ii) the ideal I is prime;
(iii) the ideal I is maximal.

Proof (iii) $\Rightarrow$ (ii) by 3.1.4, Corollary.

(ii) $\Rightarrow$ (i): Suppose that I is prime and $x = yz$. Then y or z is in I. Without loss of generality, assume $y \in I$. Then x divides y and z is invertible. Hence x is irreducible.

(i) $\Rightarrow$ (iii): Let $J = Ay$ be such that $I \underset{\neq}{\supsetneq} J \underset{\neq}{\supsetneq} A$. Then y is not invertible and x can be written yz where z is not invertible. This is impossible if x is irreducible.   □

Remark For invertible x, none of the conditions (i), (ii), (iii) are satisfied.

3.2.11

Proposition *Let A be a PID, (x_i) a finite family of elements of A. Then $x \in A$ is a g.c.d. of x_i if and only if $Ax = \sum Ax_i$.*

Proof Let $x \in A$ be such that $Ax = \sum Ax_i$. For $y \in A$, $(\forall i)\ y \langle x_i \Leftrightarrow (\forall i)\ Ay \supset Ax_i \Leftrightarrow Ay \supset \sum Ax_i \Leftrightarrow Ay \supset Ax \Leftrightarrow y \langle x$. So x is a g.c.d. of x_i. If x' is any other g.c.d. of x_i, then $x = ux$ for some invertible u. The proposition follows.   □

Corollary (Bezout's identity) *Under the assumptions of the proposition, if x is a g.c.d. of x_i, there is a family (a_i) of elements of A such that $x = \sum a_i x_i$.*

Exercises 3.2. (UFDs)

1.—Let K be a field. The subring $K[t^2, t^3]$ of $K[t]$ is not a UFD.

2.—The ring $\mathbb{R}[X, Y, Z]/(X^2 + Y^2 + Z^2)$ is a UFD, but not $\mathbb{C}[X, Y, Z]/(X^2 + Y^2 + Z^2)$ (it is isomorphic to the ring given in the counterexample in 3.2.9).

3.—Let A be an integral domain, and K its field of fractions. Then A is said to be *integrally closed* if, whenever P_1 and P_2 are monic polynomials in $K[X]$ then $P_1 P_2 \in A[X]$ implies $P_1 \in A[X]$ and $P_2 \in A[X]$. (this condition can be shown to be equivalent to the apparently weaker one: "any root of a monic polynomial of A[X] in K belongs to A".)

Show that, if A is integrally closed, then the multiplicative monoid M of monic polynomials of A[X] is a free commutative monoid.

4.—Let E be a complex Banach space, and $\mathscr{O}_E$ the ring of germs at the origin of holomorphic functions in the neighbourhood of 0 in E. The aim is to show that the ring $\mathscr{O}_E$ is a UFD by using Exercise 3 and assuming the following results, where F denotes a hyperplane in E:

REMOVABLE SINGULARITY THEOREM. *The ring $\mathscr{O}_F$ is integrally closed.*

A polynomial $P \in \mathscr{O}_F[Z]$ is *normal* if it is monic and all its coefficients except the leading one are functions vanishing at 0. Let R be the set of $f \in \mathscr{O}_{F \times \mathbb{C}}$ such that $f(0, Z) \in \mathbb{C}[Z]$ is not zero.

PREPARATION THEOREM. *Every element f of R can be uniquely written as $f = u \cdot P$, where u is an invertible element of $\mathscr{O}_{F \times \mathbb{C}}$ and $P \in \mathscr{O}_F[Z]$ is a normal polynomial.*

(a) Show that the normal polynomials of $\mathscr{O}_F[Z]$ together with multiplication form a free commutative monoid. Deduce that $M = R/G_m(\mathscr{O}_{F \times \mathbb{C}})$ is a free commutative monoid. Let $G_m(A)$ be the group of invertible elements of A.

(b) For any line D in E passing through 0, let R^D be the monoid of functions $f \in \mathscr{O}_E$ that do not vanish everywhere in the neighbourhood of 0 in D. Set $M^D = R^D/G_m(\mathscr{O}_E)$. Show that M^D is a free monoid. Deduce that $\mathscr{O}_E$ is a UFD (criterion 3.2.6 may be used).

5.—Let E be a complex Banach space, U a connected open subset of E and K a compact subset of $\mathbb{C}$. Let $\mathscr{O}(U)$ be the ring of holomorphic functions on U. A polynomial $P \in \mathscr{O}(U)[Z]$ is said to be K-*normal* if, for all $x \in U$, every root of the polynomial $P(x, Z) \in \mathbb{C}[Z]$ is in the interior $\overset{\circ}{K}$ of K. Show that the K-normal polynomials of $\mathscr{O}(U)[Z]$ together with multiplication form a free commutative monoid.

6.—Let A be a UFD all of whose nonzero prime ideals are maximal. The aim is to show that A is principal.

(a) Show that, for all $x, y \in A^*$, there exist $u, v \in A$ such that $ux + vy$ is a g.c.d. of x and y. (First consider the case where x is irreducible, then the case where $y/\text{g.c.d.}(x, y) = \text{l.c.m.}(x, y)/x$ is irreducible, and finally the general case.)

(b) Let $I \subset A$ be a non-zero ideal. Show that there is a g.c.d. $d \in A^*$ of all the elements of $I \cap A^*$. Conclude.

(c) Show that the ring A of real analytic functions with period 2π (3.1, Exercise 12) is not a UFD. Give an example of an element of $\text{Mon}^*(A)$ whose decomposition into irreducible elements is not unique.

3.3 Modules

Throughout this section, A denotes a (commutative) ring.

3.3.1

Definition A **module** over A or A-**module** is a set E together with an internal composition law called addition and written $(x, y) \mapsto x + y$ and an external law $A \times E \to E$ written $(\lambda, x) \mapsto \lambda x$, satisfying the following conditions:

(M_1) The additive set E is a commutative group;

(M_2) $(\forall \lambda, \mu \in A)\, (\forall x \in E)\quad \lambda(\mu x) = (\lambda \mu)x$

$\quad\quad (\forall \lambda, \mu \in A)\, (\forall x \in E)\quad (\lambda + \mu)x = \lambda x + \mu x$

$\quad\quad (\forall \lambda \in A)\, (\forall x, y \in E)\quad \lambda(x + y) = \lambda x + \lambda y$

$\quad\quad (\forall x \in E)\quad 1x = x.$

Examples (1) All A-modules for a field A are vectorial spaces over A.

(2) Every commutative group has a unique $\mathbb{Z}$-module structure.

(3) Let $\varphi : A \to B$ be a ring homomorphism. Setting $a \cdot b = \varphi(a)\, b$ for $a \in A$ and $b \in B$ gives B an A-module structure.

(4) The space $\mathscr{D}'$ of distributions on $\mathbb{R}$ together with multiplication is a module over the ring $\mathscr{E}$ of C^∞-functions. The space $\mathscr{E}$ together with convolution is a module over the convolution ring of distributions with compact support.

(5) Let X be a topological space and E a vector bundle over X. The space of continuous sections of E is a module over the ring of continuous functions on X.

3.3.2

Definition Let E and F be two A-modules. A map f from E to F is said to be a homomorphism or A-linear if

(L_1) $(\forall x, y \in E)\quad f(x + y) = f(x) + f(y)$;

(L_2) $(\forall x \in E)\, (\forall \lambda \in A)\quad f(\lambda x) = \lambda f(x)$.

More generally, let $h : A \to B$ be a ring homomorphism, E an A-module, F a B-module and f a map from E to F. Then f is said to be a *h-morphism* or *h-linear* if (L_1) holds and

(L_2') $(\forall x \in E)\, (\forall \lambda \in A)\quad f(\lambda x) = h(\lambda) f(x)$.

Let $h : A \to B$ be a ring homomorphism, E a B-module. Define an A-module structure on E by setting for $x \in E$ and $\lambda \in A$

$$\lambda \cdot x = h(\lambda)x.$$

The A-module structure on E is said to be obtained by **scalar restriction.** The identity map from the A-module E onto the B-module E is a *h*-morphism.

3.3.3 Submodules

Let E be an A-module. A subset E′ of E is said to be an A-*submodule* (or simply a *submodule*) of E if

(SM$_1$) E′ is an additive subgroup of E;
(SM$_2$) $(\forall x \in E')$ $(\forall \lambda \in A)$ $\lambda x \in E'$.

Every submodule E′ of an A-module E is an A-module for the induced laws, and the canonical injection $\iota : E' \to E$ is a homomorphism.

Examples (1) The A-submodules of A are the ideals of A.
(2) Let E and F be two A-modules, $f : E \to F$ a homomorphism. The *kernel* of f, written Ker f, is the submodule $f^{-1}(0)$ of E. The *image* of f, written Im f, is the submodule $f(E)$ of F.

3.3.4 Torsion

Suppose that A is integral and let E be an A-module. An element x of E is said to be a *torsion element* if the homomorphism $\lambda \mapsto \lambda x$ from A to E is not injective. The set of torsion elements of E is a submodule T of E called the *torsion submodule* of E; E is **torsion-free** if $T = \{0\}$, and a **torsion module** if $T = E$.

Let E and F be two modules over an integral domain. Every homomorphism $f : E \to F$ maps the torsion submodule of E to that of F.

Examples Suppose that A is integral and let I be a nonzero ideal in A. As an A-module, the quotient ring A/I is a torsion module.

The $\mathbb{Z}$-module $\mathbb{Q}/\mathbb{Z}$ is a torsion module.

3.3.5 Generating Families

Let E be an A-module and $(x_i)_{i \in I}$ a family of elements of E. An A-*submodule of* E *generated* by the family $(x_i)_{i \in I}$ is the smallest submodule E′ of E such that

$$(\forall i \in I) \quad x_i \in E'.$$

It is also the set of finite linear combinations of x_i with coefficients in A. If the submodule generated by $(x_i)_{i \in I}$ is E, the latter is a *generating family* of E.

The A-module E is **finitely generated** if it has a finite generating family, and it is *cyclic* if it is generated by one element.

Let (E_i) be a family of submodules of E. The submodule generated by the union of all E_i is the subgroup $\sum E_i$ generated by this union.

3.3.6   Quotient Modules

Let E be an A-module, E′ a submodule of E. A *congruence modulo* E′ is the relation
$x - y \in E'$ between elements x and y of E. It is written

$$x \equiv y \pmod{E'},$$

and is an equivalence relation. Relations $x' \equiv y'$ (mod E′) and $x'' \equiv y''$ (mod E′)
imply $x' + x'' \equiv y' + y''$ (mod E′) and $\lambda x' \equiv \lambda y'$ (mod E′) for all $\lambda \in A$.

The quotient set of E by congruence modulo E′ has a unique A-module structure
for which the quotient map is A-linear, and with this structure, is called the *quotient
module* of E by E′ and is written E/E′.

UNIVERSAL PROPERTY. *Let* E *be an* A-*module,* E′ *a submodule of* E. *For all* A-
modules F *and homomorphisms* $f : E \to F$ *such that* $f|_{E'} = 0$, *there is a unique
homomorphism* $\bar{f} : E/E' \to F$ *satisfying* $f = \bar{f} \circ \chi$, *where* χ *is the canonical map*
$E \to E/E'$.

In other words, the map $g \mapsto g \circ \chi$ is a bijection from Hom(E/E′, F) onto the set
of $f \in$ Hom(E, F) for which $f|_{E'} = 0$.

Example Every cyclic A-module is isomorphic to a module A/I, where I is an ideal
of A. Indeed, if $E = Ax$, then the map $\lambda \mapsto \lambda x$ from A to E is surjective, giving an
isomorphism $A/I \to E$, where I is the *l'annihilator* of x, i.e. the ideal consisting of
elements $\lambda \in A$ such that $\lambda x = 0$.

3.3.7   Canonical Factorization

Let E and F be two A-modules, $f : E \to F$ a homomorphism.There is a unique
isomorphism $\tilde{f} : E/\operatorname{Ker} f \to \operatorname{Im} f$ such that the diagram

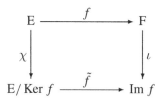

commutes.

This diagram gives the *canonical factorization* of f.

The *cokernel* of f, written Coker f, is the quotient module F/ Im f.

3.3.8 Exact Sequences

Let $E_m \to \cdots \to E_{i+1} \overset{f_{i+1}}{\to} E_i \overset{f_i}{\to} E_{i-1} \to \cdots \to E_n$ be a sequence of A-module homomorphisms (the decreasing order of indices has been chosen with homology in mind, see 3.10.2). It is said to be an **exact sequence** if Im $f_{i+1} = $ Ker f_i for $m - 1 \geqslant i \geqslant n + 1$.

Example Let E and F be two A-modules, $f : E \to F$ a homomorphism. $0 \to$ Ker $f \to E \overset{f}{\to} F \to$ Coker $f \to 0$ is an exact sequence.

In particular f is injective (resp. surjective) if and only if $0 \to E \overset{f}{\to} F$ (resp. $E \overset{f}{\to} F \to 0$) is an exact sequence.

If $0 \to E' \to E \to E'' \to 0$ is an exact sequence of A-modules, then E' can be identified with a submodule of E and E'' to the quotient module E/E'. The exact sequence is then written $0 \to E' \to E \to E/E' \to 0$.

An exact sequence of the form $0 \to E' \to E \to E'' \to 0$ is called a *short exact sequence*. If $E_m \to \cdots \to E_i \to \cdots \to E_n$ is an exact sequence, there are short exact sequences $0 \to E_i' \to E_i \to E_{i-1}' \to 0$ for $m \geqslant i \geqslant n$, with $E_i' =$ Ker $f_i =$ Im $f_{i+1} \approx$ Coker f_{i+2} for $m - 2 \geqslant i \geqslant n + 1, E_{m-1}' =$ Ker $f_{m-1} =$ Im $f_m, E_m' =$ Ker $f_m, E_n' =$ Im $f_{n-1} \approx$ Coker $f_{n+2}, E_{n-1}' =$ Coker f_{n+1}.

3.3.9

Proposition *Let* $0 \to E' \overset{u}{\to} E \overset{v}{\to} E'' \to 0$ *be an exact sequence of* A-modules, $(x_i')_{i \in I'}$ *and* $(x_i'')_{i \in I''}$ *generating families of* E' *and* E'' *respectively. Suppose that* $I' \cap I'' = \varnothing$ *and set* $I = I' \cup I''$. *For* $i \in I'$, *let* $x_i = u(x_i')$ *and, for each* $i \in I''$, *choose* $x_i \in E$ *such that* $v(x_i) = x_i''$. *Then the family* $(x_i)_{i \in I}$ *generates* E.

Proof Let $x \in E$. There is a family $(a_i)_{i \in I''}$ with finite scalar support such that $v(x) = \sum_{i \in I''} a_i x_i''$. Then $x - \sum_{i \in I''} a_i x_i$ is in Ker v, and so is of the form $u(x')$. There is a family $(a_i)_{i \in I'}$ of finite scalar support such that $x' = \sum_{i \in I'} a_i x_i'$. Then $x = \sum_{i \in I} a_i x_i$. $\square$

Corollary *Let* $0 \to E' \to E \to E'' \to 0$ *be an exact sequence of* A-modules. *If* E' *and* E'' *are of finitely generated, then so is* E.

3.3.10 Complement Submodules

Let E be an A-module, the submodules E_1 and E_2 are *complements,* or E is a *direct sum of the submodules* E_1 *and* E_2, and is written $E = E_1 \oplus E_2$, if any element of E is of the form $x_1 + x_2$, with $x_1 \in E_1$ and $x_2 \in E_2$. E_1 and E_2 are complements if and only if $E_1 + E_2 = E$ and $E_1 \cap E_2 = 0$.

A submodule of an A-module E is a **direct factor** of E if it has a complement. There are modules with submodules that are not direct factors. For example, the submodule $2\mathbb{Z}$ is not a direct factor of $\mathbb{Z}$ because the quotient being a torsion module, there are no injections from $2\mathbb{Z}$ [6] to $\mathbb{Z}$.

More generally, let (E_i) be a family of submodules of E. Then E is said to be a direct sum of the submodules E_i if all $x \in E$ can be uniquely written in the form $\sum x_i$, with $x_i \in E_i$ for all i, the family $(x_i)_{i \in I}$ having finite support.

3.3.11 Projections

Let E be an A-module. An endomorphism p of E is a *projection* if $p \circ p = p$.

Proposition *The map* $p \mapsto (\mathrm{Im}\, p, \mathrm{Ker}\, p)$ *is a bijection from the set of projections on* E *onto the set of pairs consisting of complement submodules of* E.

Proof If p is a projection of E, then $E = \mathrm{Im}\, p \oplus \mathrm{Ker}\, p$. Indeed, any $x \in E$ can be written $p(x) + (x - p(x))$, with $p(x) \in \mathrm{Im}\, p$ and $x - p(x) \in \mathrm{Ker}\, p$. So $\mathrm{Im}\, p + \mathrm{Ker}\, p = E$; on the other hand, $p(x) = x$ for $x \in \mathrm{Im}\, p$. Thus $\mathrm{Im}\, p \cap \mathrm{Ker}\, p = 0$. Next let E_1 and E_2 be two complement submodules of E. A projection having image E_1 and kernel E_2 is necessarily the map $x_1 + x_2 \mapsto x_1$; it is the desired one, proving bijectivity.   □

3.3.12 Split Exact Sequences

Proposition and Definition *Let* $0 \to E' \xrightarrow{u} E \xrightarrow{v} E'' \to 0$ *be an exact sequence. The following conditions are equivalent:*

 (i) $u(E')$ *is a direct factor of* E;
 (ii) *there is a linear retraction of* u, *i.e. a homomorphism* $\rho : E \to E'$ *such that* $\rho \circ u = 1_{E'}$;
 (iii) *there is a linear section of* v, *i.e. a homomorphism* $\sigma : E'' \to E$ *such that* $v \circ \sigma = 1_{E''}$.

If these conditions hold, the exact sequence is said to be **split.**

Proof (i) *implies* (ii). Identify E' with its image in E under u. Let S be a complement of E' in E. The projection of E onto E' with kernel S is a retraction.

---

[6]The main difference between modules and vector spaces is that there are torsion modules and submodules that are not direct factors.

(i) *implies* (iii). Let S be a complement of Ker $v = u(E')$, the homomorphism v induces an isomorphism from S onto E″ and the composite of the inverse isomorphism and the canonical injection from S to E is a section.

(ii) *implies* (i) *(resp.* (iii) *implies* (i)). If ρ is a retraction of u (resp. σ is a section of v), then $u \circ \rho$ (resp. $\sigma \circ v$) is a projection of E whose kernel (resp. image) is $u(E')$.

3.3.13 Direct Products and Sums

Let $(E_i)_{i \in I}$ be a family of A-modules. Define an A-module structure on the product set $\prod_{i \in I} E_i$ by setting

$$(x_i)_{i \in I} + (y_i)_{i \in I} = (x_i + y_i)_{i \in I}, \quad a(x_i)_{i \in I} = (ax_i)_{i \in I}.$$

The module obtained is called the *direct product* of E_i. The submodule of $\prod_{i \in I} E_i$ consisting of families with finite support is called the *direct sum* of E_i and is written $\bigoplus_{i \in I} E_i$. If I is finite, then $\bigoplus_{i \in I} E_i = \prod_{i \in I} E_i$. For $k \in I$, $p_k : \prod_{i \in I} E_i \to E_k$ defined by $p_k((x_i)_{i \in I}) = x_k$ is a module homomorphism called the *coordinate map* of index k, and the map $\iota_k : E_k \to \bigoplus_{i \in I} E_i$ defined by $\iota_k(x) = (x_i)_{i \in I}$, where $x_k = x$ and $x_i = 0$ if $i \neq k$ is a module homomorphism called the *canonical injection of index k*.

The module $\bigoplus_{i \in I} E_i$ is a direct sum of submodules $\iota_i(E_i)$. When no confusion arises, E_i is identified with $\iota_i(E_i)$. We avoid doing so when several E_i are equal.

The module $\prod_{i \in I} E_i$ (resp. $\bigoplus_{i \in I} E_i$), together with the maps p_i (resp. ι_i), is a product (resp. sum) of the modules E_i in the category of A-modules and linear maps.

In particular, let E be an A-module and $(E_i)_{i \in I}$ a family of submodules of E, the family of canonical inclusions of E_i in E defines a homomorphism f from $\bigoplus E_i$ to E given by $(x_i)_{i \in I} \mapsto \sum x_i$.

It is an isomorphism (resp. surjective) if and only if E is the direct sum of the submodules E_i (resp. generated by $(E_i)_{i \in I}$). When $I = \{1, 2\}$, f is injective if and only if $E_1 \cap E_2 = \{0\}$.

3.3.14 Direct and Inverse Limits

Let $((E_i)_{i \in I}, (f_i^j))$ be a direct (resp. inverse) system of A-modules. The set $E_\infty = \varinjlim E_i$ (resp. $\varprojlim E_i$) has a unique A-module structure with respect to which the canonical maps $E_i \to E_\infty$ (resp. $E_\infty \to E_i$) are linear. With this structure, E_∞ is the direct (resp. inverse) limit of the E_i in the category A-$\mathcal{M}od$.

Proposition *Let* I *be a directed set,* $\mathcal{E} = ((E_i), (f_i^j))$ *and* $\mathcal{F} = ((F_i), (g_i^j))$ *direct systems of modules and* (φ_i) *a morphism from* $\mathcal{E}$ *to* $\mathcal{F}$. *Let* $E = \varinjlim E_i$, $F = \varinjlim F_i$, *and $\varphi : E \to F$ be the morphism induced by φ_i by passing to the direct limit. Then* $\mathrm{Ker}\,\varphi = \varinjlim \mathrm{Ker}\,\varphi_i$, $\mathrm{Im}\,\varphi = \varinjlim \mathrm{Im}\,\varphi_i$ *and* $\mathrm{Coker}\,\varphi = \varinjlim \mathrm{Coker}\,\varphi_i$.

Proof It suffices to check that $\mathrm{Ker}\,\varphi$ (resp. $\mathrm{Im}\,\varphi$, resp. $\mathrm{Coker}\,\varphi$) with the canonical maps $\mathrm{Ker}\,\varphi_i \to \mathrm{Ker}\,\varphi$ (resp. etc.), satisfies conditions (i), (ii) and (iii) of 2.6.5. Condition (i) is immediate in all three cases. We show that $\mathrm{Ker}\,\varphi$ satisfies (ii). Let $x \in \mathrm{Ker}\,\varphi$. There exist $i \in I$ and $x_i \in E_i$ such that $\sigma_i(x_i) = x$, where $\sigma_i : E_i \to E$ is the canonical map. Then $\tau_i(\varphi_i(x_i)) = \varphi(\sigma_i(x_i)) = 0$, where $\tau_i : F_i \to F$ is a canonical map, and so there exists $j \geqslant i$ such that $g_i^j(\varphi_i(x_i)) = 0$. Hence $x_j = f_i^j(x_i) \in \mathrm{Ker}\,\varphi_j$, and $x = \sigma_j(x_j) \in \sigma_j(\mathrm{Ker}\,\varphi_j)$.

Condition (ii) for $\mathrm{Im}\,\varphi$ and $\mathrm{Coker}\,\varphi$ is immediate. So is condition (iii) for $\mathrm{Ker}\,\varphi$ and $\mathrm{Im}\,\varphi$; for $\mathrm{Coker}\,\varphi$, it follows from condition (ii) for $\mathrm{Im}\,\varphi$, $\square$

Remark Let $\mathscr{E}$ and $\mathscr{F}$ be inverse systems, (φ_i) a morphism from $\mathscr{E}$ to $\mathscr{F}$ and φ the morphism induced by φ_i by passing to the inverse limit. Then $\mathrm{Ker}\,\varphi = \varprojlim \mathrm{Ker}\,\varphi_i$, but in general $\mathrm{Im}\,\varphi \neq \varprojlim \mathrm{Im}\,\varphi_i$ and $\mathrm{Coker}\,\varphi \neq \varprojlim \mathrm{Coker}\,\varphi_i$ (2.5, Exercise 3).

Corollary *Let I be a directed set. For $m \geqslant k \geqslant n$, let $\mathscr{E}_k = ((E_{k,i})_{i\in I}, (f_{k,i}^j))$ be a direct system of modules; for $m \geqslant k > n$, let $(u_{k,i})_{i\in I}$ be a morphism from $\mathscr{E}_k$ to $\mathscr{E}_{k-1}$. If, for all i,*

$$E_{m,i} \to \cdots \to E_{k,i} \overset{u_{k,i}}{\to} E_{k-1,i} \to \cdots \to E_{n,i}$$

is an exact sequence, passing to the direct limit gives an exact sequence

$$E_m \to \cdots \to E_k \overset{u_k}{\to} E_{k-1} \to \cdots \to E_n .$$

3.3.15 Modules of Homomorphisms

Let E and F be two A-modules. The set $\mathrm{Hom}(E, F)$ of homomorphisms from E to F is a submodule of F^E. With the induced structure it is an A-module.[7]

The composition law $(f, g) \mapsto g \circ f$ from $\mathrm{Hom}(E, F) \times \mathrm{Hom}(F, G)$ to $\mathrm{Hom}(E, G)$ is bilinear.

Let $u : E' \to E$ be a homomorphism of A-modules. Define[8] a homomorphism $u^* : \mathrm{Hom}(E, F) \to \mathrm{Hom}(E', F)$ by $u^*(f) = f \circ u$. Let $v : F \to F'$ be a homomorphism of A-modules. Define a homomorphism $v_* : \mathrm{Hom}(E, F) \to \mathrm{Hom}(E, F')$ by $v_*(f) = v \circ f$. Then there is a commutative diagram

---

[7] Here the assumption of the commutativity of A is essential.

[8] We use the usual notation u^*, v_* which omits any mention of E and F (see 2.2.1).

$$\begin{array}{ccc}
\mathrm{Hom}(E, F) & \xrightarrow{u^*} & \mathrm{Hom}(E', F) \\
v_* \downarrow & & \downarrow v_* \\
\mathrm{Hom}(E, F') & \xrightarrow{u^*} & \mathrm{Hom}(E', F')
\end{array}$$

The map $(u, v) \mapsto v_* u^*$ from $\mathrm{Hom}(E', E) \times \mathrm{Hom}(F, F')$ to $\mathrm{Hom}(\mathrm{Hom}(E, F), \mathrm{Hom}(E', F'))$ is bilinear.

Then $(E, F) \mapsto \mathrm{Hom}(E, F)$ is a functor from A-$\mathcal{M}od \times$ A-$\mathcal{M}od$ to A-$\mathcal{M}od$. Let $(E_i)_{i \in I}$ and $(F_j)_{j \in J}$ be two families of A-modules. Then

$$\mathrm{Hom}\left(\bigoplus_{i \in I} E_i, \prod_{j \in J} F_j\right) = \prod_{(i,j) \in I \times J} \mathrm{Hom}(E_i, F_j).$$

Therefore, when I and J are finite, a homomorphism from $\bigoplus_{i \in I} E_i$ to $\bigoplus_{j \in J} F_j$ is represented by a matrix $(a_i^j)_{(i,j) \in I \times J}$ where, for all (i, j) the elements a_i^j are homomorphisms from E_i to F_j.

Remark It may happen that $\mathrm{Hom}(E, F) = 0$ even when $E \neq 0$ and $F \neq 0$. For example, this is the case if A is integral and E is a torsion module, and F is torsion-free. Other examples:

$$\mathrm{Hom}_{\mathbb{Z}}(\mathbb{Z}/(p), \mathbb{Z}/(q)) = 0$$

if p and q are relatively prime and $\mathrm{Hom}_{\mathbb{Z}}(\mathbb{Q}, \mathbb{Z}) = 0$.

3.3.16 Left Exactness of the Functor Hom

Proposition (a) Let E be an A-module and $0 \to F' \xrightarrow{u} F \xrightarrow{v} F''$ an exact sequence of A-modules. Then there is an exact sequence

$$0 \to \mathrm{Hom}(E, F') \xrightarrow{u_*} \mathrm{Hom}(E, F) \xrightarrow{v_*} \mathrm{Hom}(E, F'').$$

(b) Let F be an A-module and $E' \xrightarrow{u} E \xrightarrow{v} E'' \to 0$ an exact sequence of A-modules. Then there is an exact sequence

$$0 \to \mathrm{Hom}(E'', F) \xrightarrow{v_*} \mathrm{Hom}(E, F) \xrightarrow{u_*} \mathrm{Hom}(E', F).$$

Proof (a) Identifying F' with $\mathrm{Ker}\, v$, $\mathrm{Ker}\, v_*$ is the set of $f : E \to F$, where $v \circ f = 0$, i.e. $f(E) \subset \mathrm{Ker}\, v$, and so $\mathrm{Ker}\, v_* = \mathrm{Hom}(E, F')$.

(b) Identifying E'' with $\mathrm{Coker}\, u$, $\mathrm{Ker}\, u^*$ is the set of $f : E \to F$ such that $f \circ u = 0$, i.e. $f|_{\mathrm{Im}\, u} = 0$, and so $\mathrm{Ker}\, u^*$ is $\mathrm{Hom}(\mathrm{Coker}\, u, F) = \mathrm{Hom}(E'', F)$. □

3.3.17   Multilinear Maps

Let $E_1, \ldots, E_n$ and F be A-modules. A map f from $E_1 \times \cdots \times E_n$ to F is said to be *n-linear* if, for all $i \in \{1, \ldots, n\}$ and $x_1, \ldots, x_{i-1}, x_{i+1}, \ldots x_n$, the partial map $x_i \mapsto f(x_1, \ldots, x_n)$ from E_i to F is linear. The module of n-linear maps from $E_1 \times \cdots \times E_n$ to F can be identified with $\mathrm{Hom}(E_1, \mathrm{Hom}(E_2, \ldots(\mathrm{Hom}(E_n, F))\ldots))$.

A n-linear map f from E_n to F is said to be *symmetric* if $f(x_{\sigma(1)}, \ldots, x_{\sigma(n)}) = f(x_1, \ldots, x_n)$ for all permutations $\sigma \in \mathfrak{S}_n$. It is *alternating* if $x_i = x_j$ with $i \neq j$ implies $f(x_1, \ldots, x_n) = 0$.

If the n-linear map is alternating, then $f(x_2, x_1, x_3, \ldots, x_n) = -f(x_1, x_2, \ldots, x_n)$; more generally the sign of $f(x_1, \ldots, x_n)$ is reversed if any two x_i are interchanged.

3.3.18   Signature of a Permutation

Let X be a finite set, and Y the set of 2-element subsets of X. The *binary choice function* on X is a map $\tau : Y \to X$ such that $\tau(y) \in y$ for $y \in Y$. As X is finite, the axiom of choice is not needed to ensure the existence of such a choice function.

Let σ be a permutation of X, i.e. a bijection from X to itself. Write σ_* for the map $\{x, x'\} \mapsto \{\sigma(x), \sigma(x')\}$ from Y to itself; and for every binary choice function τ, define $\sigma_*\tau$ by $\sigma_*\tau = \sigma \circ \tau \circ \sigma^{-1}$. If τ and τ' are two binary choice functions, write $\tau' - \tau$ for the function $Y \to \mathbb{Z}/(2)$ with value 0 at y if $\tau'(y) = \tau(y)$ and 1 otherwise. If f is a function $Y \to \mathbb{Z}/(2)$, write $\Sigma(f)$ for $\sum_{y \in Y} f(y)$ and set $\sigma_* f = f \circ \sigma_*^{-1}$. For functions with values in $\mathbb{Z}/(2)$, the signs $+$ and $-$ are synonymous.

Let σ be a permutation of X and $\tau : Y \to X$ a binary choice function. Set $i(\sigma, \tau) = \Sigma(\sigma_*\tau - \tau)$. In fact, $i(\sigma, \tau)$ does not depend on the choice of τ. Indeed, if τ' is another binary choice function, then $(\sigma_*\tau' - \tau') - (\sigma_*\tau - \tau) = \sigma_* h - h$, where $h = \tau' - \tau$ and $\Sigma(\sigma_* h) = \Sigma(h)$. Therefore, $(-1)^{i(\sigma, \tau)}$ does not depend on τ, and is called the **signature** of the permutation σ. It is written $\varepsilon(\sigma)$.

If σ and σ' are two permutations of X, then $\varepsilon(\sigma\sigma') = \varepsilon(\sigma) \cdot \varepsilon(\sigma')$. In other words, ε is a homomorphism from the group $\mathfrak{S}(X)$ of permutations of X to the multiplicative group $\{+1, -1\}$.

Let X and X' be two finite sets and $f : X \to X'$ a bijection. Let f_* be the isomorphism $\sigma \mapsto f \circ \sigma \circ f^{-1}$ from $\mathfrak{S}(X)$ onto $\mathfrak{S}(X')$. Then $\varepsilon(f_*(\sigma)) = \varepsilon(\sigma)$. More generally, let $f : X \to X'$ be an injection, with finite X'. Define an injective homomorphism f_* from $\mathfrak{S}(X)$ to $\mathfrak{S}(X')$ by setting $f_*(\sigma)$ to be the map equal to $f_*(\sigma)$ on $f(X)$ as defined previously, and to the identity on $X' - f(X)$. Also $\varepsilon(f_*(\sigma)) = \varepsilon(\sigma)$.

A *transposition* of X is a permutation interchanging two elements of X and keeping the others unchanged. If θ is a transposition, then $\varepsilon(\theta) = -1$. The transpositions of X generate $\mathfrak{S}(X)$, but if σ is a product of k transpositions, the number k is necessarily even if $\varepsilon(\sigma) = 1$ and odd if $\varepsilon(\sigma) = -1$.

Let E and F be A-modules and $f : E^n \to F$ a n-linear alternating map. For any permutation σ of $\{1, \ldots, n\}$, $f(x_{\sigma(1)}, \ldots, x_{\sigma(n)}) = \varepsilon(\sigma) f(x_1, \ldots, x_n)$.

3.3.19 Algebras

An A-**algebra** is an A-module E with a bilinear map $\mu : E \times E \to E$, called multiplication. It is said to be an associative (resp. commutative, resp. unital) if the composition law μ on E is associative (resp. commutative, resp. has an identity element).

Examples (1) If B is a ring and $\varphi : A \to B$ a ring homomorphism, B, together with its A-module structure (3.3.1, Example 3) and multiplication, is a unital associative commutative A-algebra.

More generally, if B is a not necessarily commutative ring and $\varphi : A \to B$ a ring homomorphism such that $\varphi(A)$ is contained in the centre of B, then B is a unital associative A-algebra. All unital associative A-algebras are uniquely defined in this way.

(2) Let M be an A-module. The module $\mathrm{End}(M) = \mathrm{Hom}(M, M)$ with composition $(f, g) \mapsto f \circ g$, is a unital associative algebra, and is not generally commutative.

Exercises 3.3. (Modules)

1.—(a) Show that there are no minimal generating families in the $\mathbb{Z}$-module $\mathbb{Q}$.

(b) Give an example of a module having two minimal generating families of distinct finite cardinalities.

(c) Show that, if there is a minimal generating infinite family, then all generating families are infinite (see proof of 3.4.5).

(d) For all A-modules E, set $g(E)$ to be the minimal cardinality of a generating family of E. Show that if $0 \to E \to F \to G \to 0$ is a short exact sequence, then $g(F) \leqslant g(E) + g(G)$. Give an example where the inequality is strict.

2.—Let K be a field; set $A = K[X, Y]$, $E = A^2$ and let F be a submodule de E generated by (X, Y). Show that:

(a) E/F is torsion-free;

(b) There are no retractions from E onto F. Deduce that F is not a direct factor of E.

3.—Let E be an A-module and $\mathcal{J}$ an ideal in A. Let $\mathcal{J} \cdot E$ be the submodule of E consisting of all elements $\sum_{i \in I} a_i x_i$, where I is a finite set, $a_i \in \mathcal{J}$ and $x_i \in E$.

(a) Let F and G be two submodules of E. Does the following necessarily hold:

$$\mathcal{J} \cdot (F \cap G) = \mathcal{J} \cdot F \cap \mathcal{J} \cdot G \, ?$$

$$\mathcal{J} \cdot (F + G) = \mathcal{J} \cdot F + \mathcal{J} \cdot G \, ?$$

(b) Let (E_λ) be a family of A-modules. Show that

$$\mathcal{J} \cdot \bigoplus E_\lambda = \bigoplus \mathcal{J} \cdot E_\lambda \, .$$

(c) Show that $\mathscr{J} \cdot \prod E_\lambda \subset \prod \mathscr{J} \cdot E_\lambda$, and that the equality holds if $\mathscr{J}$ is finitely generated. Show that, if $\mathscr{J}$ is not finitely generated, then there is a family (E_λ) such that the inclusion is strict.

4.—Let E be an A-module, F and G two submodules of E. Construct an exact sequence $0 \to F \cap G \to F \oplus G \to F + G \to 0$.

5.—Let E be an A-module, F and G-submodules of E such that $G \subset F$. Show that, if G is a direct factor of E, the G is a direct factor of F. Show that, if F is a direct factor of E, then F/G is a direct factor of E/G.

6.—Let E and F be two A-modules, and $u : E \to F$ a homomorphism.

(a) Show that, Ker u and Im u are direct factors of E and F respectively if and only if there is a homomorphism $v : F \to E$ such that $u \circ v \circ u = u$.

(b) Show that $u \circ v \circ u = u$ does not imply $v \circ u \circ v = v$, but that there exists v such that $u \circ v \circ u = u$ and w such that $u \circ w \circ u = u$ and $w \circ u \circ w = w$.

7. *(Chart changes in Grassmannians)*—(a) Let F and G be two A-modules. Show that the graphs of linear maps from F to G are the complements of $0 \times G$ in $F \times G$.

(b) Let E be an A-module, G_0 a submodule of E, F_0 and F_1 two complements of G_0. Every complement F of G_0 can be considered the graph of a map $f_0 : F_0 \to G_0$ or the graph of a map $f_1 : F_1 \to G_0$. Give the relation between f_0 and f_1.

(c) Let E be an A-module, F_0 a submodule of E, G_0 and G_1 two complements of F_0. A complement F of G_0 can be considered the graph of a map $f_0 : F_0 \to G_0$. What condition should f_0 satisfy for F to be the complement of G_1? Then F can be considered the graph of a map $f_1 : F_0 \to G_1$. Express f_1 in terms of f_0.

8.—Let E, F, G be three modules, $i_1 : E \to F$ and $i_2 : E \to G$ injective homomorphisms. Embed E in $F \oplus G$ by (i_1, i_2).

(a) Show that the following are exact sequences:

$$0 \to F \to (F \oplus G)/E \to G/E \to 0, \quad 0 \to G \to (F \oplus G)/E \to F/E \to 0 \,.$$

(b) Show that the first sequence is split if and only if there is a homomorphism from G to F inducing the identity of E. Give an example where one of the sequences is split and the other one is not.

9.—For every prime p, let M_p be the subgroup in $\mathbb{Q}/\mathbb{Z}$ of elements whose orders are p-powers.

(a) Show that $\mathbb{Q}/\mathbb{Z}$ is the direct sum of the subgroups M_p (see 3.1, Exercise 6, b).

(b) Show that none of the M_p is the direct sum of two nonzero submodules.

(c) Show that $\mathrm{End}(\mathbb{Q}/\mathbb{Z})$ can be identified with the product of rings $\mathrm{End}(M_p)$.

(d) Show that $\mathrm{End}(M_p)$ can be identified with the ring $\widehat{\mathbb{Z}}_p$ of p-adic integers.

(e) Show that $\mathrm{End}(\mathbb{Q}/\mathbb{Z})$ can be identified with the profinite completion $\widehat{\mathbb{Z}}$ of $\mathbb{Z}$. We thus recover the result of (2.5, Exercise 7, e).

10.—(This exercise uses the results of the two preceding ones.) Consider the exact sequence:

$$0 \to \mathbb{Z} \xrightarrow{\iota} \mathbb{Q} \xrightarrow{\chi} \mathbb{Q}/\mathbb{Z} \to 0 \tag{E}$$

where ι and χ are the canonical maps. Applying the functor $M \mapsto \mathrm{Hom}(\mathbb{Q}, M)$ to (E) gives an exact sequence:

$$0 \to \mathrm{Hom}(\mathbb{Q}, \mathbb{Z}) \xrightarrow{\iota_*} \mathrm{Hom}(\mathbb{Q}, \mathbb{Q}) \xrightarrow{\chi_*} \mathrm{Hom}(\mathbb{Q}, \mathbb{Q}/\mathbb{Z}) \,.$$

The aim is to describe $\mathrm{Hom}(\mathbb{Q}, \mathbb{Q}/\mathbb{Z})$ and $\mathrm{Coker}\, \chi_*$.

(a) Determine $\mathrm{Hom}(\mathbb{Q}, \mathbb{Z})$ and $\mathrm{Hom}(\mathbb{Q}, \mathbb{Q})$.

(b) Show that applying the functor $M \mapsto \mathrm{Hom}(M, \mathbb{Q}/\mathbb{Z})$ to (E) gives a non-split exact sequence:

$$0 \to \widehat{\mathbb{Z}} \to \mathrm{Hom}(\mathbb{Q}, \mathbb{Q}/\mathbb{Z}) \to \mathbb{Q}/\mathbb{Z} \to 0 \,.$$

(c) Define a homomorphism $w : \mathbb{Q} \oplus \widehat{\mathbb{Z}} \to \mathrm{Hom}(\mathbb{Q}, \mathbb{Q}/\mathbb{Z})$ by setting $w(a, 0) = \chi \circ a \cdot 1_{\mathbb{Q}}$ and $w(0, u) = u \circ \chi$, where u is considered an endomorphism of $\mathbb{Q}/\mathbb{Z}$. Determine $\mathrm{Ker}\, w$ and $\mathrm{Coker}\, w$.

(d) Show that there is an exact sequence:

$$0 \to \mathbb{Q} \to \mathrm{Hom}(\mathbb{Q}, \mathbb{Q}/\mathbb{Z}) \to \widehat{\mathbb{Z}}/\mathbb{Z} \to 0 \,.$$

(Use Exercise 8.) Determine $\mathrm{Coker}\, \chi_*$. (The above exact sequence can be shown to be split.)

(e) Show that M_p can be identified with $\mathbb{Z}[1/p]/\mathbb{Z}$, or to $\widehat{\mathbb{Q}}_p/\widehat{\mathbb{Z}}_p$, where $\mathbb{Z}[1/p]$ denotes the ring of fractions whose denominator is a power of p, and $\widehat{\mathbb{Q}}_p$ the p-adic field, the field of fractions of $\widehat{\mathbb{Z}}_p$. Define $w_p : \mathbb{Z}[1/p] \oplus \widehat{\mathbb{Z}}_p \to \mathrm{Hom}(\mathbb{Q}, M_p)$ as in (c). Determine $\mathrm{Ker}\, w_p$ and $\mathrm{Coker}\, w_p$. Deduce that $\mathrm{Hom}(\mathbb{Q}, M_p)$ can be identified with $\widehat{\mathbb{Q}}_p$.

(f) Show that $\mathrm{Hom}(\mathbb{Q}, \mathbb{Q}/\mathbb{Z})$ can be identified with the set of $x = (x_p) \in \prod \widehat{\mathbb{Q}}_p$, where all but finitely many x_p are in $\widehat{\mathbb{Z}}_p$.

11.—Let A be a ring. Show that, for all ideals $\mathcal{J}$ of A, the category $(A/\mathcal{J})$-$\mathcal{M}od$ is equivalent to a complete subcategory of A-$\mathcal{M}od$. Show that the same holds for $(S^{-1}A)$-$\mathcal{M}od$ for any multiplicatively stable subset S of A, setting $S^{-1}A$ to be the ring of fractions with numerator in A and denominator in S.

12.—(a) Show that the set of endomorphisms of the identity functor of A-$\mathcal{M}od$ can be identified with A. Give an interpretation for addition (see 2.6, Exercise 2) and multiplication in A.

(b) Deduce that if A and B are two rings such that the categories A-$\mathcal{M}od$ and B-$\mathcal{M}od$ are equivalent, then A and B are isomorphic.

(c) the previous result ((b)) does not hold for non-commutative rings: let A be the ring of 2×2 matrices with entries in $\mathbb{R}$. For any vector space E, A can be made to act on E^2 by

$$\begin{pmatrix} a & b \\ c & d \end{pmatrix} \cdot \begin{pmatrix} x \\ y \end{pmatrix} = \begin{pmatrix} ax + by \\ cx + dy \end{pmatrix} .$$

Show that the functors $E \mapsto E^2$ from $\mathbb{R}$-$\mathcal{M}od$ to A-$\mathcal{M}od$ and $F \mapsto \text{Hom}_A(\mathbb{R}^2, F)$ from A-$\mathcal{M}od$ to $\mathbb{R}$-$\mathcal{M}od$ are equivalences of categories (note that, if A is not commutative, $\text{Hom}_A(F, G)$ is not a natural A-module for A-modules F and G).

13.—Let A be an integral domain where $2 \neq 0$, E and F torsion-free A-modules and $f : E^n \to F$ a n-linear map. Show that, if $f(x_{\sigma(1)}, ..., x_{\sigma(n)}) = \varepsilon(\sigma) f(x_1, ..., x_n)$ for all sequences $(x_1, ..., x_n)$ in E and all permutations σ, then the map f is alternating.

Taking for A a field of characteristic 2, construct a counterexample.

3.4   Free Modules, Matrices

Throughout this section, A denotes a (commutative) ring.

3.4.1   Free Modules

Let E be an A-module and $(x_i)_{i \in I}$ a family of elements of E. The family $(x_i)_{i \in I}$ is a *free family* if for any family $(\lambda_i)_{i \in I}$ of elements of A with finite support, $\sum_{i \in I} \lambda_i x_i = 0$ implies $\lambda_i = 0$ for all $i \in I$.

A *basis* of E is a free generating family of E, and E is a **free** A-module, if it has a basis.

It is free of rank n (or dimension n) the basis contains n elements.

Every free module over an integral domain is torsion-free. The converse does not hold. However, as will be seen, every finitely generated torsion-free module over a PID is free (3.5.8. Corollary 3.2).

3.4.2   Examples

The module A^r is a free A-module. The ring $A[X]$ is a free A-module,[9] the rings $A[X_1, ..., X_n]$ are free A-modules.[10] If K is a field, every finite dimensional vector space is a free K-module[11]; more generally:

Proposition *Every vector space is free.*

Proof Let E be a vector space. The set $\mathcal{L}$ of free subsets of E, ordered by inclusion, is direct since it is of finite character (1.4.2, Example 3).

By Zorn's theorem, there is a maximal free subset, B say, and let M be the vector subspace of E generated by B. We prove by contradiction that $M = E$. Suppose that

---

[9] Queysanne [1], §183, Lelong-Ferrand and Arnaudiès [2], 4.1, p. 123.

[10] Queysanne [1], §186, [2], 4.7, p. 151.

[11] Queysanne [1], §135, [2], th. 8.3.1, p. 253.

$M \neq E$ and let $y \in E - M$. It follows that $B \cup \{y\}$ is free, which contradicts the maximality of B. Indeed, suppose that $\sum \lambda_i x_i + \mu y = 0$ for $x_i \in B$.

If $\mu = 0$, $\sum \lambda_i x_i = 0$ implies $\lambda_i = 0$ for all i.

If $\mu \neq 0$, then $y = -\sum \frac{\lambda_i}{\mu} x_i$ and y is in M, which is impossible since $y \in E - M$. $\qquad\square$

3.4.3 Counterexamples

As a $\mathbb{Z}$-module, $\mathbb{Q}$ is torsion-free but not free.

If A is an integral domain, a non-principal ideal of A is not a free module, despite being torsion-free as well as the submodule of a free module. As will be seen, every submodule of a free module over a PID is free (3.5.1), and that every finitely generated torsion-free module over an integral domain is isomorphic to a submodule of a free module (3.8.14).

3.4.4 Alternating n-Linear Forms

An *alternating n-linear form* on an A-module E is an alternating n-linear map from E^n to A.

If E has a generating family $(e_1, ..., e_p)$ with $p < n$, every alternating n-linear form f on E is null. Indeed, if $x_i = \sum_{1 \leqslant j \leqslant p} x_i^j e_j$, then $f(x_1, ..., x_n) = \sum_{j_1,...,j_n} x_1^{j_1} \ldots x_n^{j_n} f(e_{j_1}, ..., e_{j_n}) = 0$ because for each term there exists k and k' with $k \neq k'$ and $j_k = j_{k'}$.

Proposition *Let E be a free A-module of rank n. Then the module Λ of alternating n-linear forms on E is free of rank 1. More precisely, if $(e_1, ..., e_n)$ is a basis for E, there is a unique form $\Delta \in \Lambda$ such that $\Delta(e_1, ..., e_n) = 1$, and Δ is a basis of Λ.*

Proof Let $(e_1, ..., e_n)$ be a basis for E. If $x_i = \sum x_i^j e_j$, then

$$f(x_1, ..., x_n) = \sum_{\sigma \in \mathfrak{S}_n} \varepsilon(\sigma) x_1^{\sigma(1)} \ldots x_n^{\sigma(n)} f(e_1, ..., e_n), \qquad (3.1)$$

for any alternating n-linear form f.

The map $\varphi : f \mapsto f(e_1, ..., e_n)$ from Λ to A is linear. By (1), $f(e_1, ..., e_n) = 0 \Rightarrow f = 0$, and so φ is injective.

Define $\Delta : E^n \to A$ by

$$\Delta(x_1, ..., x_n) = \sum_{\sigma \in \mathfrak{S}_n} \varepsilon(\sigma) x_1^{\sigma(1)} \ldots x_n^{\sigma(n)}.$$

The map Δ is an alternating n-linear form on E, and $\Delta(e_1, ..., e_n) = 1$, i.e. $\varphi(\Delta) = 1$. Then φ is surjective, and Λ is isomorphic to A. $\square$

3.4.5 Uniqueness of the Dimension

Proposition and Definition *Let* A *be a nonzero ring and* E *a free* A-*module. All bases for* E *have the same cardinality, called the* **dimension** (*or* **rank**) *of* E.

Proof Let $(e_i)_{i \in I}$ and $(f_j)_{j \in J}$ be two bases for E. For every $j \in J$, there is a family $(\lambda_{ij})_{i \in I}$ of elements of A such that $f_j = \sum_{i \in I} \lambda_{ij} e_i$. Let S_j be the support of $(\lambda_{ij})_{i \in I}$. For all $j \in J$, the support S_j is a finite subset of I. Set $I' = \cup_{j \in J} S_j$. All f_j are in the submodule E' of E generated by $(e_i)_{i \in I'}$, and so E' = E, whence $I' = I$, and

$$\text{Card } I = \text{Card } I' \leqslant \sum_{j \in J} \text{Card } S_j \leqslant \text{Card } J \times \aleph_0 .$$

If the set J is finite, then so is I. Let p and q be the number of elements of I and J respectively. There is an alternating p-linear form of E vanishing everywhere. Since E has a basis with q elements, $p \leqslant q$ because every alternating r-linear form on E where $r > q$ vanishes everywhere.

If the set J is infinite, then Card $J = $ Card $J \times \aleph_0 \geqslant$ Card I. Hence Card I $\leqslant$ Card J in all cases. Interchanging the roles of I and J gives Card I $=$ Card J, $\square$

Remark This result does not hold for non-commutative rings (3.4, Exercise 2), but does for skew fields.[12]

3.4.6 Free Module on a Set

Let A be a ring, I a set. Let A^I be the set of families $(\lambda_i)_{i \in I}$ of elements in A indexed by I. Define an A-module structure on A^I by setting

$$(\lambda_i)_{i \in I} + (\mu_i)_{i \in I} = (\lambda_i + \mu_i)_{i \in I}$$

$$a(\lambda_i)_{i \in I} = (a\lambda_i)_{i \in I} .$$

Write $A^{(I)}$ for the A-submodule of A^I consisting of families with finite support. For all $i \in I$, let e_i be the element $(\delta_{ij})_{j \in I}$ of $A^{(I)}$. The family $(e_i)_{i \in I}$ is a basis for $A^{(I)}$. Indeed, for all $\lambda = (\lambda_i)_{i \in I}$ in $A^{(I)}$, $\lambda = \sum \lambda_i e_i$, and so λ is in the submodule of $A^{(I)}$ generated by the elements (e_i), and the relation $\sum \lambda_i e_i = 0$ implies $(\lambda_i)_{i \in I} = 0$.

---

[12]Bourbaki [4], chap. 2, § 7, th. 3 (A II.96).

Hence $\lambda_i = 0$ for all $i \in$ I. $A^{(I)}$ is said to be the **free A-module on** I and $(e_i)_{i \in I}$ the *canonical basis* for $A^{(I)}$.

UNIVERSAL PROPERTY. *Let* E *be an arbitrary* A-*module and* $(x_i)_{i \in I}$ *an arbitrary family of elements of* E. *There is a unique homomorphism* f *from* $A^{(I)}$ *to* E *such that* $f(e_i) = x_i$ *for all* $i \in$ I.

Indeed, if f is such a homomorphism, then for all $\lambda = (\lambda_i)_{i \in I}$ of $A^{(I)}$,

$$f(\lambda) = f\left(\sum \lambda_i e_i\right) = \sum \lambda_i f(e_i) = \sum \lambda_i x_i$$

and uniqueness follows. The map f defined by $f(\lambda) = \sum \lambda_i x_i$ for $\lambda = (\lambda_i)_{i \in I}$ of $A^{(I)}$ satisfies the desired condition.

Remarks (1) The universal property can also be stated as follows: the map $f \mapsto (f(e_i))_{i \in I}$ is a bijection from $\mathrm{Hom}(A^{(I)}, E)$ onto E^I.

(2) These definitions imply that the image of f is the submodule of E generated by the family $(x_i)_{i \in I}$. Then f is surjective (resp. injective, resp. bijective) if and only if $(x_i)_{i \in I}$ is a generating family (resp. free, resp. a basis).

An A-module E is free if and only if it is isomorphic to a module of the form $A^{(I)}$. Every module is isomorphic to a quotient of a free module.

3.4.7   Projectivity of Free Modules

Proposition *Let* E *be a free* A-*module and* $0 \to F' \overset{u}{\to} F \overset{v}{\to} F'' \to 0$ *an exact sequence of* A-*modules. Then*

$$0 \to \mathrm{Hom}(E, F') \overset{u_*}{\to} \mathrm{Hom}(E, F) \overset{v_*}{\to} \mathrm{Hom}(E, F'') \to 0 \tag{3.2}$$

is an exact sequence.

Proof We may assume that $E = A^{(I)}$. Then (1) becomes: $0 \to F'^I \to F^I \to F''^I \to 0$ and $F'^I = \mathrm{Ker}(v_* : F^I \to F''^I)$ is immediate; the surjectivity of $v_* : F^I \to F''^I$ follows from the axiom of choice. □

Corollary *If, in a short exact sequence* $0 \to E' \overset{u}{\to} E \overset{v}{\to} E'' \to 0$ *of* A-*modules,* E'' *is a free* A-*module, the exact sequence splits.*

Proof The homomorphism $v_* : \mathrm{Hom}(E'', E) \to \mathrm{Hom}(E'', E'')$ is surjective, and so there exists $\sigma \in \mathrm{Hom}(E'', E)$ such that $v \circ \sigma = 1_{E''}$. □

3.4.8   Projective Modules

Definition An A-module P is said to be **projective** if, for all surjective A-module homomorphisms $f : M \to M'$ and homomorphisms $g : P \to M'$, there exists a homomorphism $h : P \to M$ making the diagram

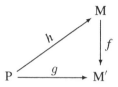

commutative.

In other words, P is projective if and only if the functor $M \mapsto \mathrm{Hom}(P, M)$ from the category A-$\mathcal{M}od$ to itself transforms surjective homomorphisms into surjective homomorphisms.

All free modules are projective (3.4.7).

Proposition *A module is projective if and only if it is the direct factor of a free module.*

Proof (a) Let P and Q be two modules such that $L = P \oplus Q$ is free, and $f : M \to M'$ a surjective module homomorphism. Then $\mathrm{Hom}(L, M) = \mathrm{Hom}(P, M) \times \mathrm{Hom}(Q, M)$. In the commutative diagram

$$\mathrm{Hom}(P, M) \times \mathrm{Hom}(Q, M) \xrightarrow{f_*} \mathrm{Hom}(P, M') \times \mathrm{Hom}(Q, M')$$
$$\mathrm{pr}_1 \Big\downarrow \qquad\qquad\qquad\qquad \Big\downarrow \mathrm{pr}_1$$
$$\mathrm{Hom}(P, M) \xrightarrow{\qquad f_* \qquad} \mathrm{Hom}(P, M')$$

the map f_* is surjective and pr_1 is surjective. So $f_* \circ \mathrm{pr}_1 = \mathrm{pr}_1 \circ f_*$ is surjective, and $f_* : \mathrm{Hom}(P, M) \to \mathrm{Hom}(P, M')$ is surjective.

(b) Conversely, suppose that P is a projective module. Let L be a free module and $f : L \to P$ a surjective homomorphism. There is a homomorphism $h : P \to L$ such that the diagram

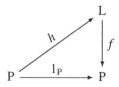

commutes. Therefore (3.3.12) P can be identified with a direct factor of L. □

For examples of non-free projective modules see 3.4, Exercise 8.

3.4.9 Matrices

Let A be a ring, I and J finite sets, $(e_i)_{i \in I}$ and $(f_j)_{j \in J}$ canonical bases for A^I and A^J respectively. For any A-linear map $\varphi : A^I \to A^J$ there is a unique family $M(\varphi) = (a_i^j)_{(i,j) \in I \times J}$ of elements of A such that for all $i \in I$, $\varphi(e_i) = \sum_{j \in J} a_i^j f_j$. Then $M(\varphi)$ is said to be the *matrix representing* φ.

For $i_0 \in I$, $(a_{i_0}^j)_{j \in J}$ is called the $i - 0$-th *column* of $M(\varphi)$; for $j_0 \in J$ the family $(a_i^{j_0})_{i \in I}$ is called the j_0-th *row*.

Assigning to each A-linear map φ its matrix $M(\varphi)$ defines a bijection from $\mathrm{Hom}(A^I, A^J)$ onto $A^{I \times J}$. Let $\alpha = (a_i^j)_{(i,j) \in I \times J}$, $\alpha' = (a'_i^{\,j})_{(i,j) \in I \times J} \in A^{I \times J}$, and let φ and φ' be the maps represented by these matrices. By definition, set $\alpha + \alpha' = M(\varphi + \varphi')$. Then,

$$\alpha + \alpha' = (a_i^j + a'_i^{\,j})_{(i,j) \in I \times J} \tag{3.3}$$

Let K be a finite set, $(g_k)_{k \in K}$ the canonical basis for A^K, $\beta = (b_j^k)_{(j,k) \in J \times K} \in A^{J \times K}$, and ψ the map represented by β. By definition, set

$$\beta\alpha = M(\psi \circ \varphi).$$

Then,

$$\beta\alpha = (c_i^k)_{(i,k) \in I \times K} \text{ with } c_i^k = \sum_{j \in J} b_j^k a_i^j. \tag{3.4}$$

In particular, the set $\mathcal{M}_I(A) = A^{I \times I} = \mathrm{Hom}(A^I, A^I)$ together with the two composition laws defined above is a ring, which is not commutative if $\mathrm{Card}\, I > 1$ and $A \neq 0$.

This ring has an identity element: the matrix $(\delta_i^j)_{(i,j) \in I \times J}$ of the identity of A^I.

3.4.10 Matrices Representing a Homomorphism with Respect to Given Bases

Let E and F be finitely generated free A-modules, $x = (x_i)_{i \in I}$ and $y = (y_j)_{j \in J}$ bases of E and F respectively, $\xi : A^I \to E$ and $\eta : A^J \to F$ isomorphisms such that for all $i \in I$ and $j \in J$, $\xi(e_i) = x_i$ and $\eta(f_j) = y_j$. For every A-linear map $\varphi : E \to F$, $M_{\mathcal{Y}}^{\mathcal{X}}(\varphi) = M(\eta^{-1} \circ \varphi \circ \xi)$ is said to be the *matrix representing* φ with respect to bases $\mathcal{X}$ and $\mathcal{Y}$. Its i-th column is the sequence of coordinates of $\varphi(x_i)$ with respect to the basis $\mathcal{Y}$. If $\mathcal{Z}$ is a basis for an A-module G and $\psi : F \to G$ a homomorphism, then $M_{\mathcal{Z}}^{\mathcal{X}}(\psi \circ \varphi) = M_{\mathcal{Z}}^{\mathcal{Y}}(\psi) \cdot M_{\mathcal{Y}}^{\mathcal{X}}(\varphi)$.

If $\mathcal{X}$ and $\mathcal{X}'$ are two bases of E, the *transition matrix* from $\mathcal{X}$ to $\mathcal{X}'$ is $M_{\mathcal{X}}^{\mathcal{X}'}(1_E)$. Its i-th column is the sequence of coordinates x_i' with respect to basis $\mathcal{X}$.

Let E and F be a finitely generated free A-modules, $\mathcal{X}$ and $\mathcal{X}'$ bases of E, $\mathcal{Y}$ and $\mathcal{Y}'$ bases of F, φ a homomorphism from E to F, $M = M_{\mathcal{Y}}^{\mathcal{X}}(\varphi)$, $M' = M_{\mathcal{Y}'}^{\mathcal{X}'}(\varphi)$, P the transition matrix from $\mathcal{X}$ to $\mathcal{X}'$ and Q the transition matrix from $\mathcal{Y}$ to $\mathcal{Y}'$. Then

$$M' = Q^{-1}MP .$$

3.4.11   Determinants

Let E be a free A-module of rank n and φ an endomorphism of E. Let Λ be the module of alternating n-linear forms on E and define the endomorphism φ^* de Λ by $\varphi^* f(x_1, \ldots, x_n) = f(\varphi(x_1), \ldots, \varphi(x_n))$. It follows from Proposition 3.4.4 that there is a unique scalar $\delta \in A$ such that $\varphi^* f = \delta f$ for all $f \in \Lambda$. It is called the *determinant* of φ and is written $\delta = \det \varphi$.

If ψ is another endomorphism of E, then $(\psi \circ \varphi)^* = \varphi^* \circ \psi^*$ and $\det(\psi \circ \varphi) = \det \psi \cdot \det \varphi$.

Let I be a finite set and $\alpha = (a_i^j)_{(i,j) \in I \times J}$ a square matrix. The determinant of α, written $\det \alpha$, is the determinant of the endomorphism of A^I represented by α. If α is the matrix representing an endomorphism φ of a free module E with respect to a basis for E (the same at both ends), then $\det \varphi = \det \alpha$.

Proposition *Let* I *be a finite set and* $\alpha = (a_i^j)_{(i,j) \in I \times J}$ *a square matrix. Then* $\det \alpha = \sum_{\sigma \in \mathfrak{S}(I)} \varepsilon(\sigma) \prod_{i \in I} a_i^{\sigma(i)}$.

Proof Suppose that $I = \{1, \ldots, n\}$. Let $(e_1, \ldots, e_n)$ be a canonical basis for A^n, Δ the alternating n-linear form on A^n such that $\Delta(e_1, \ldots, e_n) = 1$ and φ the endomorphism of A^n represented by α. Then $\det \alpha = \det \varphi = \Delta(\varphi(e_1), \ldots, \varphi(e_n)) = \sum a_1^{j_1} \cdots a_n^{j_n} \Delta(e_{j_1}, \ldots, e_{j_n})$.

Now, $\Delta(e_{j_1}, \ldots, e_{j_n}) = \varepsilon(\sigma)$ if $\sigma : k \mapsto j_k$ is a permutation of $\{1, \ldots, n\}$ and 0 otherwise; so

$$\det \alpha = \sum_{\sigma \in \mathfrak{S}_n} \varepsilon(\sigma) a_1^{\sigma(1)} \cdots a_n^{\sigma(n)} .$$

$\square$

3.4.12   Minors

Let $\alpha = (a_i^j)_{(i,j) \in I \times J}$ be a matrix, and $(i_1, \ldots, i_r)$, $(j_1, \ldots, j_r)$ sequences of elements of I and J respectively. Denote by $\mathrm{Min}_{i_1, \ldots, i_r}^{j_1, \ldots, j_r}(\alpha)$ the determinant of the matrix $(a_{i_k}^{j_l})_{k,l \in \{1, \ldots, r\}}$; it is a minor of order r of the matrix α. Such a minor is null if any two of i_k, or any two of j_l, are equal. If $I = J = \{1, \ldots, n\}$, for $i, j \in \{1, \ldots, n\}$, write $\mathrm{Min}_{\hat{i}}^{\hat{j}}(\alpha)$ for the minor $\mathrm{Min}_{1, \ldots, i-1, i+1, \ldots, n}^{1, \ldots, j-1, j+1, \ldots, n}(\alpha)$ of order $n - 1$.

Proposition (Expansion of a determinant along a row) *Let* $\alpha = (a_i^j)_{i,j \in \{1, \ldots, n\}}$ *be a square matrix of order n and* $j \in \{1, \ldots, n\}$. *Then*

$$\det \alpha = \sum_{i \in \{1,...,n\}} (-1)^{i-j} \operatorname{Min}_i^j(\alpha)\, a_i^j \,.$$

Proof

$$\det \alpha = \sum_{\sigma \in \mathfrak{S}_n} \varepsilon(\sigma) a_1^{\sigma(1)} \cdots a_n^{\sigma(n)} = \sum_{i \in \{1,...,n\}} \sum_{\substack{\sigma \in \mathfrak{S}_n \\ \sigma(i)=j}} \varepsilon(\sigma) a_1^{\sigma(1)} \cdots a_i^j \cdots a_n^{\sigma(n)} \,.$$

Let i and $j \in \{1, ..., n\}$; write $(a'^{j'}_{i'})_{i',j' \in \{1,...,n-1\}}$ for the matrix obtained by deleting the i-th column and j-th row in α and by renumbering the indices from 1 to $n - 1$. For every permutation $\sigma \in \mathfrak{S}_n$ such that $\sigma(i) = j$, define $\sigma' \in \mathfrak{S}_{n-1}$ as the composite

$$\{1, ..., n - 1\} \to \{1, ..., \widehat{i}, ..., n\} \to \{1, ..., \widehat{j}, ..., n\} \to \{1, ..., n - 1\}$$

(in this formula, the term with ⌢ is omitted). Since $\varepsilon(\sigma) = (-1)^{i-j}\varepsilon(\sigma')$,

$$\sum_{\substack{\sigma \in \mathfrak{S}_n \\ \sigma(i)=j}} \varepsilon(\sigma) a_1^{\sigma(1)} \cdots a_i^j \cdots a_n^{\sigma(n)} = (-1)^{i-j} a_i^j \sum_{\sigma' \in \mathfrak{S}_{n-1}} \varepsilon(\sigma') a'^{\sigma'(1)}_1 \cdots a'^{\sigma'(n-1)}_{n-1}$$

$$= (-1)^{i-j} a_i^j \operatorname{Min}_i^{\widehat{j}}(\alpha) \,.$$

The proposition follows.

3.4.13

Theorem (Cramer) *Let* E *be a free* A-*module of rank n. An endomorphism* φ *of* E *is an automorphism if and only if* $\det \varphi$ *is invertible.*

If φ is invertible, then $\det \varphi \cdot \det \varphi^{-1} = 1$, and so $\det \varphi$ is invertible. We give two proofs for the converse.

1st proof Let $(e_1, ..., e_n)$ be a basis for E and Δ the n-linear form on E such that $\Delta(e_1, ..., e_n) = 1$. Denote by Λ' the module of alternating $(n - 1)$-linear forms on E and let $\widetilde{\Delta} : E \to \Lambda'$ be the homomorphism defined by $\widetilde{\Delta}(x)(y_1, ..., y_{n-1}) = \Delta(x, y_1, ..., y_{n-1})$.

Lemma $\widetilde{\Delta}$ *is an isomorphism.*

Proof of the Lemma If $x = \sum x^i e_i$, then

$$\widetilde{\Delta}(x)(e_1, ..., \widehat{e_i}, ..., e_n) = (-1)^{i-1} x^i \tag{3.5}$$

(in this formula, the term with ⌢ is omitted). It follows from (1) that $\widetilde{\Delta}$ is injective. We show it is surjective. Let $f \in \Lambda'$, $x^i = (-1)^{i-1} f(e_1, ..., \widehat{e_i}, ..., e_n)$ and $x = \sum x^i e_i$. The forms f and $\widetilde{\Delta}(x)$ agree on all $(n - 1)$-uples $(e_1, ..., \widehat{e_i}, ..., e_n)$, and so are equal. $\square$

End of the 1st Proof of the Theorem Define the endomorphism φ^* of Λ' by $\varphi^* f(x_1, ..., x_{n-1}) = f(\varphi(x_1), ..., \varphi(x_{n-1}))$. In the diagram

$$
\begin{array}{ccc}
E & \xrightarrow{\tilde{\Delta}} & \Lambda' \\
\varphi \downarrow & & \uparrow \varphi^* \\
E & \xrightarrow{\tilde{\Delta}} & \Lambda'
\end{array}
$$

$\varphi^* \circ \tilde{\Delta} \circ \varphi = \det \varphi \cdot \tilde{\Delta}$.   Indeed, $(\varphi^* \circ \tilde{\Delta} \circ \varphi(x))(y_1, ..., y_{n-1}) = \Delta(\varphi(x),$ $\varphi(y_1), ..., \varphi(y_{n-1})) = \det \varphi \cdot \Delta(x, y_1, ..., y_{n-1})$.

If $\det \varphi$ is invertible, setting $\psi = (\det \varphi)^{-1}\tilde{\Delta}^{-1} \circ \varphi^* \circ \tilde{\Delta}$, $\psi \circ \varphi = 1_E$, and φ is left invertible. As $\det \psi \cdot \det \varphi = 1$, $\det \psi$ is invertible. Likewise, φ_1 is a left inverse of ψ. Now, $\varphi_1 = \varphi_1 \circ \psi \circ \varphi = \varphi$, and so ψ is an inverse of φ. $\square$

2nd Proof of the Theorem We may assume that $E = A^n$. The endomorphism φ is then represented by a matrix α. Set $b_i^j = (-1)^{j-i} \text{Min}_i^j(\alpha)$ and $\beta = (b_i^j)$. By Proposition 3.4.12, $\sum_i a_i^j b_i^j = \det \alpha$, and $\sum_i a_i^k b_i^j = 0$ for $k \neq j$ (expansion of the determinant of a matrix with two equal rows). Hence $\sum_i a_i^k b_i^j = \delta_{j,k} \det \alpha$, i.e. $\alpha \cdot \beta = \det \alpha \cdot 1_{A^n}$. Similarly, the expansion of a determinant along a column (analogous to that of Proposition 3.4.12 by interchanging rows and columns), gives $\beta \cdot \alpha = \det \alpha \cdot 1_{A^n}$. If $\det \alpha$ is invertible, $\frac{1}{\det \alpha}\beta$ is then an inverse of α. $\square$

3.4.14

Proposition *Let* E *be a free* A-*module of rank n and* φ *an endomorphism of* E. *The following conditions are equivalent:*

 (i) φ *is an automorphism;*
 (ii) φ *is surjective;*
 (iii) $\det \varphi$ *is invertible.*

Proof (iii) $\Rightarrow$ (i) by the preceding theorem. (i) $\Rightarrow$ (ii) is obvious. We show that (ii) $\Rightarrow$ (iii). Suppose that φ is surjective; by (3.4.7, Corollary), there is a homomorphism $\psi : E \to E$ such that $\varphi \circ \psi = 1_E$, and so $\det \varphi \cdot \det \psi = 1$. $\square$

3.4.15

Proposition *Let* E *be a free* A-*module of rank n and* φ *an endomorphism of* E. *The following conditions are equivalent:*

(i) φ *is injective;*

(ii) $\det \varphi$ *is not a zero divisor.*

Proof (ii) $\Rightarrow$ (i). Let $(e_1, ..., e_n)$ be a basis for E and Δ an alternating n-linear form on E such that $\Delta(e_1, ..., e_n) = 1$. For $x = \sum x^i e_i \in \mathrm{Ker}\,\varphi$, as $x^i = (-1)^{i-1}\Delta(x, e_1, ..., \widehat{e_i}, ..., e_n)$ for all i, $\det \varphi \cdot x^i = (-1)^{i-1}\Delta(\varphi(x), ...) = 0$. Hence $x^i = 0$ if $\det \varphi$ is not a zero divisor.

Not (ii) $\Rightarrow$ not (i). We show that if $\det \varphi$ is a zero divisor, then φ is not injective. Let $h \neq 0$ be such that $h \det \varphi = 0$. Let $\alpha = (a_i^j)$ be a matrix representing φ, and $\mu = \mathrm{Min}_{i_1,...,i_r}^{j_1,...,j_r}(\alpha)$ a minor such that $h\mu \neq 0$ and whose order is as maximal as possible for this property.[13] Since $h \det \varphi = 0$, $r < n$. Let $i_0 \notin \{i_1, ..., i_r\}$, and $x = \sum x^i e_i$, where $x^{i_k} = (-1)^k h\,\mathrm{Min}_{i_0,...,\hat{i}_k,...,i_r}^{j_1,...,j_r}(\alpha)$ and $x^i = 0$ if $i \neq i_k$ for any k. For all j,

$$\sum a_i^j x^i = h\,\mathrm{Min}_{i_0,...,i_r}^{j,j_1,...,j_r}(\alpha) = 0\,,$$

and so $x \in \mathrm{Ker}\,\varphi$. As $x_{i_0} = h\mu \neq 0$, $x \neq 0$ and φ is not injective. $\square$

3.4.16 Presentations

Let M be an A-module. A **presentation** of M is an exact sequence $A^{(J)} \to A^{(I)} \to M \to 0$.

Proposition *Every module has a presentation.*

Proof Let M be an A-module and $(x_i)_{i \in I}$ a generating family of M. The homomorphism $u : A^{(I)} \to M$ defined by $u(e_i) = x_i$ is surjective; let N be its kernel and $(y_j)_{j \in J}$ its generators. The homomorphism $v : A^{(J)} \to A^{(I)}$ defined by $v(e_j) = y_j$ has image N, and $A^{(J)} \xrightarrow{v} A^{(I)} \xrightarrow{u} M \to 0$ is an exact sequence. $\square$

Definition An A-module M is **finitely presented** if it has a presentation $A^{(J)} \to A^{(I)} \to M \to 0$, where I et J are finite sets.

3.4.17

Proposition *Every module is the direct limit of finitely presented modules.*

Proof Let M be an A-module, $A^{(J)} \xrightarrow{v} A^{(I)} \xrightarrow{u} M \to 0$ a presentation of M and H the set of pairs (I', J'), where I' and J' are finite subsets of I and J respectively and $v(A^{(J')}) \subset A^{(I')}$. The set H is directed with the order defined by $(I', J') \leqslant (I'', J'') \Leftrightarrow$

---

[13]$n > 0$ for otherwise $\det \varphi = 1$. If $ha_i^j = 0$, for all (i, j), $\varphi(he_i) = 0$ and φ is not injective.

$I' \subset I''$ and $J' \subset J''$. Then $\bigcup_{(I',J') \in H} I' = I$ because it is always possible to take $J' = \varnothing$, and $\bigcup_{(I',J') \in H} J' = J$ because, for $j \in J$, $v(e_j)$ has finite support in I, and so is in some $A^{I'}$. For $h = (I', J') \in H$, let $v_h : A^{J'} \to A^{I'}$ be the map induced by v and set $M_h = \text{Coker } v_h$. Then $A^{(I)} = \bigcup A^{I'} = \varinjlim A^{I'}$ and $A^{(J)} = \bigcup A^{J'} = \varinjlim A^{J'}$. Hence $M = \text{Coker } v = \varinjlim M_h$, and each M_h is finitely presented.   □

3.4.18

Proposition *Let* M *be a finitely presented* A-*module and* $((E_\lambda), (r_{\lambda,\mu}))$ *be a direct system of* A-*modules. Then*

$$\text{Hom}(M, \varinjlim E_\lambda) = \varinjlim \text{Hom}(M, E_\lambda).$$

Proof If $L = A^I$ with I finite, then for any module E, $\text{Hom}(L, E) = E^I = E^{(I)}$. Hence the functor $E \mapsto \text{Hom}(L, E)$ commutes with direct limits. If $M = \text{Coker}(v : L_1 \to L_0)$, where L_0 and L_1 are free and finitely generated, then $\text{Hom}(M, E) = \text{Ker}(v^* : \text{Hom}(L_0, E) \to \text{Hom}(L_1, E))$. So, by Proposition 3.3.14, the functor $E \mapsto \text{Hom}(M, E)$ commutes with direct limits.   □

Exercises 3.4. (Free modules and matrices)

1.—Show that, in the vector space $\mathbb{C}^{\mathbb{N}}$ of complex sequences, the sequences $(n^k r^n)_{n \in \mathbb{N}}$ with $k \in \mathbb{N}$ and $r \in \mathbb{C}$ form a free family. (For each sequence (a_n), consider the series $\sum a_n z^n$.)

2.—(a) Let A be a not necessarily commutative ring and E a left A-module. Show that, if E has a finite basis, every basis for E is finite. Show that if E has an infinite basis, all bases have the same cardinality.

(b) Let E be a vector space over a field or skew field K. Let $(a_1, ..., a_p)$ be a free family in E, and G a generating family with n elements. Show by induction that, for all $i \leqslant p$, there is a generating subset G_i of the form $\{a_1, ..., a_p\} \cup G_i'$, where G_i' has $n - i$ elements. In particular $n \geqslant p$.

Show that any two bases for E necessarily have the same number of elements.

(c) Let H be an infinite dimensional Hilbert space and A the ring $L(H, H)$ of continuous endomorphisms of H. Show that, as a left A-module, A is isomorphic to A^2 (if H_1 and H_2 are two complement subspaces of H, then $L(H, H) = L(H_1, H) \oplus L(H_2, H)$).

Show that A contains elements u_1, v_1, u_2, v_2, such that $v_1.u_1 = 1$, $v_2.u_2 = 1$ and $v_2.u_1 = 0$. Show that there is no unital homomorphism from A to a field or skew field, nor to any not necessarily commutative nonzero ring where left multiplication by a nonzero element is injective.

3.—Let A be a (commutative) ring.

(a) Let I and J be not necessarily finite sets. Show that the homomorphisms from $A^{(I)}$ to $A^{(J)}$ correspond to matrices $(a_i^j)_{(i,j) \in I \times J}$ whose columns have finite support.

Extend the results of 3.4.9 and 3.4.10 to this context. What are the homomorphisms for which the matrix only has finitely many nonzero rows?

(b) Let E be a free infinite dimensional A-module. Using the formula of Proposition 3.4.11, define $\det(1_E + u)$ for an endomorphism u of E whose image is contained in a finitely generated submodule of E. Prove the formula $\det((1_E + u)(1_E + v)) = \det(1_E + u)\det(1_E + v)$.

4.—(a) Does Proposition 3.4.11 hold when $I = \varnothing$?

(b) Does the proof of 3.4.15 still hold without the footnote?

5.—(a) Let $E = \bigoplus_{i \in I} E_i$ and $F = \bigoplus_{j \in J} F_j$ be A-modules expressible as a finite direct sum of submodules. Show that, for all homomorphisms $f : E \to F$, there is a unique family $(a_i^j)_{(i,j) \in I \times J}$ with $a_i^j \in \mathrm{Hom}(E_i, F_j)$ such that $f(x) = \sum_j a_i^j(x)$ for $x \in E_i$.

The family (a_i^j) is said to be the matrix representing f with respect to the given decompositions.

How is the matrix representing the composite of two homomorphisms computed?

(b) Let $f : E_1 \oplus E_2 \to F_1 \oplus F_2$ be a homomorphism represented by a matrix $\begin{pmatrix} a & b \\ c & d \end{pmatrix}$ with b invertible. Determine the homomorphisms $u : E_1 \to E_2$ and $v : E_1 \to F_2$ such that $\mathrm{Ker}\, f$ is the graph of the restriction from u to $\mathrm{Ker}\, v$.

6.—(a) Let $E = \left(\bigoplus E_n, d \right)$ be a bounded complex (cf. 3.10) of A-modules, and for all n, let E_n be the direct sum of three submodules T_n, F_n, S_n. The homomorphism $d_n : E_n \to E_{n-1}$ is represented by a matrix:

$$\begin{pmatrix} a_n & e_n & h_n \\ b_n & f_n & i_n \\ c_n & g_n & j_n \end{pmatrix}.$$

Assume that, for all n, $h_n : S_n \to T_{n-1}$ is an isomorphism. Show that, for all n, there is another decomposition (T_n', F_n', S_n') of E_n, where T_n', F_n', S_n' are respectively isomorphic to T_n, F_n, S_n, with respect to which the matrix representing d_n is

$$\begin{pmatrix} 0 & 0 & h_n' \\ 0 & f_n' & 0 \\ 0 & 0 & 0 \end{pmatrix},$$

h_n' being an isomorphism. (First set $T_n' = d_{n+1}(S_{n+1})$, so that a, b, c, i and j become zero, then successively modify each F_n in such a way that e_n becomes null; show that the homomorphisms g_n are not necessarily null.)

(b) Does the result of (a) extend to unbounded complexes?

7.—Let $f : A^p \to A^q$ be a A-linear map. Show that f is surjective if and only if the ideal generated by the minors of order q of the matrix representing f is the whole of the ring A.

8. *(Projective modules)*—Let E be an A-module.

(a) Show that, if E is projective and finitely generated, then it is a direct factor of a finitely generated free module.

(b) Show that, if E is projective, for all exact sequences $M' \to M \to M''$ of A-modules, $\mathrm{Hom}(E, M') \to \mathrm{Hom}(E, M) \to \mathrm{Hom}(E, M'')$ is an exact sequence.

(c) Show that the space E of analytic functions $g : \mathbb{R} \to \mathbb{R}$ such that $g(x + 2\pi) = -g(x)$ is a non-free projective module over the ring A of 3.1, Exercise 12.

Show that all maximal ideals $\mathfrak{m}_x$ in this ring are modules isomorphic to E.

9. *(Projective modules and vector bundles)*—Let X be a topological space. A real (resp. complex) **vector bundle** over X is a topological space E equipped with

(i) a continuous map $\pi : E \to X$,
(ii) the structure of a real (resp. complex) vector space on the fibre $E(x) = \pi^{-1}(x)$ for all $x \in X$,

satisfying the following condition:

(FV) For all $x \in X$, there is a neighbourhood U of x, an integer n and a homeomorphism φ from $E|_U = \pi^{-1}(U)$ onto $U \times \mathbb{R}^n$ (resp. $U \times \mathbb{C}^n$) inducing an isomorphism of vector spaces from $E(x)$ onto $\mathbb{R}^n$ (resp. $\mathbb{C}^n$) for all $x \in U$ (φ is called a *trivialization* from E onto U; E is said to be *trivial* if there is a trivialization from E over X).

A continuous section of E is a continuous map $s : X \to E$ such that $\pi \circ s = 1_X$.

If E and F are two vector bundles over X, a *morphism* from E to F is a continuous map from E to F inducing a linear map from $E(x)$ to $F(x)$ for all $x \in X$.

(a) Show that the continuous sections of E form a natural module S(E) over the ring $C(X) = C(X, \mathbb{R})$ (resp. $C(X, \mathbb{C})$), and that for every morphism $f : E \to F$, the map $f_* : s \mapsto f \circ s$ from S(E) to S(F) is C(X)-linear.

(b) Henceforth, X is assumed to be compact and metrizable. Show that, if E is a vector bundle over X passing through every point de E, then there is a continuous section (i.e. $(\forall t \in E)\ (\exists s \in S(E))\ s(\pi(t)) = t$).

(c) Let $\mathcal{U} = (U_i)_{i \in I}$ be a finite open cover of X. Show that there is a family $(f_i)_{i \in I}$ of continuous functions on X such that for all i, the function f_i has support[14] in U_i, and that $\sum f_i$ are strictly positive at all points. Deduce that there is a family of continuous functions $(\eta_i)_{i \in I}$ on X such that η_i have support in U_i for all i, and that $\sum \eta_i = 1$ (such a family is called a *partition of unity* subject to $\mathcal{U}$).

(d) Let E and F be two vector bundles over X, and $f : E \to F$ a surjective morphism. Show that, for all $x \in X$, there is a neighbourhood U of x and a morphism $g_U : F|_U \to E|_U$ such that $g_U \circ f|_U = 1_{E|_U}$. Using a partition of unity show that there is a morphism $g : F \to E$ such that $g \circ f = 1_E$. Show that $f_* : S(E) \to S(F)$ is surjective and has a C(X)-linear section.

(e) Let E be a vector bundle over X. Show that a finite family (s_i) of continuous sections of E such that, for all $x \in X$, $(s_i(x))$ generate $E(x)$. Deduce that E is a direct factor of a trivial bundle, and that S(E) is a projective C(X)-module.

---

[14]Guichardet [5], 1.2.

(f) Show that the functor $S : E \mapsto S(E)$ from the category of vector bundles over X to the category of C(X)-modules is fully faithful. (To show that $\mathrm{Hom}(E, F) \to \mathrm{Hom}(S(E), S(F))$ is bijective, E may first be considered trivial.)

(g) Let E be a vector bundle over X and p a projection of E (i.e. a morphism from E to itself such that $p \circ p = p$). Show that the image of p vector is a subbundle of E.

(h) Show that the functor S defines an equivalence between the category of vector bundles over X and the category of projective C(X)-modules.

This reduces the study and the classification of projective C(X)-modules to questions of algebraic topology. For rings that are not of the form C(X), the study of projective modules draws methods from algebraic topology.

(i) In the various parts of this problem, can the assumption of the compactness and metrizability of X be weakened? (For (c), only suppose that the open cover $\mathcal{U}$ is locally finite.)

10.—(a) Let A be a ring, and let J(A) be the set of isomorphism classes of finitely generated projective A-modules. Find an upper bound for the cardinality of J(A) in terms of that of A.

(b) The action $(E, F) \mapsto E \oplus F$ defines a structure of commutative monoid on J(A). Show that there is a commutative group K(A) and a homomorphism $\varepsilon : J(A) \to K(A)$ satisfying the following universal property: for any group G and any homomorphism $f : J(A) \to G$, there is a unique homomorphism $\bar{f} : K(A) \to G$ such that $f = \bar{f} \circ \varepsilon$.

(c) Let $\xi, \eta \in J(A)$ with representatives E and F. Show that $\varepsilon(\xi) = \varepsilon(\eta)$ if and only if $E \oplus A^n$ is isomorphic to $F \oplus A^n$ for sufficiently large n. (For an example where ε is not injective, see 4.7, Exercise 7.)

11.—(a) Show that in the Banach space ℓ^1 of real sequences x_n such that $\sum |x_n| < \infty$, the sequences $(e^{-\lambda n})_{n \in \mathbb{N}}$ form a free family as λ runs through $]0, \infty[$.

(b) Using Baire's theorem show that an infinite dimensional Banach space does not have a countable algebraic basis.

(c) Show that, for an infinite dimensional Banach space E, there is a continuous injective linear map from ℓ^1 to E.

(d) Show that, for any Hausdorff Banach space E (i.e. with a countable subset generating a dense subspace), the cardinality of E is $2^{\aleph_0}$. Deduce that all Hausdorff Banach spaces are algebraically isomorphic. Give examples where they are not isomorphic as topological vector spaces.

3.5 Modules over Principal Ideal Domains

Throughout this section, A denotes a PID; for $a \in A$, (a) the ideal generated by a, and χ the canonical map from A to Mon(A) (3.2.7).

3.5.1

> **Theorem** *Every subring of a free* A-*module is free.*

ADDENDUM. Let E be an A-module with basis $(e_i)_{i\in I}$ and F a submodule of E. Then F has a basis $(f_j)_{j\in I'}$, where the cardinality of I' is less than or equal to that of I.

For every subset J of I, let E_J be the submodule de E generated by $(e_i)_{i\in J}$ and set $F_J = F \cap E_J$. Let Ω be the set of triplets $(J, K, (f_i)_{i\in K})$, where J is a subset of I, K a subset of J, and $(f_i)_{i\in K}$ a basis for F_J. For $\alpha = (J, K, (f_i)_{i\in K})$ and $\alpha' = (J', K', (f_i')_{i\in K'})$ in Ω, write $\alpha \leqslant \alpha'$ if $J \subset J'$, $K = K' \cap J$ and $(\forall i \in K)$ $f_i' = f_i$.

This is an order relation on Ω.

Lemma *Let* $\alpha = (J, K, (f_i)) \in \Omega$. *If* α *is maximal, then* $J = I$.

Proof of the Lemma Suppose that $J \neq I$. We define $\alpha' \in \Omega$ such that $\alpha' > \alpha$. Let $i_0 \in I - J$, and set $J' = J \cup \{i_0\}$. Then $E_{J'} = E_J \oplus Ae_{i_0}$. Denote by p the projection $E_{J'} \to Ae_{i_0}$ and set $Q = p(F_{J'})$. There is an exact sequence

$$0 \to F_J \to F_{J'} \to Q \to 0. \qquad (*)$$

The submodule Q of Ae_{i_0} is of the form $\mathfrak{a}e_{i_0}$, where $\mathfrak{a}$ is an ideal in A, hence of the form Aae_{i_0} with $a \in A$ since A is principal. If $a = 0$, then $F_{J'} = F_J$ and $\alpha' = (J', K, (f_i)_{i\in K})$ is a strict upper bound of α. If $a \neq 0$, then ae_{i_0} is a basis for Q, and the sequence $(*)$ splits. Set $K' = K \cup \{i_0\}$ and choose $f_{i_0} \in F'$ such that $p(f_{i_0}) = ae_{i_0}$. Then α is strictly bounded above by $\alpha' = (J', K', (f_i)_{i\in K'})$. $\qquad\Box$

Proof of the Theorem The ordered set Ω is inductive. Indeed, let $\left(J_\lambda, K_\lambda, (f_{i,\lambda})_{i\in K_\lambda}\right)_{\lambda\in\Lambda}$ be a totally ordered family of elements of Ω; set $J = \bigcup J_\lambda$, $K = \bigcup K_\lambda$, and for $i \in K$ let f_i be the common value of $f_{i,\lambda}$ for λ such that $i \in J_\lambda$; the family $(f_i)_{i\in K}$ is a basis for F_J since all elements of F_J are in some F_{J_λ}, so are linear combinations of the elements f_i, and a relation between these f_i only involves finitely many i; therefore the family is bounded above by $(J, K, (f_i)_{i\in K})$.

By Zorn's theorem, Ω has a maximal element $(J, K, (f_i)_{i\in K})$, and $J = I$ necessarily holds. So $(f_i)_{i\in K}$ is a basis for F. $\qquad\Box$

3.5.2 Adapted Bases for Submodules

Let E be a free A-module, F a submodule of E and $(e_i)_{i\in I}$ a basis for E. This basis is said to be *adapted* F if there is a subset J of I and a family $(a_i)_{i\in J}$ in A^* such that $(a_ie_i)_{i\in J}$ is a basis for F.

If there is a basis for E adapted to F, then the quotient module E/F is the direct sum of a free module and a torsion module, the latter being the direct sum of cyclic modules. Indeed, with the above notation, E/F is isomorphic to $A^{(I-J)} \oplus \bigoplus_{i\in J} A/(a_i)$.

3.5.3 Content

We use the notation of 3.2.7. Let E be a free A-module and $x \in E$. Let (x_i) be the coordinates of x with respect to the basis (e_i). The g.c.d. d of $\chi(x_i)$ in $\mathrm{Mon}(A)$ is independent of the choice of the basis (e_i) since, for $a \in A$, $\chi(a) \langle d$ if and only if $(\exists y \in E)\ x = ay$. It is said to be the *content* of x and is written $c(x)$. An element $a \in A$ is a content of x if $\chi(a) = c(x)$.

3.5.4

Proposition *Let E be a free A-module, $x \in E$ and a a content of x. Then, for all linear forms $\varphi : E \to A$, $a \langle \varphi(x)$, and there is a linear form φ such that $\varphi(x) = a$.*

Proof There exists $y \in E$ such that $x = ay$; so $\varphi(x) = a\varphi(y)$, and $a \langle \varphi(x)$. We may assume that $E = A^{(I)}$ and $x = (x_i)$. Then a is a g.c.d. of the x_i and, by Bezout's formula (3.2.11, Corollary), there are $u_i \in A$ such that $a = \sum u_i x_i$, and the linear form $\varphi : (y_i) \mapsto \sum u_i y_i$ on E satisfies $\varphi(x) = a$. □

Remark The proposition can also be stated as follows: Let $E^\top = \mathrm{Hom}(E, A)$ and define $\delta_x : E^\top \to A$ by $\delta_x(\varphi) = \varphi(x)$. Then $\mathrm{Im}\,\delta_x = A \cdot c(x)$.

3.5.5 Elements of Content 1

Proposition *Let E be a free A-module and $x \in E$. The following conditions are equivalent:*

 (i) $c(x) = 1$;
 (ii) $(\exists \varphi : E \to A)\ \varphi(x) = 1$;
 (iii) $x \neq 0$ and Ax is a direct factor of E;
 (iv) there is a basis $(e_i)_{i \in I}$ such that $e_{i_0} = x$ for some $i_0 \in I$.

Proof By 3.5.4, (i) ⇒ (ii). If (ii) holds, then $x \neq 0$, and $y \mapsto \varphi(y) \cdot x$ is a linear retraction from E onto Ax, implying (iii). Assume (iii) holds and that F such that $E = Ax \oplus F$. By 3.5.1, F is free. Let $(e_i)_{i \in J}$ be a basis for F, and $i_0 \notin J$. Set $I = J \cup \{i_0\}$ and $e_{i_0} = x$. As x is torsion-free, $(e_i)_{i \in I}$ is a basis for E, and so (iv) follows. Finally, (iv) ⇒ (i) is immediate. □

3.5.6   Elements of Minimal Content in a Submodule

Let E be a free A-module and F a submodule of E. The ring A, being a PID, is a
a UFD, and so every decreasing sequence of elements of $\mathrm{Mon}(A)$ is stationary. By
1.6.1, there exists $f \in F$ such that $c(f)$ is a minimal element of $\{c(y)\}_{y \in F}$.

Proposition *Let* E *be a free* A-*module,* F *a nonzero submodule of* E, $f \in F$ *an
element of minimal content,* $a \in A^*$ *a content of* f *and* $e \in E$ *such that* $f = ae$. *Then*
 (a) Ae *is a direct factor of* E*;*
 (b) Let E_1 *be a complement of* Ae *in* E *and set* $F_1 = F \cap E_1$*; then* $F = Af \oplus F_1$*;*
 (c) $c(f)$ *is the smallest element of* $\{c(y)\}_{y \in F}$.

Proof (a) $c(f) = ac(e)$ and soù $c(e) = 1$, and the assertion follows from Proposition
3.5.5.

(b) Let $p : E \to Ae$ be the projection with kernel E_1. The submodule $p(F)$ of
Ae is of the form $\mathfrak{b}e$ where $\mathfrak{b}$ is an ideal, and so of the form A$\mathfrak{b}e$, where $\mathfrak{b} \in A$
since A is principal. Since $f = ae \in p(F) = A\mathfrak{b}e, \mathfrak{b}\langle a$. There exists $y \in F$ such that
$p(y) = \mathfrak{b}e$. Hence $\mathfrak{b} = \varphi(y)$ for some linear form φ on E, and so $c(y)\langle \chi(\mathfrak{b})$ by 3.5.4.
$c(y)\langle \chi(\mathfrak{b})\langle \chi(a) = c(f)$; as $c(f)$ is minimal, $c(y) = c(f)$ and $\chi(\mathfrak{b}) = \chi(a)$.
Hence $p(F) = A\mathfrak{b}e = Aae = Af \subset F$, and p induces a projection of F with image
Af and kernel $F \cap E_1 = F_1$. So $F = Af \oplus F_1$ by 3.3.11.

(c) $c(f)$ is minimal by assumption. We show that, for all $y \in F$, $c(f)\langle c(y)$.
Let $y \in F$, write $y = \lambda f + y_1$ with $y_1 \in F_1$. Then $c(y) = $ g.c.d.$(\lambda c(f), c(y_1))$ follows by considering a basis for E consisting of e and of a basis for E_1. To
show that $c(f)\langle c(y)$ it suffices to show that $c(f)\langle c(y_1)$. Set $y' = f + y_1$. Then
$c(y') = $ g.c.d.$(c(f), c(y_1))\langle c(f)$, and so $c(y') = c(f)$ since $c(f)$ is minimal. Hence
$c(f) = c(y')\langle c(y_1)$.   □

3.5.7

Proposition (Adapted basis Theorem) *Let* E *be a finitely generated free* A-*module.
For all submodules* F *of* E *there is basis for* E *adapted to* F.

ADDENDUM. There is a basis $(e_1, ...e_n)$ for E, an integer $p \leqslant n$ and elements
$a_1, ..., a_p$ of A^* such that $(a_1 e_1, ..., a_p e_p)$ is a basis for F and $a_1 \langle a_2 \langle \cdots \langle a_p$.

Proof By induction on the rank p of F. If $p = 0$, then there is nothing to show.
Suppose that F is of rank $p \neq 0$ and that $f_1 \in F$ has minimal content. Let $a_1 \in A^*$
be a content of f_1 and $e_1 \in E$ such that $a_1 e_1 = f_1$. By 3.5.6, $E = Ae_1 \oplus E_1$ and
$F = Af_1 \oplus F_1$, where E_1 is a submodule of E and $F_1 = F \cap E_1$. By 3.5.1, the modules
E_1 and F_1 are free. As $\mathrm{rk}(F) = \mathrm{rk}(F_1) + 1$, F_1 is free of rank $p - 1$. By the induction
hypothesis, there is a basis $(e_2, ..., e_n)$ for E_1 and elements $a_2, ..., a_p$ of A^* such that
$(a_2 e_2, ..., a_p e_p)$ is a basis for F_1 and $a_2 \langle \cdots \langle a_p$. Then $(e_1, ..., e_n)$ is a basis for E,
$(a_1 e_1, ..., a_p e_p)$ a basis for F, and, if $p > 1$, $a_1 \langle a_2$ by 3.5.6, since a_2 is a content of
$f_2 = a_2 e_2 \in F$.   □

3.5.8

> **Theorem** *Every finitely generated* A-*module is the direct sum of cyclic submodules.*

ADDENDUM. Every finitely generated A-module is isomorphic to a module $A^s \oplus \bigoplus_{i \in I} A/(a_i)$, where $s \in \mathbb{N}$, I is a finite set and $a_i \in A^*$ for $i \in I$. In such a module, the second direct factor is a torsion submodule.

Proof Every finitely generated A-module is isomorphic to the quotient of a finitely generated free A-module E by a submodule of E. The theorem and its addendum then follow from Proposition 3.5.7 and from 3.5.2. □

Corollary 3.1 *In a finitely generated* A-*module, the torsion submodule is a direct factor.*

Corollary 3.2 *All finitely generated torsion-free* A-*modules are free.*

Corollary 3.3 *All commutative finite groups are products of cyclic groups.*

3.5.9

Two elements $x, y \in A$ are said to be *relatively prime* if the ideal they generate is A, i.e. if 1 is a g.c.d. of x and y.

Proposition (Chinese Theorem for two elements) *Let x and y be two relatively prime elements of A. Then the module $A/(xy)$ is isomorphic to $A/(x) \oplus A/(y)$.*

ADDENDUM. Let $\chi_x : A \to A/(x)$ and $\chi_y : A \to A/(y)$ be the canonical maps. The homomorphism $\Delta : z \mapsto (\chi_x(z), \chi_y(z))$ from A to $A/(x) \times A/(y)$ factorizes naturally (3.3.7) to give an isomorphism from $A/(xy)$ onto $A/(x) \oplus A/(y)$. A more general version will be considered in (5.1.5).

Proof The kernel of Δ is the ideal of common multiples of x and y and is generated by xy since xy is a l.c.m. of x and y (3.2.5).

We show that Δ is surjective. By Bezout's formula (3.2.11, Corollary), there exist $u, v \in A$ such that $ux + vy = 1$. Let: $(\alpha, \beta) = (\chi_x(a), \chi_y(b)) \in A/(x) \times A/(y)$; set $c = vya + uxb$. As $c = a + u(b - a)x = b + v(a - b)y$,

$$\chi_x(c) = \alpha, \quad \chi_y(c) = \beta, \quad \Delta(c) = (\alpha, \beta).$$

Hence Δ is surjective, and so the proposition and its addendum follow. □

Corollary *Let $x \in A^*$. Write x as $x = u \cdot p_1^{r_1} \cdots p_k^{r_k}$, where u is invertible, all p_i irreducible and $\chi(p_i)$ distinct in* $\mathrm{Mon}^*(A)$. *Then the module $A/(x)$ is isomorphic to* $A/(p_1^{r_1}) \oplus \cdots \oplus A/(p_k^{r_k})$.

3.5.10 Elementary Divisors

Proposition and Definition *Let* E *be a finitely generated* A-*module.*
(a) The module E *is isomorphic to a module*

$$A^s \oplus \bigoplus_{i \in I} A/(p_i^{r_i}),$$

where $s \in \mathbb{N}$, I *is a finite set, and, for all* $i \in I$, $p_i \in A^*$ *is irreducible and* $r_i > 0$.
(b) If $E \approx A^s \oplus \bigoplus_{i \in I} A/(p_i^{r_i}) \approx A^{s'} \oplus \bigoplus_{i \in I'} A/(p_i'^{r_i'})$, *then as the conditions of*
(a) *hold for* s, I, p_i, r_i *and for* s', I', p_i', r_i', $s = s'$ *and there is a bijection* τ *from* I
onto I' *such that* $\chi(p_i) = \chi(p_{\tau(i)}')$ *and* $r_i = r_{\tau(i)}'$ *for all* $i \in$ I.
$(p_i^{r_i})_{i \in I}$ *is the family of* **elementary divisors** *of* E.

(a) follows from 3.5.8 and 3.5.9. The proof of (b) uses a lemma:

Lemma *Let* $E = A^s \oplus \bigoplus_{i \in I} A/(p_i^{r_i})$, *where all* p_i *are irreducibles,* $p \in A^*$ *an*
irreducible element and $r > 0$. *Then* $p^{r-1}E/p^r E$ *is isomorphic to* $(A/(p))^{s+d(p,r)}$,
$d(p, r)$ *being the number of indices i such that* $\chi(p_i) = \chi(p)$ *and* $r_i \geqslant r$.

Proof of the Lemma The general case follows by passing to the direct sum in the
following particular cases:

(i) If $E = A$, $p^{r-1}E/p^r E = Ap^{r-1}/Ap^r \approx A/Ap$.

(ii) If $E = A/(p'^{r'})$, where $\chi(p') \neq \chi(p)$, there exist $u, v \in A$ such that $up + vp'^{r'} = 1$. So the image of p in the quotient ring $A/(p'^{r'})$ is invertible, and $p^{r-1}E = p^r E = E$, and soù $p^{r-1}E/p^r E = 0$.

(iii) If $E = A/(p^{r'})$ with $r' \geqslant r$, then $p^r E = Ap^r/Ap^{r'}$ and $p^{r-1}E = Ap^{r-1}/Ap^{r'}$, and so $p^{r-1}E/p^r E \approx Ap^{r-1}/Ap^r \approx A/Ap$.

(iv) If $E = A/(p^{r'})$ with $r' < r$, then $p^{r-1}E = p^r E = 0$, and so $p^{r-1}E/p^r E = 0$.

Proof of the Proposition (b) The quotient module of E by its torsion submodule is
isomorphic to A^s and to $A^{s'}$. Thus $s = s'$ by the uniqueness of dimension (3.4.5).
Defining $d'(p, r)$ as $d(p, r)$, $s' + d'(p, r) = s + d(p, r)$: it is the dimension of
$p^{r-1}E/p^r E$ as a vector space over $A/(p)$. Hence $d'(p, r) = d(p, r)$.

The number $f(p, r)$ of indices $i \in I$ such that $\chi(p_i) = \chi(p)$ and $r_i = r$ is
$d(p, r) - d(p, r + 1)$; The same holds for the number $f'(p, r)$ defined likewise.
Henceù $f(p, r) = f'(p, r)$.   □

Remark In part (b) of the proposition, the submodules of E corresponding to $A/(p_i^{r_i})$
and to $A/(p_{\tau(i)}^{r_{\tau(i)}})$ are isomorphic, but may be distinct.

3.5.11 Primary Decomposition

Let E be a torsion A-module and $p \in A^*$ an irreducible element. The p-**primary**
component of E, written E_p, is the set of $x \in E$ such that $(\exists r \in \mathbb{N})\ p^r x = 0$. It is

a submodule of E. Then $E_p = E_{p'}$ if $\chi(p) = \chi(p')$. Hence we can talk of E_p for irreducible $p \in \mathrm{Mon}^*(A)$. When $E_p = E$, E is said to be *p-primary*.

If F is a submodule of E, then $F_p = F \cap E_p$. If E and E' are torsion A-modules and $\varphi : E \to E'$ is a homomorphism, then $\varphi(E_p) \subset E'_p$.

Theorem *Let* E *be a torsion* A*-module. Then* $E = \bigoplus_{p \in M} E_p$, *where* M *is the set of irreducible elements of* $\mathrm{Mon}^*(A)$.

Proof Let $x \in E$. The cyclic submodule Ax of E is isomorphic to a $A/(a)$, where $a \in A^*$ (3.3.6, Example). Write $a = u \cdot p_1^{r_1} \cdots p_k^{r_k}$ as in (3.5.9, Corollary); then $A/(a) \approx A/(p_1^{r_1}) \oplus \cdots \oplus A/(p_k^{r_k})$. Set $e_i = (0, \ldots, \chi_{p_i^{r_i}}(1), \ldots, 0)$ in $A/(p_1^{r_1}) \oplus \cdots \oplus A/(p_k^{r_k})$. Let φ be an isomorphism from this direct sum onto Ax and set $x_i = \varphi(e_i)$. Then $p_i^{r_i} x_i = 0$. So $x_i \in E_{p_i}$, and x is of the form $\sum \lambda_i x_i$. Hence E is generated by the submodules E_{p_i}.

Let $p_1, \ldots, p_k \in A^*$ be irreducible elements for which $\chi(p_1), \ldots, \chi(p_k)$ are distinct, and let $x_1 \in E_{p_1}, \ldots, x_k \in E_{p_k}$ be elements such that $\sum x_i = 0$. We show that $x_1 = \cdots = x_k = 0$. For $i \in \{1, \ldots, k\}$, let r_i be such that $p_i^{r_i} x_i = 0$. By Bezout's formula (3.2.11, Corollary), there exist $u, v \in A$ such that $u \cdot p_1^{r_1} + v \cdot p_2^{r_2} \cdots p_k^{r_k} = 1$. Then

$$x_1 = (1 - u p_1^{r_1}) x_1 = v \cdot p_2^{r_2} \cdots p_k^{r_k} x_1 = -v \cdot p_2^{r_2} \cdots p_k^{r_k} (x_2 + \cdots + x_k) = 0 \; ;$$

Similarly, $x_2 = \cdots = x_k = 0$. Hence E is a direct sum of the submodules E_p. $\square$

ADDENDUM. For all submodules F of E, $F = \bigoplus_{p \in M} (F \cap E_p)$.

Exercises 3.5. (Modules over principal ideal domains)
1. *(Invariant factors)* Let A be a PID and E a finitely generated A-module.

(a) Show that E is isomorphic to a module $A/(a_1) \oplus \cdots \oplus A/(a_n)$ with $a_1 \rangle a_2 \rangle \cdots \rangle a_n$, and non-invertible a_n.

(b) Show that the sequence $(a_1, \ldots, a_n)$ is uniquely determined up to multiplication of the elements a_i by invertible elements (these a_i are called invariant factors). Show that $a_1 = 0$ if and only if A is torsion-free.

(c) Find the elementary divisors and invariant factors of the $\mathbb{Z}$-module $\mathbb{Z}/(4) \oplus \mathbb{Z}/(8) \oplus \mathbb{Z}/(6) \oplus \mathbb{Z}/(18) \oplus \mathbb{Z}/(7)$.

(d) Show that n is the minimal cardinality of a generating family of M.

(e) How can the invariant factors be computed in terms of the elementary divisors and conversely?

2. *(Multiplicative groups of modulo n integers)*—Let n be an integer. Set $G(n)$ to be the finite multiplicative group of invertible elements of the ring $\mathbb{Z}/(n)$. The aim is to express $G(n)$ as a product of cyclic groups.

(a) Let G be a finite commutative group, written multiplicatively. Assume that, for every prime p, the number of elements of G of order p is less than p. Show that G is cyclic.

(b) Let p be a prime. Show that the group $G(p)$ is cyclic (use the fact that a polynomial of degree q has at most q roots in the field $\mathbb{Z}/(p)$).

(c) Let $N(p^r)$ denote the set of elements of $\mathbb{Z}/(p^r)$ with image 1 in $\mathbb{Z}/(p)$. Show that $N(p^r)$ is a multiplicative group (every element in $N(p^r)$ is of the form $1 + u$ withù u nilpotent). What is its order?

(d) Assume that $p \neq 2$. Show that $(1 + ap^k)^p \equiv 1 + ap^{k+1}$ (mod p^{k+2}). Which are the elements of order p in $N(p^r)$? Show that $N(p^r)$ is cyclic. Is the group $G(p^r)$ cyclic?

(e) For $r \geqslant 2$, let $N'(2^r)$ be the subgroup of $N(2^r) = G(2^r)$ consisting of the elements with image 1 in $\mathbb{Z}/(4)$. Show that $N'(2^r)$ is cyclic, and that it is a direct factor of $N(2^r)$. Is the group $G(2^r)$ cyclic?

(f) Show that, if $n = p_1^{r_1} \cdots p_k^{r_k}$, the group $G(n)$ is isomorphic to the product group $G(p_1^{r_1}) \times \cdots \times G(p_k^{r_k})$.

(g) Express $G(4761)$ as a product of cyclic groups.

3. *(Fitting ideal)*—Let A be a PID and E a finitely generated torsion A-module.

(a) Show that E has a presentation $A^q \xrightarrow{f} A^p \to E \to 0$, with f injective. Show that, for such a presentation, $q = p$ necessarily holds, and that, up to multiplication by an invertible element, the determinant of f only depends on E and not on the presentation. The ideal generated by $\det(f)$ is called the *Fitting ideal* of E.

(b) Compute $\det(f)$ for $E = A/(a_1) \oplus \cdots \oplus A/(a_n)$. Give an interpretation for it when $A = \mathbb{Z}$.

(c) Let $0 \to E \to F \to G \to 0$ be a short exact sequence of finitely generated torsion A-modules. What is the relation between the generators of the Fitting ideals of E, F and G?

4.—Let n be an integer and A a ring. Set $B = A[X]$. A vector of B^n (resp. a matrix with entries in B) can be written $u_0 + Xu_1 + \cdots + X^d u_d$, with $u_i \in A^n$ (resp. u_i is a matrix with entries in A).

(A) Let M be a matrix with entries in A and consider the B-linear map $\Phi_M = M - XI : B^n \to B^n$. Show that the A-submodule A^n of B^n is a complement of the image of Φ_M.

(B) Consider the B-module E_M obtained by equipping A^n with the A-bilinear map $B \times A^n \to A^n$ defined by $X.v = M(v)$ for $v \in A^n$.

(a) Show that E_M is not isomorphic to any submodule of B^n.

(b) When A is an algebraically closed field, give an interpretation of the primary decomposition of E_M, and more particularly of Proposition 3.5.10.

(c) Show that there is an exact B-linear sequence (resolution of E_M)

$$0 \to B^n \to B^n \to E_M \,,$$

with Φ_M, and a map $\varepsilon : B^n \to E_M$ which should be defined.

(C) Let M and M' be two $n \times n$ matrices with entries in A. Show that the following conditions are equivalent:

(i) The B-modules E_M and $E_{M'}$ are isomorphic;

(ii) there are invertible matrices V and W with entries in B such that $M' - XI = W(M - XI)V^{-1}$;

(iii) there is a matrix U with entries in A such that $M' = UMU^{-1}$.

(D) Show by an example (with $A = \mathbb{R}$, $n = 2$) that condition (iii) of (C) is not generally equivalent to the analogous condition (iii') where det $U = 1$. What happens when A is an algebraically closed field?

5.—Let E be a finite n-dimensional vector space over a field K and u an endomorphism of E. Endow E with the structure of a $K[T]$-module defined by $T \cdot x = u(x)$ for $x \in E$.

(a) Show that E is a finitely generated torsion $K[T]$-module.

(b) Show that the first invariant factor of the $K[T]$-module E is the minimal polynomial[15] of u. What is the Fitting ideal of E?

(c) Suppose that K is algebraically closed. Using 3.5.10 recover the fact that $E = E_1 \oplus \cdots \oplus E_k$, where all E_i are stable under u, and where, for all i, the endomorphism u_i of E_i induced by u is the sum of a homothety and a nilpotent endomorphism.[16]

(d) Show that from the ith factor, the product of invariants factors is equal to the g.c.d. of minors of order $n - i$ of a matrix representing u.

6.—Let E be the $\mathbb{Z}$-module $\mathbb{Z}^{\mathbb{N}}$ of all integers. The aim of this exercise is to determine the module $E^{\top} = \mathrm{Hom}(E, \mathbb{Z})$.

Let F be the submodule $\mathbb{Z}^{(\mathbb{N})}$ of E consisting of sequences with finite support, and $(e_n)_{n \in \mathbb{N}}$ the canonical basis for F defined by $e_n = (\delta_{n,p})_{p \in \mathbb{N}}$. For all $u \in E^{\top}$, define $\varphi(u) \in E$ by $\varphi(u) = (u(e_n))_{n \in \mathbb{N}}$. This gives a homomorphism from $E^{\top}$ to E.

Parts A an B aim to find the kernel and image of φ. They are independent.

A. *Determination of* Ker φ

(a) Show that Ker $\varphi = G^{\top}$, where $G = E/F$ (module of the "germs of sequences at infinity").

(b) Let H_2 be the set of elements of G divisible by 2^k for all k. Show that H_2 is a submodule of G. Provide a method for finding the nonzero elements of H_2.

(c) Show that all linear forms $u : G \to \mathbb{Z}$ vanish on H_2.

(d) Define H_3 similarly. Determine $H_2 + H_3$. Conclude.

B. *Determination of* Im φ

For all $x = 2^a q \in \mathbb{Z}$ with q odd, set $|x|_2 = 2^{-a}$; let $|0|_2 = 0$.

(a) Show that $(x, y) \mapsto |y - x|_2$ is a distance on $\mathbb{Z}$. Is the space $\mathbb{Z}$ complete with respect to this distance?

Show that, if $x_1, ..., x_n$ are integers such that $|x_i|_2$ are mutually distinct, then $\left| \sum x_i \right|_2$ is the largest of the $|x_i|_2$.

(b) For $x = (x_n)_{n \in \mathbb{N}} \in E$, set $\|x\|_2 = \sup |x_n|_2$. Show that:

$$(\forall u \in E^{\top}) \, (\forall x \in E) \quad |u(x)|_2 \leqslant \|x\|_2.$$

(c) Let $x = (x_n)_{n \in \mathbb{N}}$. What condition must the sequence $\left(\|x - \sum_0^n x_k e_k\|_2 \right)_{n \in \mathbb{N}}$ satisfy so as to tend to 0?

---

[15]Cf. Queysanne [1], Example 488, p. 509 or [2], def. 11.3.2, p. 343
[16]Queysanne [1], Exerc. 486 and 495, [2], § 11.4, p. 348.

(d) Let $u \in E^\top$ and set S to be the support of $\varphi(u)$, i.e. $S = \{n \mid u(e_n) \neq 0\}$. Let $x \in E$ be an element whose support is S and such that the maps

$$s \mapsto |x_s|_2 \quad \text{and} \quad s \mapsto |u(e_s)|_2 |x_s|_2$$

from S to $\mathbb{R}_+$ are strictly decreasing.

Let A be the set of sequences having support in S and which only take values 0 and 1 , i.e. $A = \{0, 1\}^S$. For $\varepsilon \in A$, set $\psi(\varepsilon) = u(\varepsilon x)$, where $\varepsilon x = (\varepsilon_n x_n)_{n \in \mathbb{N}}$. Determine $|\psi(\varepsilon) - \psi(\varepsilon')|_2$ in terms of $s_0 = \inf\{s \mid \varepsilon_s \neq \varepsilon'_s\}$. Deduce that $\psi : A \to \mathbb{Z}$ is injective.

(e) Conclude by considering the cardinality of A.

C. *Conclusion*

(a) Describe $E^\top$.

(b) Is the module E free?

(c) Do these results generalize to infinite sets other than $\mathbb{N}$, to rings other than $\mathbb{Z}$?

7.—Let A be a (not necessarily principal) ring all of whose ideals are projective A-modules (for example a Dedekind ring). Let E be a free A-module, $(e_i)_{i \in I}$ a basis for E, and F a submodule of E. Show that F is isomorphic to a module of the form $\bigoplus_{i \in I} J_i$, where J_i is an ideal of A for each i (use the Proof of Theorem 3.5.1).

8.—Let G be a countable $\mathbb{Z}$-module whose points are separated by linear forms (i.e. $(\forall x \neq 0)$ $(\exists \varphi : G \to \mathbb{Z}$ linear) $\varphi(x) \neq 0$). The aim is to show that G is free. This exercise uses some of the results of 3.8.

(a) Show that G is torsion-free. Set $V = \mathbb{Q} \otimes G$. Show that G is embedded in V, and that V is a finite dimensional or countable vector space. Show that $G^\top = \mathrm{Hom}(G; \mathbb{Z})$ is embedded in $V^* = \mathrm{Hom}(V; \mathbb{Q})$.

(b) Suppose V is finite dimensional. Show that there is a family of elements of $G^\top$ forming a basis for V^*, and that such a family separates the points of G. Deduce that G is embedded in a finitely generated $\mathbb{Z}$-module, and that it itself is finitely generated.

(c) Still supposing V to be finite dimensional, let V' be a hyperplane in V. Set $G' = G \cap V'$. Show that any basis for G' can be extended to a basis for G.

(d) Suppose V is of countably infinite dimension. Let $(e_1, ..., e_n, ...)$ be a basis for V, and V_n the vector subspace generated by $(e_1, ..., e_n)$, and set $G_n = V_n \cap G$. Construct by induction a sequence (x_n) of elements of G such that $(x_1, ..., x_n)$ is a $\mathbb{Z}$-base for G_n for all n. Show that $(x_1, ..., x_n, ...)$ is then a basis for G.

(e) Deduce that every countable subgroup of $\mathbb{Z}^\mathbb{N}$ is free.

9.—Let $G \subset \mathbb{Z}^\mathbb{N}$ be a countable subgroup that is not finitely generated. Show that G is not a direct factor (it may be shown that there is element $\xi \neq 0$ in $\mathbb{Z}^\mathbb{N}/G$ with infinitely many divisors).

10.—The aim is to find an example of a torsion-free countable $\mathbb{Z}$-module G all whose nonzero elements are only divisible by finitely many scalars, but which is not free.

Let p be a prime and A the ring $\mathbb{Z}\left[\frac{1}{p}\right] = \left\{\frac{m}{p^r}\right\}_{m \in \mathbb{Z}, r \in \mathbb{N}}$; Denote by $\widehat{\mathbb{Z}}_p$ the ring of p-adic integers (see 2.5, Exercise 7) and by $\widehat{\mathbb{Q}}_p$ its field of fractions.

Let $\alpha \in \widehat{\mathbb{Z}}_p - \mathbb{Q}$, and G the subgroup of A^2 consisting of pairs (x, y) such that $|y - \alpha x|_p \leqslant 1$ (absolute p-adic value).

(a) Show that the projection $(x, y) \mapsto x$ from G to A is surjective and that its kernel can be identified with $\mathbb{Z}$. Deduce that G is countable, but not finitely generated.

(b) Show that G is embedded in $\mathbb{Q}^2$. Deduce that G is not free.

(c) Show that the nonzero elements of G are only divisible by finitely many powers of p, and only have finitely many divisors.

3.6 Noetherian Rings

3.6.1

Theorem and Definition *Let* A *be a ring. The following properties are equivalent:*

(i) *every ideal of* A *is finitely generated;*

(ii) *the set of ideals of* A, *ordered by inclusion, is Noetherian (i.e. every increasing sequence of ideals of* A *is stationary);*

(iii) *every submodule of a finitely generated* A-*module is finitely generated;*

(iv) *For any finitely generated* A-*module* E, *the ordered set of submodules of* E *is Noetherian.*

If these conditions hold, the ring A *is said to be* **Noetherian.**

Implications (iii) $\Rightarrow$ (i) and (iv) $\Rightarrow$ (ii) are immediate (take E = A). The equivalences (i) $\Leftrightarrow$ (ii) and (iii) $\Leftrightarrow$ (iv) follow from the next lemma:

Lemma *Let* E *be an* A-*module. The following conditions are equivalent:*

(i_E) *every submodule of* E *is finitely generated;*

(ii_E) *the ordered set of submodules of* E *is Noetherian.*

Proof of the Lemma (i_E) $\Rightarrow$ (ii_E). Let (F_n) be an increasing sequence of submodules of E. Set $F = \bigcup F_n$. Let $(x_i)_{i \in I}$ be a finite generating family of F. For each i, there exists n_i such that $x_i \in F_{n_i}$. Thus, $F_n = F$ for $n = \sup(n_i)$, and so the sequence (F_n) is stationary.

(ii_E) $\Rightarrow$ (i_E). Let F be a submodule of E. The set X of finitely generated submodules of F is not empty since $\{0\} \in X$. So it contains a maximal element G (1.6.1), and G = F. Indeed were x in G − F, then G would be strictly contained in the submodule $G + Ax$. Hence F is finitely generated. $\qquad \square$

Proof of the Theorem (i) $\Rightarrow$ (iii) remains to be shown. We show by induction on n that every submodule of an A-module E generated by n elements is finitely generated. This is obvious for $n = 0$. It holds for $n = 1$. Indeed, a cyclic A-module is isomorphic to a quotient module A/I, and submodules of A/I are of the form J/I, where J is an ideal of A, and is finitely generated if J is a finitely generated ideal. Let E be an A-module generated by $x_1, ..., x_n$ with $n \geqslant 1$, and F a submodule of E. Let E′ be the submodule of E generated by $(x_1, ..., x_{n-1})$, E″ = E/E′, F′ = F ∩ E′ and F″ = F/F′.

The module F″ can be identified with a submodule of E″, and so is finitely generated, and by the induction hypothesis, F′ is finitely generated. This gives an exact sequence $0 \to F' \to F \to F'' \to 0$, and so F is finitely generated by (3.3.9, Corollary).   □

3.6.2   Examples

(1) Every PID is Noetherian.

(2) Every quotient ring of a Noetherian ring is Noetherian, Indeed, if $B = A/I$, the ideals of B correspond to ideals of A containing I.

(3) Let $\varphi : A \to B$ be a ring homomorphism. If A is Noetherian, and if φ turns B into a finitely generated A-module (see 3.3.1, Example 3), then the ring B is Noetherian. Indeed, every ideal in B is finitely generated over A, perforce over B. This will be generalized (3.7.7, Corollary 3.5).

(4) If K is a field, the ring of polynomials $K[X_1, ..., X_n]$ in n variables will be seen (3.7.7) to be Noetherian. The same holds for the ring $K[[X_1, ..., X_n]]$ of formal series, and for the ring of convergent series $\mathbb{C}\{X_1, ..., X_n\}$.

(5) Every ring of fractions of a Noetherian ring is Noetherian. In particular every localization of a Noetherian ring is Noetherian.

(6) The ring of analytic functions on a compact $\mathbb{R}$-analytic manifold is Noetherian.

3.6.3   Counterexamples

(1) Let K be a field and I an infinite set. The ring $K[(X_i)_{i \in I}]$ is not Noetherian: the ideal of polynomials with constant terms is not finitely generated.

(2) If E is an infinite dimensional Banach space, the ring of germs of analytic functions on E is not Noetherian: if (F_n) is a strictly decreasing sequence of closed vector subspaces of finite codimension in E, then we get a strictly increasing sequence (I_n) of ideals, where I_n is the ideal of germs of analytic functions vanishing on F_n in the neighbourhood of 0.

(3) A subring of a Noetherian ring is not necessarily Noetherian. Indeed, a non-Noetherian integral domain is a subring of its field of fractions.

(4) The ring $C([0, 1])$ of continuous functions on $[0, 1]$ is not Noetherian: the ideal $\mathfrak{m}$ of functions $f \in C([0, 1])$ such that $f(0) = 0$ is not finitely generated. Indeed, let $f_1, ..., f_n \in \mathfrak{m}$, and X be the set of $x \in [0, 1]$ such that $f_1(x) = \cdots = f_n(x) = 0$. If $X \neq \{0\}$, the function $x \mapsto x$ is not in the ideal generated by $(f_1, ..., f_n)$; if $X = \{0\}$, every function in the ideal generated by $(f_1, ..., f_n)$ is $O(h)$, where $h = \sup |f_i|$, and the function $\sqrt{h}$ is not in this ideal.

The ring of germs at 0 of continuous functions on $\mathbb{R}$ can likewise be shown not be Noetherian.

(5) The ring E_0 of germs at 0 of C^∞-functions on $\mathbb{R}$ is not Noetherian: the sequence of ideals (f_n), where $f_n(x) = \frac{1}{x^n} e^{-1/x^2}$, is strictly increasing.

(6) The ring $\mathcal{O}(\mathbb{R})$ of analytic functions on $\mathbb{R}$ is not Noetherian: the sequence of ideals (I_n), where I_n is the ideal of functions $f \in \mathcal{O}(\mathbb{R})$ such that $f(p) = 0$ for all integers $p \geqslant n$, is strictly increasing. Indeed the function $x \mapsto \frac{\sin \pi x}{x - n}$ is in I_{n+1} but not in I_n.

(7) An infinite dimensional Banach algebra over $\mathbb{R}$ or $\mathbb{C}$ is never a Noetherian ring (3.6, Exercise 3). However, there are Noetherian and even principal infinite dimensional Banach algebras over ultrametric fields (3.6, Exercise 4).

3.6.4 Minimal Prime Ideals

Proposition *Let* A *be a ring.*
  (a) Every prime ideal contains a minimal prime ideal..
  (b) The intersection of minimal prime ideals is the nilradical of A.
  (c) If A *is Noetherian, it only has finitely many minimal prime ideals.*

Proof (a) The set of prime ideals of A is codirect (i.e. direct for the reverse inclusion order). Indeed, if $(\mathfrak{p}_i)$ is a totally ordered family of prime ideals, then $\mathfrak{p} = \bigcap \mathfrak{p}_i$ is prime since $A/\mathfrak{p}$ can be identified with a subring of $\varprojlim A/\mathfrak{p}_i$, and so is integral.

(More fundamentally, if $x \notin \mathfrak{p}$ and $y \notin \mathfrak{p}$, there exists i such that $x \notin \mathfrak{p}_i$ and $y \notin \mathfrak{p}_i$, and so $xy \notin \mathfrak{p}_i$. Hence $xy \notin \mathfrak{p}$.) By Zorn's theorem, every prime ideal contains a minimal prime ideal.

(b) By (a), the intersection of minimal prime ideals is equal to the intersection of all prime ideals.

(c) The ordered set E of ideals in A is Noetherian. For any ideal I of A, let $\mathcal{M}(I)$ be the set of minimal prime ideals containing I and let us show by Noetherian induction (1.6.3) that, for all I, the set $\mathcal{M}(I)$ is finite. Let $I \in E$ be such that, for any ideal J strictly containing I, the set $\mathcal{M}(J)$ is finite. If I is prime, then $\mathcal{M}(I)$ has at most one element. If I is not prime, let $x, y \notin I$ such that $xy \in I$, and set $J' = I + Ax$ and $J'' = I + Ay$. Every prime ideal containing I contains xy, and hence contains J' or J''; thus $\mathcal{M}(I) = \mathcal{M}(J') \cup \mathcal{M}(J'')$ is finite. $\square$

3.6.5 Associated Prime Ideals

Definition Let A be a ring and M an A-module. The annihilator of an element of M is called an **associated prime ideal** of M.

The set of all associated prime ideals associated of M is denoted by Ass(M). For $x \in M$, write Ann(x) for the annihilator of x.

Proposition *Let* A *be a Noetherian ring and* M *a nonzero* A*-module. Then* Ass(M) *is not empty.*

The set of ideals $\mathrm{Ann}(x)$ with $x \in \mathrm{M}$, $x \neq 0$, ordered by inclusion, is Noetherian. Hence it has a maximal element $\mathrm{Ann}(x_0)$. We show by contradiction that $\mathrm{Ann}(x_0)$ is prime. Consider a and b not in $\mathrm{Ann}(x_0)$ such that $ab \in \mathrm{Ann}(x_0)$. Then $\mathrm{Ann}(x_0) \subset \mathrm{Ann}(bx_0)$, $a \in \mathrm{Ann}(bx_0)$ and $a \notin \mathrm{Ann}(x_0)$; so $\mathrm{Ann}(bx_0)$ strictly contains $\mathrm{Ann}(x_0)$, contradicting maximality.   □

3.6.6

Proposition *Let* A *be a ring and* $0 \to \mathrm{E} \xrightarrow{u} \mathrm{F} \xrightarrow{v} \mathrm{G} \to 0$ *an exact sequence of* A-*modules. Then* $\mathrm{Ass}(\mathrm{F}) \subset \mathrm{Ass}(\mathrm{E}) \cup \mathrm{Ass}(\mathrm{G})$.

Proof Let $\mathfrak{p} = \mathrm{Ann}(x) \in \mathrm{Ass}(\mathrm{F})$. We show that $\mathfrak{p} \in \mathrm{Ass}(\mathrm{E})$ or $\mathfrak{p} \in \mathrm{Ass}(\mathrm{G})$. If $x \in u(\mathrm{E})$, then $\mathfrak{p} \in \mathrm{Ass}(\mathrm{E})$; Suppose that $x \notin u(\mathrm{E})$; so $v(x) \neq 0$. Then $\mathfrak{p} \subset \mathrm{Ann}(v(x))$. If $\mathfrak{p} = \mathrm{Ann}(v(x))$, then $\mathfrak{p} \in \mathrm{Ass}(\mathrm{G})$. Suppose that $\mathfrak{p} \neq \mathrm{Ann}(v(x))$. Hence there exists $a \in \mathrm{A}$ such that $a \in \mathrm{Ann}(v(x))$ and $a \notin \mathfrak{p}$, i.e. $av(x) = 0$ and $ax \neq 0$. We show that $\mathfrak{p} = \mathrm{Ann}(ax)$. It is obvious that $\mathfrak{p} = \mathrm{Ann}(x) \subset \mathrm{Ann}(ax)$; we prove the converse inclusion. Let $b \in \mathrm{A}$ such that $bax = 0$, then $ba \in \mathfrak{p}$ and $a \notin \mathfrak{p}$, and so $b \in \mathfrak{p}$; hence $\mathrm{Ann}(ax) \subset \mathfrak{p}$.   □

3.6.7

Theorem *Let* A *be a Noetherian ring and* M *a finitely generated* A-*module. Then* $\mathrm{Ass}(\mathrm{M})$ *is finite.*

Proof We first prove the theorem when M is cyclic, i.e. $\mathrm{M} = \mathrm{A}/\mathrm{I}$ for some ideal I, by Noetherian induction on I (1.6.3). If I is prime then, for all $x \neq 0$ in M, the annihilator of x in A/I is the zero set and so the annihilator of x in A is I and $\mathrm{Ass}(\mathrm{M}) = \{\mathrm{I}\}$ contains a unique element, and so is finite. Suppose that I is not prime and that, for all ideals I' strictly containing I, the set $\mathrm{Ass}(\mathrm{A}/\mathrm{I}')$ is finite. Let $a, b \in \mathrm{A} - \mathrm{I}$ be such that $ab \in \mathrm{I}$; set $\mathrm{I}' = \mathrm{I} + \mathrm{A}a$ and let I'' be the ideal of $c \in \mathrm{A}$ such that $ac \in \mathrm{I}$. As $\mathrm{I} \subsetneq \mathrm{I}'$, $\mathrm{I} \subsetneq \mathrm{I}''$ since $b \in \mathrm{I}''$. There is an exact sequence $0 \to \mathrm{A}/\mathrm{I}'' \xrightarrow{u} \mathrm{A}/\mathrm{I} \xrightarrow{v} \mathrm{A}/\mathrm{I}' \to 0$, where u is multiplication by a. Therefore, $\mathrm{Ass}(\mathrm{A}/\mathrm{I})$ is finite by Proposition 3.6.6.

Next suppose that $x_1, ..., x_r$ generate M. The proof is by induction on r. Let M' be the submodule of M generated by $x_1, ..., x_{r-1}$. By the induction hypothesis, the set $\mathrm{Ass}(\mathrm{M}')$ is finite and $\mathrm{Ass}(\mathrm{M}/\mathrm{M}')$ is finite since M/M' is cyclic. Hence $\mathrm{Ass}(\mathrm{M})$ is finite by Proposition 3.6.6.   □

3.6.8 Remark

Let A be a Noetherian ring, M a finitely generated A-module, I the annihilator of M, i.e. $I = \bigcap_{x \in M} \text{Ann}(x)$. The inverse image of every minimal prime ideal of A/I is an element of $\text{Ass}(M)$.

Exercises 3.6. (Noetherian rings)

1.—Let U be a non-empty open subset of $\mathbb{C}$. The aim is to show that the ring $\mathcal{O}(U)$ of holomorphic functions on U is not Noetherian.

(a) Show that there is a holomorphic function on U with infinitely many simple roots. (Study the case where U contains $\mathbb{R}_+$ and is a neighbourhood of zero not containing $\mathbb{R}_+$.)

(b) Let $f \in \mathcal{O}(U)$ admit infinitely many simple roots a_n. Denote by I_n the set $h \in \mathcal{O}(U)$ such that $h(a_k) = 0$ for all $k \geqslant n$. Show that (I_n) is a strictly increasing sequence of ideals.

2.—For every open subset U of $\mathbb{C}^2$, let $\mathcal{O}(U)$ be the ring of holomorphic functions on U and for every compact subset K of $\mathbb{C}^2$, $\mathcal{O}(K) = \varinjlim \mathcal{O}(U)$ for open neighbourhoods U in K. Denote by D the closed unit disc of $\mathbb{C}$.

$\mathcal{O}(D \times D)$ may be shown to be Noetherian.[17] The aim is to show that there are convex compact subsets K_1 and K_2 in $\mathbb{C}$ such that $\mathcal{O}(K_1 \times K_2)$ is not Noetherian.

Let $K_1 \subset D$ be a convex compact subset whose intersection with the boundary of D consists of 1 and of the points $a_n = e^{\frac{1}{2n}i}$ and K_2 be the image of K_1 under conjugation $z \mapsto \bar{z}$. Set $K = K_1 \times K_2$.

(a) Describe the intersection of K with $H = \{(x, y) \in \mathbb{C}^2 \mid xy = 1\}$.

(b) Set

$$b_n = e^{i/2n+1}, \quad u_n(z) = -i \frac{z - b_n}{z + b_n}, \quad v_n(z) = i \frac{z - \bar{b}_n}{z + \bar{b}_n}.$$

Show that $u_n(z)$ (resp. $v_n(z)$) has an imaginary part $\geqslant 0$ (resp. $\leqslant 0$) for $z \in K_1$ (resp. $z \in K_2$).

Show that $u_n(a_i) = v_n(\bar{a}_i) \in \mathbb{R}$; what is the sign of this number?

(c) Define $h_n \in \mathcal{O}(K_1 \times K_2)$ by $h_n(x, y) = \frac{1}{2i\pi}(\log v_n(y) - \log u_n(x))$, log being chosen so that that the imaginary parts are in $[0, 2\pi]$. Calculate $h_n(a_k, \bar{a}_k)$. Conclude as in the previous exercise.

3.—Let A a commutative Banach algebra over $\mathbb{R}$. The aim is to show that, if A is a Noetherian ring, then it is finite dimensional as a vector space over $\mathbb{R}$.

Assume the following results:

BANACH THEOREM. *Let* E *and* F *be two Banach spaces,* $f : E \to F$ *a continuous bijective linear map. Then* f^{-1} *is continuous.*

GELFAND-MAZUR THEOREM. *Let* A *be a Banach algebra over the field* $\mathbb{R}$. *Then* A *is isomorphic to* $\mathbb{R}$ *or* $\mathbb{C}$ (4.4, Exercise 5).

(a) Let E and F be Banach spaces, $f : E \to F$ an injective closed morphism. Show that there exists $c > 0$ such that every $f' : E \to F$ with $\|f' - f\| \leqslant c$ is injective and

---

[17]Frisch [6].

closed. Show that if f is not surjective, then this is also the case for all f' sufficiently near f (extend f to a closed injective morphism $g : E \oplus \mathbb{R} \to F$).

(b) Let E and F be Banach spaces and $f : E \to F$ a surjective morphism. Show that any morphism sufficiently near f is surjective.

(c) Let A be a Banach algebra, E a Banach A-module and F a submodule of E. Show that if the closure $\overline{F}$ of F is a finitely generated A-module, then F is closed (find elements of F approaching the generators of $\overline{F}$; these elements generate $\overline{F}$). Deduce that if A is a Noetherian ring, then every ideal is closed.

(d) Show that every integral Noetherian Banach algebra is a field (consider the map $a \mapsto a\mathrm{I}$ from $A - \{0\}$ to $L(A, A)$ and show that its image is contained in the set of automorphisms of A regarded as a vector space by noting that $A - \{0\}$ is connected if $A \neq \mathbb{R}$).

(e) Prove the result when A is reduced (use Proposition 3.6.4).

(f) Let A be a Noetherian ring and $\mathfrak{N}$ its nilradical. Show that there exists r such that $\mathfrak{N}^r = 0$. Show that the quotients $\mathfrak{N}^k/\mathfrak{N}^{k+1}$ are finitely generated A/$\mathfrak{N}$-modules. Deduce that, if A is a $\mathbb{R}$-algebra such that A/$\mathfrak{N}$ is a finite dimensional vector space, then A is finite dimensional. Conclude.

4.—Let K be a complete ultrametric valued field (i.e. equipped with a map $x \mapsto |x|$ from K to $\mathbb{R}_+$ satisfying $|xy| = |x||y|, |1| = 1, |x + y| \leqslant \sup(|x|, |y|)$, and complete with respect to the distance $d(x, y) = |x - y|$). Denote by K{X} the subring of K[[X]] consisting of the formal series $\sum a_n X^n$ such that the sequence $(|a_n|)$ tends to 0.

(a) For $f = \sum a_n X^n \in$ K{X}, set $\|f\| = \sup(|a_n|)$. Show that this defines a norm on K{X} with respect to which this vector space is complete, and that multiplication is continuous.

The aim is to prove that K{X} is a PID.

(b) Let $d \in \mathbb{N}$, and E_d the vector space of polynomials of degree $< d$ with coefficients in K. Equip this space with the norm $\|f\| = \sup(|a_i|)$ for $f = \sum a_i X^i$, and the space $E_d \oplus$ K{X} with the norm defined by $\|(R, f)\| = \sup(\|R\|, \|f\|)$. Let $P \in$ K[X] be a monic polynomial of degree d. Show that the map $\varphi : (R, f) \mapsto Pf + R$ is an isometry from $E_d \oplus$ K{X} onto K{X} first show that φ is isometric, then that its image is dense, noting that the initial space is complete).

(c) Let $h = \sum a_n X^n$ be a nonzero element of K{X}, and d the largest integer i such that $|a_i| = \|h\|$. Show that $\psi : (R, f) \mapsto hf + R$ is an isomorphism from $E_d \oplus$ K{X} onto K{X} (reduce to the case $|a_d| = 1$, where $\psi = \varphi + u$ for some isometry φ and u with $\|u\| < 1$).

(d) Show that K{X} is Euclidean.

5.—Let E be a finite dimensional vector space over $\mathbb{C}$, and A the ring $\mathcal{O}_E$ of germs at the origin of holomorphic functions on E. Denote by m the maximal ideal of A consisting of f such that $f(0) = 0$. For any linear form u on E, let u^* be the homomorphism $f \mapsto f \circ u$ from $\mathbb{C}\{X\} = \mathcal{O}_\mathbb{C}$ to A, and for all A-modules M, u^*M the $\mathbb{C}\{X\}$-module obtained by restricting scalars from A to $\mathbb{C}\{X\}$ along u^*. Assume that the ring A is Noetherian.

Let M be a finitely generated A-module, N the submodule of M consisting of elements x such that $(\exists r)\ \mathfrak{m}^r x = 0$. Show that there is a linear form u on E such that the torsion submodule of u^*M is N (reduce to the case $N = 0$, and note that then for all $\mathfrak{p} \in \mathrm{Ass}(M)$, the set of $u \in E^\top$ in $\mathfrak{p}$ is a strict vector subspace, and that a finite union of such subspaces cannot be the whole of $E^\top$).

6. *(Fractional ideals)*—Let A be an integral domain and K its field of fractions. The *fractional ideal* of A (or by abuse of language K) is an A-submodule of K contained in a cyclic A-submodule of K.

(a) Let I be a fractional ideal of A and $f : I \to K$ an A-linear map. Show that f is of the form: $x \mapsto ax$, with $a \in K$. In particular f is injective or null, and its image is a fractional ideal.

(b) Show that all fractional ideals are finitely generated projective A-modules.

(c) Let I be a fractional ideal. It is said to be invertible if there is a fractional ideal I' such that $I \cdot I' = A$. Show that the following conditions are equivalent:

 (i) I is invertible;
 (ii) there are $x_1, ..., x_r \in I$ and $y_1, ..., y_r \in K$ such that $I \cdot y_1 \subset A, ..., I \cdot y_r \subset A$, and $x_1 y_1 + \cdots + x_r y_r = 1$;
 (iii) I is a projective A-module.

(d) Let I_1 and I_2 be two ideals of A. I_1 is said to divide I_2 if there is an ideal J of A such that $I_2 = J \cdot I_1$. Show that, if I_1 divides I_2, then $I_1 \supset I_2$. Give an example for which the converse does not hold. Show that, if I_1 is invertible, I_1 divides I_2 if and only if $I_1 \supset I_2$.

7. *(Dedekind rings)*—This exercise follows on the previous one.

A **Dedekind ring** is an integral domain A in which every ideal is a projective A-module.

(a) Show that all Dedekind rings are Noetherian.

(b) Let A be a Dedekind ring. Show that the multiplicative monoid of nonzero ideals in A is a free commutative monoid.

(c) Show that in a Dedekind ring, every nonzero prime ideal is maximal.

(d) Show that all PIDs are Dedekind. The ring of analytic functions of period 2π on $\mathbb{R}$ (see 3.1, Exercise 12) is Dedekind, but not principal. All Dedekind UFDs are PIDs (3.2, Exercise 6).

(e) Let A be a Dedekind ring. For any multiplicatively closed subset S in A, the ring of fractions $S^{-1}A$ is Dedekind.

An integrally closed Noetherian ring in which every nonzero prime ideal is maximal can be shown to be a Dedekind ring.

3.7 Polynomial Algebras

Throughout this section, A denotes a (commutative unital) ring.

3.7.1 Monoid Algebra

Let M be a multiplicative monoid. Consider the free A-module $A^{(M)}$ on a set M, and let $(e_m)_{m \in M}$ be its canonical basis.

Proposition and Definition *There is a unique bilinear multiplication on* $A^{(M)}$ *such that* $(\forall (m, m') \in M^2)$ $e_m \cdot e_{m'} = e_{mm'}$. *With this multiplication,* $A^{(M)}$ *is a unital associative commutative algebra if* M *is commutative. It is the* **monoid algebra** M.

Proof Such a multiplication is necessarily given by:

$$\left(\sum a_m e_m \right) \cdot \left(\sum b_{m'} e_{m'} \right) = \sum a_m b_{m'} e_{mm'} \, ,$$

i.e. $(a_m)_{m \in M} \cdot (b_{m'})_{m' \in M} = (c_{m''})_{m'' \in M}$, where $c_{m''} = \sum_{mm'=m''} a_m b_{m'}$. The multiplication is defined by this formula is easily seen to satisfy the required conditions.

□

UNIVERSAL PROPERTY. *For all unital associative A-algebras* B *and all homomorphisms* f *from* M *to the multiplicative monoid underlying* B, *there is a unique unital algebra homomorphism:* $\bar{f} : A^{(M)} \to B$ *such that* $(\forall m \in M)$ $\bar{f}(e_m) = f(m)$.

Proof It suffices to check that the linear map $\bar{f} : A^{(M)} \to B$ defined by $\bar{f}(e_m) = f(m)$ (universal property 3.4.6) is a multiplicative homomorphism. This follows from an immediate computation.

Remark The universal property can also be stated as follows: the functor from the category of monoids (resp. commutative monoids) to the category of unital associative A-algebras (resp. unital associative commutative) assigning to a monoid M its algebra $A^{(M)}$ is the left adjoint of the forgetful functor assigning to an algebra its underlying multiplicative monoid.

3.7.2 Polynomial Algebras

Definition Let S be a set and $L = \mathbb{N}^{(S)}$ the free commutative monoid on S (3.2.4). The **polynomial algebra** on S with coefficients in A, written A[S], is the algebra $A^{(L)}$ of the monoid L.

For $s \in S$, set e_s to be the basis element of L corresponding to s and ε_s the element of the canonical basis for $A[S] = A^{(L)}$ corresponding to e_s.
UNIVERSAL PROPERTY. *For any unital associative commutative* A-*algebra* B *and any family* $(x_s)_{s \in S}$ *of elements in* B, *there is a unique unital algebra homomorphism* f *from* A[S] *to* B *satisfying* $(\forall s \in S)$ $f(\varepsilon_s) = x_s$.
In other words, the functor $S \mapsto A[S]$ from $\mathscr{E}ns$ to the category $\mathscr{A}$ of unital associative commutative A-algebras is the left adjoint of the forgetful functor from $\mathscr{A}$ to $\mathscr{E}ns$.

Proof Let $\mathcal{M}$ be the category of commutative monoids. The forgetful functor ω from $\mathcal{A}$ to $\mathcal{E}ns$ is the composite of the forgetful functor $\omega_1 : \mathcal{A} \to \mathcal{M}$ assigning to an algebra its underlying multiplicative monoid, and of the forgetful functor $\omega_2 : \mathcal{M} \to \mathcal{E}ns$. The left adjoint of the functor ω_2 is the functor $\lambda_2 : S \mapsto \mathbb{N}^{(S)}$, and that of ω_1 is $\lambda_1 : M \mapsto A^{(M)}$; so the left adjoint of $\omega = \omega_2 \circ \omega_1$ is $\lambda_1 \circ \lambda_2$.   □

3.7.3   Notation

The elements of A[S] will be denoted in one of the following manner:

1st method. For $s \in S$, identify s with ε_s. The canonical basis element of A[S] corresponding to $(n_s)_{s\in S}$ is then the *monomial* $\prod_{s\in S} s^{n_s}$ and all elements of A[S] are A-linear combinations of finitely many monomials.

If, for example, S = {X, Y, Z}, then write A[X, Y, Z] for A[S], and so for instance $aX^3 + bY^2Z^4 \in$ A[X, Y, Z] if $a, b \in$ A.

2nd method. For $s \in S$, let X_s be the element ε_s and denote A[S] by $A[(X_s)_{s\in S}]$.

Then, for example, $aX_1^3 + bX_2^2X_3^4 \in A[(X_s)_{s\in S}]$. For $n = (n_s)_{s\in S} \in \mathbb{N}^{(S)}$, the canonical basis element of A[S] corresponding to n is often written X^n. Then, $X^n = \prod_{s\in S} X_s^{n_s}$, and all elements of $A[(X_s)_{s\in S}] = A[S]$ can be uniquely written as $\sum_{n\in\mathbb{N}^{(S)}} a_n X^n$.

The second method becomes necessary to avoid confusion when S is a subset of $\mathbb{N}$, or of a set equipped with a composition law.

Let B be a unital associative commutative A-algebra and $(x_i)_{i\in I}$ a family of elements of B. Let $A[(x_i)]_B$ denote the unital subalgebra of B generated by (x_i): it is the image of $A[(X_i)_{i\in I}]$ under the homomorphism $f : A[(X_i)] \to$ B defined by $f(X_i) = x_i$. Similarly, if $x, y, z \in$ B, let $A[x, y, z]_B$ denote the unital subalgebra of B generated by x, y and z; it is the image of A[X, Y, Z] under the homomorphism $f : A[X, Y, Z] \to$ B defined by $f(X) = x$, $f(Y) = y$, $f(Z) = z$. When there is no possible confusion, write $A[(x_i)]$ or $A[x, y, z]$ instead of $A[(x_i)]_B$ or $A[x, y, z]_B$, even when f is not injective, so that the polynomial algebra over I or over the set {x, y, z} cannot be identified with its image in B.

For example $\mathbb{Z}[\sqrt{3}]$ can denote the unital subalgebra of $\mathbb{R}$ generated by $\sqrt{3}$, which is identified with $\mathbb{Z}[X]/(X^2 - 3)$. To avoid confusion, the following convention is followed: When the intention is to consider the polynomial algebra over S, the elements of S are called "indeterminates" and denoted by capital letters; when S is a subset of an algebra B and A[S] denotes the subalgebra of B generated by S, the elements of S are denoted by small Latin or Greek letters, or else by the expressions they are determined by in B.

A final point: if M is a monoid, algebraists often write A[M] for the monoid algebra $A^{(M)}$. For example, if G is a group, $\mathbb{Z}[G]$ generally denotes the group algebra $\mathbb{Z}^{(G)}$.

3.7.4

Proposition *(a) If* S′ *is a subset of* S, *the algebra* A[S′] *is a subalgebra of* A[S].

(b) If S = S′ ∪ S″, *with* S′ ∩ S″ = ∅, *then the algebra* A[S] *is the same as* (A[S′])[S″].

(c) If S = $\varinjlim$ S$_i$, *then* A[S] = $\varinjlim$ A[S$_i$].

(d) Let I *be an ideal of* A *and* J *the ideal of* A[S] *generated by* I. *Then the algebra* (A/I)[S] *can be identified with* (A[S])/J.

Proof (a) $\mathbb{N}^{(S')} \subset \mathbb{N}^{(S)}$, and (a) follows.

(b) $\mathbb{N}^{(S)} = \mathbb{N}^{(S')} \times \mathbb{N}^{(S'')}$. Every element of A[S] can be uniquely written as

$$\sum_{n \in \mathbb{N}^{(S)}} a_n X^n = \sum_{(n',n'') \in \mathbb{N}^{(S')} \times \mathbb{N}^{(S'')}} a_{n',n''} X^{n'} X^{n''}$$

$$= \sum_{n'' \in \mathbb{N}^{(S'')}} \left(\sum_{n' \in \mathbb{N}^{(S')}} a_{n',n''} X^{n'} \right) X^{n''},$$

where $(a_n)_{n \in \mathbb{N}^{(S)}}$ is a family with finite support in A, and so uniquely as $\sum_{n'' \in \mathbb{N}^{(S'')}}$ $b_{n''} X^{n''}$, where $(b_{n''})_{n'' \in \mathbb{N}^{(S'')}}$ is a family with finite support in A[S′].

(c) follows from the universal property and from (2.7.5, Corollary).

(d) For any A-algebra B, set F(B) = B^S if IB = 0 and F(B) = ∅ if IB ≠ 0. The algebras (A/I)[S] and (A[S])/J represent the functor F.   □

3.7.5

Proposition *Suppose that* A *is integral. For all sets* S, *the ring* A[S] *is integral.*

Proof (a) *First suppose that* S *has a unique element.* Let P and P′ be two nonzero polynomials of respective degrees d and d′, and leading terms aX^d and a′X$^{d'}$. The leading term of PP′ is aa′X$^{d+d'}$; as it is not zero, PP′ ≠ 0.

(b) *Next suppose that* S *is finite:* the proof follows by induction from the previous case since A[X$_1$, ..., X$_n$] = A[X$_1$, ..., X$_{n-1}$][X$_n$].

(c) *General case:* it follows from the previous one, together with (3.7.4, c) and (3.1, Exercise 4).   □

3.7.6 Degree of Polynomials

For $n = (n_s)_{s \in S} \in \mathbb{N}^{(S)}$, set $|n| = \sum_{s \in S} n_s$.

Let P = $\sum a_n X^n \in$ A[S]. The *degree* of P, written d(P) is the upper bound (in $\overline{\mathbb{Z}}$) of $|n|$ for n such that $a_n \neq 0$. The polynomial P is said to be *homogeneous of degree* d if $a_n = 0$ for all n such that $|n| \neq d$.

Remark The degree of the polynomial 0 is $-\infty$, although 0 is homogeneous of degree d for all d. For any nonzero polynomial P, $d(P) \in \mathbb{N}$.

Polynomials of degree $\leqslant d$ (resp. homogeneous of degree d) form a free A-module. If S is a finite set with m elements, the dimension of this module is the binomial coefficient $\binom{m+d}{d}$ (resp. $\binom{m+d-1}{d}$).

Indeed $(n_1, \ldots, n_m) \mapsto (n_1 + 1, n_1 + n_2 + 2, \ldots, n_1 + \cdots + n_m + m)$ is a bijection from the set of $n \in \mathbb{N}^m$ such that $|n| \leqslant d$ (resp. $|n| = d$) onto the set of strictly increasing sequences with m elements in $\{1, \ldots, d + m\}$ (resp. whose last element is $d + m$).

If P and Q are homogeneous of respective degrees p and q, then PQ is homogeneous of degree $p + q$.

Proposition *Let* P, Q $\in$ A[S]. *Then* $d(PQ) \leqslant d(P) + d(Q)$. *If* A *is an integral domain, then* $d(PQ) = d(P) + d(Q)$.

Proof The inequality is immediate. We prove the equality when A is integral. We may assume that P and Q are nonzero. Write $P = P_0 + P_1$, where P_0 is homogeneous of degree $d(P)$ and P_1 of degree $< d(P)$,[18] and likewise write Q as $Q_0 + Q_1$. Then $PQ = P_0Q_0 + R$, where $R = P_0Q_1 + P_1Q_0 + P_1Q_1$ and P_0Q_0 are homogeneous of nonzero degree $d(P) + d(Q)$ by Proposition 3.7.5 and R is of degree $< d(P) + d(Q)$. Hence the degree of PQ is $d(P) + d(Q)$. $\square$

3.7.7

Theorem (Hilbert[19]) *If* A *is a Noetherian ring, then so is* A[X].

Proof Let I be an ideal of A[X]; we show that I is finitely generated. For any $P \in$ A[X] with $P \neq 0$, let $d(P)$ be the degree of P and $\alpha(P)$ the leading coefficient of P (i.e. the coefficient of $X^{d(P)}$).

Let J be the subset of A consisting of 0 and of $\alpha(P)$ for $P \in I - \{0\}$. The set J is an ideal in A. Indeed, let $a = \alpha(P)$ and $b = \alpha(Q)$ be elements of J, and suppose that $d(P) \leqslant d(Q)$; if $a + b \neq 0$, then $a + b = \alpha(X^{d(Q)-d(P)}P + Q) \in J$; for $\lambda \in A$, if $\lambda a \neq 0$, then $\lambda a = \alpha(\lambda P)$.

As A is Noetherian, J is finitely generated. Let $a_1, \ldots, a_k$ be its generators and $P_1, \ldots, P_k$ polynomials such that $a_i = \alpha(P_i)$. All P_i may be assumed to be of degree d to within multiplication by powers of X. Let L be the submodule of A[X] consisting of polynomials of degree $< d$ and set $M = L \cap I$. The module L is isomorphic to A^d, and so M is finitely generated; let $(R_1, \ldots, R_q)$ generate M as an A-module.

---

[18]P_1 not necessarily homogeneous.

[19]This theorem is known as the finite basis theorem (the term basis being at that time used for generating family).

We show that the polynomials $P_1, \ldots, P_k, R_1, \ldots, R_q$ generate the ideal I. We do this by showing by induction on $d(P)$ that all polynomials $P \in I$ are linear combinations of these polynomials, with coefficients in A[X]. If $d(P) < d$, the polynomial P is the A-linear combination of $R_1, \ldots, R_q$.

If $d(P) \geqslant d$, the scalar $\alpha(P)$ is of the form $\sum \lambda_i a_i$, and $P - X^{d(P)-d} \sum \lambda_i a_i$ is of degree $< d(P)$, and so ... $\square$

Corollary 3.4 *If* A *is a Noetherian ring then, for all n, so is* $A[X_1, \ldots, X_n]$.

This follows from the theorem by induction on n since

$$A[X_1, \ldots, X_n] = A[X_1, \ldots, X_{n-1}][X_n].$$

Corollary 3.5 *If* A *is a Noetherian ring, then so are all unital associative commutative* A-*algebras.*

Indeed, such an algebra is isomorphic to a quotient of $A[X_1, \ldots, X_n]$ for some n.

Corollary 3.6 *Every ring is the union of an increasing directed family of Noetherian subrings.*

Indeed, the subring generated by a finite subset of A is a finitely generated $\mathbb{Z}$-algebra, and so is Noetherian.

3.7.8 Polynomials with Coefficients in a UFD

Throughout this subsection, A is assumed to be a UFD, K denotes the field of fractions of A and for irreducible element $m \in A$, χ_m the canonical homomorphism from A[X] to $(A/m)[X]$.

Definition A polynomial $P = \sum a_i X^i \in A[X]$ is said to be **primitive** if 1 is a g.c.d. of the a_i.

P is primitive if and only if for all irreducible elements $m \in A$, $\chi_m(P) \neq 0$.

Lemma 3.1 *Every nonzero polynomial* $P \in K[X]$ *can be written* $P = \lambda P_1$ *for some primitive polynomial* $P_1 \in A[X]$ *and some nonzero* $\lambda \in K$.

Proof $P = \frac{1}{d} P_0$, for someù $d \in A$ with $d \neq 0$ and some $P_0 \in A[X]$. Let a be the g.c.d. of the coefficients of P_0. Then $P_0 = a P_1$, where $P_1 \in A[X]$ is primitive, and $P = \lambda P_1$ with $\lambda = \frac{a}{d}$.

Lemma 3.2 *Let* P *and* Q *be polynomials in* A[X]. *Assume* P *is primitive and that* P *divides* Q *in* K[X]. *Then* P *divides* Q *in* A[X].

Proof Suppose that $Q \neq 0$, and $Q = PR$ with $R \in K[X]$. There exist $a, b \in A$ with g.c.d.$(a, b) = 1$ such that $R = \frac{a}{b}R_1$ for some primitive $R_1 \in A[X]$. Then $bQ = aPR_1$. If $m \in A$ is irreducible element dividing b, then $\chi_m(a) \neq 0$, $\chi_m(P) \neq 0$ and $\chi_m(R_1) \neq 0$; the ring A/m is an integral domain (3.2.8, (ii)), and hence so is $(A/m)[X]$ (3.7.5), and $\chi_m(bQ) = \chi_m(aPR_1) \neq 0$, which contradicts $\chi_m(b) = 0$. Hence b is invertible in A. $\qquad\qquad\square$

Lemma 3.3 *Let* $P \in A[X]$ *be a nonzero polynomial.*

(a) *If* P *is of degree* 0, *then* P *is irreducible in* $A[X]$ *if and only if it is irreducible in* A.

(b) *If* P *is of degree* > 0, P *is irreducible in* $A[X]$ *if and only if it is primitive and irreducible in* $K[X]$.

Proof (a) If $P = QR$, with Q and R non-invertible elements of $A[X]$, the polynomials Q and R are of degree 0, and so are non-invertible elements of A. If $P = ab$, where $a, b \in A$ are non-invertible, then a, b remain non-invertible in $A[X]$.

(b) Suppose that P is irreducible in $A[X]$. Then P is primitive for otherwise $P = aP_1$, with $a \in A$ non-invertible in A and so non-invertible in $A[X]$, and P_1 of degree > 0, hence non-invertible in $A[X]$. We show that P is irreducible in $K[X]$. If $P = QR$ for non-invertible $Q, R \in K[X]$, and so of degree > 0 then, by Lemma 3.1, Q may be assumed to be a primitive polynomial of $A[X]$, and thus $R \in A[X]$ by Lemma 3.2.

Suppose that P is primitive and irreducible in $K[X]$. Then there are no $Q, R \in A[X]$ of degree > 0 such that $P = QR$ since Q and R would then be non-invertible in $K[X]$. There are also no non-invertible Q of degree 0 in $A[X]$ such that $P = QR$ since Q would then be non-invertible element in A and P would not be primitive. Therefore P is irreducible in $A[X]$. $\qquad\qquad\square$

3.7.9

***

Theorem *If* A *is a UFD then so is* $A[S]$ *for all sets* S.

***

Proof We keep the notation of 3.2.7. By 3.2.8, it amounts to showing that $A[S]$ satisfies the following two conditions:

(i) every decreasing sequence of elements of $\mathrm{Mon}^*(A[S])$ is stationary;
(ii) for every irreducible element $P \in A[S]$, the ideal generated by P is prime.

(a) *Condition* (i). Let $(P_k)_{k \in \mathbb{N}}$ be a sequence of elements of $A[S]$, with $P_k = \sum a_{k,n}X^n$ such that $(\chi(P_k))_{k \in \mathbb{N}}$ is a decreasing sequence of elements of $\mathrm{Mon}^*(A[S])$. The degrees $d(P_k)$ form a decreasing sequence of elements of $\mathbb{N}$, hence constant from some rank k_0. For $k \geqslant k_0$, $P_k = c_k P_{k+1}$, where $c_k \in A^*$. Let $n \in \mathbb{N}^{(S)}$ be such that $a_{k_0,n} \neq 0$. For $k > k_0$, $a_{k,n} = c_k a_{k+1,n}$ and $(\chi(a_{k,n}))_{k \geqslant k_0}$ is a decreasing sequence of elements of Mon* A. This sequence is therefore stationary, and all c_k are invertible beyond a certain rank, proving (i).

(b) *Condition / (ii). Suppose first that* S *has only one element* X. Let P be an irreducible element of A[X] and K the field of fractions of A. By Lemma 3.3, P is primitive, and by Lemma 3.2, $P \cdot A[X] = A[X] \cap P \cdot K[X]$. So $A[X]/P \cdot A[X]$ can be identified with a subring of $K[X]/P \cdot K[X]$. By Lemma 3.3, P is irreducible in K[X]; as K[X] is a PID, and thus a UFD, $K[X]/P \cdot K[X]$ is an integral domain, and hence so is $A[X]/P \cdot A[X]$.

(c) *Suppose next that* S *is finite.* The theorem follows by induction from the previous case.

(d) *General case.* Let P be an irreducible element of A[S]. There is a finite subset S' of S such that $P \in A[S']$. Then P is irreducible in A[S'], and so $A[S']/P \cdot A[S']$ is an integral domain. Set . Then so is $A[S]/P \cdot A[S] = (A[S']/P \cdot A[S'])[S'']$, where $S'' = S - S'$, since $A[S] = (A[S'])[S'']$ (3.7.5).   □

3.7.10 Substitution, Polynomial Functions

Let B be a unital associative commutative A-algebra, $(x_i)_{i \in I}$, a family of elements in B and $f : A[(X_i)_{i \in I}] \to B$ the homomorphism defined by $f(X_i) = x_i$.

The image of f is the unital subalgebra of B generated by $(x_i)_{i \in I}$. If f is injective, then the family $(x_i)_{i \in I}$ is said to be *algebraically free,* or that the elements x_i are *algebraically independent.* We say that $x \in B$ is **transcendental** over A if the family consisting of x is algebraically free over A. In general, the kernel of f is an ideal of $A[(X_i)_{i \in I}]$ called the *ideal of algebraic relations between the* x_i.

For $P \in A[(X_i)_{i \in I}]$, $f(P)$ is the element of B obtained by *substituting* the x_i for the indeterminates X_i in P, and it is written $P_B((x_i)_{i \in I})$ or simply $P((x_i)_{i \in I})$.[20]

The map P_B from B^I to B thus defined is the *polynomial function* defined by P on B^I. The map $P \mapsto P_B$ is a unital A-algebra homomorphism from $A[(X_i)]$ to the algebra of maps from B^I to B; it associates to X_i the projection of index i from B^I to B.

Proposition *If* B *an infinite integral domain and the homomorphism* $i : A \to B$ *defining the algebra structure of* B *is injective, then the homomorphism* $P \mapsto P_B$ *is injective.*

Proof (a) *Suppose that* I *has only one element.* Let K be the field of fractions of B and let $P \in A[X] \subset K[X]$. If P is of degree d, there are at most d roots[21] in K; hence $P(x) = 0$ for all $x \in B$ only if $P = 0$.

(b) *Suppose next that* I *is finite.* Assume $I = \{1, ..., n\}$ and use induction on n. Let $P = \sum a_k X_n^k \in A[X_1, ..., X_{n-1}][X_n] = A[X_1, ..., X_n]$, where $a_k \in A[X_1, ..., X_{n-1}]$,

---

[20]If $B = A[(X_i)_{i \in I}]$ and $\forall i$, $x_i = X_i$, then $f = 1_{A[(X_i)]}$; so $P((X_i)) = P$.

[21]Queysanne [1], § 192, Corollary 1, p. 413, Lelong-Ferrand and Arnaudiès [2], theorem 4.5.2, p. 139.

and assume the map $P_B : B^n \to B$ is zero. Let $(x_1, ..., x_{n-1}) \in B^{n-1}$, and set $Q = P(x_1, ..., x_{n-1}, X_n) = \sum a_k(x_1, ..., x_{n-1})X_n^k \in B[X_n]$. By assumption, $(\forall x_n \in B)$ $Q(x_n) = P(x_1, ..., x_{n-1}, x_n) = 0$, and so $Q = 0$ by (a), and $(\forall k) a_k(x_1, ..., x_{n-1}) = 0$. As this holds for all $x_1, ..., x_{n-1} \in B$, $(\forall k) a_k = 0$ by the induction hypothesis. Thus $P = 0$.

(c) *General case.* Let $(0 \neq) P \in A[(X_i)_{i \in I}]$. There is a finite subset J of I such that $P \in A[(X_i)_{i \in J}]$. Then the map $P_B : B^I \to B$ is the composite of the surjective projection $B^I \to B^J$ and of the nonzero map $P_B : B^J \to B$, and so in not zero. $\square$

Remarks (1) If A is an integral domain, then we may take $B = A$ in the proposition. For all integral domains A, there is an infinite integral domain B, for example $A[X]$, with a subring with which A can be identified.

(2) The assumption of the infinity of B cannot be removed, nor that of the integrality of B (see Exercise 15).

(3) A map φ from A^p to A^q is said to be polynomial if the maps $\varphi_1, ..., \varphi_q$ from A^p to A such that $\varphi = (\varphi_1, ..., \varphi_q)$ are polynomials. Let M and N be finitely generated free A-modules. A map φ from M to N is said to be polynomial with respect to the bases $e = (e_1, ..., e_p)$ of M and $f = (f_1, ..., f_q)$ of N if the *expression* of φ with respect to the bases e and f (i.e. the map from A^p to A^q induced by φ by identifying A^p and A^q with M and N via e and f respectively) is polynomial. If a map φ from M to N is polynomial with respect to one pair of bases, it remains so with respect to any pair of bases. Hence it may be called a *polynomial map* from M to N.

The composite of two polynomial maps is polynomial.

If $A = \mathbb{R}$ or $\mathbb{C}$, equip M with the topology obtained by transferring that of A^p via a basis (this topology, called the canonical *topology* is independent of the choice of basis; It is the only Hausdorff topology compatible with the vector space structure); do the same for N. Every polynomial map from M to N is continuous (and even C^∞, or better still: analytic). Two polynomial maps from M to N coinciding on a nonempty open subset of M are equal: this follows from the Taylor formula for polynomials. In particular, if φ is a nonzero polynomial map from M to N, then $\varphi^{-1}(0)$ is closed and with empty interior in M.

3.7.11 Resultants

For $p \in \mathbb{N}$, denote by $A[X]_p$ the submodule of $A[X]$ consisting of polynomials of degree $\leqslant p$. It is a free module of rank $p + 1$ as $(1, X, ..., X^p)$ is a basis.

Definition Let $P \in A[X]_p$ and $Q \in A[X]_q$. The **resultant** of P and Q for the degrees p and q, written $\mathrm{Res}_{p,q}(P, Q)$, is the determinant of the matrix representing the map $(U, V) \mapsto U.P + V.Q$ from $A[X]_{q-1} \oplus A[X]_{p-1}$ to $A[X]_{p+q-1}$, with respect to bases $((1, 0), ..., (X^{q-1}, 0), (0, 1), ..., (0, X^{p-1}))$ and $(1, ..., X^{p+q-1})$.

If p and q are the degrees of P and Q respectively, we simply write $\mathrm{Res}(P, Q)$, and call this the resultant of P and Q.

For $P = a_p X^p + \cdots + a_0$ and $Q = b_q X^q + \cdots + b_0$, the resultant $\operatorname{Res}_{p,q}(P, Q)$ is therefore the determinant of the Sylvester matrix:

$$
\begin{pmatrix}
a_0 & 0 & \cdots & 0 & b_0 & 0 & \cdots & & 0 \\
a_1 & a_0 & \ddots & 0 & b_1 & b_0 & \ddots & & \vdots \\
\vdots & & \ddots & 0 & \vdots & & \ddots & & 0 \\
 & & & a_0 & & & & & \\
 & & & & b_q & & & & b_0 \\
a_p & & & & 0 & b_q & & & \\
0 & \ddots & & & \vdots & & \ddots & \ddots & \vdots \\
\vdots & \ddots & \ddots & & \vdots & & & \ddots & \\
0 & \cdots & 0 & a_p & 0 & \cdots & \cdots & 0 & b_q
\end{pmatrix}
$$

It is in A, and $\operatorname{Res}_{q,p}(Q, P) = -1^{pq} \cdot \operatorname{Res}_{p,q}(P, Q)$.

If $\varphi : A \to A'$ is a ring homomorphism, defining $\varphi_* : A[X] \to A'[X]$ by $(\varphi_*)_{|A} = \varphi$ and $\varphi_*(X) = X$ gives

$$
\operatorname{Res}_{p,q}(\varphi_*(P), \varphi_*(Q)) = \varphi(\operatorname{Res}_{p,q}(P, Q)) .
$$

Here, two elements a and b of A will be said to be relatively prime if the ideal they generate is the whole of A (even if A is not principal).

Proposition *Let* $P, Q \in A[X]$ *be polynomials of respective degrees p and q with relatively prime leading coefficients* $a_p, b_q \in A$. *Then, P and Q are relatively prime in A[X] if and only if the resultant* $\operatorname{Res}(P, Q)$ *is invertible in A.*

Proof (a) Suppose that $\operatorname{Res}(P, Q)$ is invertible. The map $(U, V) \mapsto U.P + V.Q$ from $A[X]_{q-1} \oplus A[X]_{p-1}$ to $A[X]_{p+q-1}$ is bijective. In particular there exist U and V such that $U.P + V.Q = 1$.

(b) Conversely, suppose that P and Q are relatively prime. Let $R \in A[X]_{p+q-1}$, and h be minimal of the form $R = U.P + V.Q$ with $U \in A[X]_{h-p}$ and $V \in A[X]_{h-q}$; we show that $h < p + q$.

Assume $h \geqslant p + q$; let α and β be the coefficients of degree $h - p$ and $h - q$ of U and V respectively. Since $R = U.P + V.Q$ is of degree $< h$, $\alpha.a_p + \beta.b_q = 0$. As a_p and b_q are relatively prime, $\mu.a_p + \nu.b_q = 1$. It follows that $\alpha = \lambda.b_q$ and $\beta = -\lambda.a_p$ with $\lambda = \alpha.\nu - \beta.\mu$. Set $\tilde{U} = U - \lambda.Q.X^d$ and $\tilde{V} = V + \lambda.P.X^d$, where $d = h - (p + q)$. Then $\tilde{U}.P + \tilde{V}.Q = U.P + V.Q = R$, with $d^o(\tilde{U}) < h - p$ and $d^o(\tilde{V}) < h - q$, contradicting the minimality of h. $\square$

3.7.12 Discriminants

Let $P = a_0 + a_1 X + \cdots + a_p X^p \in A[X]$, and consider its derivative $P' = a_1 + \cdots + p a_p X^{p-1}$. The resultant $\mathrm{Res}_{p,p-1}$ can be written $a_p . \Delta$, where Δ is the determinant of

$$
\begin{pmatrix}
a_0 & \cdots & 0 & a_1 & 0 & \cdots & 0 \\
 & \ddots & \vdots & 2a_2 & \ddots & \ddots & \vdots \\
\vdots & & a_0 & \vdots & & \ddots & 0 \\
 & & & p a_p & & & a_1 \\
a_p & & \vdots & 0 & \ddots & & \vdots \\
\vdots & \ddots & & \vdots & \ddots & p a_p & \\
0 & \cdots & 1 & 0 & \cdots & 0 & p
\end{pmatrix}
$$

This determinant is called the *discriminant* of P and is written $\mathrm{discr}(P)$.

Examples For a second degree polynomial $P = aX^2 + bX + c$,

$$
\mathrm{discr}(P) = \det \begin{pmatrix} c & b & 0 \\ b & 2a & b \\ 1 & 0 & 2 \end{pmatrix} = 4ac - b^2 .
$$

For $P = X^3 + pX + q$,

$$
\mathrm{discr}(P) = \det \begin{pmatrix} q & 0 & p & 0 & 0 \\ p & q & 0 & p & 0 \\ 0 & p & 3 & 0 & p \\ 1 & 0 & 0 & 3 & 0 \\ 0 & 1 & 0 & 0 & 3 \end{pmatrix} = \det \begin{pmatrix} p & -3q & 0 \\ 0 & -2p & -3q \\ 3 & 0 & -2p \end{pmatrix} = 4p^3 + 27q^2 .
$$

3.7.13 Algebraic Sets

Let K be a field and $(P_\lambda)_{\lambda \in \Lambda}$ a family of polynomials in $K[X_1, \ldots, X_n]$. Denote by $V((P_\lambda))$ the set of $x = (x_1, \ldots, x_n) \in K^n$ such that $(\forall \lambda \in \Lambda)\, P_\lambda(x) = 0$. The subsets of K^n thus obtained are called **algebraic sets**. The set $V((P_\lambda))$ is defined by equations P_λ, and does not depend on the ideal I generated by P_λ. It will also be written $V(I)$. Since the ring $K[X_1, \ldots, X_n]$ is Noetherian, every algebraic set can be defined by finitely many equations. Let $\sqrt{I}$ be the ideal consisting of elements $P \in K[X_1, \ldots, X_n]$ with some power in I. It is also the inverse image in $K[X_1, \ldots, X_n]$ of the nilradical of $K[X_1, \ldots, X_n]/I$. Hence it is the intersection of prime ideals containing I. The following hold:

(1) $V(\sqrt{I}) = V(I)$.

Let I and J be ideals in $K[X_1, ..., X_n]$, then

(2) $I \supset J \Rightarrow V(I) \subset V(J)$;

(3) $V(I + J) = V(I) \cap V(J)$;

(4) $V(I \cap J) = V(I \cdot J) = V(I) \cup V(J)$.

If (I_λ) is a family of ideals, then

(5) $V(\sum I_\lambda) = \bigcap_\lambda V(I_\lambda)$.

Thanks to formulas (4) and (5), there is a unique topology on K^n with respect to which the closed subsets are the algebraic sets. It is called the **Zariski topology**. It is Hausdorff (except when K is finite or $n = 0$), but all points are closed.

If K is algebraically closed, then it may be shown that $V(I) = V(J)$ if and only if $\sqrt{I} = \sqrt{J}$. This result is known as the *Hilbert's Nullstellensatz* (Exercises 13, 14).

Let k be a subfield of K. A subset of K^n is a k-**algebraic set** in K^n if it is defined by equations in $k[X_1, ..., X_n]$.

3.7.14   Principle of Extension of Identities

SCHOLIUM. Let $\mathscr{R}$ be a relation involving a ring (resp. a unital associative commutative $\mathbb{Q}$-algebra), stable under taking subrings (resp. subalgebras), quotients and directed unions. If $\mathscr{R}$ holds for $\mathbb{R}$ or $\mathbb{C}$, then it holds for all rings (resp. all $\mathbb{Q}$-algebras).

More precisely:

Proposition *Let $\mathscr{A}$ be the category of commutative rings (resp. commutative, associative, unital $\mathbb{Q}$-algebras), and $\mathscr{A}'$ a complete subcategory of $\mathscr{A}$. Suppose that*

(i) *if A and B are objects of $\mathscr{A}$ for which there exists an injective homomorphism $A \to B$, then $B \in \mathscr{A}' \Rightarrow A \in \mathscr{A}'$;*

(ii) *if A and B are objects of $\mathscr{A}$ for which there exists a surjective homomorphism $A \to B$, then $A \in \mathscr{A}' \Rightarrow B \in \mathscr{A}'$;*

(iii) *if an object A de $\mathscr{A}$ is a directed union of objets of $\mathscr{A}'$, then A is an object of $\mathscr{A}'$.*

Then, $\mathscr{A}' = \mathscr{A}$ if $\mathbb{R} \in \mathscr{A}'$, or if $\mathbb{C} \in \mathscr{A}'$.

Lemma *There is an algebraically free sequence $(x_n)_{n \in \mathbb{N}}$ in $\mathbb{R}$ over $\mathbb{Q}$.*

Proof Let us construct such a sequence by induction. Let $(x_0, ..., x_{n-1})$ be a finite sequence in $\mathbb{R}$ algebraically free over $\mathbb{Q}$, and set $A_{n-1} = \mathbb{Q}[x_0, ..., x_{n-1}]$. The set A_{n-1} is countable. Let $x_n \in \mathbb{R}$. Then, $(x_0, ..., x_n)$ is algebraically free over $\mathbb{Q}$ if and only if x_n is transcendental over A_{n-1}. Elements of $\mathbb{R}$ that are not transcendental over A_{n-1} are algebraic over A_{n-1} i.e. roots of a nonzero polynomial $P \in A_{n-1}[X]$. The set $A_{n-1}[X] - \{0\}$ is countable and every polynomial in it has only finitely many roots. So the set of algebraic elements over A_{n-1} is a countable subset of $\mathbb{R}$. As $\mathbb{R}$ is not countable, there is a transcendental real x_n over A_{n-1}, enabling us to continue the construction by induction. $\square$

Proof of the Proposition For all n, there are injective homomorphisms $\mathbb{Z}[X_1, ..., X_n] \to \mathbb{Q}[X_1, ..., X_n] \to \mathbb{R} \to \mathbb{C}$, and so, if $\mathbb{R}$ or $\mathbb{C}$ is an object of $\mathscr{A}'$, $\mathbb{Z}[X_1, ..., X_n]$ (resp. $\mathbb{Q}[X_1, ..., X_n]$) is an object of $\mathscr{A}'$. By (ii), any finitely generated $\mathbb{Z}$-algebra (resp. $\mathbb{Q}$-algebra) of $\mathscr{A}$ is an object of $\mathscr{A}'$. Since any objet of $\mathscr{A}$ is an increasing directed union of finitely generated subobjects, by (iii) every object of $\mathscr{A}$ is an object of $\mathscr{A}'$.

$\square$

3.7.15 Examples

(1) EXPONENTIAL AND LOGARITHM.

Proposition *Let* A *be a* $\mathbb{Q}$-*algebra. For a nilpotent* $u \in A$, *set* $\exp(u) = \sum_0^\infty \frac{1}{n!} u^n$ *and* $\log(1 + u) = \sum_1^\infty (-1)^{n-1} \frac{1}{n} u^n$. *Then* $\log(\exp(u)) = u$ *and* $\exp(\log(1 + u)) = 1 + u$.

Scholium 3.7.14 cannot be directed applied to this proposition, but it can to the next lemma from which the latter readily follows.

Lemma *Let* A *be a* $\mathbb{Q}$-*algebra. Define the polynomials* $\lambda_p, \varepsilon_p \in A[X]$ *by* $\lambda_p(X) = \sum_1^p (-1)^{n-1} \frac{1}{n} X^n$ *and* $\varepsilon_p(X) = \sum_0^p \frac{1}{n!} X^n$. *Then* $\lambda_p(\varepsilon_p(X)) \equiv X$ (mod X^{p+1}) *and* $\varepsilon_p(\lambda_p(X)) \equiv X$ (mod X^{p+1}).

For $A = \mathbb{R}$, this lemma follows by considering the Taylor expansion of degree p of exp and log. It readily follows that the category of unital associative commutative $\mathbb{Q}$-algebras for which this lemma holds satisfies conditions (i), (ii) and (iii) of Proposition 3.7.14.

(2) HAMILTON- - CAYLEY THEOREM.

Let $\mathscr{M}_n(A)$ be the algebra of $n \times n$ square matrices with coefficients in A. For $M \in \mathscr{M}_n(A)$, consider $XI_n - M \in \mathscr{M}_n(A[X])$, where I_n is the unital element of $\mathscr{M}_n(A)$, and define the characteristic polynomial $P_M \in A[X]$ by $P_M(X) = \det(XI_n - M)$. In the unital subalgebra of $\mathscr{M}_n(A)$ generated by M, which is commutative, consider $P_M(M)$.

Theorem *For all* $M \in \mathscr{M}_n(A)$, $P_M(M) = 0$.

Proof (a) *Suppose first that* $A = \mathbb{C}$ *and that* P_M *has* n *distinct roots.* In this case, M is diagonalizable, and M may be assumed to be diagonal; let $\lambda_1, ..., \lambda_n$ be its diagonal entries. Then $P_M(X) = (X - \lambda_1) \cdots (X - \lambda_n)$, and $P_M(M)$ is diagonal with diagonal entries $P_M(\lambda_1), ..., P_M(\lambda_n)$ all zero, and so $P_M(M) = 0$.

(b) *Suppose next that* $A = \mathbb{C}$, *and that* M *is arbitrary.* For $M \in \mathscr{M}_n(\mathbb{C})$, let $\Delta(M)$ be the discriminant of P_M. The map $\Delta : \mathscr{M}_n(\mathbb{C}) \to \mathbb{C}$ is polynomial because it is the composite of the map $M \mapsto P_M$ from $\mathscr{M}_n(\mathbb{C})$ to the vector space S_{n+1} of polynomials of degree $\leqslant n$ and of the map $P \mapsto \text{discr } P$ from S_{n+1} to $\mathbb{C}$, both of which are polynomial. It is nonzero because, if M is diagonal with distinct diagonal entries, then $\Delta(M) \neq 0$. Therefore the set W of $M \in \mathcal{M}_n(\mathbb{C})$ such that $\Delta(M) \neq 0$ is open and dense in $\mathcal{M}_n(\mathbb{C})$. The map $M \mapsto P_M(M)$ from $\mathcal{M}_n(\mathbb{C})$ to itself is polynomial, and vanishes on W, and so is null.

(c) *General case.* The rings A for which the theorem holds form a subcategory $\mathscr{A}'$ of $\mathbf{Ann}$ satisfying conditions (i), (ii) and (iii) of Proposition 3.7.14, and $\mathbb{C} \in \mathscr{A}'$; hence $\mathscr{A}' = \mathbf{Ann}$.   □

Exercises 3.7. (Polynomials)

1.—What are the prime ideals of $\mathbb{C}[X]$? of $\mathbb{R}[X]$?

2.—Show that in $\mathbb{R}[X, Y]$, the ideal of polymials vanishing on the circle $S^1 = \{(x, y) \in \mathbb{R}^2 \mid x^2 + y^2 = 1\}$ is principal and prime.

3.—(a) Show that the ideal generated by $Y^2 - X^2 - 1$ in $\mathbb{R}[X, Y]$ is prime, but that the ideal generated by the same function in the ring of analytic functions on $\mathbb{R}^2$ is not prime.

(b) Give an example of a polynomial $P \in \mathbb{R}[X, Y]$ generating a prime ideal in $\mathbb{R}[X, Y]$, but an ideal that is not prime in $\mathbb{C}[X, Y]$. Show that the converse is not possible.

4.—Let K be a field. For every ideal I in $K[X_1, ..., X_n]$, define V(I) as in 3.7.13, and for every subset V of K^n, set $\mathscr{J}(V)$ to be the ideal in $K[X_1, ..., X_n]$ consisting of polynomials vanishing on V.

(a) Show that $\mathscr{J}(V \cup W) = \mathscr{J}(V) \cap \mathscr{J}(W)$ and that $\mathscr{J}(V \cap W) \supset \mathscr{J}(V) + \mathscr{J}(W)$. Give an example where inclusion is strict.

(b) Show that $V(\mathscr{J}(V(I))) = V(I)$ and $\mathscr{J}(V(\mathscr{J}(V))) = \mathscr{J}(V)$.

5.—Let $I = \{f \in \mathbb{R}[X, Y] \mid (\forall x \in \mathbb{R}) \ f(x, 0) = 0 \text{ and } f'_Y(0, 0) = 0\}$. Show that I is an ideal and give a family of generators. Is the ideal I principal? Describe V(I) and $\mathscr{J}(V(I))$.

6.—Describe V(I) and $\mathscr{J}(V(I))$ for the following ideals of $\mathbb{R}[X, Y]$: $(X^2 - Y^2)$; $(X^2 + Y^2)$; $(X^2 + Y^2 - 1)$; $(X^2 + Y^2 + 1)$; $(X(X^2 + Y^2 + 1))$; (X^2Y, X^3); $(X^2 - Y^2, X^3 + Y^3)$.

Which among these ideals can be defined by vanishing conditions on polynomials and their derivatives?

7.—(a) Show that in the space of one variable polynomials of degree d with real coefficients, the set of polynomials with d real distinct roots is open.

(b) Let $P \in \mathbb{R}[X, Y]$ be a polynomial of degree d in which the coefficient of Y^d is nonzero. Assume there exists $x \in \mathbb{R}$ such that the polynomial $P(x, Y) \in \mathbb{R}[Y]$ has d distinct real roots. Show that every polynomial vanishing on V(P) is a multiple of P (divide by P in $(A[X])[Y]$).

(c) In (b) replace the assumption that the coefficient of Y^d is nonzero with: P is primitive in $(A[X])[Y]$, i.e. $P(X, Y) = \sum a_i(X)Y^i$, where the polynomials a_i do not have any common divisors (use 3.7.8, Lemma 3.2).

(d) Generalize to more than two variables.

8. *(Continuity of roots)*—(a) Let A be the Banach algebra of series $\sum a_n X^n \in \mathbb{C}[[X]]$ such that $\sum |a_n| < \infty$. Let $P \in \mathbb{C}[X]$ be a polynomial of degree d. Show that if P has

k roots with absolute value < 1 and $d - k$ roots with absolute value > 1 (counting multiplicities), every element $f \in A$ can be uniquely written as $f = Pg + R$, with $g \in A$ and $R \in \mathbb{C}[X]$ of degree $< k$. Converse.

(b) Let $P_0 \in \mathbb{C}[X]$ be a polynomial of degree d having 0 as a root of order k, and c be the smallest absolute value of the nonzero roots. Show that, for all $\varepsilon > 0$, there exists $\alpha > 0$ such that any polynomial P of degree d satisfying $\|P - P_0\| < \alpha$ has k roots of absolute value $< \varepsilon$ and $d - k$ roots of absolute value $> c - \varepsilon$. (Use the fact that, if E and F are Banach spaces, then $\mathrm{Isom}(E, F)$ is open in $L(E, F)$.)

(c) Let $P_0 \in \mathbb{C}[X]$ be a polynomial of degree d, $x_1, ..., x_d$ its roots. Show that, for all $\varepsilon > 0$, there exists $\alpha > 0$ such that for any polynomial P of degree d satisfying $\|P - P_0\| < \alpha$, its roots $y_1, ..., y_d$ can be classified so that $|y_1 - x_1| < \varepsilon$, ..., $|y_d - x_d| < \varepsilon$.

(d) Let $d' \geqslant d$. Show that, for all $\varepsilon > 0$ and all $M \in \mathbb{R}_+$, there exists $\alpha > 0$ such that, for any polynomial P of degree $d'' \leqslant d'$ satisfying $\|P - P_0\| < \alpha$, its roots $y_1, ..., y_d$ can be classified so that $|y_1 - x_1| < \varepsilon$, ..., $|y_d - x_d| < \varepsilon$, $|y_{d+1}| > M$, ..., $|y_{d''}| > M$.

(e) Extend these results to an arbitrary algebraically closed valued field K (consider the completion $\widehat{K}$ of K and define A as in (a). $\widehat{K}$ can be shown to be algebraically closed, but it is not necessary in this exercise).

9.—Consider the rings

$$A = \mathbb{R}[X, Y]/(X^2 + Y^2 - (1)), \quad B = \mathbb{C}[X, Y]/(X^2 + Y^2 - 1).$$

(a) Show that B is isomorphic to $\mathbb{C}\left[Z, \frac{1}{Z}\right]$ (subring of $\mathbb{C}(Z)$), by setting $Z = X + iY$. What is the image of A under this isomorphism?

(b) Show that every maximal ideal of B is of the form $\mathfrak{n}_{x,y} = \{f \in B \mid f(x, y) = 0\}$, where (x, y) is a point of $\mathbb{C}^2$ such that $x^2 + y^2 = 1$ (the reader may use the injection from $\mathbb{C}[X]$ to B). Show that all nonzero ideals in B are of the form $\mathfrak{n}_{x_1,y_1}^{k_1} \cdots \mathfrak{n}_{x_r,y_r}^{k_r}$.

(c) For $(x, y) \in \mathbb{C}^2$, $x^2 + y^2 = 1$, set $\mathfrak{m}_{x,y} = \mathfrak{n}_{x,y} \cap A$. What is the codimension of $\mathfrak{m}_{x,y}$ as a real vector space in A? Show that every nonzero ideal in A is of the form $\mathfrak{m}_{x_1,y_1}^{k_1} \cdots \mathfrak{m}_{x_r,y_r}^{k_r}$, and that such an ideal is principal if and only if the sum of the k_i corresponding to real pairs is even.

(d) Show that A is not a PID, but that B is, and is even Euclidean.

(e) Consider the rings

$$A' = \mathbb{R}[X, Y]/(X^2 + Y^2 + 1), \quad B' = \mathbb{C}[X, Y]/(X^2 + Y^2 + 1).$$

Show that B' is isomorphic to B, but that A' is not isomorphic to A. Describe the ideals of A' and B'. Show that A' is a PID.

(f) What are the invertible elements of B'? of A'? Show that A' is not Euclidean (use 3.1, Exercise 14).

10.—For any subset V of $\mathbb{C}^n$, set

$$\mathscr{I}_{\mathbb{R}}(V) = \mathbb{R}[X_1, ..., X_n] \cap \mathscr{I}(V),$$

and for any ideal I of $\mathbb{R}[X_1, ..., X_n]$, set $V_{\mathbb{C}}(I) = V(\mathbb{C} \cdot I)$. Show that for any algebraic subset V of $\mathbb{C}^n$, $V_{\mathbb{C}}(\mathscr{I}_{\mathbb{R}}(V)) = V \cup \tau(V)$, where $\tau : \mathbb{C}^n \to \mathbb{C}^n$ is the conjugation $(z_1, ..., z_n) \mapsto (\overline{z_1}, ..., \overline{z_n})$.

11.—The aim is to give another proof of the fact that the ring $K[X_1, ..., X_n]$ is Noetherian when K is an infinite field.

(a) Let $P \in K[X_1, ..., X_n]$ be a polynomial of degree d. Show that there is an automorphism φ of $K[X_1, ..., X_n]$ such that the coefficient of X_n^d in $\varphi(P)$ is nonzero (take a point $x = (x_1, ..., x_n) \in K^n$ such that $P_d(x_1, ..., x_n) \neq 0$, where P_d denotes the homogeneous part of degree d of P, and effect a change of coordinates in K^n such that x becomes the last basis vector).

(b) Let $P \in K[X_1, ..., X_n]$ be a polynomial of degree d, where the coefficient of X_n^d is nonzero. Show that $K[X_1, ..., X_n]/(P)$ is a free module of rank d over $K[X_1, ..., X_{n-1}]$.

(c) Let A be a ring. Show that, if $A/(f)$ is Noetherian for all nonzero $f \in A$, then A is Noetherian.

(d) Conclude by induction on n (use 3.6.2, Example 3).

12.—(a) Let K be a field and $f \in K[[X_1, ..., X_n]]$ a formal series of degree d (i.e. such that the nonzero term of lowest degree is of degree d). Suppose that the coefficient of X_n^d is nonzero. Show that every formal series $g \in K[[X_1, ..., X_n]]$ can be uniquely written as $g = fq + R$, with $q \in K[[X_1, ..., X_n]]$ and $R \in K[[X_1, ..., X_{n-1}]][X_n]$ a polynomial of degree $< d$.

(b) Suppose that K is infinite. Show that as in the previous exercise $K[[X_1, ..., X_n]]$ is Noetherian.

(c) Show that the ring $\mathbb{R}\{X_1, ..., X_n\}$ is Noetherian.

13. *(Nullstellensatz for an uncountable field)*—Let K be an uncountable field (for example $\mathbb{C}$).

(a) Show that the field $K(X)$ (field of fractions of $K[X]$) is a K-vector space of uncountable dimension (as a runs through K, the $\frac{1}{X-a}$ are linearly independent). Deduce that if a finite dimensional K-algebras is a field then it is an algebraic extension of K.

(b) Suppose that K is algebraically closed. Show that all maximal ideals in $K[X_1, ..., X_n]$ are of the form

$$\mathfrak{m}_{(x_1, ..., x_n)} = \{P \in K[X_1, ..., X_n] \mid P(x_1, ..., x_n) = 0\}$$

with $(x_1, ..., x_n) \in K^n$. Deduce that for an ideal I in $K[X_1, ..., X_n]$, $V(I) = \varnothing \Rightarrow I = K[X_1, ..., X_n]$.

(c) Continue to assume that K is algebraically closed; let I be an ideal in $K[X_1, ..., X_n]$ and set $A = K[X_1, ..., X_n]/I$. Let $P \in K[X_1, ..., X_n]$ be a polynomial vanishing at every point of $V(I)$. Let J denote the ideal in $K[X_1, ..., X_n, Z]$ generated by I and $(1 - PZ)$. Show that $V(J)$ is empty in K^{n+1}.

Deduce that the image of $1 - PZ$ in $A[Z]$ is invertible, and that the image of P in A is nilpotent.

(d) Conclude.

14. *(Nullstellensatz, general case)*— Let k be a field. The aim is to prove that a finitely generated field extension of k (as a k-algebra) is an algebraic extension of k.

(A) Let L be a field, A a subring of L and $K \subset L$ the field of fractions of A.

(a) Show that if K is a finitely generated A-module, then $K = A$.

(b) Show that if L is a finitely generated A-module, then $K = A$ (since K is a direct factor in L as A-module).

(c) Show that if L is a finitely generated A-algebra and an algebraic extension of K, then there exists nonzero $b \in A$ such that $K = A[b^{-1}]$.

(B) (a) Set $A = k[X_1, ..., X_n]$ and $K = k(X_1, ..., X_n)$. Show that, if $n > 0$, there are no $b \in A$ such that $K = A[b^{-1}]$.

(b) Let L be a finitely generated extension de k (as a k-algebra). There are algebraically independent elements $x_1, ..., x_n$ in L such that L is an algebraic extension of $K = k(x_1, ..., x_n)$. Show that $n = 0$ necessarily holds. Conclude and continue as in the previous exercise.

15.—(a) Let $K = \{a_1, ...a_q\}$ be a finite field. Show that the polynomial $(X - a_1) ... (X - a_q) \in K[X]$ is nonzero, but that the map from K to K defined by it is null.

(b) Let I be an infinite set, K the field $\mathbb{Z}/(2)$ and B the K-algebra K^I. Show that the map $P \mapsto P_B$ from $K[X]$ to B^B is not injective, although B is infinite (consider $X^2 + X$).

3.8 Tensor Products

3.8.1

If E, F, G are three A-modules, let $B(E, F ; G)$ be the module of bilinear maps from $E \times F$ to G.

Theorem and Definition *Let* E *and* F *be* A-*modules. Then the covariant functor* $G \mapsto B(E, F ; G)$ *from* A-$\mathcal{M}od$ *to* $\mathcal{Ens}$ *is representable. The* **tensor product** *of* E *and* F *is a representative of this functor.*

Comments In other words, a tensor product of E and of F is an A-module T with a bilinear map θ from $E \times F$ to T, satisfying the following property: for all A-modules G and all bilinear maps f from $E \times F$ to G, there is a unique linear map $\bar{f} : T \to G$ such that $f = \bar{f} \circ \theta$.

Notation The module T is written $E \otimes_A F$, or simply $E \otimes F$, and the map $\theta(x, y) \mapsto x \otimes y$.

Cynical remark The different representatives of the functor $G \mapsto B(E, F ; G)$ can be mutually identified 2.4.11, hence they can be written $E \otimes F$, and this can be called "*the*" tensor product of E and F.

We give two proofs of the theorem.

1st proof Let L be the free A-module over the set E × F, and $(e_{x,y})_{(x,y) \in E \times F}$ the canonical basis for L. For all maps f from E × F to an A-module G, write $\bar{f}$ for the homomorphism from L to G defined by $\bar{f}(e_{x,y}) = f(x, y)$. The map $f \mapsto \bar{f}$ is a bijection from the set of maps from E × F to G onto Hom(L, G). The map f is bilinear if and only if

$$(\forall a \in A) \ (\forall b \in A) \ (\forall x \in E) \ (\forall y \in E) \ (\forall z \in F) \ (\forall t \in F)$$
$$\bar{f}(e_{ax+by,z}) - a\bar{f}(e_{x,z}) - b\bar{f}(e_{y,z}) = \bar{f}(e_{x,az+bt}) - a\bar{f}(e_{x,z}) - b\bar{f}(e_{x,t}) = 0,$$

or if and only if $\bar{f}$ vanishes on every submodule N of L generated by elements of the form

$$e_{ax+by,z} - ae_{x,z} - be_{y,z} \ \text{ou} \ e_{x,az+bt} - ae_{x,z} - be_{x,t} \ .$$

Hence B(E, F; G) can be identified with Hom(T, G), where T = L/N. More precisely, the functor B is represented by (T, θ), where $\theta(x, y)$ is the class of $e_{x,y}$. $\qquad \square$

2nd proof (a) *Suppose first that* E *is free.* If $E = A^{(I)}$, then $B(E, F; G) = \mathrm{Hom}(A^{(I)}, \mathrm{Hom}(F, G)) = (\mathrm{Hom}(F, G))^I = \mathrm{Hom}(F^{(I)}, G)$, and the functor $G \mapsto B(E, F; G)$ is represented by $F^{(I)}$.

(b) *General case.* Let $A^{(J)} \xrightarrow{\delta} A^{(I)} \xrightarrow{\varepsilon} E \to 0$ be a presentation of E. As B (M, F; G) = Hom(M, Hom(F, G)) for all modules M, there is an exact sequence

$$0 \to B(E, F; G) \xrightarrow{\varepsilon^*} B(A^{(I)}, F; G) \xrightarrow{\delta^*} B(A^{(J)}, F; G)$$
$$\| \qquad\qquad\qquad \|$$
$$\mathrm{Hom}(F^{(I)}, G) \quad \mathrm{Hom}(F^{(J)}, G) \ .$$

By the Yoneda lemma 2.4.3, the morphism $\delta^* : \mathrm{Hom}(F^{(I)}, G) \to \mathrm{Hom}(F^{(J)}, G)$, functorial at G, arises from a morphism $\delta_* : F^{(J)} \to F^{(I)}$. Let T be the cokernel of δ_*. For all G, there is an exact sequence

$$0 \to \mathrm{Hom}(T, G) \to \mathrm{Hom}(F^{(I)}, G) \xrightarrow{\delta^*} \mathrm{Hom}(F^{(J)}, G)$$

So Hom(T, G) = B(E, F; G), and T represents the functor $G \mapsto B(E, F; G)$. $\qquad \square$

Remark The construction given in the first proof is intrinsic. That given in the second one shows that, for example, if E and F are finitely generated (resp. have finite presentation), the same holds for $E \otimes F$.

3.8.2

The elements of $E \otimes F$ are called **tensors**. A tensor $x \otimes y$, with $x \in E$ and $y \in F$ is called a *simple tensor.*

Proposition *The module* $E \otimes F$ *is generated by the simple tensors.*

Proof Let $\theta : E \times F \to E \otimes F$ be the map $(x, y) \mapsto x \otimes y$, T the submodule of $E \otimes F$ generated by the simple tensors, and $\chi : E \otimes F \to E \otimes F/T$ the canonical homomorphism. Then $\chi \circ \theta = 0 = 0 \circ \theta$, and so $\chi = 0$ by uniqueness of the universal property, which in turn gives $E \otimes F/T = 0$. $\qquad\qquad\square$

(This also follows by considering the construction given in the First Proof of Theorem 3.8.1.)

The product of a simple tensor and of a scalar is a simple tensor: indeed $a(x \otimes y) = (ax) \otimes y$. Hence all tensors are the sum of finitely many simple ones. The **rank** of a tensor $t \in E \otimes F$ is the smallest integer r for which t is the sum of r simple tensors.

If f is a bilinear map from $E \times F$ to a module G, let $x \otimes y \mapsto f(x, y)$ denote the linear map of $E \otimes F \to G$ corresponding to it, i.e. the unique linear map $\bar{f} : E \otimes F \to G$ such that $(\forall x \in E) (\forall y \in F) \, \bar{f}(x \otimes y) = f(x, y)$. For example, if M, N, P are three A-modules, then it is the map $(f \otimes g) \mapsto g \circ f$ from $\text{Hom}(M, N) \otimes \text{Hom}(N, P)$ to $\text{Hom}(M, P)$.

Remark To show that a property holds for every element of $E \otimes F$, checking it for all simple tensors is not always sufficient. For example, $f(x \otimes y) \neq 0$ for $x \neq 0$ and $y \neq 0$ does not necessarily imply that the map $f : E \otimes F \to G$ is injective. (Example: multiplication $\mathbb{C} \otimes_{\mathbb{R}} \mathbb{C} \to \mathbb{C}$ or see 3.8, Exercise 3.)

3.8.3 Functoriality

Let E, E', F, F' be A-modules, $f : E \to E'$ and $g : F \to F'$ homomorphisms. There is a linear map $x \otimes y \mapsto f(x) \otimes g(y)$ from $E \otimes F$ to $E' \otimes F'$, written $f \otimes g$ $f \otimes g$ of $\text{Hom}(E, E') \otimes \text{Hom}(F, F')$. However these two elements can be identified when the Kronecker homomorphism (see 3.9) is injective. corresponding to $(x, y) \mapsto f(x) \otimes g(y)$ from $E \times F$ to $E' \otimes F'$. In particular, $1_E \otimes 1_F = 1_{E \otimes F}$ and, for homomorphisms $f' : E' \to E''$ and $g' : F' \to F''$, $(f' \circ f) \otimes (g' \circ g) = (f' \otimes g') \circ (f \otimes g)$.

3.8.4 Symmetry

Let E and F be A-modules. The linear map $x \otimes y \mapsto y \otimes x$ from $E \otimes F$ to $F \otimes E$ is an isomorphism, called *symmetry*. For $F = E$, the symmetry $\sigma : E \otimes E \to E \otimes E$ is not the identity, but $\sigma \circ \sigma = 1_{E \otimes E}$. In general $E \otimes F$ and $F \otimes E$ not identified under σ.

3.8.5 Identity Object

Let E be an A-module. The maps $a \otimes x \mapsto ax$ from $A \otimes E$ to E and $x \mapsto 1 \otimes x$ from E to $A \otimes E$ are converse isomorphisms which enable us to identify $A \otimes E$ with E. Similarly, $E = E \otimes A$.

3.8.6 Associativity

For finitely many A-modules $(E_i)_{i \in I}$ and an A-module G, let $B((E_i)_{i \in I} ; G)$ denote the module of multilinear maps from $\prod_{i \in I} E_i$ to G.

Proposition and Definition *(a) For every finite family $(E_i)_{i \in I}$ of A-modules, the covariant functor $G \mapsto B((E_i)_{i \in I} ; G)$ is representable. If (T, θ) is a representative of this functor, then set $\bigotimes_{i \in I} E_i = T$ and, for $(x_i)_{i \in I} \in \prod_{i \in I} E_i$, $\bigotimes_{i \in I} x_i = \theta((x_i)_{i \in I})$.*

(b) If $I = J \cup K$ with $J \cap K = \varnothing$, the linear map $\bigotimes_{i \in I} x_i \mapsto (\otimes x_i) \otimes (\otimes x_i)$ from $\bigotimes_{i \in I} E_i$ to $(\bigotimes_{i \in J} E_i) \otimes (\bigotimes_{i \in K} E_i)$ is an isomorphism.

Proof Let $I = J \cup K$ with $J \cap K = \varnothing$, and suppose that the functors $G \mapsto B((E_i)_{i \in J} ; G)$ and $G \mapsto B((E_i)_{i \in K} ; G)$ are representable. Show that the functor $G \mapsto B((E_i)_{i \in I} ; G)$ is represented by $(\bigotimes_{i \in J} E_i) \otimes (\bigotimes_{i \in K} E_i)$ equipped with the multilinear map $(x_i)_{i \in I} \mapsto (\bigotimes_{i \in J} x_i) \otimes (\bigotimes_{i \in K} x_i)$.

For all A-modules G,

$$
\begin{aligned}
B((E_i)_{i \in I} ; G) &= B((E_i)_{i \in J} ; B((E_i)_{i \in K} ; G)) \\
&= B\left((E_i)_{i \in J} ; \mathrm{Hom}\left(\bigotimes_{i \in K} E_i, G\right)\right) \\
&= \mathrm{Hom}\left(\bigotimes_{i \in J} E_i, \mathrm{Hom}\left(\bigotimes_{i \in K} E_i, G\right)\right) \\
&= B\left(\bigotimes_{i \in J} E_i, \bigotimes_{i \in K} E_i ; G\right) = \mathrm{Hom}\left(\left(\bigotimes_{i \in J} E_i\right) \otimes \left(\bigotimes_{i \in K} E_i\right), G\right),
\end{aligned}
$$

proving the existence of $\bigotimes_{i \in I} E_i$ as well as (b) if $\bigotimes_{i \in J} E_i$ and $\bigotimes_{i \in K} E_i$ are assumed to exist.

For $I = \varnothing$, $B((E_i)_{i \in \varnothing} ; G) = G = \mathrm{Hom}(A, G)$, and so $\bigotimes_{i \in \varnothing} E_i = A$.
For $I = \{a\}$, $B((E_i)_{i \in \{a\}} ; G) = \mathrm{Hom}(E_a, G)$, and so $\bigotimes_{i \in \{a\}} E_i = E_a$.
(a) now follows by induction on the number of elements of I.

Corollary *For any A-modules* E, F, G,

$$E \otimes (F \otimes G) = E \otimes F \otimes G = (E \otimes F) \otimes G.$$

3.8.7 Right Exactness, Distributivity, Passing to Direct Limits

Proposition *Let* F *be an* A-*module.*
   (a) Let $f : E' \to E$ *be a homomorphism of* A-*modules. Then*

$$\mathrm{Coker}(f \otimes 1_F) = (\mathrm{Coker}\ f) \otimes F.$$

   (b) Let $(E_i)_{i \in I}$ *be a family of* A-*modules. Then*

$$\left(\bigoplus_{i \in I} E_i \right) \otimes F = \bigoplus_{i \in I} (E_i \otimes F).$$

   (c) Let $(E_i)_{i \in I}$ *a direct system of* A-*modules. Then*

$$(\varinjlim E_i) \otimes F = \varinjlim (E_i \otimes F).$$

Proof For all A-modules G,

$$\mathrm{Hom}(E \otimes F, G) = \mathrm{Hom}(E, \mathrm{Hom}(F, G)).$$

In other words, the functor $F \mapsto E \otimes F$ from A-$\mathcal{M}od$ to A-$\mathcal{M}od$ has a right adjoint functor $G \mapsto \mathrm{Hom}(F, G)$. The proposition then follows from (2.7.3, Corollary). □

Corollary *(a) If* $E' \to E \to E'' \to 0$ *is an exact sequence of* A-*modules, then* $E' \otimes F \to E \otimes F \to E'' \otimes F \to 0$ *is an exact sequence.*
   (b) Let $f : E \to E'$ *and* $g : F \to F'$ *be surjective homomorphisms, then* $f \otimes g : E \otimes F \to E' \otimes F'$ *is surjective.*
   (c) Let $(E_i)_{i \in I}$ *and* $(F_j)_{j \in J}$ *be two families of* A-*modules. Then*

$$\left(\bigoplus_{i \in I} E_i \right) \otimes \left(\bigoplus_{j \in J} F_j \right) = \bigoplus_{(i,j) \in I \times J} E_i \otimes F_j.$$

   (d) Let $(E_i)_{i \in I}$ *and* $(F_j)_{j \in J}$ *be direct systems of* A-*modules. Then*

$$(\varinjlim E_i) \otimes (\varinjlim F_j) = \varinjlim_{(i,j) \in I \times J} E_i \otimes F_j.$$

Remarks (1) Let F be an A-module and $f : E' \to E$ an injective homomorphism. The homomorphism $f \otimes 1_F : E' \otimes F \to E \otimes F$ is not in general injective. For example, suppose that A is integral and take $E = E' = A$. For all nonzero $a \in A$, $a \cdot 1_E$ is injective, but if F has torsion, then there exists $a \neq 0$ such that $a \cdot 1_E \otimes 1_F = a \cdot 1_F$ is not injective.

In particular, if E' is a submodule of E, then in general $E' \otimes F$ cannot be identified with a submodule of $E \otimes F$. For example, for $A = \mathbb{Z}$, $E = \mathbb{Q}$, $E' = \mathbb{Z}$, $F = \mathbb{Z}/(n)$, $E \otimes F = 0$ but $E' \otimes F \neq 0$.

However, if E' is a direct factor of E, the module $E' \otimes F$ is a direct factor of $E \otimes F$. For example, if $f : E' \to E$ and $g : F' \to F$ are injective *vector space* homomorphisms over a field K, the map $f \otimes g : E' \otimes F' \to E \otimes F$ is injective, since f and g identify E' and F' with direct factors of E and F respectively.

(2) Let $f : E \to E'$ and $g : F \to F'$ be homomorphisms. Then the modules $\mathrm{Coker}(f \otimes g : E \otimes F \to E' \otimes F')$ and $(\mathrm{Coker}\, f) \otimes (\mathrm{Coker}\, g)$ are not in general isomorphic. For example, for $A = \mathbb{R}$, $E = \mathbb{R}^n$, $E' = \mathbb{R}^{n+1}$, $F = \mathbb{R}^p$, $F' = \mathbb{R}^{p+1}$, f and g injective, the above modules are vector spaces of dimension $n + p + 1$ and 1 respectively.

3.8.8 Tensor Product of Free Modules

Proposition *Let* E *and* F *be free* A-*modules with respective bases* $(e_i)_{i \in I}$ *and* $(f_j)_{j \in J}$. *Then* $E \otimes F$ *is a free module and* $(e_i \otimes f_j)_{(i,j) \in I \times J}$ *is a basis for* $E \otimes F$.

Proof Taking $E_i = F_j = A$ in Corollary (3.8.7, c), $A^{(I)} \otimes A^{(J)} = A^{(I \times J)}$, and the proposition follows. □

Corollary 3.7 *Let* E *and* F *be* A-*modules*, $(x_i)_{i \in I}$ *and* $(y_j)_{j \in J}$ *generators of* E *and* F *respectively. Then* $(x_i \otimes y_j)_{(i,j) \in I \times J}$ *generate* $E \otimes F$.

Corollary 3.8 *The tensor product of two projective modules is projective.*

Proof Let E and F be two projective A-modules. There are A-modules E' and F' such that $E \oplus E'$ and $F \oplus F'$ are free. Then, $(E \oplus E') \otimes (F \oplus F') = (E \otimes F) \oplus (E \otimes F') \oplus (E' \otimes F) \oplus (E' \otimes F')$ is free. So $E \otimes F$ is projective. □

3.8.9 Tensor Product of Matrices

Proposition and Definition *Let* E, E', F, F' *be finitely generated free* A-*modules*, $(e_i)_{i \in I}$, $(e'_{i'})_{i' \in I'}$, $(f_j)_{j \in J}$, $(f'_{j'})_{j' \in J'}$ *bases for* E, E', F, F' *respectively. Let* $u : E \to E'$ *and* $v : F \to F'$ *be homomorphisms*, $M = (m_i^{i'})$ *and* $N = (n_j^{j'})$ *their respective matrices with respect to the given bases. Then the matrix of the homomorphism* $u \otimes v : E \otimes F \to E' \otimes F'$ *with respect to the bases* $(e_i \otimes f_j)$ *and* $(e'_{i'} \otimes f'_{j'})$ *is* $(t_{i,j}^{i',j'})$, *where* $t_{i,j}^{i',j'} = m_i^{i'} n_j^{j'}$. *This matrix is called the* **tensor product** *of* M *and* N, *and is written* $M \otimes N$.

Proof Since $u(e_i) = \sum m_i^{i'} e'_{i'}$ and $v(f_j) = \sum n_j^{j'} f'_{j'}$, $u \otimes v\,(e_i \otimes f_j) = u(e_i) \otimes v(f_j) = \sum_{i',j'} m_i^{i'} n_j^{j'} e'_{i'} \otimes f'_{j'}$. □

Corollary *Let* K *be a field*, M *and* N *matrices with entries in* K. *Then the rank of* $M \otimes N$ *is the product of the ranks of* M *and* N.

Proof We may assume M and N are as follows to within base changes

$$\begin{pmatrix} 1 & 0 & \ldots & 0 & 0 & \ldots & 0 \\ 0 & 1 & \ldots & 0 & 0 & \ldots & 0 \\ \vdots & & & \vdots & \vdots & & \vdots \\ 0 & & \ldots & \ldots & 1 & 0 & \ldots & 0 \\ 0 & & \ldots & \ldots & 0 & 0 & \ldots & 0 \\ 0 & & \ldots & \ldots & 0 & 0 & \ldots & 0 \end{pmatrix}$$

$\square$

3.8.10 Tensor Product of Cyclic Modules

Proposition *Let* I *and* J *be two ideals of* A. *Then*

$$(A/I) \otimes (A/J) = A/(I+J).$$

Proof Let e and f be the images of 1 in A/I and A/J. The module $(A/I) \otimes (A/J)$ is generated by $e \otimes f$ (3.8.8, Corollary 3.7), and the annihilator of $e \otimes f$ contains I and J, and so $I + J$. On the other hand, the map $(a, b) \mapsto ab$ defined by passing to the quotient is a bilinear map from $(A/I) \times (A/J)$ to $A/(I+J)$, and hence a homomorphism of $(A/I) \otimes (A/J)$ mapping $e \otimes f$ onto the image g of 1. Therefore the annihilator of $e \otimes f$ is contained in that of g, i.e. in $I + J$; and so is equal to it. $\square$

3.8.11 Examples of Tensor Products over $\mathbb{Z}$

Consider the following $\mathbb{Z}$-modules: $\mathbb{Z}^r, \mathbb{Z}/(n), \mathbb{Q}, \mathbb{Q}/\mathbb{Z}, \mathbb{Z}^{(N)}, \mathbb{Z}^N$. Their mutual tensor products are given by the following table,

$\otimes$	$\mathbb{Z}^r$	$\mathbb{Z}/(n)$	$\mathbb{Q}$	$\mathbb{Q}/\mathbb{Z}$	$\mathbb{Z}^{(N)}$	$\mathbb{Z}^N$
$\mathbb{Z}^{r'}$	$\mathbb{Z}^{rr'}$					
$\mathbb{Z}/(n')$	$(\mathbb{Z}/(n'))^r$	$\mathbb{Z}/(d)$ $d = \text{g.c.d.}(n, n')$				
$\mathbb{Q}$	$\mathbb{Q}^r$	0	$\mathbb{Q}$			
$\mathbb{Q}/\mathbb{Z}$	$(\mathbb{Q}/\mathbb{Z})^r$	0	0	0		
$\mathbb{Z}^{(N)}$	$\mathbb{Z}^{(N \times \{1,\ldots,r\})}$	$(\mathbb{Z}/(n))^{(N)}$	$\mathbb{Q}^{(N)}$	$(\mathbb{Q}/\mathbb{Z})^{(N)}$	$\mathbb{Z}^{(N \times N)}$	
$\mathbb{Z}^N$	$\mathbb{Z}^{N \times \{1,\ldots,r\}}$	$(\mathbb{Z}/(n))^N$	E	F	G	H

where $E \subset \mathbb{Q}^N$ is the set of sequences of the form $\left(\frac{p_i}{q_i}\right)_{i \in \mathbb{N}}$, the sequence (q_i) being bounded; F is the image of E in $(\mathbb{Q}/\mathbb{Z})^N$; $G \subset E^{\mathbb{N} \times \mathbb{N}}$ is the set of double sequences whose support is contained in a "strip", i.e. in a set $\{1, \ldots, k\} \times \mathbb{N}$; $H \subset \mathbb{Z}^{\mathbb{N} \times \mathbb{N}}$ is the set of double sequences $(n, p) \mapsto u_{n,p}$ such that there exists a finitely generated submodule M of $\mathbb{Z}^{\mathbb{N}}$ for which, for all $n \in \mathbb{N}$, the partial sequence $u_n : p \mapsto u_{n,p}$ is in M.

Proof (a) *Column of $\mathbb{Z}^r$:* for all $\mathbb{Z}$-modules M, $\mathbb{Z}^r \otimes M = M^r$ by distributivity (3.8.7, b).

(b) *Row of $\mathbb{Z}^{(\mathbb{N})}$:* for all $\mathbb{Z}$-modules M, $\mathbb{Z}^{(\mathbb{N})} \otimes M = M^{(\mathbb{N})}$ by distributivity. The tensor products $\mathbb{Z}^{(\mathbb{N})} \otimes \mathbb{Z}^r$ and $\mathbb{Z}^{(\mathbb{N})} \otimes \mathbb{Z}^{(\mathbb{N})}$ are given by Proposition 3.8.8.

(c) *Column of $\mathbb{Z}/(n)$:* For all Z-modules M, $\mathbb{Z}/(n) \otimes M = M/nM$.

Indeed, the exact sequence $\mathbb{Z} \xrightarrow{n} \mathbb{Z} \to \mathbb{Z}/(n) \to 0$ gives

$$M \xrightarrow{n} M \to \mathbb{Z}/(n) \otimes M \to 0.$$

For $M = \mathbb{Z}/(n')$, let $\chi : \mathbb{Z} \to \mathbb{Z}/(n')$ be the canonical homomorphism. Then $nM = \chi(n\mathbb{Z}) = \chi(n\mathbb{Z} + n'\mathbb{Z}) = (n\mathbb{Z} + n'\mathbb{Z})/n'\mathbb{Z} = d\mathbb{Z}/n'\mathbb{Z}$ where $d = $ g.c.d.(n, n'), and so $M/nM = (\mathbb{Z}/(n'))/(d\mathbb{Z}/(n')) = \mathbb{Z}/(d)$.

For $M = \mathbb{Q}, \mathbb{Q} \otimes \mathbb{Z}/(n) = \mathbb{Q}/n\mathbb{Q} = 0$ since $n\mathbb{Q} = \mathbb{Q}$. Similarly, $\mathbb{Q}/\mathbb{Z} \otimes \mathbb{Z}/(n) = 0$ since $n(\mathbb{Q}/\mathbb{Z}) = \mathbb{Q}/\mathbb{Z}$.

For $M = \mathbb{Z}^N$, $\mathbb{Z}^N \otimes \mathbb{Z}/(n) = \mathbb{Z}^N/n\mathbb{Z}^N = (\mathbb{Z}/(n))^N$.

(d) $\mathbb{Q} \otimes \mathbb{Q} = \mathbb{Q}$. For $q \in \mathbb{N}^*$, consider the submodule $\mathbb{Z}\frac{1}{q}$ de $\mathbb{Q}$. Then $\mathbb{Q} = \bigcup_{q \in \mathbb{N}^*} \mathbb{Z}\frac{1}{q} = \varinjlim \mathbb{Z}\frac{1}{q}$, the direct limit being indexed by $\mathbb{N}^*$ ordered by divisibility. For all $q \in \mathbb{N}^*$, the homomorphism $x \otimes y \xrightarrow{\alpha_q} xy$ from $\mathbb{Q} \otimes (\mathbb{Z}\frac{1}{q})$ to $\mathbb{Q}$ is an isomorphism thanks to the diagram

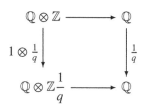

These isomorphisms form a coherent system, i.e. for all q and q' such that q divides q', the following diagrams

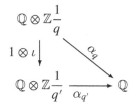

are commutative.

Hence, passing to the direct limit, we get an isomorphism $\alpha : \mathbb{Q} \otimes \mathbb{Q} \to \mathbb{Q}$ characterized by $\alpha(x \otimes y) = xy$.

(e) The tensor product of the exact sequence $0 \to \mathbb{Z} \xrightarrow{\iota} \mathbb{Q} \to \mathbb{Q}/\mathbb{Z} \to 0$ with $\mathbb{Q}$ is the following exact sequence $\mathbb{Q} \to \mathbb{Q} \to \mathbb{Q} \otimes \mathbb{Q}/\mathbb{Z} \to 0$, where the homomorphism $1 \otimes \iota : \mathbb{Q} \to \mathbb{Q}$ is the identity since the isomorphisms identifying $\mathbb{Q} \otimes \mathbb{Z} \to \mathbb{Q}$ with $\mathbb{Q} \otimes \mathbb{Q} \to \mathbb{Q}$ are defined by $1 \otimes 1 \mapsto 1$.

(f) The homomorphism $x \otimes (u_i)_{i \in \mathbb{N}} \mapsto (xu_i)_{i \in \mathbb{N}}$ from $\mathbb{Q} \otimes \mathbb{Z}^{\mathbb{N}}$ to $\mathbb{Q}^{\mathbb{N}}$ induced for all $q \in \mathbb{N}^*$ is an isomorphism from $\mathbb{Z} \cdot \frac{1}{q} \otimes \mathbb{Z}^{\mathbb{N}}$ onto the submodule $\mathbb{Z}^{\mathbb{N}} \cdot \frac{1}{q}$ of $\mathbb{Q}^{\mathbb{N}}$ consisting of the sequences in $\mathbb{Z} \cdot \frac{1}{q}$. Then

$$\mathbb{Q} \otimes \mathbb{Z}^{\mathbb{N}} = \varinjlim \mathbb{Z}^{\mathbb{N}} \frac{1}{q} = \bigcup_{q \in \mathbb{N}^*} \mathbb{Z}^{\mathbb{N}} \frac{1}{q} = \mathrm{E} .$$

(g) *Column of* $\mathbb{Q}/\mathbb{Z}$. The module $\mathbb{Q}/\mathbb{Z} \otimes \mathbb{Q}/\mathbb{Z}$ is a quotient of $\mathbb{Q} \otimes \mathbb{Q}/\mathbb{Z} = 0$ and so is zero.

The tensor product of the exact sequence $0 \to \mathbb{Z} \to \mathbb{Q} \to \mathbb{Q}/\mathbb{Z} \to 0$ with $\mathbb{Z}^{\mathbb{N}}$ is $\mathbb{Z}^{\mathbb{N}} \to \mathrm{E} \to \mathbb{Z}^{\mathbb{N}} \otimes \mathbb{Q}/\mathbb{Z} \to 0$. As $\mathbb{Z}^{\mathbb{N}} \to \mathrm{E}$ is injective, $\mathbb{Z}^{\mathbb{N}} \otimes \mathbb{Q}/\mathbb{Z} = \mathrm{E}/\mathbb{Z}^{\mathbb{N}}$.

(h) $\mathbb{Z}^{\mathbb{N}} \otimes \mathbb{Z}^{(\mathbb{N})} = (\mathbb{Z}^{\mathbb{N}})^{(\mathbb{N})} = \mathrm{G}$.

(i) For every $\mathbb{Z}$-module M, let α_{M} denote the homomorphisms $(u_i) \otimes m \mapsto (u_i m)$ from $\mathbb{Z}^{\mathbb{N}} \otimes \mathrm{M}$ to $\mathrm{M}^{\mathbb{N}}$. These define a morphism of functors, and α_{M} is an isomorphism if M is free and finitely generated. Every finitely generated submodule M of $\mathbb{Z}^{\mathbb{N}}$ is free (3.5.8, Corollary 3.2) and for such a submodule, $\mathrm{M}^{\mathbb{N}}$ can be identified with a submodule of $(\mathbb{Z}^{\mathbb{N}})^{\mathbb{N}} = \mathbb{Z}^{\mathbb{N} \times \mathbb{N}}$. As $\mathbb{Z}^{\mathbb{N}}$ is the direct limit of its finitely generated submodules M, $\mathbb{Z}^{\mathbb{N}} \otimes \mathbb{Z}^{\mathbb{N}} = \varinjlim \mathbb{Z}^{\mathbb{N}} \otimes \mathrm{M} = \bigcup_{\mathrm{M}} \mathrm{M}^{\mathbb{N}} = \mathrm{H}$. $\square$

3.8.12 Extension of Scalars

Let $\varphi : \mathrm{A} \to \mathrm{B}$ be a ring homomorphism. Consider B as an A-module (3.3.1, Example 3). For all A-modules E, $\mathrm{B} \otimes_{\mathrm{A}} \mathrm{E}$ can be endowed with a structure of B-module by $b \cdot (b' \otimes m) = bb' \otimes m$. With this structure, $\mathrm{B} \otimes_{\mathrm{A}} \mathrm{E}$, is said to be the B-module obtained by the **extension of scalars** of A to B along φ. This B-module is written $\varphi_*(\mathrm{E})$.

If $f : \mathrm{E} \to \mathrm{E}'$ is an A-module homomorphism, then the map $f_* = 1_{\mathrm{B}} \otimes_{\mathrm{A}} f : \mathrm{B} \otimes_{\mathrm{A}} \mathrm{E} \to \mathrm{B} \otimes_{\mathrm{A}} \mathrm{E}'$ is B-linear. This defines a functor $\varphi_* : \mathrm{A}\text{-}\mathcal{M}od \to \mathrm{B}\text{-}\mathcal{M}od$ which commutes with cokernels, direct sums, and direct limits.

For a B-module F, let $\varphi^*(\mathrm{F})$ be the A-module obtained by restricting scalars from B to A along φ (3.3.2). This gives a functor $\varphi^* : \mathrm{B}\text{-}\mathcal{M}od \to \mathrm{A}\text{-}\mathcal{M}od$. The modules $\varphi^*(\mathrm{F})$ and F have the same underlying additive group.

Proposition *The extension of scalars functor* $\varphi_* : \mathrm{A}\text{-}\mathcal{M}od \to \mathrm{B}\text{-}\mathcal{M}od$ *is the left adjoint of the restriction of scalars functor* φ^*.

In other words, for all A-modules E and all B-modules F, $\text{Hom}_B(B \otimes_A E, F) = \text{Hom}_A(E, F)$. This is the set of φ-linear maps from E to F.

Proof For an A-linear map $f : E \to F$, set $\bar{f}(b \otimes x) = bf(x)$ for all $b \in B$ and $x \in E$. For $g : B \otimes_A E \to F$, set $\bar{g}(x) = g(1 \otimes x)$ for all $x \in E$. The maps $f \mapsto \bar{f}$ and $g \mapsto \bar{g}$ are mutual inverses.   □

Corollary *If $\varphi : A \to B$ and $\psi : B \to C$ are ring homomorphisms then, for all A-modules E, $(\psi \circ \varphi)_*(E) = \psi_*(\varphi_*(E))$.*
   In other words, $C \otimes_B (B \otimes_A E) = C \otimes_A E$.

Proof The functor $\psi_* \circ \varphi_*$ is a left adjoint of $\varphi^* \circ \psi^* = (\psi \circ \varphi)^*$, and so may be identified with $(\psi \circ \varphi)_*$.   □

Examples (1) If E is a real vector space, then $\mathbb{C} \otimes_{\mathbb{R}} E$ is a complex vector space called the *complexification* of E whose $\mathbb{C}$-dimension is equal to the $\mathbb{R}$-dimension of E.
   (2) Let E be an A-module and I an ideal in A; the (A/I)-module $A/I \otimes E$ can be identified with E/IE, where IE is the submodule generated by the elements $ax, a \in I$ and $x \in E$. Indeed, IE is the image of $I \otimes E$ in E.

3.8.13 Extension of Scalars to the Field of Fractions

Suppose that A is an integral domain, K its field of fractions and $\iota : A \to K$ the canonical injection.

Proposition *For all A-modules E, the kernel of $\iota \otimes 1_E : E \to K \otimes_A E$ is the torsion submodule of E.*

Proof For $q \in A^*$, consider the submodule $A\frac{1}{q}$ of K and let $\iota_q : A \to A\frac{1}{q}$ be the inclusion homomorphism. Tensoring the commutative diagram

with E gives

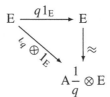

and so $\text{Ker}(\iota_q \otimes 1_E) = \{x \in E \mid qx = 0\}$.

Passing to the direct limit over the ordered set $\text{Mon}^*(A)$, we get $\text{Ker}(\iota \otimes 1_E) = \bigcup_q \{x \in E \mid qx = 0\}$, and this set is the torsion submodule of E. $\qquad\square$

3.8.14 Rank of a Module

Let A be an integral domain and K its field of fractions.

The *rank* of a finitely generated A-module M, written rk M, is the dimension of the K-vector space $K \otimes_A M$. If A is principal, M is isomorphic to some module $A^r \oplus T$, where T is a torsion module (3.5.8), and the rank of M is r.

If $0 \to M \to N \to P \to 0$ is a short exact sequence of finitely generated A-modules, then rk N = rk M + rk P. Indeed, K being the direct limit of free modules of rank 1, there is an exact sequence

$$0 \to K \otimes_A M \to K \otimes_A N \to K \otimes_A P \to 0.$$

If $f : M \to N$ is a finitely generated A-module homomorphism, let rk f denote the rank of Im f. Then rk N = rk f + rk(Coker f) and, assuming Ker f is finitely generated (which is always the case if A is Noetherian), rk M = rk(Ker f) + rk f.

Corollary *Let A be an integral domain. Every finitely generated torsion free A-module is isomorphic to a submodule of a finitely generated free module of the same rank.*

Proof Let K be the field of fractions of A, E a finitely generated torsion free A-module generated by the finite family (x_i) in E, and (e_j) a basis for the finite dimensional K-vector space $K \otimes_A E$. Define the scalars $a_i^j \in K$ by $\iota(x_i) = \sum a_i^j e_j$, and let d be their common denominator. The injection $\iota : E \to K \otimes_A E$ maps E into the free A-submodule $K \otimes_A E$ generated by $\frac{1}{d} e_j$. $\qquad\square$

3.8.15 Tensor Products of Algebras

Let E and F be A-algebras. There is a unique A-algebra structure on the A-module $E \otimes F$ such that $(x \otimes y)(x' \otimes y') = xx' \otimes yy'$. If E and F are associative (resp. unital, resp. commutative), the same holds for $E \otimes F$. If E and F are unital, the maps $\iota_1 : E \to E \otimes F$ and $\iota_2 : F \to E \otimes F$ defined by $\iota_1(x) = x \otimes 1$ and $\iota_2(y) = 1 \otimes y$ are algebra homomorphisms.

Let A-$\mathcal{Alg}$ denote the category of associative unital commutative, A-algebras.

Proposition *Let E and F be objects of A-$\mathcal{Alg}$. The algebra $E \otimes F$, equipped with ι_1 and ι_2, is the sum of E and F in the category A-$\mathcal{Alg}$.*

Proof We show that, for any A-algebra C and all morphisms $f : E \to C$, $g : F \to C$ there is a unique morphism $h : E \otimes F \to C$ such that, for all $x \in E$ and $y \in F$, $h(x \otimes 1) = f(x)$ and $h(1 \otimes y) = g(y)$, or such that

$$h(x \otimes y) = f(x)g(y) \tag{*}$$

since $x \otimes y = (x \otimes 1)(1 \otimes y)$. We know that there is a unique A-linear map $h : E \otimes F \to C$ satisfying (∗); moreover, for $x, x' \in E$ and $y, y' \in F$ $h((x \otimes y)(x' \otimes y')) = h(xx' \otimes yy') = f(xx')g(yy') = f(x)g(y)f(x')g(y') = h(x \otimes y)h(x' \otimes y')$ since C is commutative.   □

Corollary *Under the same assumptions,* $E \otimes_A F$ *is an amalgamated sum of* E *and* F *over* A *in the category of commutative rings.*

3.8.16

Let E, B, C be algebras, $f : B \to E$ and $g : C \to E$ algebra homomorphisms, then the image of the homomorphism $x \otimes y \mapsto f(x)g(y)$ of $B \otimes C \to E$ is the subalgebra of E generated by $f(B)$ and $g(C)$. In particular, if B and C are unital subalgebras of an algebra E, the subalgebra $B \cdot C$ of E generated by B and C is identified with a quotient algebra of $B \otimes C$.

3.8.17

Let A and B be two rings, $f : A \to B$ a ring homomorphism and E an A-algebra. The B-module $B \otimes_A E$, equipped with the multiplication defined in 3.8.15, is a B-algebra, said to be the B-algebra obtained from E by extension of scalars from A to B along f. For any B-algebra F, F can be regarded as an A-algebra by restriction of scalars, and

$$\mathrm{Hom}_{B\text{-}\mathscr{A}lg}(B \otimes_A E, F) = \mathrm{Hom}_{A\text{-}\mathscr{A}lg}(E, F) .$$

3.8.18   Examples

(1) $A[X, Y] = A[X] \otimes_A A[Y]$.

Similarly $A[X_1, ..., X_p, Y_1, ..., Y_q] = A[X_1, ..., X_p] \otimes_A A[Y_1, ..., Y_q]$. More generally, if I is an arbitrary set and (I', I'') a partition of I, then $A[(X_i)_{i \in I}] = A[(X_i)_{i \in I'}] \otimes_A A[(X_i)_{i \in I''}]$.

(2) Let $P_1, ..., P_k \in A[X_1, ..., X_p]$, $Q_1, ..., Q_l \in A[Y_1, ..., Y_q]$. Then

$$(A[X_1, ..., X_p]/(P_1, ..., P_k)) \otimes_A (A[Y_1, ..., Y_q]/(Q_1, ..., Q_l))$$
$$= A[X_1, ..., X_p, Y_1, ..., Y_q]/(P_1, ..., P_k, Q_1, ..., Q_l).$$

(3) By extension of scalars from A to B along f, the algebra $A[(X_i)_{i \in I}]$ gives $B[(X_i)_{i \in I}]$ and $A[(X_i)_{i \in I}]/((P_j)_{j \in J})$ gives $B[(X_i)_{i \in I}]/((f_* P_j)_{j \in J})$.

(4) The algebra $\mathbb{C} \otimes_{\mathbb{R}} \mathbb{C}$ is isomorphic to the product algebra $\mathbb{C} \times \mathbb{C}$. Indeed $\mathbb{C} = \mathbb{R}[X]/(X^2 + 1)$, and so $\mathbb{C} \otimes_{\mathbb{R}} \mathbb{C} = \mathbb{C}[X]/(X^2 + 1) = \mathbb{C}[X]/((X - i)(X + i))$. The substitutions $X \mapsto i$ and $X \mapsto -i$ give two homomorphisms from $\mathbb{C}[X]$ to $\mathbb{C}$, and hence there is a homomorphism from $\mathbb{C}[X]$ to $\mathbb{C} \times \mathbb{C}$ whose kernel is the ideal $(X - i) \cap (X + i) = ((X - i)(X + i))$. This in turn gives an injective homomorphism from $\mathbb{C}[X]/((X - i)(X + i))$ to $\mathbb{C} \times \mathbb{C}$. This homomorphism is bijective since both these algebras have dimension 4 over $\mathbb{R}$. (We will resume this argument in the more general framework of Galois theory.)

(5) If X is a topological space, $\mathbb{C} \otimes_{\mathbb{R}} C(X, \mathbb{R}) = C(X, \mathbb{C})$.

Exercises 3.8. (Tensor product)

1.—Let E and F be two A-modules, $L_1 \xrightarrow{\delta} L_0 \xrightarrow{\varepsilon} E \to 0$ and $M_1 \xrightarrow{\delta'} M_0 \xrightarrow{\varepsilon'} F \to 0$ presentations of E and F respectively. Show that

$$(L_1 \otimes M_0) \oplus (L_0 \otimes M_1) \xrightarrow{(\delta \otimes 1, 1 \otimes \delta')} L_0 \otimes M_0 \xrightarrow{\varepsilon \otimes \varepsilon'} E \otimes F \to 0$$

is a presentation of $E \otimes F$.

2.—Let K be a field and $E_1 \xrightarrow{u} E \xrightarrow{f} F \xrightarrow{v} F_1$ and $E_1' \xrightarrow{u'} E' \xrightarrow{f'} F' \xrightarrow{v'} F_1'$ two exact sequences of vector spaces. Show that $(E_1 \otimes E') \oplus (E \otimes E_1') \xrightarrow{(u \otimes 1, 1 \otimes u')}$

$$E \otimes E' \xrightarrow{f \otimes f'} F \otimes F' \xrightarrow{\begin{pmatrix} v \otimes 1 \\ 1 \otimes v' \end{pmatrix}} (F_1 \otimes F') \oplus (F \otimes F_1')$$ is an exact sequence. In particular $\operatorname{Ker}(f \otimes f') = (\operatorname{Ker} f \otimes E') + (E \otimes \operatorname{Ker} f')$.

3.—Let E and F be K-vector spaces and $t \in E \otimes F$. Set $G = E \otimes F/K \cdot t$ and let $f : E \otimes F \to G$ be the canonical map. Show that $(f(x \otimes y) = 0) \Rightarrow (x \otimes y = 0)$ if and only if t is of rank $\geqslant 2$ and that the restriction of f to the set of simple tensors is injective if and only if t is of rank $\geqslant 3$.

4.—Let E and F be modules over an integral domain A, $x \in E$ and $y \in F$ non-torsion elements. Show that $x \otimes y \neq 0$ (extend the scalars to the field of fractions of A).

5.—Let M and N be square matrices with entries in a field K.

(a) Show that if M and N are diagonal (resp. triangular, resp. symmetric), then the same holds for $M \otimes N$.

(b) Show that if M_1 and N_1 are respectively similar to M and N, $M_1 \otimes N_1$ is similar to $M \otimes N$.

(c) Write $\det(M \otimes N)$ in terms of $\det(M)$ and $\det(N)$.

(d) Determine the eigenvalues of $M \otimes N$ with their multiplicities assuming known those of M and N.

6.—Let K be a field.

(a) Let X be a set and E a K-vector space. Show that the map $f \otimes t \mapsto (x \mapsto f(x) \cdot t)$ from $K^X \otimes E$ to E^X is bijective if either X is finite or E is finite dimensional, and is always injective (consider E as the increasing union of finite dimensional vector spaces). What is its image?

(b) Let X and Y be two sets. Show that the map $f \otimes g \mapsto h$ from $K^X \otimes K^Y$ to $K^{X \times Y}$ defined by $h(x, y) = f(x) \cdot g(y)$ is injective.

(c) Let X and Y be topological spaces. Show that the map $C(X, \mathbb{R}) \otimes_{\mathbb{R}} C(Y, \mathbb{R})$ assigning to every vector $f \otimes g\, h$ defined by $h(x, y) = f(x) \cdot g(y)$ is injective. Can $\mathbb{R}$ be replaced by $\mathbb{C}$?

(d) Supposing X and Y to be compact and metrizable, show that this map is image-dense.

7.—Let $h : A \to B$ be a ring homomorphism, M and N two B-modules. Show that there is an exact sequence:

$$M \otimes_A B \otimes_A N \xrightarrow{d} M \otimes_A N \to M \otimes_B N \to 0 \,,$$

where $d(x \otimes b \otimes y) = bx \otimes y - x \otimes by$.

8.—Let K be a field. For every algebraic subset X of K^n, let A(X) be the algebra of functions on X induced by polynomials.

(a) Let $X \subset K^m$ and $Y \subset K^n$ be algebraic sets. Show that $A(X \times Y) = A(X) \otimes_K A(Y)$.

(b) Let $p : K^{n+r} = K^n \times K^r \to K^n$ and $q : K^{n+s} \to K^n$ be the first projections, $S \subset K^n$, $X \subset K^{n+r}$ and $Y \subset K^{n+s}$ algebraic sets such that $p(X) \subset S$, $q(Y) \subset S$. Define $p^* : A(S) \to A(X)$ by $p^*(f) = f \circ p$, and $q^* : A(S) \to A(Y)$ likewise. Show that $A(X \times_S Y) = A(X) \otimes_{A(S)} A(Y)$.

9.—(a) Let X and Y be two compact metrizable spaces, $f : Y \to X$ a continuous map, and E a vector bundle over X (3.4, Exercise 9). Define a bundle $F = f^*(E)$ over Y by equipping the fibre product en $Y \times_X E$ with its first projection onto Y and, for all $y \in Y$, with the structure obtained on $F(y)$ by identifying it with $E(f(y))$. Show that the module S(F) of continuous sections of F can be identified with a module obtained from S(E) by extension of scalars along $f^* : C(X) \to C(Y)$.

(b) Let E and F be two vector bundles over X. Define a vector bundle $E \otimes F$ over X such that $(E \otimes F)(x) = E(x) \otimes F(x)$ for all $x \in X$, and show that $S(E \otimes F) = S(E) \otimes_{C(X)} S(F)$.

10.—Let K be a field, A the ring $K[X, Y]$ and $\mathfrak{m}$ the maximal ideal in A consisting of polynomials without constant terms. The aim is to determine the module $\mathfrak{m} \otimes_A \mathfrak{m}$.

(a) Show that there is an exact sequence $0 \longrightarrow A \xrightarrow{\begin{pmatrix} -Y \\ X \end{pmatrix}} A^2 \xrightarrow{(X, Y)} \mathfrak{m} \longrightarrow 0$.

(b) Show that this sequence induces an exact sequence

$$0 \to A \to \mathfrak{m} \oplus \mathfrak{m} \to \mathfrak{m}^2 \to 0 \,,$$

where $\mathfrak{m}^2$ denotes the ideal of A generated by the monomials of degree 2.

(c) Deduce an exact sequence $0 \to K \to \mathfrak{m} \otimes \mathfrak{m} \to \mathfrak{m}^2 \to 0$, where K is endowed with the A-module structure given by its identification with $A/\mathfrak{m}$.

(d) Show that the element $x \otimes y - y \otimes x$ of $\mathfrak{m} \otimes \mathfrak{m}$ is nonzero and generates the torsion submodule.

(e) Show that the submodule of $\mathfrak{m} \otimes \mathfrak{m}$ generated by $x \otimes x$, $x \otimes y$ and $y \otimes y$ is isomorphic to $\mathfrak{m}^2$, and finally that $\mathfrak{m} \otimes \mathfrak{m}$ is isomorphic to $\mathfrak{m}^2 \oplus K$.

11.—In what follows, A and B denote two unitary commutative rings, and $\alpha : A \to B$ a unitary ring homomorphism. The A-module $B \otimes_A B$ being equipped with its canonical ring structure (multiplication being given by: $(x \otimes y)(x' \otimes y') = xx' \otimes yy')$, let $p : B \otimes_A B \to B$ be the homomorphism given by: $p(x \otimes y) = xy$, and $j_1, j_2 : B \to B \otimes_A B$ the homomorphisms defined by: $j_1(x) = x \otimes 1$, $j_2(x) = 1 \otimes x$.

(a) Show that the kernel I of p is generated by the elements $1 \otimes x - x \otimes 1$ ($x \in B$).

(b) For any B-module M, let $\mathscr{D}_A(B, M)$ be the set of "A-derivation from B to M", defined by:

$$\mathscr{D}_A(B, M) = \{D \in \mathrm{Hom}_{A\text{-}\mathscr{M}od}(B, M) \mid \forall x, y \in B, D(xy) = xD(y) + yD(x)\}.$$

(α) Show that $\mathscr{D}_A(B, M)$ is a B-submodule of $\mathrm{Hom}_{A\text{-}\mathscr{M}od}(B, M)$.

(β) Show that $D \circ \alpha = 0$, for any A-derivation D from B to M.

(c) The ring $B \otimes_A B$ being equipped with its B-algebra structure defined by j_1 (i.e. B acts on $B \otimes_A B$ by: $b \cdot (x \otimes y) = j_1(b)(x \otimes y)$), consider the isomorphism of B-modules:

$$v : \mathrm{Hom}_{A\text{-}\mathscr{M}od}(B, M) \to \mathrm{Hom}_{B\text{-}\mathscr{M}od}(B \otimes_A B, M)$$

given by: $v(f)(x \otimes y) = xf(y)$. Show that $v(\mathscr{D}_A(B, M))$ is the B-submodule of $\mathrm{Hom}_{B\text{-}\mathscr{M}od}(B \otimes_A B, M)$ whose elements vanish on $j_1(B)$ and on I^2.

(d) Deduce that there exists a B-module $\Omega_{B/A}$ (called the *module of* A-*differentials of* B) and an A-derivation $d_{B/A} : B \to \Omega_{B/A}$ such that, for all A-modules M and A-derivations $D : B \to M$, there is a unique homomorphism of B-modules $u : \Omega_{B/A} \to M$ such that: $D = u \circ d_{B/A}$.

(e) Describe $\Omega_{B/A}$ in the following cases:

(α) $A = \mathbb{R}$, $B = \mathbb{C}$, α is the canonical injection;

(β) $A = \mathbb{Z}$, $B = \mathbb{Q}$, α is the canonical injection;

(γ) $A = F_p$, $B = F_q$, α is the canonical injection (F_p and F_q are respectively the fields of order p and $q = p^f$, p being prime and $f \in \mathbb{N}^*$);

(δ) $A = k[X^p]$, $B = k[X]$, where p is prime and k a field, α being the canonical injection;

(ε) $B = A^n$, $n \geqslant 1$, α being defined by: $\alpha(a) = (a, a, ..., a)$;

(ζ) $\alpha : A \to B$ is surjective;

(η) B is an A-algebra of polynomials with indeterminates in an arbitrary set L, the set $\alpha : A \to A[X_e]_{e \in L}$ being defined by $\alpha(a) = a$.

12.—Let E and F be real Euclidean spaces.

(a) Show that there is a unique symmetric bilinear form on $E \otimes F$ satisfying $\langle x \otimes y, x' \otimes y' \rangle = \langle x, x' \rangle \cdot \langle y, y' \rangle$.

(b) Let $(e_1, ..., e_n)$, $(f_1, ..., f_p)$ be orthonormal bases for E and F respectively. Show that the elements $e_i \otimes f_j$ form an orthonormal basis for $E \otimes F$. Deduce that $E \otimes F$ is Euclidian.

(c) Generalize these results to Hermitian spaces over $\mathbb{C}$.

13.—Let K be a field. For every vector space E, denote by P(E) the projective space of E, i.e. the set of vector subspaces of rank 1 of E and ψ_E the map from $E - \{0\}$ to P(E) assigning to every vector the line it generates.

(a) Let E and F be two vector spaces. Show that the map $(L, M) \mapsto L \otimes M$ from $P(E) \times P(F)$ to $P(E \otimes F)$ is injective.

(b) Let $x_0 \in E$ and $y_0 \in F$. Determine the tangent linear map $T_{x_0, y_0} \theta$ to θ : $(x, y) \mapsto x \otimes y$ from $E \times F$ to $E \otimes F$ at (x_0, y_0) (i.e. the degree 1 part of the expansion of $(u, v) \mapsto (x_0 + u) \otimes (y_0 + v)$).

(c) Suppose $x_0 \neq 0$ and $y_0 \neq 0$. Show that, by passing to quotients, $T_{x_0, y_0} \theta$ gives an injective map from $E/Kx_0 \times F/Ky_0$ to $E \otimes F/K(x_0 \otimes y_0)$.

(d) Suppose that $K = \mathbb{R}$ or $\mathbb{C}$ and that E and F are finite dimensional. Show that $P(E) \times P(F)$ can be identified with a closed submanifold of $P(E \otimes F)$.

3.9   The Kronecker Homomorphism

3.9.1

Let E, F, E', F' be A-modules. The map $(f, f') \mapsto f \otimes f'$ from $\text{Hom}(E, F) \times \text{Hom}(E', F')$ to $\text{Hom}(E \otimes E', F \otimes F')$ is bilinear. Hence there is a homomorphism

$$\varphi_{F,F'}^{E,E'} : f \otimes f' \mapsto f \otimes f'$$

from $\text{Hom}(E, F) \otimes \text{Hom}(E', F')$ to $\text{Hom}(E \otimes E', F \otimes F')$, called the **Kronecker homomorphism.**

In particular, this gives morphisms $\alpha_F^E = \varphi_{A,F}^{E,A} : E^\top \otimes F \to \text{Hom}(E, F)$ and $\beta^{E,F} = \varphi_{A,A}^{E,F} : E^\top \otimes F^\top \to (E \otimes F)^\top$, where $E^\top = \text{Hom}(E, A)$.

3.9.2   Functoriality

Let $u : E_1 \to E$, $u' : E_1' \to E'$, $v : F \to F_1$, and $v' : F' \to F_1'$ be homomorphisms of A-modules. These give a homomorphism $v_* u^* = u^* v_* : f \mapsto v \circ f \circ u$ from $\text{Hom}(E, F)$ to $\text{Hom}(E_1, F_1)$. The commutativity of the following diagram follows readily:

$$\begin{array}{ccc}
\text{Hom}(E, F) \otimes \text{Hom}(E', F') & \xrightarrow{\;v_* u^* \otimes v'_* u'^*\;} & \text{Hom}(E_1, F_1) \otimes \text{Hom}(E'_1, F'_1) \\
\varphi \downarrow & & \downarrow \varphi \\
\text{Hom}(E \otimes E', F \otimes F') & \xrightarrow{\;(v \otimes v')_* (u \otimes u')^*\;} & \text{Hom}(E_1 \otimes E'_1, F_1 \otimes F'_1)
\end{array}$$

This says that φ is a morphism from the functor $(E, E', F, F') \mapsto \text{Hom}(E, F) \otimes \text{Hom}(E', F')$ to the functor $(E, E', F, F') \mapsto \text{Hom}(E \otimes E', F \otimes F')$, both these functors being defined on $A\text{-}\mathcal{M}od^0 \times A\text{-}\mathcal{M}od^0 \times A\text{-}\mathcal{M}od \times A\text{-}\mathcal{M}od$, with values in $A\text{-}\mathcal{M}od$.

3.9.3

For homomorphisms $f : M \to M'$ and $g : N \to N'$, let $f \oplus g$ be the homomorphism $(x, y) \mapsto (f(x), g(y))$ from $M \oplus N$ to $M' \oplus N'$.

Proposition *If* $E = E_1 \oplus E_2$, *then* $\varphi_{F,F'}^{E,E'} = \varphi_{F,F'}^{E_1,E'} \oplus \varphi_{F,F'}^{E_2,E'}$. *Similarly, if* $F = F_1 \oplus F_2$, *then* $\varphi_{F,F'}^{E,E'} = \varphi_{F_1,F'}^{E,E'} \oplus \varphi_{F_2,F'}^{E,E'}$.

Proof Apply functoriality to $\iota_1 : E_1 \to E$ and $\iota_2 : E_2 \to E$ (resp. to $\pi_1 : E \to E_1$ and $\pi_2 : E \to E_2$).

Corollary *If* $E = E_1 \oplus E_2$, *then* $\varphi_{F,F'}^{E,E'}$ *is an isomorphism if and only if so are* $\varphi_{F,F'}^{E_1,E'}$ *and* $\varphi_{F,F'}^{E_2,E'}$. *Similarly, if* $F = F_1 \oplus F_2$, *then* $\varphi_{F,F'}^{E,E'}$ *is an isomorphism if and only if so are* $\varphi_{F_1,F'}^{E,E'}$ *and* $\varphi_{F_2,F'}^{E,E'}$.

This follows from the fact that $f \oplus g$ is an isomorphism if and only if so are f and g.

3.9.4 Specific Cases When φ is an Isomorphism

Proposition *If* E *and* E' *(resp.* E *and* F*) are finitely generated and projective (i.e. isomorphic to direct factors of a finitely generated free module),* $\varphi_{F,F'}^{E,E'}$ *is an isomorphism.*

Proof It is immediate if $E = E' = A$, or if $E = F = A$. The case where E and E' (resp. F and F') are finitely generated and free follows by iteration from (3.9.3, Corollary). The case where ù E and E' (resp. E and F) are finitely generated and projective then follows by applying the converse part of the same corollary.

Corollary *If either* E *or* F *is finitely generated and projective, then* $\alpha_F^E : E^\top \otimes F \to \text{Hom}(E, F)$ *is an isomorphism.*

3.9.5   Specific Cases When α is Bijective

Proposition *If E is free and F has finite presentation, then $\alpha_F^E : E^\top \otimes F \to \mathrm{Hom}(E, F)$ is an isomorphism.*

Proof Let $L_1 \to L_0 \to F \to 0$ be a presentation of F, where L_0 and L_1 are are finitely generated and free. By Corollary 3.9.4, the morphisms $\alpha_{L_0}^E$ and $\alpha_{L_1}^E$ are bijective. As the diagram

$$
\begin{array}{ccccccc}
E^\top \otimes L_1 & \longrightarrow & E^\top \otimes L_0 & \longrightarrow & E^\top \otimes F & \longrightarrow & 0 \\
\downarrow{\scriptstyle \alpha_{L_1}^E} & & \downarrow{\scriptstyle \alpha_{L_0}^E} & & \downarrow{\scriptstyle \alpha_F^E} & & \\
\mathrm{Hom}(E, L_1) & \to & \mathrm{Hom}(E, L_0) & \to & \mathrm{Hom}(E, F) & \longrightarrow & 0
\end{array}
$$

where the rows are exact, is commutative, $E^\top \otimes F$ and $\mathrm{Hom}(E, F)$ can be identified with the cokernel of the same homomorphism.   □

3.9.6   The Noetherian Case

Proposition *If A is Noetherian and E is free, then $\alpha_F^E : E^\top \otimes F \to \mathrm{Hom}(E, F)$ is injective. Its image is the set of $f : E \to F$ for which $f(E)$ is contained in a finitely generated submodule of F.*

Proof Let (F_i) be an increasing directed family of finitely generated submodules of F such that $F = \bigcup F_i$. For all i, the module F_i has a finite presentation since A is Noetherian, and so $\alpha_{F_i}^E$ is bijective. As $F = \varinjlim F_i$, passing to the direct limit, gives an isomorphism from $E^\top \otimes F = \varinjlim E^\top \otimes F_i$ onto $\varinjlim \mathrm{Hom}(E, F_i)$. This direct limit is just the union of submodules $\mathrm{Hom}(E, F_i)$ of $\mathrm{Hom}(E, F)$.   □

Corollary 3.9 *If A is Noetherian and E est free, then $\beta^{E,F} : E^\top \otimes F^\top \to (E \otimes F)^\top$ is injective.*

Indeed, $\beta^{E,F} = \alpha_{F^\top}^E : E^\top \otimes F^\top \to \mathrm{Hom}(E, F^\top) = B(E, F; A) = (E \otimes F)^\top$.

Corollary 3.10 *If A is Noetherian, E F free, then*

$$
\varphi_{F,F'}^{E,E'} : \mathrm{Hom}(E, F) \otimes \mathrm{Hom}(E', F') \to \mathrm{Hom}(E \otimes E', F \otimes F')
$$

is injective.

Proof If $F = A$, the homomorphism $\varphi_{A,F'}^{E,E'} = \alpha_{\mathrm{Hom}(E',F')}^E : \mathrm{Hom}(E, A) \otimes \mathrm{Hom}(E', F') \to \mathrm{Hom}(E, \mathrm{Hom}(E', F')) = B(E, E'; F') = \mathrm{Hom}(E \otimes E', F')$ is injective by Corollary 1. The result follows by passing to the direct sum when F is free and finitely generated, then by passing to the direct limit when F is free. Hence it is the increasing directed union of finitely generated free modules.

3.9.7 The Principal Ideal Domain Case

Proposition *If A is principal, $\alpha_F^E : E^\top \otimes F \to \mathrm{Hom}(E, F)$ is injective for all E and F.*

Proof First suppose that F is finitely generated. By 3.5.8, it reduces to the case when F is cyclic. If $F = A$, then the morphism α_F^E is the identity of $E^\top$. If $F = A/(a)$ with $a \neq 0$, the exact sequence $0 \to E^\top \xrightarrow{a\cdot 1} E^\top \to \mathrm{Hom}(E, F)$ and $E^\top \xrightarrow{a\cdot 1} E^\top \to E^\top \otimes F \to 0$ associated to the exact sequence $0 \to A \xrightarrow{a\cdot 1} A \to F \to 0$ show that α identifies $E^\top \otimes F$ with a submodule de $\mathrm{Hom}(E, F)$.

The general case follows as in Proposition 3.9.6. $\qquad\square$

3.9.8 Image of α

Proposition *The image of $\alpha_F^E : E^\top \otimes F \to \mathrm{Hom}(E, F)$ is the set of homomorphisms $f : E \to F$ factorizing through a finitely generated free A-module.*

Proof If $t = \sum_1^r \xi_i \otimes y_i \in E^\top \otimes F$, then $\alpha(t) = g \circ f$, where $f : E \to A^r$ and $g : A^r \to F$ are defined by $f(x) = (\xi_1(x), ..., \xi_r(x))$ and $g(a_1, ..., a_r) = a_1 y_1 + \cdots + a_r y_r$. Conversely, if $h = g \circ f$ with $f : E \to A^r$ and $g : A^r \to F$, then $h = \sum \xi_i \otimes y_i$, where the ξ_i are the coordinates of f and the y_i are the images under g of the elements of a canonical basis. $\qquad\square$

Corollary *Let E be an A-module. The following conditions are equivalent:*

(i) 1_E belongs to the image of $\alpha_E^E : E^\top \otimes E \to \mathrm{End}(E)$;
(ii) α_E^E is bijective;
(iii) E is finitely generated and projective.

Proof (i) $\Rightarrow$ (iii) By the above proposition and Proposition 3.4.8, (iii) $\Rightarrow$ (ii) by (3.9.4, Corollary), and (ii) $\Rightarrow$ (i) is then immediate.

3.9.9 Contraction and Composition

Proposition *Let E, F, G be three A-modules. The diagram*

$$
\begin{array}{ccc}
E^\top \otimes F \otimes F^\top \otimes G & \xrightarrow{\;\alpha_F^E \otimes \alpha_G^F\;} & \mathrm{Hom}(E, F) \otimes \mathrm{Hom}(F, G) \\
\Big\downarrow{k} & & \Big\downarrow{c} \\
E^\top \otimes G & \xrightarrow{\quad\alpha_G^E\quad} & \mathrm{Hom}(E, G)
\end{array}
$$

where k is the **contraction** *$\xi \otimes y \otimes \eta \otimes z \mapsto \eta(y) \cdot \xi \otimes z$ and c the* **composition** *$(f \otimes g) \mapsto g \circ f$, is commutative.*

Proof The image of $\xi \otimes y \otimes \eta \otimes z$ under the two composite maps in $\mathrm{Hom}(E, G)$ is the map $x \mapsto \xi(x) \cdot \eta(y) \cdot z$.   □

3.9.10 Trace

Let E be an A-module. Consider the Kronecker homomorphism $\alpha_E = \alpha_E^E : E^\top \otimes E \to \mathrm{End}(E)$ and the contraction $\kappa_E : E^\top \otimes E \to A$ defined by $\kappa_E(\xi \otimes x) = \xi(x)$. Suppose that $\mathrm{Ker}\,\alpha_E \subset \mathrm{Ker}\,\kappa_E$. Then, for all $f : E \to E$ in the image of α_E, define $\mathrm{Tr}(f) \in A$ by $\mathrm{Tr}(f) = \kappa(t)$, where $t \in \alpha_E^{-1}(f)$, a scalar independent of the choice of t. The scalar $\mathrm{Tr}(f)$ is called the **trace** of f. The map $\mathrm{Tr} : \mathrm{Im}\,\alpha_E \to A$ is A-linear.

Proposition *Suppose that* E *is finitely generated and free and let* (e_i) *be a basis for* E. *For all endomorphisms* f *of* E, *the trace of* f *is the sum* $\sum a_i^i$ *of the diagonal entries of the matrix representing* f.

Proof As α_E is bijective, $\mathrm{Tr}(f)$ is defined for all $f \in \mathrm{End}(E)$. Let (e_i') be the dual basis $E^\top$ of (e_i). For all $x \in E$, $x = \sum e_i'(x)e_i$; so $f(x) = \sum e_i'(x)f(e_i)$. Thus $f = \alpha(\sum e_i' \otimes f(e_i))$ and $\mathrm{Tr}(f) = \sum e_i'(f(e_i)) = \sum a_i^i$.   □

3.9.11

Proposition *Let* E *and* F *be two A-modules,* $f : E \to F$ *a homomorphism in the image of* α_F^E *and* $g : F \to E$ *an arbitrary homomorphism. Assume that* $\mathrm{Ker}\,\alpha_E \subset \mathrm{Ker}\,\kappa_E$ *and* $\mathrm{Ker}\,\alpha_F \subset \mathrm{Ker}\,\kappa_F$. *Then* $\mathrm{Tr}(g \circ f)$ *and* $\mathrm{Tr}(f \circ g)$ *are well defined, and* $\mathrm{Tr}(g \circ f) = \mathrm{Tr}(f \circ g)$.

Proof Let $t = \sum x_i' \otimes y_i \in E^\top \otimes F$ be such that $f = \alpha_F^E(t)$. Then $g \circ f = \alpha_E(\sum x_i' \otimes g(y_i))$ and $f \circ g = \alpha_F(\sum(x_i' \circ g) \otimes y_i)$, so that $\mathrm{Tr}(g \circ f)$ and $\mathrm{Tr}(f \circ g)$ are defined and equal to $\sum x_i'(g(y_i))$.   □

3.9.12 Trace and Extension of Scalars

Let $h : A \to A'$ be a ring homomorphism, E and F A-modules, E′ and F′ the A′-modules resulting from E and F by extension of scalars. Denote by h_* the $x \mapsto 1 \otimes x$ from E to E′ (resp. from F to F′), and by h_* the map $\xi \mapsto h \circ \xi$ from $E^\top$ to $\mathrm{Hom}_A(E, A') = \mathrm{Hom}_{A'}(E', A') = E'^\top$. For all A-linear maps $f : E \to F$, let f_* be the A′-linear map $1_{A'} \otimes f : E' \to F'$.

Proposition *The following diagrams are commutative:*

(a)

$$
\begin{array}{ccc}
\mathrm{E}^{\top} \otimes_A \mathrm{F} & \xrightarrow{\ \alpha^{\mathrm{E}}_{\mathrm{F}}\ } & \mathrm{Hom}_A(\mathrm{E}, \mathrm{F}) \\
{\scriptstyle h_* \otimes h_*}\big\downarrow & & \big\downarrow{\scriptstyle f \mapsto f_*} \\
\mathrm{E}'^{\top} \otimes_{A'} \mathrm{F}' & \xrightarrow{\ \alpha^{\mathrm{E}'}_{\mathrm{F}'}\ } & \mathrm{Hom}_{A'}(\mathrm{E}', \mathrm{F}')
\end{array}
$$

(b)

$$
\begin{array}{ccc}
\mathrm{E}^{\top} \otimes_A \mathrm{E} & \xrightarrow{\ \kappa_{\mathrm{E}}\ } & A \\
{\scriptstyle h_* \otimes h_*}\big\downarrow & & \big\downarrow{\scriptstyle h} \\
\mathrm{E}'^{\top} \otimes_{A'} \mathrm{E}' & \xrightarrow{\ \kappa_{\mathrm{E}'}\ } & A'
\end{array}
$$

Proof (a) Let $\xi \in \mathrm{E}^{\top}$ and $y \in \mathrm{F}$. Then $\alpha^{\mathrm{E}}_{\mathrm{F}}(\xi \otimes y) = (x \mapsto \xi(x) \cdot y)$, which gives a map $a' \otimes x \mapsto a' \otimes \xi(x) \cdot y$ in $\mathrm{Hom}(\mathrm{E}', \mathrm{F}')$. In other words, given the definition of the A-module structure on A', $\xi \otimes y$ gives the element $h_*(\xi) \otimes (1 \otimes y)$ in $\mathrm{E}'^{\top} \otimes \mathrm{F}'$, whose image under the Kronecker homomorphism is the map $a' \otimes x \mapsto a' \cdot h(\xi(x)) \cdot 1 \otimes y = a' \otimes \xi(x) \cdot y$.

(b) For $\xi \in \mathrm{E}^{\top}$ and $x \in \mathrm{E}$, $\kappa \circ (h_* \otimes h_*)(\xi \otimes x) = \kappa(h_*(\xi) \otimes (1 \otimes x)) = h(\xi(x))$. $\qquad\square$

Corollary *Suppose that* $\mathrm{Ker}\,\alpha_{\mathrm{E}} \subset \mathrm{Ker}\,\kappa_{\mathrm{E}}$ *and* $\mathrm{Ker}\,\alpha_{\mathrm{E}'} \subset \mathrm{Ker}\,\kappa_{\mathrm{E}'}$ *hold. Then, for* $f \in \mathrm{Im}(\alpha_{\mathrm{E}})$, $\mathrm{Tr}(f_*) = h(\mathrm{Tr}(f))$.

3.9.13

Proposition *Suppose that the ring A is reduced. Then,* $\mathrm{Ker}\,\alpha_{\mathrm{E}} \subset \mathrm{Ker}\,\kappa_{\mathrm{E}}$ *for all A-modules E.*

Proof Let $t \in \mathrm{E}^{\top} \otimes \mathrm{E}$ be such that $\alpha_{\mathrm{E}}(t) = 0$. We show that $\kappa_{\mathrm{E}}(t) = 0$. By 3.1.7, all we need to show is that, for all homomorphisms h from A to a *field* A', $h(\kappa_{\mathrm{E}}(t)) = 0$. With the notation of 3.9.12 and assuming A' is a field, the homomorphism $\kappa_{\mathrm{E}'}$: $\mathrm{E}'^{\top} \otimes_{A'} \mathrm{E}' \to \mathrm{End}(\mathrm{E}')$ is injective by Proposition 3.9.6, and so the image of t under $h_* \otimes h_* : \mathrm{E}^{\top} \otimes \mathrm{E} \to \mathrm{E}'^{\top} \otimes \mathrm{E}'$ is zero, and $h(\kappa_{\mathrm{E}}(t)) = \kappa_{\mathrm{E}'} \circ (h_* \otimes h_*)(t) = 0$. $\qquad\square$

3.9.14 Computation of the Rank of a Tensor

Proposition *Let K be a field, E and F vector spaces over K and* $t \in \mathrm{E}^{\top} \otimes \mathrm{F}$. *The rank (see 3.8.2) of t equals the rank (i.e. the dimension of the image) of the linear map* $\alpha^{\mathrm{E}}_{\mathrm{F}}(t) : \mathrm{E} \to \mathrm{F}$.

Proof Let $t \in E^\top \otimes F$ and set $f = \alpha_F^E(t)$. If $t = \sum_i x_i \otimes y_i$, then the image of f is contained in the vector subspace of F generated by the y_i, and so $\mathrm{rk}(f) \leqslant r$. Hence rk $f \leqslant \mathrm{rk}(t)$.

Let d be the dimension of the image V of f, $(y_1, ..., y_d)$ a basis for V and define the linear forms $u_1, ..., u_d$ on E by $f(x) = \sum u_i(x) \cdot y_i$. Then $f = \alpha(\sum u_i \otimes y_i)$, and as α_F^E is injective (Proposition 3.9.6), $t = \sum u_i \otimes y_i$. Thus $\mathrm{rk}(t) \leqslant d = \mathrm{rk}(f)$.
□

Corollary *Let* E *and* F *be vector spaces over* K, E' *a vector space of linear forms on* E *separating the points of* E *(i.e. such that* $(\forall x \in E, x \neq 0)$ $(\exists f \in E')$ $f(x) \neq 0)$. *Let* $t = \sum x_i \otimes y_i \in E \otimes F$. *Then the rank of* t *equals the rank of the map* $f \mapsto \sum f(x_i) \cdot y_i$ *from* E' *to* F.

Proof The map $\delta : E \to E'^\top$ defined by $\delta(x) = (f \mapsto f(x))$ is injective, and identifies E with a direct factor of $E'^\top$. As a result, the rank of t equals the rank of $(\delta \otimes 1_F)(t) \in E'^\top \otimes F$. The rank of the latter tensor equals the rank of its image in $\mathrm{Hom}(E', F)$ under $\alpha_F^{E'}$, and this image is the map $f \mapsto \sum f(x_i) \cdot y_i$.
□

Exercises 3.9. (Kronecker Homomorphism)

1. *(Norm ε)*— Let E and F be normed vector spaces over $\mathbb{R}$ or $\mathbb{C}$. Denote by $E^\top$ the space of continuous linear forms on E.

(a) Define injections from $E \otimes F$ to the space $L(E^\top, F)$ of continuous linear forms from $E^\top$ to F, to $L(F^\top, E)$ and to the space $B(E^\top, F^\top)$ of continuous bilinear forms on $E^\top \times F^\top$. Show that the norms induced on $E \otimes F$ by these three injections agree (use the Hahn-Banach theorem). Let $x \mapsto \|x\|_\varepsilon$ be the norm thus defined on $E \otimes F$.

(b) Let (ξ_i) be a family of elements of $E^\top$ such that $(\forall x \in E)$ $\|x\| = \sup |\xi_i(x)|$. Show that, for $t \in E \otimes F$, $\|t\|_\varepsilon = \sup \|(\xi_i \otimes 1_F)(t)\|$. If (η_j) is a family of elements of $F^\top$ satisfying the same property, then $\|t\|_\varepsilon = \sup_{i,j} |(\xi_i \otimes \eta_j)(t)|$.

(c) Let X be a compact metric space and suppose that F is complete. Show that the space $C(X, F)$ of continuous maps from X to F can be identified with the completion relative to the norm ε of $C(X) \otimes F$.

(d) Let X and Y be compact metric spaces. Show that $C(X \times Y)$ is identified with the completion with respect to the norm ε of $C(X) \otimes C(Y)$.

(e) Let E' and F' be subspaces of E and F respectively, equipped with the induced norms. Show that the norm ε on $E' \otimes F'$ is induced by that of $E \otimes F$. Assuming that E' and F' closed, is the norm ε on $(E/E') \otimes (F/F')$ the quotient of the norm ε on $E \otimes F$?

(f) Show that the seminorm π defined in [2.4, Exercise 3] is a norm.

2. *(Nuclear norm)*—Let E and F be finite dimensional normed spaces over $\mathbb{R}$ or $\mathbb{C}$. For all $t \in E^\top \otimes F$, let $\|t\|_\pi$ denote the lower bound of $\sum \|\xi_i\| \|y_i\|$ for the finite families $((\xi_i, y_i))$ such that $t = \sum \xi_i \otimes y_i$. The nuclear norm, written $f \mapsto \|f\|_\nu$, is the norm obtained by transferring the norm π on $L(E, F)$ via the Kronecker isomorphism.

(a) Show that for all $f \in L(E, E)$, $\|\mathrm{Tr}(f)\| \leqslant \|f\|_\nu$.

(b) Show that $\|1_E\|_\nu = \dim(E)$ (use (a) gives one of the inequalities; to prove the other one, first study the dimension 2 case by considering a parallelogram with maximal area among all those with sides of length 1).

(c) Let F be the space $\mathbb{R}^3$ equipped with the norm $x \mapsto \sup |x_i|$, and E the plane $\{x \in \mathbb{R}^3 \mid x_1 + x_2 + x_3 = 0\}$ equipped with the induced norm. Draw a unit ball of E. Compute the nuclear norm of the canonical injection from E to F.

(d) Let $f \in L(E, F)$ and F_1 a subspace de F containing the image of f (resp. N a subspace de E contained in the kernel of f). Is the nuclear norm of f in $L(E, F_1)$ (resp. in $L(E/N, F)$) necessarily equal to the nuclear norm of f in $L(E, F)$?

(e) Take $f : \mathbb{R}^3 \to \mathbb{R}^3$ to be the map represented by the matrix

$$\begin{pmatrix} 0 & 1 & 1 \\ 1 & 0 & 1 \\ 1 & 1 & 0 \end{pmatrix}$$

the space $\mathbb{R}^3$ being equipped with the norm $x \mapsto \sup |x_i|$. Compare the sum of the absolute values of the eigenvalues with the nuclear norm.

(f) Show that, for $f \in L(E, F)$, the nuclear norm of f is the upper bound of $\text{Tr}(g \circ f)$ for $g : F \to E$ such $\|g\| \leqslant 1$, i.e. $(\forall y \in D) \|g(y)\| \leqslant \|y\|$.

3.—Let A be a ring, E and F finitely generated projective A-modules. Show that the bilinear form $(f, g) \mapsto \text{Tr}(g \circ f)$ enables the identification of each of the modules $\text{Hom}(E, F)$ and $\text{Hom}(F, E)$ with the dual of the other.

4.—Let V be the real vector space of $n \times n$ matrices. Denote by $u^\top$ the transpose of the matrix u, and by $\text{Tr}(u)$ its trace. Set $p(u) = \text{Tr}(u \cdot u)$ and $q(u) = \text{Tr}(u^\top \cdot u)$.

(a) Check that p and q are quadratic forms on V.

(b) If these quadratic forms are decomposed as sums of squares with signs of linearly independent forms, then in the case of q, how many have a $+$ sign and how many a $-$ one? In the case of p? (Start with the case $n = 2$.)

(c) Is there a nonzero matrix u such that $p(u) = 0$? such that $q(u) = 0$?

5.—Let M be a square matrix with real entries. Show that

$$\frac{d}{dt} \det(1 + tM)_{t=0} = \text{Tr}(M)$$

(the computation may be done in $\mathbb{C}$ by reducing to the case when M is triangular). Compute the derivative for arbitrary t.

6.—Let K be a field and E a finite dimensional K-vector space. Show that every linear form $\theta : \text{End}(E) \to K$ such that $(\forall f, g)\, \theta(f \circ g) = \theta(g \circ f)$ can be written λTr. Can K be replaced by a ring by taking E to be a finitely generated free A-module? a finitely generated projective module?

7.—(a) Let E and F be finite dimensional vector spaces over a field K, $f : E \to F$ and $g : F \to E$ two homomorphisms such that $u = g \circ f$ and $v = f \circ g$ are of finite rank. Show that the nonzero eigenvalues of u are eigenvalues of v with the same multiplicity. Deduce that $\mathrm{Tr}(v) = \mathrm{Tr}(u)$. (If necessary, extend the scalars to the algebraic closure of K.)

(b) Let E and F be two A-modules, $f : E \to F$ and $g : F \to E$ homomorphisms such that the traces of $u = g \circ f$ and $v = f \circ g$ are well defined. Assume A is reduced. Show that $\mathrm{Tr}(u) = \mathrm{Tr}(v)$.

(c) In question (b), can the assumption that A is reduced be omitted?

8.—(a) Let $P \in \mathbb{C}[Z]$ be nonzero and A the algebra $\mathbb{C}[Z]/(P)$. For $f \in A$, let $\mathrm{Tr}_{A:\mathbb{C}}(f)$ be the trace of the endomorphism $\mu_f : g \mapsto f \cdot g$ of the vector space A. What are the eigenvalues of μ_f? Show that if F is a representative of f, then $\mathrm{Tr}_{A:\mathbb{C}}(f)$ is the sum of the values of F at the roots of P.

(b) Let $P \in \mathbb{C}[X_1, \ldots, X_n, Z]$ be a monic polynomial in Z (i.e. in $\mathbb{C}[X_1, \ldots, X_n][Z]$). Let $F \in \mathbb{C}[X_1, \ldots, X_n, Z]$, and for $(x_1, \ldots, x_n) \in \mathbb{C}^n$ set $S(x_1, \ldots, x_n) = \sum_i F(x_1, \ldots, x_n, z_i)$, where the z_i are the roots of $P(x_1, \ldots, x_n, Z) \in \mathbb{C}[Z]$, counted with their multiplicities. Show that the function $S : \mathbb{C}^n \to \mathbb{C}$ thus defined is polynomial.

9.—Let A be a ring and B an A-algebra, finitely generated and projective as A-module. For all $f \in B$, let $\mathrm{Tr}_{B:A}(f)$ be the trace of the endomorphism $g \mapsto f \cdot g$ of the A-module B.

(a) Compute $\mathrm{Tr}_{\mathbb{C}:\mathbb{R}}(z)$ for $z \in \mathbb{C}$.

(b) Show that, if $B = A[X]/(X^d - a)$, with $a \in A$, then $\mathrm{Tr}(x^k) = 0$ for $0 < k < d$. Give an example where $\mathrm{Tr}_{B:A} : B \to A$ is zero, while $B \neq 0$.

(c) Let C be a B-algebra, finitely generated and projective as B-module. Show that C is a finitely generated A-module and that, if $h \in C$, then $\mathrm{Tr}_{C:A}(h) = \mathrm{Tr}_{B:A}(\mathrm{Tr}_{C:B}(h))$. Give a slightly more general statement.

10.—Let K be a field, $A = K[X]$, $B = K[Y]$, and h a polynomial of degree $d > 0$. consider the algebra homomorphism $h^* : A \to B$ defined by $h^*(X)=h(Y)$, whereù $h^*(u) = u \circ h$ for $u \in A$.

Equip B with the A-module structure defined by this homomorphism.

(a) Show that B is a free A-module of rank d with $(1, \ldots, Y^{d-1})$ as basis.

(b) For $f \in B$, let $h_!(f)$ be the trace of the endomorphism $g \mapsto fg$ of the A-module B. Show that $h_!(f)$ is a polynomial of degree $\leqslant \frac{n}{d}$, where n is the degree of f. Show that, for $u \in A$, $h_!(h^*(u)) = du$.

(c) Suppose that K is algebraically closed. Show that, for $x \in K$ and $f \in B$, the scalar $h_!(f)(x)$ is the sum of $f(y)$ for $y \in K$ such that $h(y) = x$ (counted with their multiplicities as roots of $h(Y) - x$).

(d) Let $f \in B$. Show that there is a polynomial $g \in A$ such that, for all $x \in K$, the scalar $g(x)$ is the product of all $f(y)$ for $y \in K$ such that $h(y) = x$ (resp. $g(x) = \sum_{i<j} f(y_i)f(y_j)$, whereù $y_1, \ldots, y_d$ are the roots of $h(Y) - x$).

3.10   Chain Complexes

3.10.1   Graded Module

Definition A **graded A-module** is an A-module E. equipped with a sequence $(E_n)_{n \in \mathbb{Z}}$ of submodules such that E. $= \bigoplus E_n$.

Let E. be a graded A-module. Equipping the module E. with the sequence E'_n defined by $E'_n = E_{n-k}$ gives a graded A-module E.(k) obtained by diminishing the degrees by k.

The A-module E. is called positively (resp. negatively) graded if $E_n = 0$ for all $n < 0$ (resp. > 0). It is said to be bounded below (resp. above) if there exists $n_0 \in \mathbb{Z}$ such that $E_n = 0$ for $n < n_0$ (resp. $n > n_0$); If it is bounded both below and above, it is said to be bounded.

Let E. and F. be graded A-modules. A *degree k morphism* from E. to F. is a homomorphism f from the module E. to the module F. such that $f(E_n)$ is contained in F_{n+k} for all n. A degree k morphism induces a homomorphism f_n from E_n to F_{n+k} for all n, and $f \mapsto (f_n)_{n \in \mathbb{Z}}$ is a bijection of the set $\mathrm{Hom}_k(E., F.)$ of degree k morphisms from E. to F. onto $\prod_{n \in \mathbb{Z}} \mathrm{Hom}(E_n, F_{n+k})$. Clearly,

$$\mathrm{Hom}_k(E., F.) = \mathrm{Hom}_0(E.(k), F.) = \mathrm{Hom}_0(E., F.(-k)).$$

3.10.2   Chain Complexes

Definition A descending (resp. ascending) **chain complex** of A-modules is a graded A-module equipped with a degree -1 (resp. $+1$) endomorphism d such that $d \circ d = 0$.

Descending chain complexes are usually written with lower indices and ascending ones with upper ones. A descending (resp. ascending) complex is said to be right bounded if it is bounded below (resp. above), left bounded if its bounded above (resp. below). If E˙ is a chain complex, a descending complex can be obtained by setting $E_n = E^{-n}$. This remark enables us to reduce the study of ascending complexes to that of complexes and conversely. In what follows, we study ascending complexes.

Let $(E., d)$ be a descending chain complex. The endomorphism d is called the *differential* of the complex. Set $Z_n(E.) = \mathrm{Ker}\, d_n$ and $B_n(E.) = \mathrm{Im}\, d_{n+1}$; the elements de $Z_n(E.)$ are the degree n *cycles*, those of $B_n(E.)$ the degree n *boundaries*, and $B_n(E.) \subset Z_n(E.)$. Set $H_n(E.) = Z_n(E.)/B_n(E.)$.

Definition $H_n(E.)$ is said to be the *n*-**th homology module** of E. .

We have the following exact sequences

$$0 \to Z_n \to E_n \to B_{n-1} \to 0$$

$$0 \to B_n \to Z_n \to H_n \to 0 .$$

The module E. is said to be **acyclic** in degree n if $H_n(E.) = 0$, in other words if $E_{n+1} \xrightarrow{d} E_n \xrightarrow{d} E_{n-1}$ is an exact sequence.

3.10.3   Morphisms of Chain Complexes

Let E. and F. be complexes. A *morphism of complexes* from E. to F. is a degree 0 morphism f from the graded module E. to the graded module F. such that $d \circ f = f \circ d$, where d denotes the differential of E. as well as that of F. If $f : E. \to F.$ is a morphism of chain complexes, f induces a homomorphism from $Z_n(E.)$ to $Z_n(F.)$ and from $B_n(E.)$ to $B_n(F.)$. Passing to the quotient, this gives a homomorphism $f_* : H_n(E.) \to H_n(F.)$.

More generally, a degree k morphism of complexes from E. to F. is a graded degree k morphism $f : E. \to F.$ such that $d \circ f = (-1)^k f \circ d$. This gives a homomorphism $f_* : H_n(E.) \to H_{n+k}(F.)$ for all n. The morphisms of chain complexes are the degree 0 morphisms.

The chain complex E.(k) is defined by equipping the graded module E.(k) with the differential $(-1)^k d$. The degree k morphisms of chain complexes from E. to F. are the (degree 0) morphisms of chain complexes from E.(k) to F. .

3.10.4   Homotopic Morphisms

Let E. and F. be two chain complexes.

Definition The degree k morphisms f and g of complexes from E. to F. are said to be **homotopic** if there is a graded degree $k + 1$ morphism $s : E. \to F.$ such that $g - f = d \circ s + (-1)^k s \circ d$.

Proposition *If f and g are homotopic degree k morphisms from E. to F., then the homomorphisms f_* and g_* from $H_n(E.)$ to $H_{n+k}(F.)$ are equal.*

Proof For $x \in Z_n(E.)$, $g(x) - f(x) = d(s(x)) + (-1)^k s(d(x)) = d(s(x)) \in B_{n+k}(F.)$, and so, denoting the canonical homomorphism by χ, $Z_n \to H_n$, $g_*(\chi(x)) - f_*(\chi(x)) = \chi(d(s(x))) = 0$.

Remark Set $\mathrm{Hom}_k(E., F.)$ to be the module of graded degree k morphisms from E. to F., and $\mathrm{Hom}(E., F.)$ the graded module $\bigoplus_k \mathrm{Hom}_k(E., F.)$. The module underlying $\mathrm{Hom}(E., F.)$ is a submodule of the module of module homomorphisms from E. to F.: a module homomorphism $h : E. \to F.$ can be represented by a matrix $(h_{n,p})$, where $h_{n,p}$ is a homomorphism from E_n to F_p. $h \in \mathrm{Hom}(E., F.)$ if and only if the support of the family $(h_{n,p})$ is contained in a ' strip 'parallel to the diagonal'', i.e. in some set $\{(n, p) \mid a \leqslant p - n \leqslant b\}$.

For $f \in \mathrm{Hom}_k(E_., F_.)$, set $D(f) = d \circ f + (-1)^{k-1} f \circ d$. We check that $D(D(f)) = 0$, and $\mathrm{Hom}(E_., F_.)$ equipped with the differential D is a chain complex. Degree k cycles of this complex are degree k morphisms of complexes from $E_.$ to $F_.$, and the k-th homology module of $\mathrm{Hom}(E_., F_.)$ is the module of homotopy classes of degree k morphisms of complexes from $E_.$ to $F_.$.

3.10.5

Proposition *Let* $E_., F_., G_.$ *be chain complexes.*

(a) The homotopy relation is an equivalence relation between degree k morphisms of complexes from $E_.$ to $F_.$;

(b) let f_0 and f_1 be degree p morphisms of complexes from $E_.$ to $F_.$, g_0 and g_1 degree q morphisms of complexes from $F_.$ to $G_.$. If f_1 is homotopic to f_0 and g_1 homotopic to g_0, then $g_1 \circ f_1$ is homotopic to $g_0 \circ f_0$.

Proof (a) follows from the previous remark. We prove (b).

1. Suppose that $g_0 = g_1 = g$: if $f_1 - f_0 = d \circ s + (-1)^p s \circ d$, then $g \circ f_1 - g \circ f_0 = g \circ d \circ s + (-1)^p g \circ s \circ d = (-1)^q d \circ g \circ s + (-1)^p g \circ s \circ d = d \circ s' + (-1)^{p+q} s' \circ d$, with $s' = (-1)^q g \circ s$.

2. Next suppose that $f_0 = f_1 = f$: if $g_1 - g_0 = d \circ s + (-1)^q s \circ d$, then $g_1 \circ f - g_0 \circ f = d \circ s \circ f + (-1)^q s \circ d \circ f = d \circ s \circ f + (-1)^{p+q} s \circ f \circ d$.

3. The general case follows from the previous ones.   □

Definition A morphism of chain complexes $f : E_. \to F_.$ is a **homotopy equivalence** if there is a morphism of chain complexes $g : F_. \to E_.$ such that $g \circ f$ and $f \circ g$ are respectively homotopic to the identity of $E_.$ and $F_.$.

Two complexes $E_.$ and $F_.$ are said to be *homotopy equivalent* if there is a homotopy equivalence from $E_.$ to $F_.$. The relation thus defined is an equivalence relation between complexes.

3.10.6 Connecting Homomorphism

Let $0 \to E_. \xrightarrow{f} F_. \xrightarrow{g} G_. \to 0$ be a short exact sequence, where f and g are degree 0 morphisms of complexes. We next define for all n a homomorphism $\delta : H_n(G_.) \to H_{n-1}(E_.)$.

Let $\gamma \in H_n(G_.)$. Choose $z \in Z_n(G_.)$ such that $\chi(z) = \gamma$, where $\chi : Z_n \to H_n$ is a canonical homomorphism. Choose $y \in F_n$ such that $g(y) = z$; as $d(y) \in F_{n-1}$ and $g(d(y)) = 0$, there is a unique $x \in E_{n-1}$ such that $f(x) = d(y)$.

Since $f(d(x)) = d(d(y)) = 0, d(x) = 0$ as f is injective. Set $\eta = \chi(x) \in H_{n-1}(E_.)$.

Proposition and Definition *The element η is independent of the choices made. The map $\delta : \gamma \mapsto \eta$ from $H_n(G.)$ to $H_{n-1}(E.)$ thus defined is a homomorphism called the* **connecting homomorphism.**

Proof (a) *Independence from the choice of* y: let $y' \in F_n$ such that $g(y') = z$; then $y' = y + f(u)$ for some $u \in E_n$. Then $d(y') = d(y) + d(f(u))$, and $d(y') = f(x')$ with $x' = x + d(u)$; so $\chi(x') = \chi(x)$.

(b) *Independence from the choice of* z: let $z' \in Z_n(G.)$ be such that $\chi(z') = \chi(z) = \gamma$, then z' is of the form $z + d(w)$, with $w \in G_{n+1}$, and w is of the form $g(v)$, with $v \in F_{n+1}$. For y' such that $g(y') = z'$, we may choose $y' = y + d(v)$. Then $d(y') = d(y) = f(x)$.

(c) *Linearity of* δ: follows immediately.

3.10.7 Diagramme du serpent

Théorème. *Soit $0 \to E. \xrightarrow{f} F. \xrightarrow{g} G. \to 0$ une suite exacte courte où f et g sont des morphismes de complexes de degré 0. Alors*

est une suite exacte.

Proof (a) *Exactness in $H_n(F.)$:* $g_* \circ f_* = (g \circ f)_* = 0$. Let $\beta \in H_n(F.)$ be such that $g_*(\beta) = 0$. We show that $\beta = f_*(\alpha)$ for some $\alpha \in H_n(E.)$. Let $y \in Z_n(F)$ be such that $\chi(y) = \beta$. As $g_*(\beta) = 0$, $g(y) = d(w)$ for some $w \in G_{n+1}$ and $w = g(v)$ for some $v \in F_{n+1}$ since g is surjective. Set $y' = y - d(v)$. As $\chi(y') = \chi(y) = \beta$, and $g(y') = 0$, $y' = f(x)$ for some $x \in E_n$. As $f(d(x)) = d(y') = 0$, $d(x) = 0$ since f is injective, and $x \in Z_n(E.)$. Setting $\chi(x) = \alpha$, $f(\alpha) = \beta$.

(b) $\delta \circ g_* = 0$: if $\gamma = g_*(\beta)$ with $\beta \in H_n(F.)$, choose $y_1 \in Z_n(F.)$ such that $\chi(y_1) = \beta$. With the notation of 3.10.6, we may take $z = g(y_1)$ and $y = y_1$. Then $d(y) = 0$, and so $\delta(\gamma) = 0$.

(c) *Exactness in $H_n(G.)$:* let $y \in H_n(G.)$ be such that $\delta(\gamma) = 0$. We show that $\gamma = g_*(\beta)$ for some $\beta \in H_n(F.)$. With the notation of 3.10.6, $x = d(u)$ for some $u \in E_n$. Set $y' = y - f(u)$; as $d(y') = 0$, $y' \in Z_n(F.)$, and $g(y') = g(y) = z$; so $\gamma = g_*(\chi(y'))$.

(d) $f_* \circ \delta = 0$: with the notation of 3.10.6,

$$f_*(\delta(\gamma)) = f_*(\chi(x)) = \chi(f(x)) = \chi(d(y)) = 0.$$

(e) *Exactness in* $H_{n-1}(E.)$: let $\alpha \in H_{n-1}(E.)$ be such that $f_*(\alpha) = 0$. Choose $x \in Z_{n-1}(E.)$ such that $\chi(x) = \alpha$. As $f(x) \in B_{n-1}(F.)$, $f(x) = d(y)$ for some $y \in F_n$. Set $g(y) = z$. As $d(z) = g(d(y)) = g(f(x)) = 0$, $z \in Z_n(G.)$, and $\alpha = \delta(\gamma)$ with $\gamma = \chi(z)$.   □

3.10.8 Mapping Cylinder

Let E. and F. be two chain complexes and $f : E. \to F.$ a morphism of chain complexes. Set $M_n = F_n \oplus E_{n-1}$ and let $d : M_n \to M_{n-1}$ be the homomorphism defined by the matrix

$$\begin{pmatrix} d & f \\ 0 & -d \end{pmatrix}.$$

We check that $d \circ d = 0 : M_n \to M_{n-2}$. The complex $M. = \bigoplus M_n$ thus constructed is called the **mapping cylinder** of f.

Proposition *With the above notation,*

(a) there is a short exact sequence of complexes

$$0 \to F. \xrightarrow{u} M. \xrightarrow{v} E.(1) \to 0,$$

where $u_n : F_n \to M_n$ *and* $v_n : M_n \to E_{n-1}$ *are the canonical injection and projection;*

(b) the connecting homomorphism $\delta : H_{n-1}(E.) \to H_{n-1}(F.)$ *of this exact sequence is* f_*.

Proof (a) is immediate; we show (b). Let $\gamma \in H_{n-1}(E.)$ and choose $z \in Z_{n-1}(E.)$ such that $\chi(z) = \gamma$. For $y \in M_n$ such that $v(y) = z$ we may choose $y = \begin{pmatrix} 0 \\ z \end{pmatrix}$. Then $d(y) = \begin{pmatrix} f(z) \\ 0 \end{pmatrix} = u(x)$ with $x = f(z)$, and by construction $\delta(\gamma) = \chi(x) = \chi(f(z)) = f_*(\gamma)$.   □

Corollary 3.11 *There is an exact sequence*

$$\cdots \to H_n(E.) \xrightarrow{f_*} H_n(F.) \xrightarrow{u_*} H_n(M.) \xrightarrow{v_*} H_{n-1}(E.) \to \cdots$$

Corollary 3.12 *Let* $a, b \in \mathbb{Z}$ *be such that* $a \leqslant b$. $H_n(M.) = 0$ *for* $a \leqslant n \leqslant b$ *if and only if* $f_* : H_n(E.) \to H_n(F.)$ *is an isomorphism for* $a \leqslant n < b$, *surjective for* $n = b$ *and injective for* $n = a - 1$.

3.10.9   Resolutions

Definition Let M be an A-module. A left (right) **resolution** of M is a descending positively graded chain complex L. (resp. ascending L'), acyclic in degrees $\neq 0$ and equipped with an isomorphism $\iota : H_0(L.) \to M$ (resp. $\iota : M \to H_0(L')$).

A left (resp. right) resolution of M gives an exact sequence

$$\cdots \to L_n \to L_{n-1} \to \cdots \to L_1 \to L_0 \to M \to 0$$

$$(\text{resp. } 0 \to M \longrightarrow L^0 \to L^1 \to \cdots \to L^n \to L^{n+1} \to \cdots).$$

We say that L. is a *free resolution* (resp. finite free) of M if L. is a *left* resolution of M such that L_n is free (resp. finite free) for all n. Projective and flat resolutions are defined likewise; however an injective resolution is a *right* resolution such that L^n is injective for all n.

Proposition *(a) All A-modules admit a free resolution.*

(b) If A is a Noetherian ring, then all finitely generated A-modules admit a finite free resolution.

Proof (a) (resp. (b)): by (3.4.6, Remark 2), there is a pair (L_0, ε) where L_0 is a free (resp. finite free) module libre and $\varepsilon : L_0 \to M$ a surjective morphism, and a sequence of pairs $((L_n, d_n))_{n \geqslant 1}$ where all L_n are free (resp. finite free) modules can be inductively constructed and so can homomorphisms $d_n : L_n \to L_{n-1}$ such that $\operatorname{Im} d_1 = \operatorname{Ker} \varepsilon$ and $\operatorname{Im} d_n = \operatorname{Ker} d_{n-1}$ for $n > 1$ ($\operatorname{Ker} d_{n-1}$ is a finitely generated submodule of L_{n-1} in case (b)).   □

Let L. and L'. be two resolutions of M. A *resolution morphism* from L. to L'. is a morphism of chain complexes $h : L. \to L'.$ such that $\iota' \circ h_* = \iota$. More generally, let $f : M \to M'$ be an A-module homomorphism, L. a resolution of M and L'. a resolution of M'. A *f-morphism* from L. to L'. is a morphism of chain complexes $h : L. \to L'.$ for which the diagram

$$
\begin{array}{ccc}
H_0(L.) & \xrightarrow{\;h_*\;} & H_0(L'.) \\
\left\downarrow{\iota}\right. & & \left\downarrow{\iota'}\right. \\
M & \xrightarrow{\;f\;} & M'
\end{array}
$$

commutes.

3.10.10   Projective Resolutions

Let M be an A-module. A *projective resolution* of M is a left resolution L. of M such that L_n is a projective module for all n, or, equivalently, such that the module $\bigoplus L_n$ is projective.

A free resolution is projective, and so all modules admit a projective resolution.

> **Theorem** *Let* M *and* M′ *be two* A-*modules*, L. *and* L′. *projective resolutions of* M *and* M′ *respectively, and* $f : M \to M'$ *a homomorphism. There exists a* f-*morphism* $h : L. \to L'.$, *unique up to homotopy.*

This theorem is a particular case of the following proposition:

Proposition *Let* L. *and* E. *be positively graded complexes. Suppose that* L. *is projective and* E. *acyclic in degrees* $\neq 0$. *Then, for all homomorphisms* $f : H_0(L.) \to H_0(E.)$, *there is a morphism of chain complexes* h, *unique up to homotopy, such that* $f = h_*$.

Proof (a) *Existence.* Construct by induction a sequence $(h_n)_{n \in \mathbb{N}}$ of homomorphisms $h_n : L_n \to E_n$ for which the diagrams

$$
\begin{array}{ccc}
L_0 & \xrightarrow{\ h_0\ } & E_0 \\
{\scriptstyle\varepsilon}\downarrow & & \downarrow{\scriptstyle\varepsilon} \\
H_0(L.) & \xrightarrow{\ f\ } & H_0(E.)
\end{array}
\qquad
\begin{array}{ccc}
L_n & \xrightarrow{\ h_n\ } & E_n \\
{\scriptstyle d}\downarrow & & \downarrow{\scriptstyle d} \\
L_{n-1} & \xrightarrow{\ h_{n-1}\ } & E_{n-1}
\end{array}
\qquad (n \geqslant 1)
$$

commute. This is possible since L_0 being projective and $\varepsilon : E_0 \to H_0(E.)$ being surjective, we can complete the diagram

$$
\begin{array}{ccc}
L_0 & & E_0 \\
{\scriptstyle\varepsilon}\downarrow & & \downarrow{\scriptstyle\varepsilon} \\
H_0(L.) & \xrightarrow{\ f\ } & H_0(E.)
\end{array}
$$

to obtain a commutative one with $h_0 : L_0 \to E_0$; suppose that h_n has been defined such that $d_n \circ h_n = h_{n-1} \circ d_n$. Then h_n maps $Z_n(L.)$ to $Z_n(E.)$ and, as L_{n+1} is projective and $d_{n+1} : E_{n+1} \to Z_n(E.)$ surjective, we can complete the diagram

$$
\begin{array}{ccc}
L_{n+1} & & E_{n+1} \\
{\scriptstyle d_{n+1}}\downarrow & & \downarrow{\scriptstyle d_{n+1}} \\
Z_n(L.) & \xrightarrow{\ h_n\ } & Z_n(E.)
\end{array}
$$

to obtain a commutative one with $h_{n+1} : L_{n+1} \to E_{n+1}$.

(b) *Uniqueness up to homotopy.* By difference, it suffices to show that, if $h : L. \to E.$ is a morphism of chain complexes such that $h_* = 0 : H(L.) \to H_0(E.)$, and by induction there is a sequence $(s_n)_{n \in \mathbb{N}}$ of homomorphisms $s_n : L_n \to E_{n+1}$ such that $h_0 = d_1 \circ s_0$ and $h_n = d_{n+1} \circ s_n + s_{n-1} \circ d_n$ for $n \geqslant 1$. Since $h_* = 0$, the homomorphism h_0 maps L_0 to $B_0(E.)$, and the diagram

$$E_1$$
$$\downarrow d_1$$
$$L_0 \xrightarrow{\;h_0\;} B_0(E.)$$

where L_0 is projective and d_1 surjective, can be completed to obtain a commutative diagram with $s_0 : L_0 \to E_1$. Suppose that s_n has been defined such that $h_n = d_{n+1} \circ s_n + s_{n-1} \circ d_n$, and set s_{n+1} such that $h_{n+1} = d_{n+2} \circ s_{n+1} + s_n \circ d_{n+1}$, i.e. $d_{n+2} \circ s_{n+1} = h_{n+1} - s_n \circ d_{n+1}$. As $d_{n+1} \circ s_n \circ d_{n+1} = 0$, $h_{n+1} - s_n \circ d_{n+1}$ maps L_{n+1} to $Z_{n+1}(E.) = \operatorname{Im} d_{n+2}$ and this homomorphism can be written $d_{n+2} \circ s_{n+1}$. □

Corollary *Any two projective resolutions of an* A-*module* M *are homotopy equivalent.*

3.10.11 Injective Modules

Definition An A-module E is said to be **injective** if, for all injective A-module homomorphisms $f : M' \to M$ and all homomorphisms $g : M' \to E$, there is a homomorphism $h : M \to E$ making the diagram

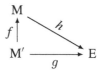

commutative.

In other words, E is injective if and only if the contravariant functor $M \mapsto \operatorname{Hom}(M, E)$ from the category $A\text{-}\mathcal{M}od$ to itself transforms injective homomorphisms into surjective ones.

Proposition *Let* E *be an* A-*module. The following conditions are equivalent:*

(i) the module E *is injective;*
(ii) for all ideals I *of* A, *and all homomorphisms* $f : I \to E$, *there exists* $x \in E$ *such that, for all* $a \in I$, $f(a) = ax$.

Proof (i) $\Rightarrow$ (ii): follows readily by applying the definition of the canonical injection $I \to A$ and by noting that $\operatorname{Hom}(A, E) = E$.

(ii) $\Rightarrow$ (i): let $f : M' \to M$ be an injective homomorphism and $g : M' \to E$ a homomorphism. Let Ω be the set of submodules N of M equipped with a homomorphism $h : N \to E$. Define an order on Ω by $(N, h) \leqslant (N', h')$ if and only if $N \subset N'$ and $h'|_N = h$. Equipped with this order, Ω is inductive. Indeed, if $((N_\lambda, h_\lambda))_{\lambda \in \Lambda}$ is a totally ordered family of elements of Ω, then the pair (N, h), where $N = \bigcup_{\lambda \in \Lambda} N_\lambda$ and h is obtained by gluing together all h_λ, is an upper bound for the family. The set of upper bounds for (M', g) in Ω is also inductive. Hence, by Zorn's theorem, there is a maximal upper bound (N, h) for (M', g) in Ω. Assuming that (ii) holds, we show by contradiction that, for such an element, $N = M$. Let $x \in M - N$, and I be the ideal consisting of $a \in A$ such that $ax \in N$. There exists $y \in E$ such that, for all $a \in I$, $h(ax) = ay$. The homomorphism $\bar{h} : N \oplus A \to E$ defined by $\bar{h}(n, a) = h(n) + ay$ can be factorized as

$$
\begin{array}{ccc}
N \oplus A & & \\
\chi \downarrow & \searrow \bar{h} & \\
N' & \xrightarrow{\; h' \;} & E
\end{array}
$$

where $N' = N + Ax$ and $\chi(n, a) = n + ax$. The element (N', h') is a strict upper bound for (N, h), which is impossible. $\qquad\square$

Definition Suppose that A is an integral domain and E an A-module. An element $x \in E$ is said to be **divisible** if for all $a \in A - \{0\}$, there exists $y \in E$ such that $ay = x$. The module E is said to be **divisible** if all elements of E are divisible.

The above proposition has the following corollary:

Corollary *Suppose that* A *is an integral domain.*
(a) All injective A*-modules are divisible.*
(b) If A *is a PID, all divisible modules are injective.*

Proof Condition (ii) of the proposition always holds for the 0 ideal. As E is divisible it also holds for every nonzero principal ideal.

(a) If the module E is injective, (ii) holds for all ideals, and hence in particular for all principal ideals.

(b) If A is a PID, and (ii) holds for all principal ideals, it holds for all ideals. $\quad\square$

3.10.12

The results of this subsection will be needed in the following one.

Let $\varphi : A \to B$ be a ring homomorphism, and E an A-module. Set $\widetilde{E} = \mathrm{Hom}_A (B, E)$. Define a B-module structure on $\widetilde{E}$ by setting $bf(x) = f(xb)$ for $b, x \in B$ and $f \in \widetilde{E}$.

Proposition *For all* B*-modules* M,

$$
\mathrm{Hom}_B (M, \widetilde{E}) = \mathrm{Hom}_A (M, E) .
$$

In other words, the functor $E \mapsto \widetilde{E}$ *is the right adjoint of the scalar restriction functor.*

More precisely, define $\alpha : \widetilde{E} \to E$ by $\alpha(f) = f(1)$. The map α is A-linear, and so $\alpha_* : \mathrm{Hom}_B (M, \widetilde{E}) \to \mathrm{Hom}_A (M, E)$. For A-linear maps $h : M \to E$, define $\tilde{h} : M \to \widetilde{E}$ by $\tilde{h}(x)(b) = h(bx)$, and define $\beta : \mathrm{Hom}_A (M, E) \to \mathrm{Hom}_B (M, \widetilde{E})$ by $\beta(h) = \tilde{h}$. The maps α_* and β are mutually inverses.

Proof For $h \in \text{Hom}_A(M, E)$, $\alpha_*(\beta(h))(x) = \alpha_*(\tilde{h})(x) = \tilde{h}(x)(1) = h(1.x) = h(x)$. For $u \in \text{Hom}_B(M, \tilde{E})$, set $v = \alpha_*(u)$, i.e. $v(x) = u(x)(1)$. As $\tilde{v}(x)(b) = v(bx) = u(bx)(1) = bu(x)(1) = u(x)(1 \cdot b) = u(x)(b)$, $\tilde{v} = u$, i.e. $\beta \circ \alpha_* = 1$. $\square$

Corollary *If* E *is an injective* A-*module, then* $\tilde{E}$ *is an injective* B-*module.*

Proof The functor $M \mapsto \text{Hom}_B(M, \tilde{E}) = \text{Hom}_A(M, E)$ is the composition of the scalar restriction functor $M \mapsto M$ from B-***Mod*** to A-***Mod***, which transforms injections into injections, and of the functor $M \mapsto \text{Hom}_A(M, E)$ which transforms injections into surjections. $\square$

3.10.13

Theorem *Any module is isomorphic to a submodule of an injective module.*

Proof (a) *Suppose first that the ring* A *is principal.* Let K be the field of fractions of A, M an A-module, $A^{(J)} \xrightarrow{u} A^{(I)} \to M$ a presentation of M, and $\iota : A^{(I)} \to K^{(I)}$ the canonical injection. Set $E = \text{Coker}(\iota \circ u)$. The module E is divisible since it is the quotient of $K^{(I)}$ which is divisible, and so E is injective, and M is isomorphic to a submodule of E.

(b) *Now let* A *be arbitrary.* Let M be an A-module. As $\mathbb{Z}$ is a PID, there is a $\mathbb{Z}$-linear injection f from M to an injective $\mathbb{Z}$-module E. Set $\tilde{M} = \text{Hom}_{\mathbb{Z}}(A, M)$ and $\tilde{E} = \text{Hom}_{\mathbb{Z}}(A, E)$, and let $h : M \to \tilde{M}$ be the map defined by $h(x)(a) = ax$. The maps h and $f_* : \tilde{M} \to \tilde{E}$ are A-linear injections, and so $f_* h : M \to \tilde{E}$ is an A-linear injection and by (3.10.12, corollary), E is injective. $\square$

3.10.14 Injective Resolutions

Let M be an A-module. An *injective resolution* of M is a right resolution E˙ of M such that E^n is an injective module for all n.

Remark This does not imply the injectivity of the module E˙: a direct sum of injective modules is generally not injective. As can be checked, this is however the case when the ring A is Noetherian.

Theorem *(a) All* A-*modules admit an injective resolution.*

(b) Let M *be an* A-*module. Any two injective resolutions of* M *are homotopy equivalent.*

(c) Let M *and* M' *be two* A-*modules,* E˙ *and* E'˙ *injective resolutions of* M *and* M' *respectively, and* $f : M \to M'$ *a homomorphism. Then there is a* f-*morphism* $h : E˙ \to E'˙$. *It is unique up to homotopy.*

Proof It is similar to those of (3.10.9, Proposition, a), (3.10.10, Corollary and Theorem), by reversing the direction of the arrows.

References

1. M. Queysanne, *Algèbre*. Collection U (Armand Colin, Paris, 1964)
2. J. Lelong-Ferrand, J.-M. Arnaudis, *Cours de mathématiques*, tome 1. Algèbre (Dunod, Paris, 1971)
3. J. Lelong-Ferrand, J.-M. Arnaudis, *Cours de matéhématiques*, tome 2. Analyse (Dunod, Paris, 1972)
4. N. Bourbaki, *Algèbre, ch. 1 à 3*, Hermann (1970); *Algèbre, ch. 10, algèbre homologique*, Masson (1980)
5. A. Guichardet, *Calcul intégral*. Collection U (Armand Colin, Paris, 1969)
6. J. Frisch, Points de platitude d'un morphisme. Invent. Math. (1967)

Chapter 4
Coverings

4.1 Spaces over B

4.1.1

Definition Let B be a topological space. A **space over** B is a topological space X together with a continuous map $\pi : X \to B$ called **projection**.

If X is a space over B, a continuous section or simply a **section** of X is a continuous map $\sigma : B \to X$ such that $\pi \circ \sigma = 1_B$. We say that σ **passes** through a point $x \in X$ if $\sigma(\pi(x)) = x$.

For all $b \in B$, a **fiber** of X at b, written $X(b)$, is the subspace $\pi^{-1}(b)$ of X.

Let X and Y be spaces over B, π and π' their projections. A B-morphism, or simply a **morphism** from X to Y is a continuous map $f : X \to Y$ for which the diagram

commutes. If $f : X \to Y$ is a B-morphism, the map f induces a map $f_b : X(b) \to Y(b)$ for all $b \in B$. Define a category B-$\mathcal{T}\!op$ by taking its objects to be the spaces over B and its morphisms the B-morphisms.

Let X and Y be spaces over B, π and π' their projections. The *fiber product* of X and Y, written $X \times_B Y$, is the subspace of the topological space $X \times Y$ consisting of pairs (x, y) such that $\pi(x) = \pi'(y)$. The space $X \times_B Y$ equipped with the map $(x, y) \mapsto \pi(x)$ is the product of X and Y in B-$\mathcal{T}\!op$. For all $b \in B$, $(X \times_B Y)(b) = X(b) \times Y(b)$.

The sum of X and Y in B-$\mathcal{T}\!op$ is the disjoint union $X \sqcup Y$ equipped with the projection inducing π on X and π' on Y. So

$$(X \sqcup Y)(b) = X(b) \sqcup Y(b).$$

© Springer Nature Switzerland AG 2020
R. Douady and A. Douady, *Algebra and Galois Theories*,
https://doi.org/10.1007/978-3-030-32796-5_4

4.1.2   Change of Basis

Let B and B′ be topological spaces, X a space over B and $h : B' \to B$ a continuous map. The topological space $X' = B' \times_B X$, equipped with the first projection $X' \to B'$, is a space over B′ written $h^*(X)$ and called the *space over* B′ *obtained from X by a change of basis from* B *to* B′ *along* h. The fibre of $h^*(X)$ at a point $b' \in B'$ can be identified with the fibre of X at $h(b')$.

Let A be a subspace of B, and $\iota : A \to B$ the canonical injection. The space $\iota^*(X)$ can be identified with $\pi^{-1}(A)$, equipped with the projection onto A induced by π. This space is written $X|_A$.

4.1.3   Hausdorff Spaces over B

A space X over B is said to be *Hausdorff over* B if the diagonal $\Delta_X = \{(x, x) \in X \times X\}$ is closed in $X \times_B X$, i.e. if for all pairs (x, y) of distinct elements lying in the same fibre of X, there are two neighbourhoods of x and y respectively, open in X and disjoint.

Remark If X is a Hausdorff space over B, then the fibres of X are Hausdorff.

If B is Hausdorff, the conditions "X is Hausdorff over B" and "X is Hausdorff" are equivalent.

4.1.4   Etale Spaces over B

Let X and Y be topological spaces. A continuous map $f : X \to Y$ is a *local homeomorphism* if, for all $x \in X$, there is an open neighbourhood U of x such that f induces a homeomorphism from U onto an open subset of Y.

A space X over B is said to be *etale over* B if the projection is a local homeomorphism.

Example (*Implicit function theorem*) Let X and Y be $\mathscr{C}^1$-manifolds, $f : X \to Y$ a $\mathscr{C}^1$-map. If, for all $x \in X$, the tangent linear map

$$T_x f : T_x X \to T_{f(x)} Y$$

is an isomorphism, then f is a local homeomorphism.

4.1.5 Proper Spaces

The related concept of a proper space over B corresponds to that of a quasi-compact space.

Proposition and Definition *Let* X *be a space over* B. *The following conditions are equivalent:*

(i) *Every filter* $\mathscr{F}$ *on* X *whose image* $\pi_*\mathscr{F}$ *in* B *converges to a point* $b \in$ B, *has a cluster point* x *such that* $\pi(x) = b$;

(ii) *for all* $b \in$ B, *and open subsets* $(U_i)_{i \in I}$ *of* X *satisfying*

$$\bigcup_{i \in I} U_i \supset X(b),$$

there is a finite subset J *of* I *and a neighbourhood* V *of* b *such that*

$$\bigcup_{i \in J} U_i \supset \pi^{-1}(V) ;$$

(iii) *The fibres of* X *are quasi-compact and the map* π *is closed.*

If these conditions hold, then X *is* **proper** *over* B.

Proof (i) $\Rightarrow$ (iii). Let $b \in$ B, ι the injection $X(b) \to X$, and $\mathscr{G}$ a filter on $X(b)$. Set $\mathscr{F} = \iota_*\mathscr{G}$, the filter $\pi_*\mathscr{F}$ is defined by b, and so $\mathscr{F}$ admits a cluster point $x \in X(b)$. Then x is a cluster point of $\mathscr{G}$. Since this holds for all filters $\mathscr{G}$, the space $X(b)$ is quasi-compact.

Let Y be a closed subset of X and $b \in$ B a cluster point of $\pi(Y)$. Set $\mathscr{V}_b$ to be the filter of neighbourhoods of b in B. For all U in $\mathscr{V}_b$, $\pi^{-1}(U) \cap Y$ generated a filter $\mathscr{F}$ on X. The filter $\pi_*\mathscr{F}$ converges to b, and so $\mathscr{F}$ admits a cluster point $x \in X(b)$. Then x is a cluster point of Y, and so $x \in$ Y and $b \in \pi(Y)$. Therefore, $\pi(Y)$ is closed.

(iii) $\Rightarrow$ (ii). Let $b \in$ B, and $(U_i)_{i \in I}$ open subsets in X whose union cover $X(b)$. As $X(b)$ is quasi-compact, there is a finite subset J of I such that $\bigcup_{i \in J} U_i \supset X(b)$. Then $Y = X - \bigcup_{i \in J} U_i$ is closed in X and does meet $X(b)$, the set $\pi(Y)$ is closed in B and does not contain b, and $V = B - \pi(Y)$ is an open neighbourhood of b with the desired property.

(ii) $\Rightarrow$ (i). Let $\mathscr{F}$ be a filter on X such that $\pi_*\mathscr{F}$ converges to $b \in$ B.

Suppose that, for all $x \in \pi^{-1}(b)$ the filter $\mathscr{F}$ is incompatible with the filter of neighbourhoods of x, i.e., there exists an open neighbourhood U_x of x and a subset $A_x \in \mathscr{F}$ such that $A_x \cap U_x = \varnothing$. The subsets U_x form an open cover of $\pi^{-1}(b)$. Let J be a finite subset of X and V a neighbourhood of b such that $\bigcup_{x \in J} U_x \supset \pi^{-1}(V)$. Set $A = \bigcap_{x \in J} A_x$. Then $\pi(A) \in \pi_*\mathscr{F}$ and $\pi(A) \cap V = \varnothing$; so $\pi_*\mathscr{F}$ does not converge to b, contradicting assumption. $\qquad\square$

Corollary *If* X *is proper over* B *then for all* $b \in$ B, *all neighbourhoods* X(b) *in* X *contain a subset* $\pi^{-1}(V)$, *where* V *is a neighbourhood of b in* B.

Condition (ii) for I reduces to one element.

4.1.6

Proposition *Let* X *be a Hausdorff space over* B, *where* B *is assumed to be locally compact. Then* X *is proper over* B *if and only if the inverse image of every compact subset of* B *is compact. If this is the case, then* X *is locally compact.*

Proof Suppose first that X is proper over B. We first show that if B is compact, then so is X. Let $\mathscr{F}$ be a filter on X. The filter $\pi_*\mathscr{F}$ admits a cluster point in B since B is compact. Let $\mathscr{G}$ be a filter on B finer than $\pi_*\mathscr{F}$ and convergent. Then the filter $\mathscr{F} \wedge \pi^*\mathscr{G}$ admits a cluster point in X since π is proper, and so $\mathscr{F}$ admits a cluster point.

In the general case, for any compact subset K of B, the space $\pi^{-1}(K)$ is proper over K, and hence compact.

We now prove the converse. Let $b \in$ B and K a compact neighbourhood of b. Every filter $\mathscr{F}$ on X such that $\pi_*\mathscr{F}$ converges to b induces a compact filter on $\pi^{-1}(K)$, and thus admits a cluster point. The projection of this value must be b since B is Hausdorff. So π is proper.

X is locally compact. Let $x \in$ X and set $b = \pi(x)$. Let K be a compact neighbourhood of b. Then $\pi^{-1}(K)$ is a compact neighbourhood of x.   □

4.1.7 *Fibres*

Let X be a space over B. A *trivialization* of X is an isomorphism $\tau : X \rightarrow B \times F$ of spaces over B, where F is a topological space, the product $B \times F$ being equipped with its projection onto the first factor. We say that X is a *trivial fibre bundle* if it admits a trivialization. It is a *a locally trivial fibre bundle* or simply a *fibre bundle* if for all $b \in$ B, there is a neighbourhood U of b in B such that $X|_U$ is a trivial fibre bundle, or equivalently, if there exists an open cover $(U_i)_{i \in I}$ of B such that, for all i, $X|_{U_i}$ is a trivial fibre bundle. It is a *fibre bundle with fibre* F if all the fibres are homeomorphic to F.

Remark Let X be a fibre bundle over B. If B is connected, then the fibres of X are mutually homeomorphic. Indeed, the equivalences classes of the equivalence relation "X(b) is homeomorphic to X(b')" between elements $b, b' \in$ B are open; hence there is a unique class.

4.1.8 Example. The Möbius Strip

Consider the set X of pairs (L, x) where L $\subset \mathbb{R}^2$ is a 1-dimensional vector subspace and $x \in$ L. Let π be the projection (L, x) $\mapsto$ L of X onto the set B of 1-dimensional vector subspaces. Identify B with $\mathbb{R}/\pi.\mathbb{Z}$, and X with the quotient of $\mathbb{R}^2$ by the equivalence relation identifying (θ, ρ) with $(\theta + k\pi, (-1)^k \rho)$ for $k \in \mathbb{Z}$. Equip X and B with the topologies defined by these identifications, and X with the projection π. Then X is a fibre space with fibre $\mathbb{R}$ over B.

This fibre space is not trivial. Indeed a continuous section σ of X is given by a continuous function $h : \theta \mapsto \rho$ from $\mathbb{R}$ to $\mathbb{R}$ satisfying $h(\theta + k\pi) = (-1)^k h(\theta)$. Given two continuous sections σ_1 and σ_2 represented by two functions h_1 and h_2, $g = h_2 - h_1$ vanishes at some point: indeed, if $g(\theta) > 0$ the $g(\theta + \pi) < 0$ and conversely. Hence there exists $b \in$ B such that $\sigma_2(b) = \sigma_1(b)$.

The set X can be identified with the set of affine lines in $\mathbb{R}^2$ by assigning to (L, x) the line D perpendicular to L at x.

This set can in turn be identified with the set of 1-dimensional vector spaces that are not vertical in $\mathbb{R}^3$, by assigning to every affine line D in the horizontal plane H with side 1 the 1-dimensional vector space Δ perpendicular to the plane containing O and D. In other words, X may be considered a projective plane $\mathbf{P}^2\mathbb{R}$ without a point. All these identifications are compatibles with the natural topologies.

On the other hand, the space X can be topologically embedded in $\mathbb{R}^3$ by

$$(\theta, \rho) \mapsto \begin{pmatrix} \cos 2\theta(1 + u(\rho)\cos\theta) \\ \sin 2\theta(1 + u(\rho)\cos\theta) \\ u(\rho)\sin\theta \end{pmatrix}$$

where $u(\rho) = \frac{1}{\pi}\operatorname{Arctg}\rho$ varies from $-\frac{1}{2}$ to $\frac{1}{2}$ as ρ varies from $-\infty$ to $+\infty$.

The closure M of the image of X in $\mathbb{R}^3$ is the image of $\mathbb{R} \times \left[-\frac{1}{2}, \frac{1}{2}\right]$ under

$$(\theta, s) \mapsto \begin{pmatrix} \cos 2\theta(1 + s\cos\theta) \\ \sin 2\theta(1 + s\cos\theta) \\ s\sin\theta. \end{pmatrix}$$

It is the *Möbius strip*, which may be regarded as the (non trivial) fibre bundle over the circle S^1 with fibre $\left[-\frac{1}{2}, \frac{1}{2}\right]$.

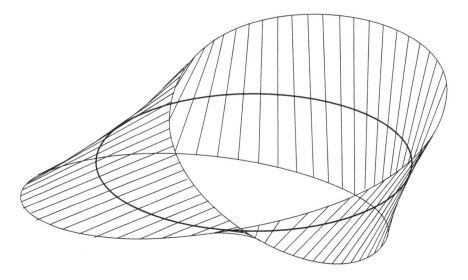

4.1.9   Fibre Bundle over a Segment

Theorem *Every fibre bundle over a compact interval* $[a, b]$ *of* $\mathbb{R}$ *is trivial.*

The proof of this theorem will be given in 4.1.11.

Remark This result holds for any interval in $\mathbb{R}$ (4.1, Exercise 12), not just a compact one.

4.1.10   Partition Subordinate to a Cover

Lemma and Definition *Let* X *be a compact metric space and* $\mathcal{U} = (U_i)_{i \in I}$ *an open cover of* X*. There exists* $\mu > 0$ *such that every ball of radius* $< \mu$ *is contained in some* U_i*. Every such* μ *is called a* **mesh** *of* $\mathcal{U}$*.*

Proof The distance $\rho_i(x)$ from x to $X - U_i$ (and ∞ if $U_i = X$) for $i \in I$ and $x \in X$ is a Lipschitz function with constant 1 or identically $+\infty$. So is $\rho(x) = \sup_i(\rho_i(x))$; it is >0 at all points since (U_i) is a cover of X. A lower bound $\mu > 0$ of the function ρ is a mesh of $\mathcal{U}$. $\square$

Proposition and Definition *Let* $[a, b]$ *be a compact interval in* $\mathbb{R}$ *and* $\mathcal{U} = (U_i)_{i \in I}$ *an open cover of* $[a, b]$*. Then there is an increasing finite sequence* $(t_0, ..., t_n)$ *with* $t_0 = a, t_n = b$ *such that*

$$(\forall k \in \{1, ..., n\})\, (\exists i \in I)\, [t_{k-1}, t_k] \subset U_i \, .$$

The intervals $[t_{k-1}, t_k]$ *are said to form a* **partition subordinate** *to* $\mathcal{U}$*.*

Proof Choose n such that $\frac{b-a}{n}$ are less than a mesh of $\mathscr{U}$ and set $t_k = a + k\frac{b-a}{n}$.

4.1.11 Gluing Trivializations

Lemma 4.1 *Let* E *and* F *be two topological spaces,* f *a map from* E *to* F *and* (A_k) *a finite closed cover of* E. *If* $f|_{A_k}$ *continuous for all* k, *then* f *is continuous.*

Proof of Lemma 4.1. Let G be a closed subset of F. For all k, $f^{-1}(G) \cap A_k = (f|_{A_k})^{-1}(G)$ is closed in A_k, hence in E, and $f^{-1}(G) = \bigcup_k (f^{-1}(G) \cap A_k)$ is closed.   □

Lemma 4.2 *Let* a, b, $c \in \mathbb{R}$ *with* $a < b < c$, *and* X *a space over* $[a, c]$. *If* $X|_{[a,b]}$ *and* $X|_{[b,c]}$ *are trivial fibre bundles, then so is* X *over* $[a, c]$.

Proof Let $\tau_1 : X|_{[a,b]} \to [a, b] \times F_1$ and $\tau_2 : X|_{[b,c]} \to [b, c] \times F_2$ be trivializations. Define a homeomorphism $\gamma : F_1 \to F_2$ by $(b, \gamma(y)) = \tau_2(\tau_1^{-1}(b, y))$. Setting $\tau(x) = (\mathrm{Id} \times \gamma)(\tau_1(x))$ if $x \in X|_{[a,b]}$ and $\tau_2(x)$ if $x \in X|_{[b,c]}$ gives a bijection $X \to [a, c] \times F_2$. Applying Lemma 4.1 to τ and τ^{-1} shows that τ is a homeomorphism, hence a trivialization.   □

Proof of Theorem 4.1.9. Let X be a fibre bundle over $[a, b]$, and $\mathscr{U} = (U_i)$ an open cover of $[a, b]$ such that $X|_{U_i}$ is trivial for all i. Let $(t_0, ..., t_n)$ be such that $[t_{k-1}, t_k]$ form a partition subordinate to $\mathscr{U}$. Applying Lemma 4.2, induction on k shows that $X|_{[a,t_k]}$ is trivial for $k = 1, ..., n$. For $k = n$, X is a trivial fibre bundle.   □

Exercises 4.1. (Spaces over B)
1.—Let B be a topological space.

(a) Show that if X is a Hausdorff (resp. quasi-compact) space, then $B \times X$ is Hausdorff (resp. proper) over B.

(b) Let X be a space over B, Hausdorff (resp. etale, resp. proper) over B, B' a topological space and $f : B' \to B$ a continuous map. Show that $f^*(X)$ is Hausdorff (resp. etale, resp. proper) over B'.

(c) Let X and Y be spaces Hausdorff (resp. etale, resp. proper) over B. Show that $X \times_B Y$ is Hausdorff (resp. etale, resp. proper) over B.

(d) Let X be a space over B and (U_i) an open cover of B. Show that, if $X|_{U_i}$ is Hausdorff (resp. etale, resp. proper) over U_i for all i, then the space X is Hausdorff (resp. etale, resp. proper) over B. Does the same hold for a locally finite closed cover?
2.—Let X be a space over B.

(a) If X is proper over B, The fibre $X(b)$ is quasi-compact for all $b \in B$.

(b) If X is quasi-compact, then X is proper over B.

(c) If B is quasi-compact, then X is proper over B if and only if X is quasi-compact.

(d) If X is proper over B then for any quasi-compact $K \subset B$, the space $\pi^{-1}(K)$ is quasi-compact.

3.—Let X be a space proper over B.

(a) Every closed subset of X is proper over B.

(b) Let Y be a space over B and $f : X \to Y$ a B-morphism. Then $f(X)$ is proper over B and, if Y is Hausdorff over B, $f(X)$ is closed in Y. If f is bijective and Y Hausdorff over B, then the map f is a homeomorphism.

(c) Let $f : X \to \mathbb{R}$ be a continuous map. If the projection $\pi : X \to B$ is surjective, then the function $g : B \to \mathbb{R}$ defined by $g(b) = \sup_{x \in \pi^{-1}(b)} f(x)$ is upper semicontinuous. If moreover the projection $\pi : X \to B$ is an open map, then g is continuous.

4.—A map $f : X \to Y$ is said to be proper (resp. Hausdorff) if X, equipped with f, is proper (resp. Hausdorff) over Y. Let $f : X \to Y$ and $g : Y \to Z$ be continuous maps.

(a) If f and g are proper, then so is $g \circ f$.

(b) If $g \circ f$ is proper and g is Hausdorff, then f is proper.

5.—Let B be a topological space. If K is a compact set, then all closed subsets of $B \times K$ are proper over B. Conversely, let X be a proper space over B. Show that if X is completely regular (2.6, Exercise 2), then there is a compact set K and a homeomorphism over B from X onto a closed subset of $B \times K$ (take the Čech compactification of X [2.6, Exercise 2]).

6.—Let B be a topological space, X a locally compact space and Y a subspace of $B \times X$. Show that the following conditions are equivalent:

(i) Y is proper over B;

(ii) Y is closed in $B \times X$ and, for all $b \in B$, there is a compact subset $K \subset X$ and a neighbourhood V of b in B such that $Y|_V \subset V \times K$.

7.—Let X be a space over B, and π its projection. Assume that X and B are metrizable (or simply that X is metrizable and that all points of B admit a countable fundamental system of neighbourhoods). The aim is to show that the following conditions are equivalent:

(i) X is proper over B;

(ii) every sequence (x_n) in X such that the sequence $(\pi(x_n))$ converges to a point $b \in B$ admits an a cluster point in $X(b)$.
(a) Show that (i) $\Rightarrow$ (ii).

(b) Choose a distance on X. Show that the following conditions are equivalent:

(α) every Cauchy filter of $\mathscr{F}$ over X such that $\pi_* \mathscr{F}$ converges to a point $b \in B$ converges to a point of $X(b)$;

(β) every Cauchy sequence (x_n) in X such that the sequence $(\pi(x_n))$ converges to a point $b \in B$ converges to a point of X(b).

If these conditions hold, then X will be said to be B-*complete*.

(c) X will be said to be *pre-proper* over B if, for all $b \in B$ and all $\varepsilon > 0$, there is a neighbourhood V of b such that $\pi^{-1}(V)$ has a finite cover consisting of sets having diameter $\leqslant \varepsilon$. Show that if X is pre-proper over B, then every ultrafilter $\mathcal{U}$ over X such that $\pi_* \mathcal{U}$ converges in B is a Cauchy filter. Deduce that, if X is pre-proper over B and B-complete, then X is proper over B.

(d) Show that if X is not pre-proper over B, then there is a sequence (x_n) in X and $a > 0$ such that $d(x_p, x_q) \geqslant a$ for $p \neq q$, and the sequence $(\pi(x_n))$ converges B. Conclude.

8.—Let X and Y be topological spaces and $f : X \to Y$ a continuous map. The map f is said to be *closed* if, for any closed subset A of X, the image $f(A)$ is closed in Y.

(a) Show that f is closed if and only if for all $y \in Y$, every neighbourhood of $f^{-1}(y)$ contains a set $f^{-1}(V)$, where V is a neighbourhood of y in Y.

(b) Show that every proper map is closed. Give an example of a non-proper closed map and an example of a non-proper map with compact fibres.

(c) Let X be a space over B. Show that X is proper over B if and only if for every space B$'$ and continuous map $h : B' \to B$, the projection of $h^*(X)$ onto B$'$ is closed. (To show X is proper over B, the previous exercise may be used and B$'$ taken to be $\mathbb{N} \cup \{\infty\}$ if X and B are metrizable; in the general case, B$'$ may be taken to be $X \sqcup \{\infty\}$, equipped with a topology inducing the discrete topology on X and with a carefully chosen filter of neighbourhoods of ∞.)

9.—Let X be a topological space, (A_i) subspaces of X, and $f : \bigsqcup A_i \to X$ the map inducing the canonical injection on each A_i. Show that, f is proper if and only if (A_i) is a locally finite family of closed sets. When is f etale? open?

10. *(Descent)*—Let B and B$'$ be topological spaces and $f : B' \to B$ a continuous map. Let X be a space over B and set $X' = f^*(X)$. Assume that f is open and surjective, or else proper and surjective. Show that, if X$'$ is proper (resp. etale, resp. Hausdorff) over B$'$, then the space X is proper (resp. etale, resp. Hausdorff) over B. Show that, if $\pi' : X' \to B'$ is an open map, then the same holds for $\pi : X \to B$.

11. *(Classical fibrations)*—Consider $\mathbb{R}^n$ (resp. $\mathbb{C}^n$) equipped with the Euclidean (resp. Hermitian) norm. Define the following spaces:

- S^{n-1} (resp. S^{2n-1}) is the unit sphere of $\mathbb{R}^n$ (resp. $\mathbb{C}^n$);
- O_n (resp. U_n) is the space of linear isometries of $\mathbb{R}^n$ (resp. $\mathbb{C}^n$);
- SO_n (resp. SU_n) is the subspace of O_n (resp. U_n) of maps having determinant $+1$;
- for $p \leqslant n$, the Stiefel manifold $V_p(\mathbb{R}^n)$ (resp. $V_p(\mathbb{C}^n)$) is the space of orthonormal families $(x_1, \ldots, x_p)$ of vectors of $\mathbb{R}^n$ (resp. $\mathbb{C}^n$), the Grassmannian $G_p(\mathbb{R}^n)$ (resp. $G_p(\mathbb{C}^n)$) is the set of p-dimensional subspaces of $\mathbb{R}^n$ (resp. $\mathbb{C}^n$);

- for integers p_1, p_2, p_3 such that $p_1 \leqslant p_2 \leqslant p_3 \leqslant n$, D_{p_1,p_2,p_3} is the set of families (F_1, F_2, F_3) of subspaces of $\mathbb{R}^n$ (resp. $\mathbb{C}^n$), where F_i has dimension p_i, such that

$$F_1 \subset F_2 \subset F_3 .$$

Consider the surjective map $\psi : V_p(\mathbb{R}^n) \to G_p(\mathbb{R}^n)$ which assigns to every family the subspace it generates. Equip $G_p(\mathbb{R}^n)$ with the quotient topology of $V_p(\mathbb{R}^n)$ and $D_{p_1,p_2,p_3}(\mathbb{R}^n)$ with the topology induced by that of $G_{p_1} \times G_{p_2} \times G_{p_3}$.

Show that the map ψ is a fibration. What is its fibre? Same questions for the following maps:

- $(x_1, \ldots, x_{p_3}) \mapsto (F_1, F_2, F_3)$ from V_{p_3} to $D_{p_1\,p_2\,p_3}$, where F_i is the subspace generated by $x_1, \ldots, x_{p_i}$;
- $(F_1, F_2, F_3) \mapsto F_i$ from D_{p_1,p_2,p_3} to G_{p_i} for $i = 1, 2, 3$;
- for $q \leqslant p$, $(x_1, \ldots, x_p) \mapsto (x_1, \ldots, x_q)$ from $V_p(\mathbb{R}^n)$ (resp. $V_p(\mathbb{C}^n)$) onto $V_q(\mathbb{R}^n)$ (resp. $V_q(\mathbb{C}^n)$).

This in particular gives a fibration of $V_p(\mathbb{R}^n)$ over S^{n-1} with fibre $V_{p-1}(\mathbb{R}^{n-1})$, of $V_p(\mathbb{R}^n)$ over $V_{p-1}(\mathbb{R}^n)$ with fibre S^{n-p}, of O_n over S^{n-1} with fibre O_{n-1}, of U_n over S^{2n-1} with fibre U_{n-1}, of S^{2n-1} over $P_{n-1}(\mathbb{C}) = G_1(\mathbb{C}^n)$ with fibre $U_1 \approx S^1$. More particularly, since $P_1(\mathbb{C}) \approx S^2$ (Riemann sphere), this gives a fibration of S^3 over S^2 with fibre S^1, called the *Hopf fibration*. Show that S^3 is not homeomorphic to $S^2 \times S^1$.

12.—(a) Let $\mathscr{U} = (U_i)$ be an open cover of $\mathbb{R}$. Show that there is an increasing family $(t_k)_{k \in \mathbb{Z}}$ of points of $\mathbb{R}$ such that for all k, the interval $[t_{k-1}, t_k]$ is contained in some U_i and such that the union of these intervals is $\mathbb{R}$.

(b) Show that a fibre bundle over a not necessarily compact interval in $\mathbb{R}$ is trivial.

4.2 Locally Connected Spaces

4.2.1

Definition A topological space X is said to be **locally connected** if all points of X admit a fundamental system of connected neighbourhoods.

This definition is equivalent to the following condition:

(CCOO) *Every connected component of open subsets of X is open in X.*

Indeed, suppose that X is locally connected. Let U be an open subset of X and U′ a connected component of U. For all $x \in U'$, the set U contains a connected neighbourhood V of x. Then $U' \supset V$, and so U′ is a neighbourhood of x, proving (CCOO). Conversely, suppose that condition (CCOO) holds. Then, for all $x \in X$ and all open neighbourhoods U of X, the connected component of U containing x is open, and so is a neighbourhood of x contained in U.   □

4.2.2 Examples

The space $\mathbb{R}^n$ is locally connected. Every topological manifold, i.e. every space locally homeomorphic to $\mathbb{R}^n$, is locally connected.

A finite union of locally connected closed sets is locally connected. More generally:

Proposition *Let* X *be a topological space and* Y *a proper space over* X. *If* Y *is locally connected and the projection* $\pi : Y \to X$ *is surjective, then the space* X *is locally connected.*

Proof Let $x \in X$ and U be a neighbourhood of x in X; set $F = \pi^{-1}(x)$. The intersection of the union W of connected components of $\pi^{-1}(U)$ and of F is a neighbourhood of F. By 4.1.5, corollary, it contains a set $\pi^{-1}(V)$, where V is a neighbourhood of x. Then $\pi(W)$ contains V; so it is a connected neighbourhood of x in U.   □

4.2.3 Counterexamples

A totally discontinuous space is locally connected only if it is discrete. Thus $\mathbb{Q}$ is not locally connected. Non-finite profinite sets are not locally connected. A compact space is locally connected only if it has finitely many connected components.

In $\mathbb{T}^2 = \mathbb{R}^2/\mathbb{Z}^2$, the image of a line with irrational slope in $\mathbb{R}^2$ is an arc-connected set, but is not locally connected.

Let $\mathbf{m} = (m_n)$ be a sequence of numbers tending to infinity, such that m_n divides m_{n+1} for all n. The *solenoid* $S_{\mathbf{m}} = \varprojlim \mathbb{R}/\mathbb{Z}.m_n$ is a connected compact space, but is neither locally connected, nor arc-connected.

We next give three examples of connected compact subsets of $\mathbb{R}^2$, but not locally connected.

Examples (1) *Comb Space.* It is the set

$$K_1 = ([0, 1] \times \{0\}) \cup (A \times [0, 1]),$$

where $A = \{0\} \cup \left\{\frac{1}{n}\right\}_{n \in \mathbb{N}*}$.

(2) *Topologists' sine curve.* It is the set

$$K_2 = \left\{(x, y) \in \mathbb{R}^2 \mid x \in {]}0, 1], y = \sin\frac{1}{x}\right\} \cup \{0\} \times [-1, 1].$$

(3) *Spiral around circle.* It is the set

$$K_3 = \left\{\left(1 + \frac{1}{t}\right).e^{it}\right\}_{t \in [1, \infty[} \cup S^1 \subset \mathbb{C},$$

where S^1 is the unit circle.

Note that K_1 is arc-connected, whereas K_2 and K_3 are not.

4.2.4

Theorem *Every connected and locally connected compact metrizable space is arc-connected.*

To prove this theorem, we introduce the notion of uniformly locally connected metric space. Then the theorem follows from Propositions 4.2.6 and 4.2.8 given below.

4.2.5

Definition Let X be a metric space and $B(x, r)$ the closed ball in X centered at x with radius r. Then X is **uniformly locally connected** if for all $\varepsilon > 0$, there exists $\alpha > 0$ such that, for all $x \in X$, X contains a connected set L with $B(x, \alpha) \subset L \subset B(x, \varepsilon)$.

Or equivalently, there is a function $h : [0, a[\to \mathbb{R}$, continuous at 0 with $h(0) = 0$ and $a > 0$, such that, for all $x, x' \in X$ with $d(x, x') < a$, there is a connected set L having diameter $\leqslant h(d(x, x'))$ containing x and x'. Such a function is called a *modulus of local connectivity* (an allusion to uniform continuity and to the notion of the modulus of continuity).

4.2.6

Proposition *Every locally connected metric space is uniformly locally connected.*

Proof Let $\varepsilon > 0$. All $x \in X$ have a connected neighbourhood L contained in $B\left(x, \frac{\varepsilon}{2}\right)$, and hence having diameter $\leqslant \varepsilon$; For such L, there is open subset $U \leqslant L$ containing x. Hence there is an open cover $\mathcal{U} = (U_i)$ of X such that each U_i is contained in a connected set having diameter $\leqslant \varepsilon$. Let μ be a mesh of $\mathcal{U}$ (4.1.10) and $\alpha < \mu$. Then

$$(\forall x \in X) \, (\exists i) \, B(x, \alpha) \subset U_i \subset L_i \subset B(x, \varepsilon) .$$

$\square$

4.2.7 *Embeddings in a Banach Space*

The above Proposition 4.2.8 is easier to prove for a closed subset of a Banach space. The next lemma shows that this does not lessen its generality.

Lemma *For every metric space* X, *there is an isometric embedding* ι *of* X *in a Banach space* E.

Proof Suppose first that E is the vector space of bounded continuous functions on X, equipped with the norm $\|f\| = \sup_{x \in X} |f(x)|$. If X has finite diameter, then we may take $\iota(x) = h_x$, where $h_x(x') = d(x, x')$. In the general case, assuming $X \neq \varnothing$, $\iota(x) = h_x - h_{x_0}$ for some $x_0 \in X$. In all cases, this gives an isometric embedding of X in E. $\blacksquare$

4.2.8

Proposition *Every complete uniformly locally connected metric space is arc-connected.*

Proof Thanks to Lemma 4.2.7, X may be assumed to be in some Banach space E, in which it is closed since it is complete. Let $x, y \in X$. Set $t_{0,0} = 0$, $t_{0,1} = 1$, $x_{0,0} = x$, $x_{0,1} = y$.

Let (ε_n) be a sequence of reals > 0 such that $\sum \varepsilon_n < \infty$, and for all n let $\alpha_n > 0$ such that any two points at a distance $\leqslant \alpha_n$ are in the same connected set having diameter $\leqslant \varepsilon_n$. For all n, we inductively construct a finite sequence $(t_{n,0}, ..., t_{n,k(n)})$ in $I = [0, 1]$ with $0 = t_{n,0} < t_{n,1} < \cdots < t_{n,k(n)} = 1$, and a finite sequence $(x_{n,0}, ..., x_{n,k(n)})$ in X in such a way that, setting $\sigma_n = \{t_{n,i}\}_{0 \leqslant i \leqslant k(n)}$:

- $\|x_{n,i} - x_{n,i-1}\| \leqslant \alpha_n$ for $i = 1, ..., k(n)$, $n \geqslant 1$.
- $\sigma_{n+1} \supset \sigma_n$
- $x_{n+1,j} = x_{n,i}$ if $t_{n+1,j} = t_{n,i}$
- $\|x_{n+1,j} - x_{n,i-1}\| \leqslant \varepsilon_n$ and $\|x_{n+1,j} - x_{n,i}\| \leqslant \varepsilon_n$ if $t_{n,i-1} \leqslant t_{n+1,j} \leqslant t_{n,i}$.

$t_{n,i}$ and $x_{n,i}$ being defined for $0 \leqslant i \leqslant k(n)$, the points $x_{n,i-1}$ and $x_{n,i}$ are in a connected set $L_{n,i}$ having diameter $\leqslant \varepsilon_n$. There is a finite sequence $(x_{n+1,i,s})_{0 \leqslant s \leqslant k(n+1,i)}$ with $x_{n+1,i,0} = x_{n,i-1}$, $x_{n+1,i,k(n+1,i)} = x_{n,i}$, $\|x_{n+1,i,s} - x_{n+1,i,s-1}\| \leqslant \alpha_n$. Choose an arbitrary sequence $(t_{n+1,i,s})$ with

$$t_{n,i-1} = t_{n+1,i,0} < t_{n+1,i,1} < \cdots < t_{n+1,i,k(n+1,i)} = t_{n,i} \ .$$

Set

$$\sigma_{n+1,i} = \{t_{n+1,i,s}\}_{0 \leqslant s \leqslant k(n+1,i)} \ ,$$

$$\sigma_{n+1} = \bigcup \sigma_{n+1,i} = \{t_{n+1,0}, ..., t_{n+1,k(n+1)}\}$$

with $0 = t_{n+1,0} < \cdots < t_{n+1,k(n+1)} = 1$, $k(n+1) = \sum_i k(n+1,i)$. This ends the inductive construction.

Then define an affine function $\gamma_n : I \to E$ over $[t_{n,i-1}, t_{n,i}]$ by $\gamma_n(t_{n,i}) = x_{n,i}$. So $\gamma_n(0) = x$, $\gamma_n(1) = y$, $\|\gamma_{n+1}(t) - \gamma_n(t)\| \leqslant \varepsilon_n$ and $(\forall t \in I)\, d(\gamma_n(t), X) \leqslant \alpha_n$.

Therefore the functions γ_n form a Cauchy sequence in the space $C(I, E)$, converging to a path in X from x to y. $\square$

Exercises 4.2. (Locally connected spaces)

1.—Let X be a topological space. Show that there is a surjective continuous map $h : [0, 1] \to X$ if and only if X is compact, metrizable, connected and locally connected.

2.—A space X is *locally connected* (resp. *openly locally connected*) at a point x_0 if x_0 admits a fundamental system of connected (resp. open connected) neighbourhoods. The aim here is to construct an example of a locally connected space at a point but not openly locally connected at that point.

Let $I_{p,q}$ be the segment in $\mathbb{R}^2$ having as endpoints $\left(\frac{1}{p}, 0\right)$ and $\left(\frac{1}{p+1}, \frac{1}{(p+1)pq}\right)$. Set

$$X = [0, 1] \times \{0\} \cup \bigcup_{p \geqslant 1, q \geqslant 1} I_{p,q}$$

and $x_0 = (0, 0)$.

(a) Show that X is locally connected at x_0.

(b) Show that $[0, 1] \times \{0\}$ is in every open connected subset of X containing x_0.

(c) Show that a locally connected space (i.e. locally connected at all points) is openly locally connected at all points.

3.—(a) Given two metric spaces X and Y, a map $f : X \to Y$ and an increasing continuous function $h : [0, a[\to \mathbb{R}_+$, with $h(0) = 0$ and $a > 0$, f is said to admit h as a *modulus of continuity* if $d(f(x), f(x')) \leqslant h(d(x, x'))$ for all pairs (x, x') such that $d(x, x') < a$.

Show that a map admits a modulus of continuity if and only if it is uniformly continuous.

(b) Given a metric space X and a function h as above, X is said to admit h as a *modulus of local connectivity* if, for all (x, x') such that $d(x, x') < a$, there is a connected subset L of X having diameter $\leqslant h(d(x, x'))$ and containing x and x'.

Show that a metric space admits a modulus of local connectivity if and only if it is uniformly locally connected.

4.—(a) Let X be a connected compact metric space. Assuming that the diameter D of X and a modulus of local connectivity of X are known, is it possible to determine a modulus of continuity of a path from x to x' in X for $r \leqslant D$, for all (x, x') such that $d(x, x') = r$?

(b) Let X be a metric space. Assume there is a surjective continuous map $g : [0, 1] \to X$. Then g is uniformly continuous and X compact and locally connected, and so uniformly locally connected.

If a modulus of continuity of g is known, is it possible to find a modulus of local connectivity of X?

5.—Show that none of the three spaces given in 4.2.3 as examples of compact non locally connected subsets of $\mathbb{R}^2$ are homeomorphic to subspaces of the other two.

Is there a surjective continuous map from one of these compact subsets to any of the others?

6.—Show that the solenoid S_m defined in 4.2.3 can be embedded in $\mathbb{R}^3$, but not in $\mathbb{R}^2$.

4.3 Coverings

4.3.1

Definition Let B be a topological space. A **covering** of B is a fibre bundle with discrete fibres over B.

In other words, a covering of B is a space X together with a map $\pi : X \to B$, satisfying the following condition:

For all $b \in B$, there is a neighbourhood U of b in B, a discrete space F and a homeomorphism φ from $\pi^{-1}(U)$ onto $U \times F$ such that the diagram

$$\pi^{-1}(U) \longrightarrow U \times F$$
$$\searrow \qquad \swarrow$$
$$U$$

commutes.

If B is locally connected, then the condition can be restated as follows:

There is an open connected cover $(U_i)_{i \in I}$ *of B such that, for all i and all connected components V of* $\pi^{-1}(U_i)$, *the projection* π *induces a homeomorphism from V onto U.*

A covering X of B is *trivial* if it is isomorphic to a covering $B \times F$, where F is a discrete space; an isomorphism $\varphi : X \to B \times F$ is called a *trivialization* of X. Let $f : B' \to B$ be a continuous map and X a covering of B. Then (B', f) *trivializes* X if the covering $X' = f^*(X)$ of B' is trivial. In particular, if A is a subspace of B, A is said to trivialize X if $X|_A$ is a trivial covering of A.

The coverings of B form a full subcategory $\mathscr{C}ov_B$ of the category of spaces over B.

4.3.2

A covering X of B is etale and Hausdorff over B. Indeed these conditions are local in B and if F is discrete, then $U \times F$ is etale and Hausdorff over U.

Therefore, if s and s' are two continuous sections of X, the set of $b \in B$ such that $s(b) = s'(b)$ is clopen[1] in B.

The disjoint union and the fibre product of two coverings of B are coverings of B.

If X is a covering and $f : B' \to B$ a continuous map, then the space $f^*(X)$ obtained from X by a change of basis from B to B' along f is a covering of B'.

Let X be a covering of B and $b \in B$. The *degree* of X in b, written $\deg_b(X)$, is the cardinal of the fibre $X(b)$.[2] The function $b \mapsto \deg_b(X)$ is locally constant: indeed, if $X_U \approx U \times F$, then $\deg_b(X) = \operatorname{Card} F$ for all $b \in U$. The image of X in B is clopen.

If $B \neq \varnothing$ and if $\deg_b(X)$ does not depend on b (for example if B is connected), then $\deg(X)$ or $\deg_B(X)$ denote its value.

The covering X is *finite* of B if, for all $b \in B$, $\deg_b(X)$ is finite.

4.3.3

Proposition *Suppose that B is connected and the space X etale and Hausdorff over B. X is a trivial covering of B if and only if, for all $x \in X$, there is a continuous section $s : B \to X$ such that $s(\pi(x)) = x$.*

Proof Suppose that such continuous sections exist. Let Γ be the set of sections $B \to X$, equipped with the discrete topology, and define the map $\varepsilon : B \times \Gamma \to X$ by $\varepsilon(b, s) = s(b)$. The map ε is obviously continuous, it is etale since $B \times \Gamma$ and X are etale over B, it is surjective by assumption. It is injective. Indeed, if $s(b) = s'(b')$, then $b = b' = \pi(s(b))$, and the set of points where s and s' agree is closed since X is Hausdorff over B, closed since π is locally injective, nonempty since it contains b, and so is equal to B since B is connected. Hence ε, being etale and bijective, is a homeomorphism. The converse is obvious.   □

4.3.4 *Examples*

(1) Let X be clopen in a space B. Then X, equipped with inclusion, is a covering of B.

(2) Let $S^1 = \{z \in \mathbb{C}, |z| = 1\}$. Then $\mathbb{R}$ together with the map $\varphi : t \mapsto e^{it}$ is a covering of S^1. Indeed φ induces a homeomorphism from $\mathbb{R}/2\pi\mathbb{Z}$ to S^1, and $\mathbb{R}$ equipped with the quotient map is a covering of $\mathbb{R}/2\pi\mathbb{Z}$ as will be seen in (4.3.14).

---

[1] Both open and closed, Translator.

[2] We speak of an n-fold covering when $\deg_b(X) = n$, Translator.

Similarly, $\mathbb{C}$ equipped with $z \mapsto e^z$ is a covering of $\mathbb{C}^*$.

(3) The space S^1 together with $z \mapsto z^d$ is a d-fold covering of S^1. Similarly, $\mathbb{C}^*$ together with $z \mapsto z^d$ is a covering of $\mathbb{C}^*$.

(4) The action of $\mathbb{R}^*$ identifies the set $\mathbf{P}^n\mathbb{R}$ of 1-dimensional vector subspaces of $\mathbb{R}^{n+1}$ with the quotient of $\mathbb{R}^{n+1} - \{0\}$, while that of $\{+1, -1\}$ identifies it with the unit sphere S^n of $\mathbb{R}^{n+1}$. Equip $\mathbf{P}^n\mathbb{R}$ with the quotient topology (both quotient topologies agree). The space S^n, equipped with the quotient map, is a 2-fold covering of $\mathbf{P}^n\mathbb{R}$. This covering is non trivial for $n > 1$, since S^n is connected.

(5) More generally, let $\mathbf{G}_p(\mathbb{R}^n)$ be the set of p-dimensional vector subspaces of $\mathbb{R}^n$ (*Grassmannian*), and $\widetilde{\mathbf{G}}_p(\mathbb{R}^n)$ the set of *oriented* p-dimensional vector subspaces. Equip $\mathbf{G}_p(\mathbb{R}^n)$ (resp. $\widetilde{\mathbf{G}}_p(\mathbb{R}^n)$) with the topology obtained by considering it as the quotient of the *Stiefel manifold* $\mathbf{V}_p(\mathbb{R}^n) \subset \mathbb{R}^{np}$ consisting of the orthonormal families $(x_1, ..., x_p)$ in $\mathbb{R}^n$ with respect to the action of the group O_p (resp. SO_p) (4.1, Exercise 11). Then $\widetilde{\mathbf{G}}_p(\mathbb{R}^n)$ is a 2-fold covering of $\mathbf{G}_p(\mathbb{R}^n)$, non trivial if $1 \leqslant p \leqslant n-1$.

(6) Let B be a topological space and $b \mapsto P_b$ a continuous map from B to the vector space of polynomials of $\mathbb{C}[\mathbb{Z}]$ of degree $\leqslant d$. Assume that, for all $b \in$ B, the polynomial P_b admits d distinct roots in $\mathbb{C}$. Then the subspace X of B $\times \mathbb{C}$ consisting of pairs (b, z) such that $P_b(z) = 0$, equipped with the projection $(b, z) \mapsto b$, is a d-fold covering of B.

Indeed, let E be the vector space of polynomials of degree $\leqslant d$, H the subspace of E $\times \mathbb{C}$ consisting of pairs (P, z) such that $P(z) = 0$, and W the set of $P \in$ E having d distinct roots. Changing bases, it becomes a matter of showing that $H|_W$ is a d-fold covering of W. Let $P_0 \in$ W; the roots of P_0 are simple and, by the implicit function theorem, for each root z_i of P_0, there is neighbourhood U_i of P_0 in E and a neighbourhood V_i of z_i in $\mathbb{C}$ such that $G_i = H \cap (U_i \times V_i)$ is the graph of a continuous map from U_i to V_i. The sets V_i may be assumed to be disjoint, and the sets U_i to be equal to a neighbourhood U of P_0. Then $H|_U$ contains the union of the disjoint sets G_i. Since every $P \in$ U has at least d roots, $H|_U = \bigcup G_i$, and so $H|_U$ is a trivial d-fold covering of U. It follows that W is open in E and that $H|_W$ is a d-fold covering of W.

4.3.5

Proposition *Let* B *be a topological space. Let* X *and* Y *be coverings of* B, f *and* g B-*morphisms from* Y *to* X. *The set of* $y \in$ Y *such that* $f(y) = g(y)$ *is clopen in* Y. *In particular, if* Y *is connected and if there exists* $y \in$ Y *such that* $f(y) = g(y)$, *then* $f = g$.

Proof The set of points where f and g agree is closed since X is Hausdorff over B, and it is opens since the projection X $\mapsto$ B is locally injective. $\square$

Corollary *With the same notation, if* B *is connected and there exists b* ∈ B *such that* $f_b = g_b$, *then* $f = g$.

Proof Each connected component of Y contains a point of Y(b).

4.3.6

Proposition *Let* B *be a locally connected space,* X *and* Y *be coverings of* B, *and* $f : Y \to X$ *a morphism. Then* Y *together with* f *is a covering of* X.

Lemma *Assume* B *is connected, and that* X *and* Y *are trivial coverings of* B *and* $f : Y \to X$ *a covering map. Then* Y *together with* f *is a covering of* X, *trivial on each connected component of* X.

The proposition follows from the lemma by covering B with open and connected sets trivializing both X and Y.

Proof of Lemma We may assume that $X = B \times F$ and $Y = B \times G$, with F and G are discrete. Then f is of the form $(b, u) \mapsto (b, \alpha(b, u))$, where $\alpha : B \times G \to F$ is a continuous map. Since B is connected and F discrete, $\alpha(b, u)$ does not depend on b, and so $\alpha(b, u) = \varphi(u)$, where φ is a map from G to F. Then, for $t \in F$, $f^{-1}(B \times \{t\}) = B \times \varphi^{-1}(t)$, and this space is a trivial covering of $B \times \{t\}$. □

Corollary 4.1 *Let* $f : Y \to X$ *be a covering map. Then* $f(Y)$ *is clopen in* X. *In particular, if* X *is connected and* $Y \neq \varnothing$, *then* f *is surjective.*

Corollary 4.2 *Let* $f : Y \to X$ *be a covering map. If* B *is connected and there exists* $b \in B$ *such that* $f_b : Y(b) \to X(b)$ *is bijective, then* f *is an isomorphism.*

4.3.7

Let B be a locally connected space.

Proposition *Assume that* B *is connected; let* X *be a covering of* B. *If there exists* $b \in B$ *such that there is a continuous section passing through all* $x \in X(b)$, *then* X *is trivial.*

Proof Let Γ be the set of continuous sections of X equipped with the discrete topology. The map $\varepsilon : B \times \Gamma \to X$ defined by $\varepsilon(b, s) = s(b)$ is a covering map and $(B \times \Gamma, \varepsilon)$ is a covering of X (4.3.6). We show that ε is a isomorphism: for this it suffices that, for every connected component Y of X, $\deg_Y(B \times \Gamma, \varepsilon) = 1$. Let Y be a connected component of X. It is a covering of B, and its projection $\Pi(Y)$ in B is clopen and so equal to B; in particular $Y \cap X_b \neq \varnothing$: for $y \in Y \cap X_b$, $\deg_y(B \times \Gamma, \varepsilon) = \deg_Y(B \times \Gamma, \varepsilon) > 0$. If s and s' are two continuous sections X passing through y, then $s = s'$ by 4.3.5, and so $\deg_Y(B \times \Gamma, \varepsilon) = 1$. □

4.3.8 Subcoverings, Quotient Coverings

Proposition *Let* B *be a locally connected space and* X *a covering of* B.

(a) A subset Y *of* X *is a covering of* B *if and only if* Y *is clopen in* X.
(b) If R *be an equivalence relation on* X, *then* X/R *is a covering of* B *if and only if the graph of* R *is clopen in* X $\times_B$ X.

Proof

(a) Suppose that $Y \subseteq X$ is a covering of B. Let U be a connected open subset of B trivializing X and Y. Let W be a connected component of $Y|_U$. Then W is contained in a connected component V of $X|_U$. As $\pi|_W : W \to U$ and $\pi|_V : V \to U$ are bijective, W = V. In other words, $Y|_U$ is the union of the connected components of $X|_U$, and the same holds for $(X - Y)|_U$. Hence $Y|_U$ and $(X - Y)|_U$ are open in $X|_U$. Since B can be covered with connected open subsets trivializing X and Y, Y and (X − Y) must be open.

Conversely, let Y be clopen in X, U an open connected subset in B trivializing X, and $\varphi : X|_U \to U \times F$ a trivialization. The sets $\varphi^{-1}(U \times \{t\})$ for $t \in F$ are the connected components of $X|_U$. The space $Y|_U$ is clopen in $X|_U$, and so $Y|_U = \varphi^{-1}(U \times G)$, for some subset G of F. Hence $Y|_U$ is a trivial covering of U, and since all points of B admit a connected open neighbourhood trivializing X, the space Y is a covering of B.

(b) Suppose R is as stated and that X/R is a covering of B. Let $\chi : X \to X/R$ be the canonical morphism. The graph of R is the set of points of X $\times_B$ X, where $\chi \circ \mathrm{pr}_1$ and $\chi \circ \mathrm{pr}_2 : X \times_B X \to X/R$ agree. It is therefore clopen by Proposition 4.3.5. Conversely, let U be a connected open subset in B trivializing X. Then $X/R|_U = (X|_U)/R$, and the topologies agree since $X|_U$ is a saturated set for R open in X. Let $\varphi : X|_U \to U \times F$ be a trivialization and let R′ be the equivalence relation on U × F obtained by transferring via φ the relation induced by R onto $X|_U$. The graph of R′ is clopen in

$$(U \times F) \times_U (U \times F) = U \times F \times F,$$

and so can be written as U × G, where G is a subset of F × F. The set G is the graph of an equivalence relation on F since, if $x \in U$, then

$$(x, t) \sim_{R'} (x, t') \Longleftrightarrow (t, t') \in G.$$

The canonical bijection $(U \times F)/R' \to U \times (F/G)$ is a homeomorphism since the canonical map $U \times F \to U \times (F/G)$ is open. Hence $(X/R)|_U$ is isomorphic to U × (F/G), and so is a trivial covering of U. Since B can be covered by a family of nonempty connected open sets trivializing X, the space X/R is a covering of B. $\qquad \square$

4.3.9 Examples

For $f \in \mathbb{C}[z]$ considered a map $\mathbb{C} \to \mathbb{C}$, let f^k be $f \circ \cdots \circ f$, where f appears k times. Then z is said to be a *periodic point with period k* if $f^k(z) = z$ and $z_i = f^i(z) \neq z$ for $0 < i < k$. In this case, the set $\{z_0, ..., z_{k-1}\}$ is a *k-cycle* of f.

Fix $d \geqslant 2$ and $k \geqslant 1$. Let E denote the vector space of polynomials of degree $\leqslant d$, B the set of $f \in$ E for which the equation $f^k(z) - z = 0$ has d^k distinct roots and set $X = \{(f, z) \in B \times \mathbb{C} \mid f^k(z) = z\}$. In other words, X is the set of $(f, z) \in B \times \mathbb{C}$ such that z is periodic point of f with period dividing k.

The set B is open and dense in E since it is the set of f of degree d for which the discriminant of $f^k(z) - z$ is nonzero. Equipped with the projection $\pi : (f, z) \mapsto f$, the space X is a covering of B (4.3.4, Example 5).

The set Y of (f, z), for z with period k, is clopen in X. Indeed, it is clearly open; its complement is the set of (f, z), where z is periodic with period dividing some k' strictly dividing k. Hence it is the union of subcoverings, and so also open.

The set Z of (f, ζ), where $f \in$ B and ζ is a k-cycle of f may be considered a covering of B. Indeed Z is the quotient of Y by the equivalence relation identifying (f, z) and $(f, f^i(z))$ for $0 \leqslant i \leqslant k - 1$. The graph of this relation is the union of the images of the morphisms $(f, z) \mapsto (f, z, f^i(z))$ from Y to Y $\times_B$ Y. Hence it is clopen and Proposition 4.3.8. can be applied.

4.3.10

Definition A space X is **simply connected** if $X \neq \varnothing$ and every covering of X is trivial.[3]

All simply connected spaces are connected. Indeed, if Y is clopen in X and different from $\varnothing$ and X, then Y is a non trivial covering of X.

The space X is called *locally simply connected* if all points of X admit a fundamental system of simply connected neighbourhoods.

Examples (1) All intervals of $\mathbb{R}$ are simply connected. For compact intervals this follows from (4.1.9); otherwise see 4.1, Exercise 12, or 4.3.17, Corollary 4.2.

(2) More generally, in a topological vector space, every nonempty convex subset is simply connected (4.3.17, Corollary 4.2).

---

[3] The reader should be aware of possible conflicts with the usual meaning of the term "simply", for example: I do not assume X simply connected, but simply to be connected.

4.3.11

If X is a covering of B and Y a covering of X, then the space Y, equipped with the composite function onto B, is not necessarily a covering of B (4.3, Exercise 7). This holds in most cases met with in practice. The next proposition provides a criterion for this to be so.

If Y is a space over B and $b \in B$, Y is said to be a *b-neighbourhood covering* if there is neighbourhood V of b in B such that $Y|_V$ is a covering of V.

Proposition *Let* B *be a locally connected space,* X *a covering of* B *and* Y *a covering of* X. *Let* $b \in B$ *and* U *a connected neighbourhood of* b *trivializing* X. *Let* $(U_t)_{t \in F}$ *denote the family of connected components of* $X|_U$, *and for open subsets* V *of* U, *set* $V_t = U_t \cap X|_V$. *Then,* Y *is a b-neighbourhood covering of* B *if and only if there is an open neighbourhood* V *of* b *in* U *such that, for all* $t \in F$, *the open subset* V_t *trivializes* Y.

Proof Suppose that V_t trivializes Y for all $t \in F$. Then $Y|_V$ is the disjoint union of the trivial coverings $f^{-1}(V_t)$ of V, and so is a trivial covering of V. The converse follows by (4.3.1).

Corollary 4.3 *All coverings of a finite cover of* B *are coverings of* B.

Corollary 4.4 *If* B *is locally simply connected, all coverings of a covering of* B *are coverings of* B.

4.3.12

Proposition *A space* X *over* B *is a finite covering of* B *if and only if the projection* $\pi : X \to B$ *is etale, proper and Hausdorff.*

Proof Suppose the projection is etale, proper and Hausdorff, and let $b \in B$. The fibre $X(b)$ is discrete since the projection is etale; it is compact since it is is proper and Hausdorff, and hence finite. Let $X(b) = \{x_1, \ldots, x_d\}$. There are open neighbourhoods $U_1, \ldots, U_d$ of $x_1, \ldots, x_d$ such that π induces respectively a homeomorphism from U_i onto the open subsets V_i of B. Since π is Hausdorff, the sets U_i may be assumed to be disjoint. Set

$$V = \bigcap V_i - \pi \left(X - \bigcup U_i \right) .$$

Then $\pi^{-1}(V) = U'_1 \cup \cdots \cup U'_d$, where $U'_i = U_i \cap \pi^{-1}(V)$, and $X|_V$ is isomorphic to $V \times \{1, \ldots, d\}$. Conversely, "etale", "proper" and "Hausdorff" are local properties of B, and hence obviously hold for a finite covering.   □

4.3.13

Theorem *Let* X *be a topological space and* G *a group acting on* X *such that, for all* $g \in G$, *the map* $x \mapsto gx$ *is continuous. We assume the following:*

(L) For all $x \in X$, *there is a neighbourhood* U *of* x *such that, for all* $g \neq e$ *in* G, $g\mathrm{U} \cap \mathrm{U} = \varnothing$.

Then X *is a covering of* X/G.

Proof Let U be open in X and $\forall g \neq e$, $g\mathrm{U} \cap \mathrm{U} = \varnothing$. For $g \neq g'$,

$$g\mathrm{U} \cap g'\mathrm{U} = \varnothing, \text{ as } g' = gh$$

with $h \neq e$ and $g\mathrm{U} \cap g'\mathrm{U} = g(\mathrm{U} \cap h\mathrm{U}) = \varnothing$. So GU is a disjoint union of all $g\mathrm{U}$ for $g \in G$. The canonical map $\chi : \mathrm{X} \to \mathrm{X}/\mathrm{G}$ is open since the saturation of all open subsets of X, i.e. the union of the orbits, is open. In particular $\mathrm{V} = \chi(\mathrm{U})$ is open in X/G. The assumption $g\mathrm{U} \cap \mathrm{U} = \varnothing$ implies that χ induces a bijection from U onto V. This bijection is continuous and open, and hence a homeomorphism. For all $g \in G$, the restriction of χ to $g\mathrm{U}$ is the composition of $g^{-1} : g\mathrm{U} \to \mathrm{U}$ and $\chi : \mathrm{U} \to \mathrm{V}$. So it is a homeomorphism from $g\mathrm{U}$ onto V. This gives a partition of $\chi^{-1}(\mathrm{V}) = \mathrm{GU}$ into open subsets homeomorphic to V under a projection. $\qquad\square$

Remark The assumptions of the theorem hold if X is a Hausdorff space and G a finite group acting continuously and freely on X, i.e.

$$(\forall g \in G), \quad x \mapsto gx \text{ is continuous and } (\forall x \in X)\ (\forall g \neq e)\ \ g \cdot x \neq x.$$

4.3.14

Corollary *Let* G *be a topological group and* H *a discrete subgroup of* G. *Then* G *is a covering of* G/H.

Proof There is a neighbourhood W of e in G such that $\mathrm{W} \cap \mathrm{H} = \{e\}$, and a neighbourhood U of e in G such that $\mathrm{U}^{-1} \cdot \mathrm{U} \subset \mathrm{W}$. Then $\mathrm{U}h \cap \mathrm{U} = \varnothing$ for $h \neq e$ in H, and for all $x \in G$, $x\mathrm{U}h \cap x\mathrm{U} = \varnothing$. Hence the theorem can be applied to the right action of H on G. $\qquad\square$

4.3.15

Theorem *Let* B *be a topological space,* ρ *the map* $(t, b) \mapsto (0, b)$ *from* $[0, 1] \times B$ *to* $[0, 1] \times B$, *and* X *a covering of* $[0, 1] \times B$. *Then there is a unique covering isomorphism* $\varphi : X \to \rho^*(X)$ *inducing the identity over* $\{0\} \times B$.

Proof Suppose first that X is trivial. We may then assume that $X = [0, 1] \times B \times F$, where F is discrete. So $\rho^*(X) = X$, and 1_X is the desired isomorphism. If φ is another isomorphism with the required property, then φ is of the form

$$(t, b, x) \mapsto (t, b, u(t, b, x))$$

where $u : [0, 1] \times B \times F \rightarrow F$ is continuous. For fixed b and x, $t \mapsto u(t, b, x)$ is a constant continuous map from $[0, 1]$ to F since $[0, 1]$ is connected and F is discrete, so that $u(t, b, x) = u(0, b, x) = x$. Hence $\varphi = 1_X$. ☐

Lemma 4.3 *Let* $\alpha, \beta, \gamma \in \mathbb{R}$ *be such that* $\alpha \leqslant \beta \leqslant \gamma$, *and* X *a covering of* $[\alpha, \gamma] \times B$ *such that* $X_1 = X|_{[\alpha,\beta] \times B}$ *and* $X_2 = X|_{[\beta,\gamma] \times B}$ *are trivial. Then* X *is trivial.*

Proof of Lemma 4.3 Let $\tau_1 : X_1 \rightarrow [\alpha, \beta] \times B \times F$ and $\tau_2 : X_2 \rightarrow [\beta, \gamma] \times B \times G$ be trivializations. Let θ the homeomorphism from $B \times F$ onto $B \times G$ induced by $\tau_2 \circ \tau_1^{-1}$ over $\{\beta\} \times B$, and

$$\theta_* = 1_{[\beta,\gamma]} \times \theta : [\beta, \gamma] \times B \times F \rightarrow [\beta, \gamma] \times B \times G.$$

Then $\tau_2' = \theta_*^{-1} \circ \tau_2 : X_2 \rightarrow [\beta, \gamma] \times B \times F$ is a trivialization agreeing with τ_1 over $\{\beta\} \times B$. The trivializations τ_1 and τ_2' can be glued together to give a continuous bijection $\tau : X \rightarrow [\alpha, \gamma] \times B \times F$, which is continuous. So is its inverse by (4.1.11, Lemma 4.4). Hence τ is a trivialization of X. ☐

Lemma 4.4 *Keeping the notation of the theorem, for all* $b \in B$, *there is a neighbourhood* U *of* b *in* B *such that* $X|_{[0,1] \times U}$ *is trivial.*

Proof of Lemma 4.4 For $b \in B$, $[0, 1] \times \{b\}$ can be covered by a finite family of open subsets $J_i \times U_i$ of $[0, 1] \times B$, where J_i is open in $[0, 1]$ and U_i a neighbourhood of b. Set $U = \bigcap U_i$. By Proposition 4.1.10, there is an increasing sequence $(t_0, \ldots, t_n)$ with

$$t_0 = 0, \quad t_n = 1 \quad \text{and} \quad (\forall k \in \{1, \ldots, n\}) \, (\exists i) \, [t_{k-1}, t_k] \subset J_i \,.$$

Then $X|_{[t_{k-1}, t_k] \times U}$ is trivial for all $k \in \{1, \ldots, n\}$. Using Lemma 4.3, it follows by induction that $X|_{[0, t_k] \times U}$ is trivial; for $k = n$, this means that $X|_{[0,1] \times U}$ is trivial. ☐

Proof of Theorem (general case) By Lemma 4.4, there is an open cover (U_i) of B such that $X|_{[0,1] \times U_i}$ is trivial for all i. Then, for all i, there is an isomorphism $\varphi_i : X|_{[0,1] \times U_i} \rightarrow \rho^*(X|_{[0,1] \times U_i})$ inducing the identity over $\{0\} \times U_i$, and by uniqueness in the trivial case, φ_i and φ_j agree over $[0, 1] \times U_i \cap U_j$. The maps φ_i can be glued together to give an isomorphism φ with the desired property. Its uniqueness follows from that of the maps φ_i. ☐

4.3.16

Let X and Y be topological spaces. Two continuous maps f_0 and f_1 from X to Y are said to be **homotopic** if there is a continuous map $F : [0, 1] \times X \to Y$ such that $F(0, x) = f_0(x)$ and $F(1, x) = f_1(x)$ for all $x \in X$. Then the relation $\simeq$ defined by $f_0 \simeq f_1$ is an equivalence relation on the set $C(X, Y)$ of continuous maps from X to Y (for transitivity, use 4.1.11, Lemma 4.3). The topological spaces and homotopy classes of continuous maps form the quotient category $\mathscr{H}\!omotop$ of $\mathscr{T}\!op$.

A continuous map $f : X \to Y$ is called a **homotopy equivalence** if the homotopy class of f is an isomorphism in $\mathscr{H}\!omotop$, i.e. if there is a continuous map $g : Y \to X$ such that $g \circ f \simeq 1_X$ and $f \circ g \simeq 1_Y$. If there is a homotopy equivalence $X \to Y$ X and Y are said to be *homotopy equivalent*, or to have the same *homotopy type*.

A space X is said to be *contractible* if it is homotopy equivalent to a point, i.e. if $X \neq \varnothing$ and 1_X is homotopic to a constant map.

Let X be a topological space and A a subset of X. A *retraction* from X onto A is a map $r : X \to A$ such that $r(x) = x$ for $x \in A$. If there is a continuous retraction from X onto A, A is said to be a *retract* of X (this requires A to be closed if X is Hausdorff).

A continuous retraction $r : X \to A$ is called a deformation retraction if $\iota \circ r : X \to X$, where $\iota : A \to X$ is the canonical injection, is homotopic to 1_X. If there is a deformation retraction $X \to A$, A is a *deformation retract* of X. This implies that X and A are homotopy equivalent.

4.3.17

Theorem 4.3.15 admits the following corollaries:

> **Corollary 4.5** *Let* B *and* B' *be topological spaces,* X *a covering of* B, f *and* g *continuous maps from* B' *to* B. *If* f *and* g *are homotopic, then the coverings* $f^*(X)$ *and* $g^*(X)$ *of* B' *are isomorphic.*

Proof Let $h : [0, 1] \times B' \to B$ be a homotopy from f to g, r the injection $b' \mapsto (1, b')$ from B' to $[0, 1] \times B'$ and ρ the map $(t, b') \mapsto (0, b')$ from $[0, 1] \times B'$ to itself. Then

$$f = h \circ \rho \circ r \text{ and } g = h \circ r, \text{ and } \rho^*(h^*(X)) \approx h^*(X)$$

by the theorem; thus $f^*(X) \approx g^*(X)$. $\square$

Corollary 4.6 *Every contractible space is simply connected.*

Remark In Theorem 4.3.15, assuming the paracompactness of B, the existence (but not uniqueness) of φ generalizes to fibre spaces with arbitrary fibre. The proof is different. Assuming the paracompactness of B', Corollary 4.5 generalizes likewise. All fibre spaces over a contractible paracompact space are trivial.

4.3.18   Coverings of Products

Proposition *Let* X *and* Y *be topological spaces. Assume* Y *is connected and locally connected. Let* E *be a covering of* X × Y. *If* $E|_{\{x\}\times Y}$ *is trivial for all* $x \in X$ *and if there exists* $y_0 \in Y$ *such that* $E|_{X\times\{y_0\}}$ *is trivial, then* E *is trivial.*

Lemma *Let* X *and* Y *be topological spaces. Assume* Y *is connected and locally connected. Let* E *be a covering of* X × Y, *and* σ *a section of* E, *which is not assumed to be continuous. If* $\sigma|_{\{x\}\times Y}$ *is continuous for all* $x \in X$ *and if there exists* $y_0 \in Y$ *such that* $\sigma|_{X\times\{y_0\}}$ *is continuous, then* σ *is continuous.*

Proof of Lemma Let $\tau : E|_{U\times V} \to U \times V \times F$ be the trivialization with open $U \subset X$ and connected, open $V \subset Y$. For $(x, y) \in U \times V$, we may write $\tau(\sigma(x, y)) = (x, y, s(x, y))$. This defines a function $s : U \times V \to F$ continuous at y, and hence independent from y since V is connected and F discrete. Hence, if σ is continuous (resp. non continuous) at $(x, y) \in U \times V$, then it is continuous (resp. non continuous) at (x, y') for all $y' \in V$. If $x \mapsto \sigma(x, y)$ is continuous on U for some $y \in V$, then σ is continuous on $U \times V$.

It follows that, for each $x \in X$, the set of y such that σ is continuous at (x, y) is clopen in Y, and contains y_0. As Y is connected, this set is the whole of Y. ☐

Proof of Proposition Let $\tau_0 : E|_{X\times\{y_0\}} \to X \times \{y_0\} \times F$ be a trivialization. For $t \in F$ and $x \in X$, there is a unique continuous section $\sigma_{t,x}$ of $E|_{\{x\}\times Y}$ passing through $\tau_0^{-1}(x, y_0, t)$. For each t, the section $\sigma_t : (x, y) \mapsto \sigma_{t,x}(y)$ satisfies the assumptions of the lemma, and so is continuous. The map $\alpha : (x, y, t) \mapsto \sigma_t(x, y)$ is a covering map.

The morphism α is bijective. Indeed, for $z \in E(x, y)$, there is a unique section σ_z of $E|_{\{x\}\times Y}$ passing through z. Then $\tau(\sigma_z(x, y_0))$ is of the form (x, y_0, t_z) and $\alpha^{-1}(z)$ is reduced to (x, y, t_z).

Hence, α is a covering isomorphism $X \times Y \times F \to E$. ☐

4.3.19   Product of Simply Connected Spaces

Corollary *Let* X *and* Y *be two simply connected spaces, at least one of which is locally connected. Then* X × Y *is simply connected.*

Exercises 4.3. (Coverings)

1.—Show that through a change of basis the covering $\widetilde{G}_p(\mathbb{R}^n) \to G_p(\mathbb{R}^n)$ (4.3.4, Example 5) is induced by the covering of $S^N \to P^N(\mathbb{R})$ for sufficiently large N.

2.—Let X and Y be topological spaces, and $f : Y \to X$ a proper map. Consider the transition functor f^* from the category $\mathscr{C}ov_X$ of coverings of X to $\mathscr{C}ov_Y$.

   (a) Show that, if f is surjective with connected fibres, then the functor f^* is fully faithful.

   (b) Deduce that, if f is surjective with connected fibres and Y simply connected, X is simply connected.

   (c) Show that, if the fibres of f are simply connected, then the functor f^* is an equivalence of categories. Deduce that, with this assumption, if X is simply connected, then so is Y.

3.—Let X be a topological space, A and B two open (or two closed) subsets of X such that $A \cup B = X$ and $A \cap B \neq 0$ is connected.

   (a) Show that any covering E of X such that $E|_A$ and $E|_B$ are trivial is trivial.

   (b) Deduce that if A and B are simply connected, so is X.

   (c) Show that the sphere S^n is simply connected for $n \geqslant 2$.

4.—Show that the Stiefel manifolds $V_p(\mathbb{R}^n)$ (4.1, Exercise 11) are simply connected for $p \leqslant n - 2$. Deduce that so are the oriented Grassmannians $\widetilde{G}_p(\mathbb{R}^n)$ (4.3, Exercise 1).

5.—Let Φ be a functor from the category of sets and bijections into itself, B a locally connected space and X a covering of B. Define the structure of a B-covering on the set $\Phi_B(X) = \bigsqcup_{b \in B} \Phi(X(b))$. Study the properties of this action.

6.—Set C_n to be the circle of diameter $\left[(0, 0), \left(\frac{1}{n}, 0\right)\right]$ in $\mathbb{R}^2$, and

$$B = \bigcup_{n \in \mathbb{N}^*} C_n .$$

   (a) Show that B is compact.

   (b) Set $B_+ = B \cap (\mathbb{R} \times \mathbb{R}_+)$ and $B_- = B \cap (\mathbb{R} \times \mathbb{R}_-)$. Show that B_+ and B_- are simply connected (they are contractible).

   (c) Find a covering E_n of B such that $E_n|_{C_n}$ is not trivial.

   (d) Show that the disjoint union E of all E_n is a covering of $B \times \mathbb{N}^*$, which is itself a trivial covering of B, but is not a covering of B.
   Show that nonetheless $E|_{B_+}$ and $E|_{B_-}$ are trivial coverings.

7.—Let B and B' be topological spaces. Describe the maps $f : B' \to B$ for which a space X over B is a covering of B when f^*X is a covering of B'.

8.—Consider the *non locally connected* space $\overline{\mathbb{N}} = \mathbb{N} \cup \{\infty\}$.

(a) Show that the subspace $\overline{\mathbb{N}} \times \mathbb{N} - \{(\infty, 0)\}$ of $\overline{\mathbb{N}} \times \mathbb{N}$ is also a trivial covering of $\overline{\mathbb{N}}$.

(b) Find a clopen subset in $\overline{\mathbb{N}} \times \mathbb{N}$ which is not a covering of $\overline{\mathbb{N}}$.

9.—Let B be a not necessarily locally connected topological space. Define $\mathscr{Rig}$ to be the category of what are called *rigid coverings* of B as follows. An object of $\mathscr{Rig}$ is a covering X of B together with sets U_i forming an open cover of B and with trivializations τ_i $X|_{U_i} \to U_i \times F_i$, $i \in I$, satisfying the condition below. Over $U_{i,j} = U_i \cap U_j$, the map $\tau_j \circ \tau_i^{-1}$ is of the form $(b, y) \mapsto (b, \gamma_{i,j}(b)(y))$, where $\gamma_{i,j}$ is a map from $U_{i,j}$ to the set $S_{i,j} = \text{Isom}_{\mathscr{Ens}}(F_i, F_j)$. The required condition is:

(1c) *For all (i, j), the map $\gamma_{i,j} : U_{i,j} \to S_{i,j}$ is locally constant.*

(a) Note that this condition is automatically satisfied if B is locally connected. Give an example with B compact but not locally connected where it is is not satisfied.

Given two objects $\mathbf{X} = (X, (U_i, \tau_i)_{i \in I})$ and $\mathbf{X}' = (X', (U_j, \tau_j)_{j \in I'})$ of $\mathscr{Rig}$, define $\text{Hom}_{\mathscr{Rig}}(\mathbf{X}, \mathbf{X}')$ as the set of covering morphisms $h : X \to X'$ such that, for all $(i, j) \in I \times I'$ the map $\eta_{i,j} : U_i \cap U'_j \to \text{Isom}_{\mathscr{Ens}}(F_i, F'_j)$ expressing h is locally constant.

(b) Show that, if B is locally connected, then the forgetful functor $\mathscr{Rig} \to \mathscr{Cov}_B$ is an equivalence of categories. Give examples where it is not fully faithfully, or essentially surjective.

10.—Let Ω be the set of $(z_1, ..., z_d) \in \mathbb{C}^d$, with distinct z_i.

(a) Show that Ω is connected.

(b) Denote by $\mathscr{P}_d$ the set of monic polynomials $z^d + a_{d-1}z^{d-1} + \cdots + a_0$ of degree d, and $\mathscr{B}$ the set of $P \in \mathscr{P}_d$ with d distinct roots. Define $\pi : \Omega \to \mathscr{B}$ by $\pi(z_1, ..., z_d) = \prod(z - z_i)$.
Show that Ω together with π is a $d!$-fold covering of $\mathscr{B}$ and is not trivial.

(c) Denote by X the set of (P, z) such that $P \in \mathscr{B}$ and $P(z) = 0$. Define $\varpi : X \to \mathscr{B}$ by $\varpi(P, z) = P$ and $\omega : \Omega \to X$ by $\omega(z_1, ..., z_d) = (\pi(z_1, ..., z_d), z_1)$. Show that (Ω, ω) is a non trivial covering of X and that (X, ϖ) is a non trivial covering of $\mathscr{B}$.

11. (*A proof of d'Alembert's theorem.*)—Let $f \in \mathbb{C}[z]$ be a polynomial of degree d and F a primitive of f (i.e. a polynomial such that $F' = f$).

(a) Show that $F : \mathbb{C} \to \mathbb{C}$ is a proper map.

(b) Assume that f does not vanish at any point. Show that F is a $d + 1$-fold covering. Given that $\mathbb{C}$ is simply connected, show that $d = 0$.

4.4 Universal Coverings

4.4.1

Let (B, b_0) be a pointed space, i.e. a topological B with a distinguished point $b_0 \in B$ called a *basepoint*. A *pointed covering* of (B, b_0) is a covering E of B with a point $t_0 \in E(b_0)$.

Definition Let (B, b_0) be a pointed space. If the functor $X \mapsto X(b_0)$ from the category of coverings of B to the category of sets is representable, then a pair (E, t_0) representing this functor is called **a pointed universal covering** of (B, b_0).

In other words, (E, t_0) is a pointed universal covering of (B, b_0) if and only if for all coverings X of B and all $x \in X(b_0)$, there is a unique B-morphism $f : E \to X$ such that $f(t_0) = x$.

A universal pointed covering of (B, b_0) is an initial object (2, Exercise 2) in the category of pointed coverings of (X, x_0).

Two universal pointed coverings of (B, b_0) are uniquely isomorphic.

4.4.2

Let (B, b_0) be a pointed locally connected space.

Proposition *A pointed covering* (E, t_0) *of* (B, b_0) *is universal if and only if* E *is connected and trivializes all coverings of* B.

Proof

(a) *If* (E, t_0) *is universal, then* E *is connected.*
  Let F be clopen in E and $t_0 \in F$. Define f and g from E to $B \times \{0, 1\}$ by $f(t) = (\pi(t), 0)$ for all $t \in E$ and $g(t) = (\pi(t), 1)$ for $t \notin F$ and $g(t) = (\pi(t), 0)$ for $t \in F$. Then $f(t_0) = g(t_0)$; so $f = g$ by uniqueness of the universal property; hence $F = E$. Therefore E is connected.

(b) *If* (E, t_0) *is universal, then* E *trivializes every covering of* B.
  Let X be a covering of B. We show that $E \times_B X$ is trivial. For this it suffices (4.3.7) to show that for $x \in X(b_0)$, there is a continuous section $E \to E \times_B X$ passing through (t_0, x). By the universal property of E, there is a B-morphism $f : E \to X$ such that $f(t_0) = x$. Then $t \mapsto (t, f(t))$ is the desired section.

(c) *Converse.*
  Let X be a covering of B and $x \in X(b_0)$. There is a continuous section $E \to E \times_B X$ passing through (t_0, x) since E trivializes X; it is unique since E is connected. A unique B-morphism $f : E \to X$ such that $f(t_0) = x$ corresponds to it.   □

Remark The assumption that B is locally connected was only needed to prove *"necessity"*. *"Sufficiency"* holds without this assumption.

4.4.3

Let (B, b_0) be a pointed locally connected space.

Corollary *Let* (E, t_0) *be a universal pointed covering of* (B, b_0), $b \in B$ *and* $t \in E(b)$. *Then* (E, t) *is a universal pointed covering of* (B, b).

Indeed, the characterization given in Proposition 4.4.2 does not involve basepoints. (Not that the assumption $t \in E(b)$ only implies that b and b_0 belong to the same connected component.)

A covering E of B is said to be a *universal covering* of B if there are $b \in B$ and $t \in E(b)$ such that (E, t) is a universal pointed covering of (B, b). This holds for all (b, t) with $t \in E(b)$.

If E is a universal covering of B, the image B' of E in B is a connected component of B, and E a universal covering of B'.

If B is connected, then two universal coverings of B are isomorphic, but not uniquely. Worse, there is no natural way of choosing an isomorphism, and so the universal coverings of B cannot be mutually identified, nor is it possible to speak of "the universal covering of B".

4.4.4

Proposition *Every simply connected covering is universal.*

Proof Let E be a simply connected covering of B. Then E is connected and every covering of E is trivial, and so E trivializes every covering of B, and E is universal by 4.4.2.

Remark Let B be a connected space; if B admits a simply connected covering, then every universal covering of B is simply connected. There are spaces B with universal coverings but without any simply connected one (4.4, Exercise 2).

4.4.5

Theorem *Every connected, simply connected space has a simply connected covering.*

The proof requires the notion of product covering of an infinite family of coverings. For an example of locally connected space without any universal covering, see 4.4, Exercise 3.

4.4.6 Product Covering

Let B be a totally connected space and $(X_\lambda)_{\lambda \in \Lambda}$ a family of coverings of B. If Λ is finite, then the fibre product X of X_λ is a covering of B. But this is not always the case when Λ is infinite; then X may not in general be locally connected nor discrete since its fibres are infinite products of discrete spaces.

Proposition *If there is an open cover of* B *trivializing all* X_λ, *then the set* X *can be equipped with a unique topology* $\mathcal{T}$, *finer than the product topology, turning* X *into a cover of* B. *The cover thus obtained is a* **product** *of* X_λ *in the category of coverings of* B.

Proof

(a) Suppose first that B is connected and all X_λ are trivial. For all $\lambda \in \Lambda$, assume $X_\lambda = B \times F_\lambda$. Set $F = \prod_\lambda F_\lambda$. Let $\mathcal{T}_0$ be the product topology on $B \times \prod_\lambda F_\lambda$ (which is not generally locally connected), and $\mathcal{T}$ the product topology of the topology on B and of the *discrete* topology on F. The topology $\mathcal{T}$ is finer than $\mathcal{T}_0$ and turns X into a covering of B.

We show that $(X, \mathcal{T})$ is a product of X_λ in $\mathcal{C}ov_B$. For this, it suffices to show that, if Y is a covering of B and for all λ, f_λ is a morphism from Y to X_λ, then the map f from Y to X defined by f_λ is continuous from Y to $(X, \mathcal{T})$. If necessary by restricting to an open connected subset of B trivializing Y, we may assume that $Y = B \times G$, where G is discrete. Then, B being connected and F_λ discrete, $f_\lambda : (b, g) \mapsto (b, h_\lambda(g))$ for some map h_λ from G to F_λ. The map $h : G \to F$ defined by h_λ is continuous since G is discrete. Hence $f : (b, g) \mapsto (b, h(g))$ is continuous, which shows that X is the product of the X_λ in $\mathcal{C}ov_B$.

Let $\mathcal{T}'$ be another topology finer than $\mathcal{T}_0$ and turning X into a covering of B. The projection $(X, \mathcal{T}') \to X_\lambda$ is continuous for all λ, and so the identity $(X, \mathcal{T}') \to (X, \mathcal{T})$ is continuous. It is a bijective covering morphism, hence an isomorphism, and $\mathcal{T}' = \mathcal{T}$. This proves the proposition in case (a).

(b) In the general case, let (U_i) be open connected sets covering B and trivializing all X_λ. For all i, let $\mathcal{T}_i$ be the topology on the product covering $X_i = X|_{U_i}$ of $X_\lambda|_{U_i}$. By uniqueness in (a), for all (i, j) the topologies $\mathcal{T}_i$ and $\mathcal{T}_j$ agree over each connected component of $U_{i,j}$. The following lemma shows that $\mathcal{T}_i$ and $\mathcal{T}_j$ agree over $U_{i,j}$, then that the $\mathcal{T}_i$ can be glued together to give the desired topology $\mathcal{T}$.

Gluing Lemma *Let* B *be a topological space and* X *a set with a map* $\pi : X \to B$. *Let* (U_i) *be an open cover of* B. *Set* $X_i = \pi^{-1}(U_i)$ *and for each* i *let* $\mathcal{T}_i$ *be a topology on* X *making* $\pi_i = \pi|_{X_i}$ *continuous. Assume that* $\mathcal{T}_i$ *and* $\mathcal{T}_j$ *induce the same topology on*

$X_i \cap X_j$ for all i and j. Then there is a unique topology $\mathscr{T}$ on X making π continuous and inducing $\mathscr{T}_i$ on X_i for all i.

Proof Let $\mathscr{T}$ be the topology on X whose open sets are the subsets V of X such that for all i, $V \cap X_i$ is open with respect to $\mathscr{T}_i$. Then π is continuous with respect to $\mathscr{T}$. We show that $\mathscr{T}|_{U_i} = \mathscr{T}_i$ for all i. It readily follows that $\mathscr{T}_i$ is finer than $\mathscr{T}|_{U_i}$. Let W be an open subset in X_i for $\mathscr{T}_i$. For all j, $W \cap X_j$ is open in $X_{ij} = X_i \cap X_j$ with respect to $\mathscr{T}_i$, hence with respect to $\mathscr{T}_j$. Since π_j is continuous with respect to $\mathscr{T}_j$, X_{ij} is open in X_j with respect to $\mathscr{T}_j$, and so $W \cap X_j$ is open with respect to $\mathscr{T}_j$. Therefore W is open with respect to $\mathscr{T}$, and $\mathscr{T}_i = \mathscr{T}|_{U_i}$. If $\mathscr{T}'$ is any other topology making π continuous and inducing the $\mathscr{T}_i$, it readily follows that $\mathscr{T}'$ is coarser than $\mathscr{T}$. Let V be an open subset of W with respect to $\mathscr{T}$. Then $V \cap X_i$ is open in X_i with respect to $\mathscr{T}'$, and so $V \cap X_i$ is open in X with respect to $\mathscr{T}'$ and $V = \bigcup (V \cap X_i)$ is open with respect to $\mathscr{T}'$. Hence $\mathscr{T} = \mathscr{T}'$. □

4.4.7

Cardinality Lemma *Let* B *be a topological space,* Ω *an open basis for* B *having infinite cardinality* $\mathfrak{w}$, *and* X *a connected etale Hausdorff space over* B. *Then* $(\forall b \in B)$ *Card* $X(b) \leqslant \mathfrak{w}$.

Proof The open sets in Ω may be assumed to be connected without changing $\mathfrak{w}$. A finite sequence $(U_0, \ldots, U_n)$ of open connected sets in Ω such that $U_{i-1} \cap U_i \neq \varnothing$ for $i = 1, \ldots, n$ will be called a *chain* in B. If $(U_0, \ldots, U_n)$ is a chain in B, a finite sequence $(V_0, \ldots, V_n)$ of open sets in X will be called a chain in X over $(U_0, \ldots, U_n)$ if π induces a homeomorphism from V_i onto U_i for $i \in \{0, \ldots, n\}$ and $\pi(V_{i-1} \cap V_i) = U_{i-1} \cap U_i$ for $i \in \{1, \ldots, n\}$. If $(V_0, \ldots, V_n)$ and $(V'_0, \ldots, V'_n)$ are two chains over the same chain $(U_0, \ldots, U_n)$ of B, and if $V_0 \cap V'_0 \neq \varnothing$, then these two chains are equal. Indeed, writing s_i for the section $U_i \to V_i$ and s'_i for the section $U_i \to V'_i$, induction shows that s_i and s'_i agree at least at one point of U_i, and so are equal.

If there is a chain $\mathscr{U}$ in B such that $\mathscr{V}$ is a chain over $\mathscr{U}$, $\mathscr{V}$ will be said to be a chain in X. For $x, x' \in X$, let $\mathscr{C}_x$ (resp. $\mathscr{C}_{x,x'}$) be the set of chains $(V_0, \ldots, V_n)$ of X such that $x \in V_0$ (resp. $x \in V_0$ and $x' \in V_n$). The cardinality of the set of chains of B is $\leqslant \mathfrak{w}$, and over each chain of B there is at most one chain in $\mathscr{C}_x$, and so Card $\mathscr{C}_x \leqslant \mathfrak{w}$.

The relation $\mathscr{C}_{x,x'} \neq \varnothing$, where $x, x' \in X$ is an equivalence relation. Indeed, if

$$(V_0, \ldots, V_n) \in \mathscr{C}_{x,x'} \text{ and } (V'_0, \ldots, V'_n) \in \mathscr{C}_{x',x''},$$

writing W for the connected component of x' in $V_n \cap V'_0$,

$$(V_0, \ldots, V_n, W, V'_0, \ldots, V'_n) \in \mathscr{C}_{x',x''};$$

reflexivity and symmetry are immediate. The equivalence classes are open. Hence, since X is connected, all points are equivalent. Let $b \in B$ and $x_0 \in X$. As the restriction of π to the last open set of a chain in X is injective, for $x \in X(b)$, $\mathscr{C}_{x_0,x}$ form a disjoint family of nonempty subsets of $\mathscr{C}_{x_0}$, and so Card $X(b) \leqslant \mathfrak{w}$. □

Proof of Theorem 4.4.5 Let (B, b) be a pointed space. If (X, x) and (Y, y) are pointed connected coverings of (B, b), (Y, y) will be said to *dominate* (X, x) if there is a pointed covering morphism from (Y, y) to (X, x). Such a morphism is unique by (4.3.5). Hence, if (X, x) and (Y, y) mutually dominate one other, then they are isomorphic.

If (E, t) is a pointed connected covering of (B, b) dominating all others, then it is a universal covering by Definition (4.4.1). If B is moreover locally simply connected, then E is simply connected. Indeed, if (Y, y) is a pointed connected covering of (E, t), then Y is a covering of B by (4.3.11, Corollary 4.6), and so (E, t) dominates (Y, y) and the projection $Y \to E$ is an isomorphism. Hence every connected covering of E is trivial, and every sum of connected coverings is trivial.

Let B be a connected, locally simply connected space. Choose a point $b \in B$, an open basis Ω for B and a set Ξ with cardinality $\geqslant$ Card B $\times$ Card Ω. By the cardinality lemma, every covering of B has cardinality $\leqslant$ Card Ξ, and so is isomorphic to a covering whose underlying set is a subset of Ξ. Let $(X_\lambda, x_\lambda)_{\lambda \in \Lambda}$ be a family of pointed coverings of (B, b) containing every connected pointed coverings of (B, b) whose underlying set is in Ξ. A simply connected open cover of B trivializes all X_λ. Let X be the product covering of X_λ, set $x = (x_\lambda)$ and let E be the connected component of x in X. The pointed covering (E, x) is connected and dominates all X_λ, hence all connected pointed coverings of (B, b). Thus it is simply connected. □

Exercises 4.4. (Universal coverings)

1. *(Non-functoriality of the universal covering.)*—(a) Let E be the universal covering of $S^1 = \{z \in \mathbb{C} \mid |z| = 1\}$, and f the $z \mapsto -z$ from S^1 to itself. Show that there is a f-morphism $g : E \to E$, but none satisfying $g \circ g = 1_E$.

(b) Let $\mathscr{C}$ be the category of locally simply connected topological spaces. Show that there is no functor $R : \mathscr{C} \to \mathscr{C}$ and nor any functorial morphism $\pi : R \to 1_{\mathscr{C}}$ such that, for all $X \in \mathscr{C}$, the space $R(X)$ with $\pi_X : R(X) \to X$ is a universal covering of X.

(c) What do you think of the following argument: "Consider the Hopf fibre bundle $H = (S^3, \pi : S^3 \to S^2)$ with fibre S^1 (4.1, Exercise 11). Replacing each fibre by its universal covering gives a fibre bundle with fibre $\mathbb{R}$. However, every fibre bundle with fibre $\mathbb{R}$ has a continuous section; the image of such a section is a continuous section of H, and so H has a continuous section"?

2.—Let C_n be the circle of diameter $\left[(0, 0), \left(\frac{1}{2n+1}, 0\right)\right]$ for $n \in \mathbb{Z}$ in $\mathbb{R}^2$, and set $M = \bigcup C_n$. Let f be a homeomorphism from M onto itself inducing a homeomorphism from C_n onto C_{n+1} for all n. Let X be the quotient space of $\mathbb{R} \times M$ by the equivalence relation identifying (t, m) to $(t + 1, f(m))$ for $t \in \mathbb{R}$ and $m \in M$.

(a) Show that X is a fibre bundle on $S^1 = \mathbb{R}/\mathbb{Z}$ with fibre M. Show that every covering of X is trivial on each fibre of this fibration.

(b) Show that X has a non simply connected universal covering.

3.—Show that the space B of 4.3, Exercise 6, does not have a universal covering.

4.—(A) Let B be a topological space. Consider the following properties:

   (i) there is an open cover (U_i) of B such that every cover of B induces a trivial covering of each U_i;

  (ii) Every covering of a covering of B is a covering of B.

Do any one of these properties imply the other?

(B) Let (B, b_0) be a pointed connected space.

(a) Let $\mathscr{U} = (U_i)$ be a connected open cover of B, and $\mathscr{C}ov_{B,U}$ the category of coverings of B that are trivial over each U_i. Show that B has a universal covering in $\mathscr{C}ov_{B,U}$, i.e. that the functor $X \mapsto X(b_0)$ from $\mathscr{C}ov_{B,U}$ to $\mathscr{E}ns$ is representable.

(b) Show that B has a universal covering if and only if B satisfies condition (i) of part A, and that this covering is simply connected if B satisfies (ii).

5. *(Another proof of d'Alembert's theorem.)*—Let A be a commutative Banach algebra over $\mathbb{R}$ or $\mathbb{C}$, and let G be the group of invertible elements of A. Define the map $\exp : A \to A$ by

$$\exp(x) = \sum \frac{1}{n!} x^n .$$

(a) Show that A with exp is a universal covering of G.

(b) Show that if G is simply connected, then the map exp is injective.

(c) Deduce that a commutative complex Banach algebra is a field only if it has dimension 1 (Gelfand–Mazur theorem).

(d) Deduce that $\mathbb{C}$ is algebraically closed (d'Alembert's theorem).

(e) Extend the result of (c) to a not necessarily commutative algebra over $\mathbb{C}$ (consider a maximal commutative subalgebra).

For other proofs of d'Alembert's theorem, see (5.2.4), (4.6, Exercise 4) and (6.1, Exercise 3).

4.5 Galois Coverings

4.5.1

Theorem and Definition *Let B be a connected space and E a connected covering of B. The following conditions are equivalent:*

  (i) E *trivializes itself (i.e.* E $\times_B$ E *with* $\mathrm{pr}_1 : E \times_B E \to E$ *is a trivial covering of*
      E*);*
  (ii) *the group* $\mathrm{Aut}_B(E)$ *of automorphisms of* E *acts transitively on each fibre of* E.

  If these conditions hold, then E *is said to be a* **Galois covering** *of* B, *and the group*
$\mathrm{Aut}_B(E)$ *is called the* **Galois group** *of* E.

Proof By (4.3.3), condition (i) is equivalent to the following:

  (i′) There is a continuous section E $\to$ E $\times_B$ E passing through all points of E $\times_B$ E.
        Condition (i′) is equivalent to:
  (i″) For all $(x, x') \in E \times_B E$, there is a B-morphism $f : E \to E$ such that $f(x) = x'$.

  (ii) $\Rightarrow$ (i″) is obvious; we show the converse. If (i″) holds, for x and x' in the
same fibre, there are B-morphisms f and f' from E to itself, such that $f(x) = x'$
and $f'(x') = x$. Then $f' \circ f$ and $f \circ f'$ have a fixed point, and so agree with 1_E by
(4.3.5).   □

Remark If E is a Galois covering of B, then by (4.3.5) the Galois group acts simply
transitively on the fibres.

4.5.2 Examples

  (1) Under the assumptions of Theorem 4.3.13, if X is connected, then X is a
Galois covering of X/G. Indeed, every element of G defines an automorphism of
X as a covering map of X/G, and by definition G acts transitively on the fibres of
$X \to X/G$. This applies to Example (4) of (4.3.4): $S^n \to P^n \mathbb{R}$.

  (2) Similarly, under the assumptions of Corollary 4.3.14, if G is connected, then
G is a Galois covering of G/H. This applies to Examples (2) and (3) of (4.3.4).

  (3) Let B be a locally connected space, E a Galois covering of B, X a connected
cover of B and $f : E \to X$ a B-morphism. Then E together with f is a Galois covering
of X. Indeed if $t, t' \in E$ such that $f(t) = f(t')$, then there exist $g \in \mathrm{Aut}_B(E)$ such
that $g(t) = t'$. As f and $f \circ g$ are two morphisms de E in X agreeing at t, $f = f \circ g$
by (4.3.5), and $g \in \mathrm{Aut}_X(E)$. The Galois group $\mathrm{Aut}_X(E)$ is the subgroup of $\mathrm{Aut}_B(E)$
consisting of all g such that $f \circ g = f$ (4.5.7).

  (4) Let B be a connected space and E a universal covering of B. Then E is a Galois
covering of B. Indeed, E trivializes all coverings of B, in particular itself.

  (5) If X and X′ are Galois coverings of B and B′ with Galois groups G and G′
respectively, then $X \times X'$ is a Galois covering of $B \times B'$, with group $G \times G'$. Indeed
$G \times G'$ may be checked to act simply transitively on the fibres of $X \times X'$.

  For an example of a non Galois covering, see 4.5, Exercise 3.

4.5.3 The Functor S

Let B be a connected and locally connected space, E a Galois covering of B and $G = \text{Aut}_B(E)$ its Galois group. Let G-$\mathscr{E}ns$ denote the category of sets on which G acts and $\mathscr{D}$ the category of coverings of B trivialized by E. For every object X of $\mathscr{D}$, let S(X) be the set $\text{Hom}_B(E, X)$ on which G acts by $(g, f) \mapsto f \circ g^{-1}$. This defines a functor S from $\mathscr{D}$ to G-$\mathscr{E}ns$.

For $X \in \mathscr{D}$, $\deg_B(X) = \text{Card } S(X)$. Indeed, Card $S(X)$ is equal to the number of sections $E \to E \times_B X$ equal to $\deg_E(E \times_B X)$ since $E \times_B X$ is a trivial covering of E which is connected and

$$\deg_E(E \times_B X) = \deg_B(X)$$

since the fibres are the same. In particular, $S(X) = \varnothing \Leftrightarrow X = \varnothing$.

Theorem *The functor* $S : \mathscr{D} \to G\text{-}\mathscr{E}ns$ *is an equivalence of categories.*

This theorem amounts to the above Propositions 4.5.5 and 4.5.6.

4.5.4

Proposition *Let X be an object of* $\mathscr{D}$*; equip S(X) with the discrete topology. The map* $\varepsilon_X : E \times S(X) \to E \times_B X$ *defined by* $\varepsilon_X(t, s) = (t, s(t))$ *a covering isomorphism of E compatible with the actions* $\perp$ *and* $\cdot$ *of G on* $E \times S(X)$ *and* $E \times_B X$ *respectively defined by*

$$g\perp(t, s) = (g(t), s \circ g^{-1}) \qquad g \cdot (t, x) = (g(t), x).$$

Proof Clearly, ε_X is a covering morphism of E; in particular it is open. We show it is bijective. If $(t, s(t)) = (t', s'(t'))$, then $t = t'$, and $s = s'$ by (4.3.5); so ε_X is injective. For all $(t, x) \in E \times_B X$, there is a section $\sigma : E \to E \times_B X$ passing through (t, x) since $E \times_B X$ is a trivial covering of E; it corresponds to a B-morphism $s : E \to X$ such that $s(t) = x$. Hence ε_X is surjective, and it is a homeomorphism, and

$$\varepsilon_X(g\perp(t, s)) = \varepsilon_X(g(t), s \circ g^{-1}) = (g(t), s(g^{-1}(g(t))))$$
$$= (g(t), s(t)) = g \cdot \varepsilon_X(t, s).$$

$\square$

4.5.5

Proposition *The functor* S *is fully faithful.*

In other words, if X and Y are coverings of B trivialized by E, then the map $f \mapsto f_*$ from $\mathrm{Hom}_B(X, Y)$ to $\mathrm{Hom}_G(S(X), S(Y))$ is bijective. Here, $\mathrm{Hom}_B(X, Y)$ denotes the set of covering morphisms from X to Y, and $\mathrm{Hom}_G(S(X), S(Y))$ the set of maps φ from S(X) to S(Y) compatible with the action of G, i.e. such that

$$\varphi(g \cdot s) = g \cdot \varphi(s) \text{ for } g \in G \text{ and } s \in S(X),$$

and f_* is defined by $f_*(s) = f \circ s$.

Lemma *Let* X *be a covering of* B. *For the action of* G *on* E $\times_B$ X *given by the action of* G *on* E, $(E \times_B X)/G = X$.

Proof of Lemma The fibres of E $\times_B$ X $\to$ X are the same as the fibres of E $\to$ B, and so are not empty and G acts transitively on them. Hence the set $(E \times_B X)/G$ can be identified with X. The topology of X agrees with the quotient topology of E $\times_B$ X, since the map E $\times_B$ X $\to$ X defines a covering, and so is open.   □

Proof of Proposition Let $\varphi : S(X) \to S(Y)$ be a morphism in G-$\mathscr{Ens}$. There is morphism $\varphi_* : E \times S(X) \to E \times S(Y)$ defined by $\varphi_*(t, s) = (t, \varphi(s))$ corresponding to it. The morphism φ_* is compatible with the action $\perp$ of G.

As ε_X and ε_Y are isomorphisms, there is a unique E-morphism $\varphi_!$ from E $\times_B$ X to E $\times_B$ Y making the diagram

$$
\begin{array}{ccc}
E \times S(X) & \xrightarrow{\;\varepsilon_X\;} & E \times_B X \\
\downarrow{\scriptstyle \varphi_*} & & \downarrow{\scriptstyle \varphi_!} \\
E \times S(Y) & \xrightarrow{\;\varepsilon_Y\;} & E \times_B Y
\end{array}
$$

commutative. This morphism is compatible with the actions of G. It follows that there is a continuous map f from $X = (E \times_B X)/G$ to $Y = (E \times_B Y)/G$, and f is a B-morphism. We show that $f_* : S(X) \to S(Y)$ is equal to φ, i.e. that $f(s(t)) = (\varphi(s))(t)$ for $s \in S(X)$ and $t \in E$. This follows from the commutativity of the diagram

$$
\begin{array}{ccc}
E \times S(X) & \xrightarrow{\;\varphi_*\;} & E \times S(Y) \\
\downarrow{\scriptstyle \varepsilon} & & \downarrow{\scriptstyle \varepsilon} \\
E \times_B X & \xrightarrow{\;\varphi_!\;} & E \times_B Y \\
\downarrow & & \downarrow \\
X & \xrightarrow{\;\;f\;\;} & Y
\end{array}
$$

Indeed $f_* = \varphi$, which shows that the map $f \mapsto f_*$ is surjective. We show that, if $\varphi = u_*$, for some B-morphism $u : X \to Y$, then $f = u$. The injectivity of $u \mapsto u_*$ will follow. For all $s \in S(X)$, $f \circ s = u \circ s$, and so f and u agree on $s(E)$. For all $s \in S(X)$, there is a section $\tilde{s} : E \to E \times_B X$, and $E \times_B X = \bigcup_{s \in S(X)} \tilde{s}(E)$. Hence $X = \bigcup_{s \in S(X)} s(E)$, and so $f = u$. $\square$

4.5.6

Proposition *The functor* S *is essentially surjective.*

In other words, for all sets A on which G acts, there is a covering $X \in \mathscr{D}$ such that $S(X)$ is isomorphic to A in G-$\mathscr{Ens}$.

Proof (a) Assume $A = G/H$ for some subgroup H of G. Set $X = E/H$ and let ξ denote the canonical map $E \to E/H$. We show that X is an object of $\mathscr{D}$. The map ε_E (4.5.4) is a homeomorphism from $E \times G$ onto $E \times_B E$. The graph of the equivalence relation induced on E by H is the image under ε_E of $E \times H$ which is clopen in $E \times G$; so this graph is clopen in $E \times_B E$. Hence by (4.3.8, (b)) X is a covering of B. On the other hand, by passing to the quotient, ε_E gives a morphism $\varepsilon' : E \times (H\backslash G) \to E \times_B X$, where $H\backslash G$ is the set of classes Hg. This morphism is bijective, and so is an isomorphism, and $X \in \mathscr{D}$.

We show that the map $g \mapsto g \cdot \xi = \xi \circ g^{-1}$ from G to $S(X)$ is surjective. Choose $t_0 \in E$. For $f \in S(X)$, let $t \in E$ such that $\xi(t) = f(t_0)$. There exists $g \in G$ such that $g(t) = t_0$, and the morphisms f and $\xi \circ g^{-1}$ agree at t_0, and so are equal (4.3.5, Corollary).

Therefore the group G acts transitively on $S(X)$. We show that the stabilizer of ξ is H. $A \approx S(X)$ since

$$\xi \circ h^{-1} = \xi \iff \xi = \xi \circ h \iff \xi(h(t_0)) = \xi(t_0)$$
$$\iff \exists h' \in H \; h(t_0) = h'(t_0) \iff h \in H.$$

(b) If (X_i) is a family of objects of $\mathscr{D}$, then $S\left(\bigsqcup X_i\right) = \bigsqcup S(X_i)$. Indeed, for all morphisms from E to $\bigsqcup X_i$, the image of E is in some X_i since E is connected.

(c) Every $A \in$ G-$\mathscr{Ens}$ is of the form $\bigsqcup A_i$, with $A_i \approx G/H_i$, where H_i is a subgroup of G. By (a), for each i, A_i is isomorphic to some $S(X_i)$ and (b) $A \approx S(X)$, where $X = \bigsqcup X_i$. $\square$

4.5.7

Proposition *Let* B *be a locally connected space,* X *and* E *connected coverings of* B *and* $f : E \to X$ *a morphism. Assume* E *is Galois and let* G *be the Galois group* $\mathrm{Aut_B(E)}$. *Then,*

(a) E *trivializes* X *over* B*;*

(b) (E, f) *is a Galois covering of* X*;*

(c) X *is a Galois covering of* B *if and only if the group* $H = \mathrm{Aut_X}(E, f)$ *consisting of* $g \in G$ *such that* $f \circ g = f$ *is a normal subgroup of* G.

Proof (a) By (4.3.1) it suffices to show that for $b \in B$, $t \in E(b)$ and $x \in X(b)$, there is a morphism $h : E \to X$ such that $h(t) = x$. By (4.3.6, Corollary 4.5), $f(E) = X$, and so there exists $t' \in E(b)$ such that $f(t') = x$. As E is Galois, there exists $g \in G$ such that $g(t) = t'$. Then $h = f \circ g$ satisfies $h(t) = x$.

(b) Let $t, t' \in f^{-1}(x)$. As E is Galois, there exists $g \in \mathrm{Aut_B}(E)$ such that $g(t) = t'$. Then $f(g(t)) = f(t)$, and by (4.3.5) $f \circ g = f$. In other words, $g \in \mathrm{Aut_X}(E, f)$. Hence (E, f) is a Galois covering of X.

To prove (c), a lemma is needed.

Lemma X *is a Galois covering of* B *if and only if* $\mathrm{Aut_B}(X)$ *acts transitively on* $S(X) = \mathrm{Hom_B}(E, X)$.

Proof of Lemma Suppose that X is Galois, and let f and f' be two morphisms $E \to X$. Choose $t \in E$ and set $x = f(t)$ and $x' = f'(t)$. There exists $h \in \mathrm{Aut_B}(X)$ such that $h(x) = x'$. Then $h \circ f$ and f' agree at t, and so $h \circ f = f'$ by (4.3.5).

Conversely, let x and x' be in the same fibre $X(b)$. Choose $t \in E(b)$; there are morphisms f and f' from E to X such that $f(t) = x$, $f'(t) = x'$. If $\mathrm{Aut_B}(X)$ acts transitively on $S(X)$, there exists $h \in \mathrm{Aut_B}(X)$ such that $h \circ f = f'$. Then $h(x) = x'$. □

End of the Proof of the Proposition (c) The G-set $S(X)$ is isomorphic to G/H under $g \mapsto f \circ g^{-1}$. The stabilizer of $f \circ g^{-1}$ is gHg^{-1}; if the automorphisms of $S(X)$ act transitively, these stabilizers are all equal, and H is normal.

Conversely, if H is normal, then the right translations in G act transitively on G/H under G-automorphisms, and so X is Galois. □

Exercises 4.5. (Galois coverings)

1.—Let E be a 2-fold covering of B. Show that the map $\tau : E \to E$ assigning to each point the other point of the same fibre is an automorphism of E. Deduce that, if B and E are connected, then E is a Galois covering.

2.—Let B be a connected space and E a covering of B such that $\mathrm{Aut_B}E$ acts transitively on the fibres of E. Show that each connected component of E is a Galois covering of B.

3.—Let X and Y be curves having double points represented in the figure below and $\pi : Y \to X$ a continuous map sending bijectively each (open) arc A_i onto A and each B_i onto B by preserving the direction of the arrows.

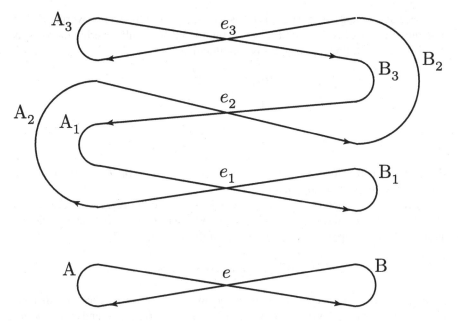

(a) Show that Y together with π is a covering of X. Give its degree.

(b) Determine $\mathrm{Aut}_X Y$. Is the covering Y Galois?

(c) Construct a 2-fold covering $\widetilde{Y}$ of Y such that $\widetilde{Y}$ is a Galois covering of X.

(d) Construct a 2-fold covering $\widetilde{X}$ of X such that $\widetilde{Y}$ is a 3-fold Galois covering of $\widetilde{X}$.

(e) Show (using the results of 4.7) that if B is a connected and locally connected space and Z a connected 3-fold non Galois covering of B, then there is a continuous map $f : X \to B$ such that $f^*(Z)$ is isomorphic to Y.

4.—Let B be a locally connected space and X a covering of B. Assume X has degree 3 and is not Galois. Show that there is a 2-fold covering E of X which is a Galois covering of B, and that there is 2-fold covering Y of B such that E is a 3-fold Galois covering of Y. How many quotient coverings of E are there?

4.6 Fundamental Groups

4.6.1

Definition Let (B, b_0) be a pointed space, Φ the functor $X \mapsto X(b_0)$ from the category $\mathcal{C}ov_B$ of coverings of B to $\mathcal{E}ns$. The **fundamental group** of B with respect to b_0, written $\pi_1(B, b_0)$, is the automorphism group of the functor Φ.

Hence an element $\gamma \in \pi_1(B, b_0)$ consists in being given for every covering X of B a permutation γ_X of $X(b_0)$, subject to the condition that, for all covering morphisms $f : X \to Y$, the following diagram is commutative.

$$
\begin{array}{ccc}
X(b_0) & \xrightarrow{\gamma_X} & X(b_0) \\
\downarrow{\scriptstyle f_{b_0}} & & \downarrow{\scriptstyle f_{b_0}} \\
Y(b_0) & \xrightarrow{\gamma_Y} & Y(b_0)
\end{array}
$$

Remark (1) If B is locally connected, then $\pi_1(B, b_0) = \pi_1(B_0, b_0)$ where B_0 is the connected component of B containing b_0.

(2) If $\mathscr{U}$ is an open cover of B, let $\mathscr{C}ov_{B,\mathscr{U}}$ denote the category of coverings of B trivialized by the open subsets of $\mathscr{U}$. We may define $\pi_1(B, \mathscr{U}, b_0)$ as the automorphism group of the restriction $\Phi_{\mathscr{U}}$ of Φ to $\mathscr{C}ov_{B,\mathscr{U}}$. Then

$$
\pi_1(B, b_0) = \varprojlim \pi_1(B, \mathscr{U}, b_0).
$$

If B is locally simply connected, then the inverse limit is reached by all simply connected open covers, in other words $\mathscr{C}ov_B = \mathscr{C}ov_{B,\mathscr{U}}$ and $\pi_1(B, b_0) = \pi_1(B, \mathscr{U}, b_0)$ if $\mathscr{U}$ is such a cover. In other cases, the pro-object of the category of groups defined by $d\pi_1(B, \mathscr{U}, b_0)$ is not all that interesting.

4.6.2

Proposition *Let* (B, b_0) *be a pointed space and* E *a universal covering of* B *such that* $E(b_0) \neq \varnothing$. *Then* $\pi_1(B, b_0)$ *acts simply transitively on* $E(b_0)$.

Proof Let $t, u \in E(b_0)$. We first show that there is at most one element $\gamma \in \pi_1(B, b_0)$ such that $\gamma_E(t) = u$. If γ has this property, then for all coverings X of B and all $x \in X(b_0)$, there is a morphism $f : E \to X$ such that $f(t) = x$ and $\gamma_X(x) = \gamma_X(f(t)) = f(\gamma_E(t)) = f(u)$, which defines $\gamma_X(x)$ for all X and $x \in X(b_0)$.

We next show existence. Define $\gamma_{ut} \in \pi_1(B, b_0)$ by setting for all coverings X of B and all $x \in X(b_0)$, $(\gamma_{ut})_X(x) = f(u)$ where $f : E \to X$ is the unique morphism such that $f(t) = x$. We check that γ_{ut} is a functorial morphism from Φ to Φ. Then $\gamma_{tt} = 1$ and $\gamma_{vu} \circ \gamma_{ut} = \gamma_{vt}$; so γ_{ut} is an isomorphism, and we also have $\gamma_{ut}(t) = u$.   □

4.6.3 *Comparison of* $\pi_1(B, b_0)$ *and* $\mathrm{Aut}_B(E)$

Let E be a universal covering of B such that $E(b_0) \neq \varnothing$. The group $\mathrm{Aut}_B(E)$ also acts simply transitively on $E(b_0)$ since E is Galois (4.5.1, Remark). By definition of $\pi_1(B, b_0)$, for $\gamma \in \pi_1(B, b_0)$ and $g \in \mathrm{Aut}_B(E)$, the permutations $\sigma_\gamma = \gamma_E$ and $\rho_g = g_{|E(b_0)}$ of $E(b_0)$, respectively defined by γ and g, commute.

Let $t_0 \in E(b_0)$. The inverse of the bijection $\theta \mapsto \theta \cdot t_0$ identifies $E(b_0)$ with $\pi_1(B, b_0)$ and the action σ_γ becomes the left translation by γ. The action ρ_g becomes the right translation by some $\alpha_{t_0}(g) \in \pi_1(B, b_0)$. Indeed, every permutation of $\pi_1(B, b_0)$ commuting with left translations is a right translation.

Proposition *The map* $\alpha_{t_0} : \mathrm{Aut}_B(E) \to \pi_1(E, b_0)$ *is an anti-isomorphism.*

This means that α_{t_0} is bijective and that $\alpha_{t_0}(gg') = \alpha_{t_0}(g') \cdot \alpha_{t_0}(g)$.

Proof The map α_{t_0} is bijective since $\mathrm{Aut}_B(E)$ acts simply transitively on $E(b_0)$, and

$$x \cdot \alpha_{t_0}(gg') = gg'(x) = g(g'(x)) = g(x \cdot \alpha_{t_0}(g')) = x \cdot \alpha_{t_0}(g') \cdot \alpha_{t_0}(g).$$

$\square$

Corollary *The groups* $\pi_1(B, b_0)$ *and* $\mathrm{Aut}_B(E)$ *are isomorphic.*

Remark The anti-isomorphism α_{t_0} depends on the choice of t_0. Indeed, if t_1 is another point of $E(b_0)$, then let $\beta \in \pi_1(B, b_0)$ be the unique element such that $t_1 = \beta \cdot t_0$; so α_{t_1} is the composition of α_{t_0} and of an inner automorphism of $\pi_1(B, b_0)$ defined by β. Indeed,

$$g(\theta \cdot t_1) = \theta \cdot \alpha_{t_1}(g) \cdot t_1 \text{ for } \theta \in \pi_1(B, b_0) \text{ and } g \in \mathrm{Aut}_B(E),$$

where $g(\theta \cdot \beta \cdot t_0) = \theta \cdot \alpha_{t_1}(g) \cdot \beta \cdot t_0$, but $g(\theta \cdot \beta \cdot t_0) = \theta \cdot \beta \cdot \alpha_{t_0}(g) \cdot t_0$, where

$$\alpha_{t_1}(g) = \beta \cdot \alpha_{t_0}(g) \cdot \beta^{-1}.$$

Hence, except when $\pi_1(B, b_0)$ is commutative, $\mathrm{Aut}_B(E)$ should not be identified with $\pi_1(B, b_0)$.

4.6.4 Functoriality

Let (B, b_0) and (B', b_0') be two pointed spaces and $f : (B, b_0) \to (B', b_0')$ a continuous pointed map, i.e. a continuous map $f : B \to B'$ such that $f(b_0) = b_0'$. Let $\gamma \in \pi_1(B, b_0)$. For every covering X' of B', $f^*(X')$ is a covering of B, and $\gamma_{f^*(X')}$ is a permutation of $f^*(X')(b_0) = X'(b_0')$. When X' varies, these $\gamma_{f^*(X')}$ form an automorphism of the functor $X' \mapsto X'(b_0')$, i.e. an element of $\pi_1(B', b_0')$, which we will denote $f_*(\gamma)$.

The map $f_* : \pi_1(B, b_0) \to \pi_1(B', b_0')$ thus defined is a group homomorphism. If $g : B' \to B''$ is a continuous map such that $g(b_0') = b_0''$, then $(g \circ f)_* = g_* \circ f_*$. In other words, π_1 is a functor from the category of pointed spaces to the category of groups.

Let B_0 be the connected component of b_0 in B and $\iota : B_0 \to B$ the canonical injection. Then $\iota_* : \pi_1(B_0, b_0) \to \pi_1(B, b_0)$ is an isomorphism. Indeed, apart from the functor $\iota^* : \mathscr{C}\!\mathit{ov}_B \to \mathscr{C}\!\mathit{ov}_{B_0}$, there is a functor $\iota_* : \mathscr{C}\!\mathit{ov}_{B_0} \to \mathscr{C}\!\mathit{ov}_B$ assigning to all coverings X of B_0 the space X over B. These functors can be checked to define mutual inverses between $\pi_1(B_0, b_0)$ and $\pi_1(B, b_0)$.

4.6.5

Proposition *Let $f : (B, b_0) \to (B', b_0')$ be a pointed space morphism, X and X' respective coverings of B and B', and $h : X \to X'$ an f-morphism. Then the map $h : X(b_0) \to X'(b_0')$ is an f_*-morphism in the following sense: for $g \in \pi_1(B_0, b_0)$ and $x \in X(b_0)$, $h(g \cdot x) = f_*(g) \cdot h(x)$.*

Proof Let $\tilde{h} : X \to B \times_{B'} X'$ be the B-morphism $x \mapsto (p(x), h(x))$, where p is the projection $X \to B$.

The map $\tilde{h} : X(b_0) \to (B \times_{B'} X')(b_0) = \{b_0\} \times X'(b_0') = X'(b_0')$ is a $\pi_1(B, b_0)$-morphism, and by definition of f_*, the identity of $X'(b_0')$. The proposition follows. $\square$

4.6.6 Basepoint-Change

Proposition *Let B be a connected space admitting a universal covering, and $b_0, b_1 \in$ B. Then the groups $\pi_1(B, b_0)$ and $\pi_1(B, b_1)$ are isomorphic.*

Proof If E is a universal covering, then $\pi_1(B, b_0)$ and $\pi_1(B, b_1)$ are isomorphic to $\mathrm{Aut}_B(E)$ by (4.6.3, Corollary).

Remark (1) The assumption that B admits a universal covering can be weakened, but not removed. All that is needed is the existence of a simply connected space J and of a continuous map $g : J \to B$ whose image contains b_0 and b_1. For example, it is sufficient for b_0 and b_1 to be in the same arc-connected component of B.

For an example of a compact, connected non locally connected space B with two points b_0 and b_1 such that $\pi_1(B, b_0) \not\approx \pi_1(B, b_1)$, see (4.6, Exercise 15, B). We do not know if there is such an example when B is locally connected.

(2) There is no natural isomorphism from $\pi_1(B, b_0)$ onto $\pi_1(B, b_1)$, and hence these cannot be identified, and "the fundamental group of B" is not well defined even when B is connected and admits a universal covering. More precisely, let $\mathscr{C}$ be the category of connected spaces admitting a universal covering and $\mathscr{C}^{\scriptscriptstyle\bullet}$ the category of pointed spaces (B, b_0) such that $B \in \mathscr{C}$. We defined a functor $\pi_1 : \mathscr{C}^{\scriptscriptstyle\bullet} \to \mathscr{G}\!\mathit{r}$. There is no functor $F : \mathscr{C} \to \mathscr{G}\!\mathit{r}$ such that π_1 is isomorphic to the composition of the forgetful functor $\mathscr{C}^{\scriptscriptstyle\bullet} \to \mathscr{C}$ and of F (cf. 4.6, Exercise 8).

4.6.7

Let (B, b_0) and (B', b_0') be pointed topological spaces, f and g pointed continuous maps from (B, b_0) to (B', b_0'). A *pointed homotopy* from f to g is a continuous map $h : [0, 1] \times B \to B'$ such that

$$(\forall t \in [0, 1]) \ h(t, b_0) = b_0', \ (\forall b \in B) \ h(0, b) = f(b) \text{ and } h(1, b) = g(b).$$

Theorem *If there is a pointed homotopy from f to g, then the homomorphisms f_* and g_* from $\pi_1(B, b_0)$ to $\pi_1(B', b_0')$ agree.*

Proof Set $I = [0, 1]$, let ι_0 and ι_1 be the injections $b \mapsto (0, b), b \mapsto (1, b)$ from B to $I \times B$, and ρ the map $(t, b) \mapsto (0, b)$ from $I \times B$ to itself. Let h be a pointed homotopy from f to g and X' a covering of B'. Set $X = h^*(X')$, $X_0 = \iota_0^*(X) = f^*(X')$ and $X_1 = \iota_1^*(X) = g^*(X')$. Then $X_0(b_0) = X_1(b_0) = X'(b_0') = X(t, b_0)$ for all $t \in I$, and

$$(h \circ \rho)|_{I \times \{b_0\}} = h|_{I \times \{b_0\}}, \text{ and so } \rho^* X|_{I \times \{b_0\}} = X|_{I \times \{b_0\}}.$$

By Theorem 4.3.15, there is an isomorphism $\varphi : X \to \rho^*(X)$ agreeing with 1_X over $\{0\} \times B$. By (4.3.5, Corollary), φ agrees with 1_X on $I \times \{b_0\}$, in particular $\varphi_{(1, b_0)} = 1_{X'(b_0')}$. Then $\alpha = \iota_1^*(\varphi)$ is an isomorphism from X_1 onto X_0 such that $\alpha_{b_0} = 1_{X'(b_0')}$.

Let $\gamma \in \pi_1(B, b_0)$. Since γ is a functorial morphism, there is a commutative diagram

$$
\begin{array}{ccc}
X_1(b_0) & \xrightarrow{\gamma X_1} & X_1(b_0) \\
\alpha_{b_0} \downarrow & & \downarrow \alpha_{b_0} \\
X_0(b_0) & \xrightarrow{\gamma X_0} & X_0(b_0)
\end{array}
$$

which can also be written as

$$
\begin{array}{ccc}
X'(b_0') & \xrightarrow{\gamma X_1} & X'(b_0') \\
1 \downarrow & & \downarrow 1 \\
X'(b_0') & \xrightarrow{\gamma X_0} & X'(b_0')
\end{array}
$$

and $\gamma X_0 = \gamma X_1$. As this holds for all X', $f_*(\gamma) = g_*(\gamma)$. As this holds for all γ, $f_* = g_*$. □

4.6.8

Theorem *Let* B *be a connected, locally connected space admitting a universal covering and* $b_0 \in$ B. *Then the functor* $\Phi : X \mapsto X(b_0)$ *from* $\mathscr{Cov}_B$ *to* $\pi_1(B, b_0)$-$\mathscr{Ens}$ *is an equivalence of categories.*

Proof Let (E, t_0) be a pointed universal covering of (B, b_0). Set $G = \mathrm{Aut}_B(E)$. The isomorphism $g \mapsto \alpha_{t_0}(g^{-1})$ (4.6.3) identifies G with $\pi_1(B, b_0)$. Since E is a Galois covering of B and trivializes every covering X of B, by Theorem 4.5.3 the functor $S : X \mapsto \mathrm{Hom}(E, X)$ from $\mathscr{Cov}_B$ to G-$\mathscr{Ens}$ is an equivalence of categories. We show that the functors S and Φ are isomorphic. By the universal property of (E, t_0), for all coverings X of B, the map $\delta_{t_0} : f \mapsto f(t_0)$ from $\mathrm{Hom}(E, X)$ onto $X(b_0)$ is bijective. Hence it is a G-set isomorphism. Functoriality is obvious and so δ_{t_0} is an isomorphism of functors.   □

4.6.9 Dictionary

Let B be a connected, locally connected space admitting a universal covering, $b_0 \in$ B and X a covering of B.

(a) If X is a disjoint union (resp. product covering) of all X_λ, then the $\pi_1(B, b_0)$-set $X(b_0)$ is the sum (resp. product) of $X_\lambda(b_0)$.

(b) Assuming $X \neq \varnothing$, X is connected if and only if $\pi_1(B, b_0)$ acts transitively on $X(b_0)$. Indeed, X est connected if and only if $X = X_1 \sqcup X_2$ for some $X_1 \neq \varnothing$ and $X_2 \neq \varnothing$, and a $\pi_1(B, b_0)$-set F is non-homogenous if $F = F_1 \sqcup F_2$ for some $F_1 \neq \varnothing$ and $F_2 \neq \varnothing$.

More generally, let (X_λ) be the connected components of X. Then $X_\lambda(b_0)$ are the orbits in $X(b_0)$ of the action of $\pi_1(B, b_0)$.

(c) Assigning to every pointed covering (X, x_0) of (B, b_0) the stabilizer of x_0 gives a bijection between the isomorphic classes of connected pointed coverings of (B, b_0) and the subgroups of $\pi_1(B, b_0)$. Indeed, for every group G this gives a bijection between the isomorphic classes of homogenous pointed G-sets and the subgroups of G mapping (F, x_0) onto the stabilizer of x_0; the inverse is given by $H \mapsto G/H$.

(d) Let (X, x_0) and (X', x_0') be two connected pointed coverings of (B, b_0). The coverings X and X' are isomorphic if and only the stabilizers S and S' of x_0 and x_0' are conjugate. Indeed, for $g \in \pi_1(B, b_0)$, the stabilizer of gx_0 is gSg^{-1}; the coverings X and X' are isomorphic if an only if there exists $g \in \pi_1(B, b_0)$ such that all pointed coverings (X, gx_0) and (X', x_0') are isomorphic, i.e. $gSg^{-1} = S'$.

(e) Let (X, x_0) be a connected pointed covering of (B, b_0). The covering X is Galois if and only if the stabilizer of x_0 is normal. Indeed, X is Galois if and only if, for all $g \in G$, the pointed coverings of (X, x_0) and (X, gx_0) are isomorphic.

4.6.10

Proposition *Let B be a connected space admitting a simply connected covering, $b_0 \in B$, (X, x_0) a pointed covering of (B, b_0) and $p : X \to B$ the projection. Then $p_* : \pi_1(X, x_0) \to \pi_1(B, b_0)$ is injective and its image is the stabilizer of x_0.*

Proof Let E be a simply connected pointed covering of B and $t_0 \in E(b_0)$. The pointed covering (E, t_0) is universal by (4.4.4), and so there is a morphism $f : (E, t_0) \to (X, x_0)$. The space E together with f is a covering of X by (4.3.6), which is universal by (4.4.4).

Consider the diagram

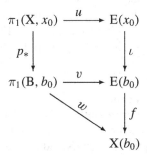

where ι is the canonical injection, $u(g) = g \cdot t_0$, $v(g') = g' \cdot t_0$ and $w(g') = g' \cdot x_0$. This diagram is commutative: by applying (4.6.5) to the p-morphism 1_E, this holds for the upper square, and is obvious for the lower triangle.

The maps u and v are bijective by (4.6.2), ι is injective and its image is $f^{-1}(x_0)$. Hence p_* is injective and its image is $w^{-1}(x_0)$. $\qquad\square$

4.6.11

Corollary *Let (B, b_0) be a pointed space admitting a simply connected pointed covering.*

(a) *For all subgroups H of $\pi_1(B, b_0)$, there is a connected pointed covering (X, x_0) of (B, b_0) such that the image of $\pi_1(X, x_0)$ in $\pi_1(B, b_0)$ is H, and it is unique up to isomorphism.*

(b) *Assume that B is connected and let (X, x_0) be a connected pointed covering of (B, b_0). Then X is Galois if and only if the image of $\pi_1(X, x_0)$ is a normal subgroup of $\pi_1(B, b_0)$.*

If X *is Galois, then* $\mathrm{Aut_B(X)}$ *is isomorphic to the quotient group*

$$\pi_1(\mathrm{B}, b_0)/\pi_1(\mathrm{X}, x_0).$$

4.6.12   Fundamental Group of a Product

Proposition *Let* B *(resp.* B'*) be a space admitting a connected pointed covering* E *(resp.* E'*),* $b_0 \in \mathrm{B}$ *and* $b_0' \in \mathrm{B}'$ *such that* $\mathrm{E}(b_0)$ *and* $\mathrm{E}'(b_0')$ *are nonempty. Assume that* B *or* B' *is locally connected. Then the homomorphism*

$$\pi_1(\mathrm{B} \times \mathrm{B}', (b_0, b_0')) \to \pi_1(\mathrm{B}, b_0) \times \pi_1(\mathrm{B}', b_0')$$

given by the projections is an isomorphism.

Proof The groups $\mathrm{G} = \pi_1(\mathrm{B}, b_0)$ and $\mathrm{G}' = \pi_1(\mathrm{B}', b_0')$ act simply transitively on $\mathrm{E}(b_0)$ and $\mathrm{E}'(b_0')$ respectively, and so $\mathrm{G} \times \mathrm{G}'$ acts simply transitively on $\mathrm{E} \times \mathrm{E}'(b_0, b_0') = \mathrm{E}(b_0) \times \mathrm{E}'(b_0')$. If B (resp. B') is locally connected, then so is E (resp. E'). Thanks to (4.3.19, Corollary), the space $\mathrm{E} \times \mathrm{E}'$ is simply connected, and so is a universal covering of $\mathrm{B} \times \mathrm{B}'$. It follows that $\pi_1(\mathrm{B} \times \mathrm{B}', (b_0, b_0'))$ can be identified with $\mathrm{G} \times \mathrm{G}'$.   □

Exercises 4.6. (Fundamental group)

1.—Let (B, b_0) be a pointed space admitting a pointed universal covering. Show that every endomorphism of the functor Φ of (4.6.1) is an automorphism. Does this result still hold without the assumption that (B, b_0) admits a pointed universal covering?

2.—Let X be a topological space and G a group acting on X in such a way that assumption (L) of Theorem 4.3.13 is satisfied. Set $\mathrm{B} = \mathrm{X}/\mathrm{G}$. Let $x_0 \in \mathrm{X}$ and b_0 be its image in B.

(a) Show that if X is simply connected, then $\pi_1(\mathrm{B}, b_0)$ is isomorphic to G.

(b) Show that if X is connected and admits a simply connected covering, then X is a Galois covering of B and G is isomorphic to the quotient group of $\pi_1(\mathrm{B}, b_0)$ by the image of $\pi_1(\mathrm{X}, x_0)$.

(c) Show that if X is locally simply connected, then each connected component of X is a Galois covering of a connected component of B, and the quotient of $\pi_1(\mathrm{B}, b_0)$ by the image of $\pi_1(\mathrm{X}, x_0)$ is isomorphic to the stabilizer of the connected component of b_0 in B.

3. *(Degree of a map from* S^1 *to* S^1 *.)*—Set $\mathrm{S}^1 = \{z \in \mathbb{C} \mid |z| = 1\}$ and let ε be the map $t \mapsto e^{2i\pi t}$ from $\mathbb{R}$ to S^1.

(a) Show that $\pi_1(\mathrm{S}^1, 1)$ is isomorphic to $\mathbb{Z}$. Since this group is commutative, $\pi_1(\mathrm{S}^1)$ is well defined.

(b) Let $f : S^1 \to S^1$ be a continuous map, define $d°(f) \in \mathbb{Z}$ by $f_*(\gamma) = d°(f) \cdot \gamma$ for $\gamma \in \pi_1(S^1)$; this number is called the *degree* of f. Show that there is a continuous map $\tilde{f} : \mathbb{R} \to \mathbb{R}$ such that $\varepsilon \circ \tilde{f} = f \circ \varepsilon$ and that for any $\tilde{f}$ with this property and all $t \in \mathbb{R}$,

$$\tilde{f}(t + 1) = \tilde{f}(t) + d°(f).$$

(c) Show that any two continuous maps f and g from S^1 to S^1 are homotopic if and only if $d°(f) = d°(g)$.

(d) Let $f : S^1 \to S^1$ be a differentiable map (i.e. continuous and such that $\tilde{f}$ is differentiable). For $z \in S^1$, let $f'(z)$ be the common value of all $\tilde{f}'(t)$ for $t \in \varepsilon^{-1}(z)$. Let $a \in S^1$; assume that $f'(z) \neq 0$ for all $z \in f^{-1}(a)$. Show that the set $f^{-1}(a)$ is finite and that

$$d°(f) = \mathrm{Card}\{z \in f^{-1}(a) \mid f'(z) > 0\} - \mathrm{Card}\{z \in f^{-1}(a) \mid f'(z) < 0\}.$$

(e) Let f be a continuous map from S^1 to $\mathbb{C}^* = \mathbb{C} - \{0\}$; set $d°(f) = d°\left(\frac{f}{|f|}\right)$. Assuming that f is infinitely differentiable, express $d°(f)$ as an integral.

4. *(A proof of d'Alembert's theorem.)*—(a) Calculate the degree of the map $z \mapsto z^d$ from S^1 to S^1.

(b) Show that two sufficiently near (relative to uniform convergence) maps from S^1 to S^1 are homotopic.

(c) Let P be a complex polynomial of degree d, and for $r \in \mathbb{R}_+$ let f_r be the map $z \mapsto P(rz)$ from S^1 to $\mathbb{C}$. Show that, for sufficiently large r, f_r is a map from S^1 to $\mathbb{C}^*$ of degree d.

(d) Show that, if P has no roots z such that $r \leqslant |z| \leqslant r'$, then the maps f_r and $f_{r'}$ from S^1 to $\mathbb{C}^*$ have the same degree. What is the degree of f_0 (assuming $P(0) \neq 0$)? Show that P has at least one root if $d > 0$.

5.—Let G be a finite group. The aim is to show that there is a compact manifold whose fundamental group is isomorphic to G.

(a) Show that, for sufficiently large n, G is isomorphic to a subgroup of the group O_n of isometries of Euclidean $\mathbb{R}^n$.

(b) Show that, for $p \geqslant n + 2$, the space V_n^p of isometric embeddings of $\mathbb{R}^n$ in $\mathbb{R}^p$ is a simply connected compact manifold (4.1, Exercise 11 and 4.3, Exercise 4).

(c) Make G act freely on V_n^p. Conclude.

6.—Let G be an arbitrary group. The aim is to show that there is a connected space B admitting a contractible universal covering and a point $b_0 \in B$ such that $\pi_1(B, b_0)$ is isomorphic to G.

(a) Let E be the space of maps $f : [0, 1] \to G$ such that there is an increasing finite sequence $(t_0, \ldots, t_n)$ with $t_0 = 0, t_n = 1, f(0) = 0$ and f constant on $]t_{i-1}, t_i]$ for $i \in \{1, \ldots, n\}$. For $f, g \in E$, let $d(f, g)$ be the sum of the lengths of intervals on which they differ. Show that E becomes a metric group and that G can be identified to a discrete subgroup of E.

(b) For $t \in [0, 1]$ and $f \in E$, define f_t by $f_t(s) = f(s)$ if $s \leqslant t$ and $f_t(s) = e$ if $s > t$. Show that the map $(t, f) \mapsto f_t$ from $[0, 1] \times E$ to E is continuous. Deduce that E is contractible.

(c) Conclude.

7.—Let K be a connected space admitting a universal contractible covering E, $k_0 \in K$ and $e_0 \in E(k_0)$. The aim is to show that, for any metrizable compact pointed space (B, b_0) admitting a universal covering, and for any homomorphism $\varphi : \pi_1(B, b_0) \to \pi_1(K, k_0)$, there is a pointed continuous map $f : (B, b_0) \to (K, k_0)$ such that $f_* = \varphi$, and that it is unique up to pointed homotopy.

(a) Let G be a group. A **principal G-covering** of B is a covering of B equipped with a continuous G-action inducing a simply transitive action of G on each fibre. Show that every principal G-cover admitting a section is trivial.

(b) Set $G = \pi_1(K, k_0)$, and consider E as a principal G-covering. Prove that the problem amounts to showing that, for any principal pointed G-covering (X, x_0) of (B, b_0), there is, up to pointed homotopy, a unique pointed continuous map $f : (B, b_0) \to (K, k_0)$ such that all principal pointed G-coverings (X, x_0) and $f^*(E, e_0)$ of (B, b_0) are isomorphic.

(c) Let F be a contractible space, B a compact metrizable space, A a closed subset of B, U a neighbourhood of A and $f : U \to F$ a continuous map. Show that there is a continuous map $g : B \to F$ agreeing with f in the neighbourhood of A (if F is a convex containing 0 in a normed space, multiply f by a function with support in U).

(d) Let B be a compact metrizable space, A a closed subset of B, U a neighbourhood of A and X a trivial principal G-covering of B. Show that, for all continuous $f : U \to K$ such that $f^*(E)$ is isomorphic to $X|_U$, there is a continuous map $g : B \to K$ such that $g^*(E)$ is isomorphic to X. More precisely, if $\alpha : X|_U \to f^*(E)$ is an isomorphism, then g may be chosen so that there is an isomorphism $\beta : X \to g^*(E)$ agreeing with α over a neighbourhood of A.

(e) Prove the result without assuming that X is trivial (take a finite open cover $(U_i)_{1 \leqslant i \leqslant n}$ of B trivializing X, a shrinking (V_i), i.e. an open cover such that $\overline{V_i} \subset U_i$, a shrinking (W_i) of (V_i), set $A_i = A \cup \overline{W_1} \cup \cdots \cup \overline{W_i}$, and proceed by induction by applying (d) to $(\overline{V_i}, \overline{V_i} \cap A_{i-1})$).

(f) Conclude: for existence, take $A = \{b_0\}$; for uniqueness, consider

$$\left([0, 1] \times B, \{0, 1\} \times B \ \cup \ [0, 1] \times \{b_0\} \right).$$

(g) Can the assumption that B is compact and metrizable be replaced by a weaker assumption? Can the result be formulated in terms of the representability of some functor?

8. (*Non-functoriality of* π_1 *without basepoint.*)—With the notation of (4.6.6, Remark 2), assume there is a functor F : $\mathscr{C}$ → $\mathscr{Gr}$ such that the diagram

commutes.

(a) Show that, for any space B, the maps ι_{0*} and ι_{1*} from F(B) to F([0, 1] × B) agree. Deduce that, if f and g are two pointed continuous maps from (B, b_0) to (B', b_0') such there is a non-pointed homotopy from f to g, then the maps f_* and g_* from π_1(B, b_0) to π_1(B', b_0') agree.

(b) Let f and g are two pointed continuous maps from (B, b_0) to (B', b_0'), and assume there is non-pointed homotopy h from f to g. Show that g_* : π_1(B, b_0) → π_1(B', b_0') is the composition of f_* with the inner automorphism defined by the element $\gamma \in \pi_1$(B', b_0') corresponding to the loop $t \mapsto h(t, b_0)$ (see 4.9.9). Deduce a contradiction.

(c) Can there be a functor G : $\mathscr{C}$ → $\mathscr{Gr}$ such that the group G(B) is isomorphic to π_1(B, b_0) for all (B, b_0) ∈ $\mathscr{C}$?

9.—In Proposition 4.6.10, instead of a simply connected covering, can B be assumed to admit a universal covering? (study the example of 4.3, Exercise 2).

10. Let U be the set of all $(p, q) \in \mathbb{C}^2$ such that $4p^3 + 27q^2 \neq 0$. We remind the reader that, for $(p, q) \in$ U, the equation $X^3 + pX + q = 0$ has three distinct roots. Set V = $\{(p, q, x) \in$ U × $\mathbb{C}$ | $x^3 + px + q = 0\}$ and E = $\{(x, y, z) \in \mathbb{C}^3$ | $x + y + z = 0, x \neq y \neq z \neq x\}$. Equip V with the projection $(p, q, x) \mapsto (p, q)$ from V to U and E with the map $(x, y, z) \mapsto (p, q, x)$ from E to V, where p and q are defined by

$$X^3 + pX + q = (X - x)(X - y)(X - z).$$

(a) Show that E is a covering of V, that V is a covering of U and that E is a covering of U.

What are the degrees of these coverings?

(b) Show that E is connected. Deduce that V and U are connected.

(c) Let $a = (-1, 0) \in$ U. Show that the action of π_1(U, a) on the fibre V_a defines a surjective homomorphism from π_1(U, a) onto $\mathfrak{S}_3$.

(d) Which of the coverings E → V, V → U, E → U, are Galois?

(e) Let S^3 be the unit sphere of $\mathbb{C}^2$, i.e. $S^3 = \{(p, q) \mid p\bar{p} + q\bar{q} = 1\}$, and set $\Gamma = \{(p, q) \in S^3 \mid 4p^3 + 27q^2 = 0\}$. Show that $S^3 - \Gamma$ has the same homotopy type as U. Deduce that $\pi_1(S^3 - \Gamma, a)$ is not commutative.

(f) Let $\Gamma_0 = \{(\lambda, 0) \mid |\lambda| = 1\}$. Show that Γ is homeomorphic to Γ_0, but that there is no homeomorphism $f : S^3 \to S^3$ such that $f(\Gamma) = \Gamma_0$.

(g) The space $S^3 - \{a\}$ is homeomorphic to $\mathbb{R}^3$. Represent with an iron wire the image of Γ under a homeomorphism from $S^3 - \{a\}$ onto $\mathbb{R}^3$.

11.—Let G be a topological group and e its identity element. Assume that G admits a universal covering.

(a) Show that $\pi_1(G, e)$ is commutative (2.5, Exercise 2, (d)).

(b) Let (H, e) be a connected pointed covering of (G, e). Show that there is a unique topological group law on H having e as its identity, such that the projection $H \to G$ is a homomorphism. (The case of a universal covering may be first considered.)

(c) Can the space $\mathbb{C} - \{+1, -1\}$ be endowed with a composition law making it a topological group?

12.—Let (X, x_0) and (Y, y_0) be two pointed spaces and let $f : (X, x_0) \to (Y, y_0)$ be a pointed continuous map. Let (E, e_0) be a pointed universal covering of (Y, y_0). Show that the image of $\pi_1(f^*E, (x_0, e_0))$ in $\pi_1(X, x_0)$ is the kernel of the homomorphism $f_* : \pi_1(X, x_0) \to \pi_1(Y, y_0)$.

13.—Let V be a manifold and ω a real-valued closed differential form of degree 1 on V. A subset in $V \times \mathbb{R}$ will be called a ω-*slice* if it is the graph of a function $f : U \to \mathbb{R}$ such that $df = \omega|_U$, where U is open in V. The slices form a basis with respect to the topology $\mathscr{T}$ on $V \times \mathbb{R}$.

(a) Show that $V \times \mathbb{R}$, equipped with the topology $\mathscr{T}$ and the projection pr_1 onto V, is a covering of V.

(b) Let $b \in V$. Show that there is a homomorphism $\varphi : \pi_1(V, b) \to \mathbb{R}$ such that the action of $\pi_1(V, b)$ on the fibre of this covering, namely on $\mathbb{R}$ equipped with the discrete topology, is given by $\gamma \cdot x = x + \varphi(\gamma)$.

(c) Let $\gamma \in \pi_1(V, b)$ be represented by a differentiable loop C. Show that

$$\varphi(\gamma) = \int_C \omega .$$

(d) Show that the following conditions are equivalent:

(i) ω is exact, i.e. $(\exists f)\, \omega = df$;
(ii) $\varphi = 0$;
(iii) the covering $(V \times \mathbb{R}, \mathscr{T})$ is trivial.

(e) Assume V is connected and let E be a connected component of $(V \times \mathbb{R}, \mathscr{T})$ and $\pi : E \to V$ the projection. Show that E is a Galois covering of V and that the differential form $\pi^*\omega$ on E is exact.

(f) Show that $\omega \mapsto \varphi$ is a linear map from the space of closed differential forms on V to the vector space $\mathrm{Hom}(\pi_1(V, b), \mathbb{R})$. What is its kernel?

(g) Generalize the results to complex valued differential forms.

14.—(a) Let G be a group and H a subgroup of G. Show that $\mathrm{Aut}_G\, G/H$ is isomorphic to N/H, where N is the normalizer of H in G.

(b) Let (B, b_0) be a connected, locally simply connected pointed space, and let (X, x_0) be a connected pointed covering of (B, b_0). Describe $\mathrm{Aut}_B\, X$.

15.—(A) Let B be a topological space, $b_0, b_1 \in B$. Assume that b_0 and b_1 are in the same arc-connected component, or more generally, that there is a simply connected space J and a continuous map $g : J \to B$ such that $g(J) \supset \{b_0, b_1\}$. Let Φ_0 (resp. Φ_1) denote the functor $\mathscr{C}\!ov_B \to \mathscr{E}\!ns$ assigning the fibre $X(b_0)$ (resp. $X(b_1)$) to the covering X of B. Show that the functors Φ_0 and Φ_1 are isomorphic (they are both isomorphic to the functor $\mathrm{Sec} \circ g^*$ assigning to X the set of sections de $g^*(X)$).

Deduce that the groups $\pi_1(B, b_0)$ and $\pi_1(B, b_1)$ are isomorphic.

(B) Let p be a (not necessarily prime) integer $\geqslant 2$, and take B to be the *solenoid* $\widehat{\mathbb{R}} = \varprojlim \mathbb{R}/p^n\mathbb{Z}$ with its group structure, the projections $\pi_n : \widehat{\mathbb{R}} \to \mathbb{R}/p^n\mathbb{Z}$, in particular $\pi_0 : \widehat{\mathbb{R}} \to \mathbb{R}/\mathbb{Z}$, and the injection $\iota : \mathbb{R} \to \widehat{\mathbb{R}}$. Let $b_0, b_1 \in B$.

The aim is to prove that if b_0 and b_1 are not in the same arc-connected component of B, then the functors Φ_0 and Φ_1 are not isomorphic.

(a) Show that b_0 and b_1 are in the same arc-connected component of $B = \widehat{\mathbb{R}}$ if and only if $b_1 - b_0 \in \iota(\mathbb{R})$.
(b) Let $\widehat{\mathbb{Z}}$ be the subgroup $\varprojlim \mathbb{Z}/p^n.\mathbb{Z}$ of $\widehat{\mathbb{R}}$, and by $\psi : (s, t) \mapsto s + \iota(t)$, consider $E = \widehat{\mathbb{Z}} \times \mathbb{R}$ as a space over $\widehat{\mathbb{R}}$.
   Show that E is a covering of $\widehat{\mathbb{R}}$ isomorphic to $\pi_0^*(\mathbb{R})$, where $\mathbb{R}$ is considered the universal covering of $\mathbb{R}/\mathbb{Z}$.
   For all n, consider $E_n = p^n\mathbb{Z} \times \mathbb{R} \subset E$. Show that E_n, equipped with the induced projection is a covering of $\widehat{\mathbb{R}}$, isomorphic to $\pi_n^*(\mathbb{R})$.
(c) Let $b \in \widehat{\mathbb{R}}$. Show that $\bigcap E_n(b) = \{0\}$ if $b = 0$, and is empty if $b \notin \iota(\mathbb{R})$. Conclude.

(C) Keep $B = \widehat{\mathbb{R}}$.

(a) Show that, for all $b \in B$, the fundamental group $\pi_1(B, b)$ acts trivially on $E(b)$, although the fiber bundle E is not trivial.

(b) Show that there is some n such that every covering of B is induced by a covering of $\mathbb{R}/p^n.\mathbb{Z}$ by change of basis by π_n. Show that for all b, $\pi_1(B, b) = \{0\}$ although B is connected and not simply connected.

(D) Let B′ be a connected space consisting of the union of two closed subsets of B and Γ, with $B = \widehat{\mathbb{R}}$, Γ homeomorphic to S^1 and $B \cap \Gamma = \{0\}$.

Let X be a covering of B' inducing a universal covering over Γ and a trivial covering over B. Denote by π_X the projection $X \to B'$. Consider X as the real line $\mathbb{R}$ with a copy B_k of B at each point $k \in \mathbb{Z}$; for each k, there exists $b_k \in B_k$ over b.

(a) Show that $\pi_1(B', 0)$ acts non trivially on $X(0)$. Deduce that $\pi_1(B', 0) \neq \{0\}$.

(b) The aim is to show that, for all $b \in B$, the group $\pi_1(B', b)$ acts trivially on $X(b)$ if $b \notin \iota(\mathbb{R})$.

Let $\mathscr{C}ov_{B',X}$ be the category of coverings Y of B' together with a morphism $\varphi : Y \to X$. If Y is an object of $\mathscr{C}ov_{B',X}$, then the fibre $Y(b)$ is the disjoint union of $Y(b_k) = \varphi^{-1}(b_k)$. Let $\alpha \in \pi_1(B', b)$. Then α acts on $X(b)$ and on $Y(b)$, and $\alpha.Y(b_k) = Y(\alpha.b_k)$.

Assuming that α acts non trivially on $X(b)$, let k be such that $\alpha(b_k) = b_{k'}$ with $k' \neq k$. Restricting to the objects of $\mathscr{C}ov_{B',X}$ inducing a trivial covering over B_k (but an arbitrary one over $B_{k'}$), and starting with α, define an isomorphism between the functors Φ_0 and $\Phi_b : \mathscr{C}ov_B \to \mathscr{E}ns$. Conclude.

4.7 Van Kampen's Theorem

The aim of this section is to find the fundamental group of a space B when $B = U_1 \cup U_2$ and the fundamental groups of U_1, U_2 and $U_1 \cap U_2$ are known.

4.7.1 Categorical Preliminaries

Let G and H be two groups and $f : H \to G$ a homomorphism. For all G-sets X, let $f^*(X)$ be the H-set obtained by making H act on X by $h \cdot x = f(h)x$. This gives a functor $f^* : G\text{-}\mathscr{E}ns \to H\text{-}\mathscr{E}ns$.

Proposition *If f^* is an equivalence of categories, then f is an isomorphism.*

Proof

(a) f^* *essentially surjective* $\Rightarrow$ f *injective*. For any set X, all actions of H on X stems from an action of G, and so the elements of $\mathrm{Ker}\, f$ act trivially. Taking $X = H$ and making H act by left translation, $\mathrm{Ker}\, f = \{e\}$ follows.

(b) f^* *fully faithful* $\Rightarrow$ f *surjective*. For all G-sets X and Y, every H-morphism from X to Y is a G-morphism. Take $X = \{x\}$ and $Y = G/f(H)$. The map $\varphi : X \to Y$ such that $\varphi(X)$ is the class of the identity e is a H-morphism, hence a G-morphism, and e is fixed by G. As G acts transitively on $G/f(H)$,

$$G/f(H) = \{e\}, \quad \text{and} \quad f(H) = G. \qquad \square$$

4.7.2

If $\mathscr{C}_1$, $\mathscr{C}_2$, $\mathscr{D}$ are categories, $R_1 : \mathscr{C}_1 \to \mathscr{D}$ and $R_2 : \mathscr{C}_2 \to \mathscr{D}$ functors, define a category $\mathscr{C} = \mathscr{C}_1 \times_{\mathscr{D}} \mathscr{C}_2$ as follows: an object of $\mathscr{C}$ is a triplet (X_1, X_2, f) where $X_1 \in \mathscr{C}_1$, $X_2 \in \mathscr{C}_2$, and $f : R_1(X_1) \to R_2(X_2)$ is an isomorphism of $\mathscr{D}$. A morphism from $\mathscr{C}$ to (X_1, X_2, f) in (Y_1, Y_2, g) is a pair (h_1, h_2), with $h_1 : X_1 \to Y_1$ a morphism in $\mathscr{C}_1$, $h_2 : X_2 \to Y_2$ a morphism in $\mathscr{C}_2$, and with commuting diagram

$$
\begin{array}{ccc}
R_1(X_1) & \xrightarrow{\ f\ } & R_2(X_2) \\
{\scriptstyle h_{1*}}\downarrow & & \downarrow{\scriptstyle h_{2*}} \\
R_1(Y_1) & \xrightarrow{\ g\ } & R_2(Y_2)
\end{array}
$$

If

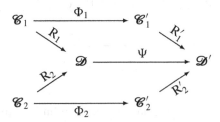

is a commutative diagram of functors, assign to it a functor $\Phi : \mathscr{C}_1 \times_{\mathscr{D}} \mathscr{C}_2 \to \mathscr{C}'_1 \times_{\mathscr{D}'} \mathscr{C}'_2$. If Φ_1, Φ_2 are ψ equivalences of categories, then so is Φ.

4.7.3

Let B be a locally connected space, U_1, U_2 subspaces of B, both open or both closed. Assume that $B = U_1 \cup U_2$ and set $V = U_1 \cap U_2$ (if both U_i are open, then they are locally connected and so is V, if they are closed, then U_1, U_2, and V have to be assumed to be locally connected).

Proposition *The functor* $\alpha : \mathcal{Cov}_B \to \mathcal{Cov}_{U_1} \times_{\mathcal{Cov}_V} \mathcal{Cov}_{U_2}$ *defined by* $\alpha(X) = (X|_{U_1}, X|_{U_2}, 1_X|_V)$ *is an equivalence of categories.*

Proof The functor α *is fully faithful.* Indeed, if X and Y are coverings of B and if $f_1 : X|_{U_1} \to Y|_{U_1}$, $f_2 : X|_{U_2} \to Y|_{U_2}$ are morphisms agreeing over V, then gluing together f_1 and f_2 gives a morphism f from X to Y. (When U_1 and U_2 are closed, the continuity of f follows from (4.1.11, Lemma 4.3).)

The functor α *is essentially surjective.* Let $(X_1, X_2, f) \in \mathcal{Cov}_{U_1} \times_{\mathcal{Cov}_V} \mathcal{Cov}_{U_2}$. Denote by X the space over B consisting of the quotient of the disjoint union $X_1 \sqcup X_2$

by the equivalence relation identifying x with $f(x)$ for $x \in X_1|_V$. The canonical injection $\iota_1 : X_1 \to X$ is a homeomorphism from X_1 onto $X|_{U_1}$. Indeed the canonical injection $X_1 \to X_1 \sqcup X_2$ is clopen, and the canonical map $X_1 \sqcup X_2 \to X$ is open if both U_i are open, and closed if they are both closed. Similarly for ι_2.

We show that X is a covering of B. This is obvious if both U_i are open. Assume they are closed and let $b \in B$. If $b \in B - V$, b is an interior point of one of the U_i, and X is a covering over a neighbourhood of b. Assume $b \in V$. Since V is locally connected, there is a neighbourhood S of b in B such that $T = S \cap V$ is connected, and such that $X_1|_{S_1}$ and $X_2|_{S_2}$ are trivial, with $S_i = S \cap U_i$. Let $\tau_i : X_i|_{S_i} \to S_i \times F_i$ be trivializations. We may assume that $F_1 = F_2 = F$ and that the diagram

$$
\begin{array}{ccc}
X_1(b) & \xrightarrow{\;f_b\;} & X_2(b) \\
& \searrow^{\tau_{1,b}} \quad \nearrow_{\tau_{2,b}} & \\
& F &
\end{array}
$$

commutes within to identification of F_1 with F_2 by $\tau_{2,b} \circ f_b \circ \tau_{1,b}^{-1}$. Then the diagram

$$
\begin{array}{ccc}
X_1|_T & \xrightarrow{\;f\;} & X_2|_T \\
& \searrow^{\tau_1} \quad \nearrow_{\tau_2} & \\
& T \times F &
\end{array}
$$

commutes. Indeed $\tau_2 \circ f$ and τ_1 are morphisms from $X_1|_T$ to $T \times F$ agreeing over b, hence at least at one point in each of the connected components (4.3.5). So gluing together τ_1 and τ_2 gives a trivialization $\tau : X_S \to S \times F$; the continuity of τ and of its inverse follows from (4.1.11, Lemma 4.3).

Then $X \in \mathscr{C}\!ov_B$ and $\alpha(X) \approx (X_1, X_2, f)$. $\square$

4.7.4   Free Groups on Sets

Let X be a set. The free product $L(X) = \star_{x \in X} L_x$ (2.5.2, Example 3), where, for all $x \in X$, $L_x = \mathbb{Z}$, is called the **free group** on X. For all $x \in X$, set $e_x = \iota_x(1) \in L_x$. The group $L(X)$, equipped with the family $(e_x)_{x \in X}$, represents the functor $G \to G^X$, in other words:

UNIVERSAL PROPERTY. *For every group* G *and family* $(g_x)_{x \in X}$ *of elements in* G, *there is a unique homomorphism* $\varphi : L(X) \to G$ *such that, for all* $x \in X$, $\varphi(e_x) = g_x$.
   Indeed

$$
\mathrm{Hom}_{\mathscr{G}\!r}(L(X) ; G) = \prod_{x \in X} \mathrm{Hom}_{\mathscr{G}\!r}(L_x ; G) = G^X .
$$

The homomorphism φ is surjective if and only if $(g_x)_{x \in X}$ generates G. If φ is bijective (resp. injective), $(g_x)_{x \in X}$ is said to be a basis (resp. a free family), and G a **free group** if it has a basis.

4.7.5

We keep the notation of 4.7.3. Let $b_0 \in V$, $(V_\lambda)_{\lambda \in \Lambda}$ the family of connected components of V, with V_0 the connected component of b_0, and $\Lambda^* = \Lambda - \{0\}$. Assume that there is a universal covering of B.

Theorem *If U_1 and U_2 are simply connected, then $\pi_1(B, b_0)$ is isomorphic to the free group over Λ^*.*

More precisely, let X be a covering of B. Set $F = X(b_0)$ and let $\tau_1 : X|_{U_1} \to U_1 \times F$ and $\tau_2 : X|_{U_2} \to U_2 \times F$ be the trivializations inducing the identity over b_0. Over V, $\tau_2 \circ \tau_1^{-1}$ is an automorphism of $V \times F$ defining a permutation g_b^X of F for all $b \in V$. This permutation only depends on the connected component of b in V, and we get a family $(g_\lambda^X)_{\lambda \in \Lambda}$ of permutations of F. For all $\lambda \in \Lambda$, as X varies, g_λ^X gives an automorphism g_λ of the functor $X \mapsto X(b_0)$ and hence an element of $\pi_1(B, b_0)$. Obviously $g_0 = e$. We prove that:

Proposition *Under the assumptions of the theorem, $(g_\lambda)_{\lambda \in \Lambda^*}$ is a basis for $\pi_1(B, b_0)$ as a (not necessarily commutative) group.*

Proof Consider the diagram of functors

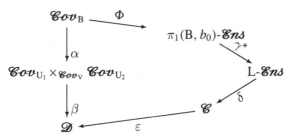

where L is the free group over Λ^*, $\gamma : L \to \pi_1(B, b_0)$ the homomorphism defined by $\gamma(e_\lambda) = g_\lambda$, $\mathscr{C}$ the category of pairs $(F, (f_\lambda)_{\lambda \in \Lambda^*})$, where F is a set and f_λ a permutation of F for all λ, $\mathscr{D}$ the category of triplets $(F_1, F_2, (f_\lambda)_{\lambda \in V})$, where F_1 and F_2 are sets and $f_\lambda : F_1 \to F_2$ a bijection for λ; the functor δ assigns to a L-set F the set F equipped with the permutations defined by e_λ, the functor ε assigns to $(F, (f_\lambda))$ the object $(F, F, (f_\lambda))$, with $f_0 = 1_F$. The functor α is the one defined in 4.7.3, and the functor β is defined as follows:

$$\beta(X_1, X_2, f) = (F_1, F_2, (f_\lambda)), \text{ where } F_1 = X_1(b_0), F_2 = X_2(b_0),$$

f_λ is the map $x \mapsto \tau_2 \circ \tau_1^{-1}(b, x)$ for $b \in V_\lambda$, where $\tau_1 : X_1 \to U_1 \times F_1$ and $\tau_2 : X_2 \to U_2 \times F_2$ are the trivializations inducing the identity over b_0.

This diagram is easily seen to be commutative. By Proposition (4.7.3), the functor α is an equivalence of categories. That δ and ε are equivalences follows readily, and it is easily seen that so is β. As a result, $\gamma^* \circ \Phi$ is an equivalence of categories.

The functor Φ is an equivalence by (4.6.8), and hence so is γ^*, and by Proposition (4.7.1) γ is an isomorphism.   □

4.7.6   Amalgamated Sums

Let G_1, G_2, H be groups, $u_1 : H \to G_1$ and $u_2 : H \to G_2$ homomorphisms. The sum of G_1 and G_2 *amalgamated by* H, written $G_1 \overset{H}{\star} G_2$, is the quotient $Q = (G_1 \star G_2)/N$, where $G_1 \star G_2$ is the summation group of G_1 and G_2 and N the normal subgroup generated by $u_1(h)u_2(h^{-1})$, $h \in H$. Equip Q with the homomorphisms $j_1 : G_1 \to Q$ and $j_2 : G_2 \to Q$ composed of the canonical injections $G_1 \to G_1 \star G_2$ and $G_2 \to G_1 \star G_2$ and the canonical quotient map.

UNIVERSAL PROPERTY.
(a) *The diagram*

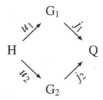

is commutative.
(b) *Let* Γ *be a group,* f_1 *and* f_2 *homomorphisms such that the diagram*

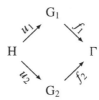

commutes. Then there is a unique homomorphism $g : Q \to \Gamma$ *such that* $g \circ j_1 = f_1$ *and* $g \circ j_2 = f_2$.

Proof
(a) For all $h \in H$, $u_1(h)u_2(h^{-1}) \in N$, and so $j_1 \circ u_1 = j_2 \circ u_2$.

(b) Define $g : G_1 \star G_2 \to \Gamma$ by $g(x) = f_1(x)$ if $x \in G_1$ and $g(x) = f_2(x)$ if $x \in G_2$. For $h \in H$, $g(u_1(h)u_2(h^{-1})) = f_1 \circ u_1(h) f_2 \circ u_2(h^{-1}) = e$, where e is the identity element of Γ. Hence there is a homomorphism f from Q to Γ. As G_1 and G_2

generate $G_1 \star G_2$, their images $j_1(G_1)$ and $j_2(G_2)$ generate Q and the homomorphism $f : Q \to \Gamma$ thus constructed is the only one with the desired property. $\qquad\square$

The universal property is equivalent to the statement that $G_1 \overset{H}{\star} G_2$, equipped with (j_1, j_2) represents the functor

$$\Gamma \mapsto \mathrm{Hom}(G_1, \Gamma) \times_{\mathrm{Hom}(H,\Gamma)} \mathrm{Hom}(G_2, \Gamma).$$

4.7.7

We keep the notation of 4.7.3. Let $b_0 \in B$. Assume that there are universal pointed coverings of (B, b_0), (U_1, b_0), (U_2, b_0) and of (V, b_0).

Theorem (Van Kampen) *If V is connected, then $\pi_1(B, b_0)$ is the sum of $\pi_1(U_1, b_0)$ and $\pi_1(U_2, b_0)$ amalgamated by $\pi_1(V, b_0)$.*

More precisely, set $G = \pi_1(B, b_0)$, $G_i = \pi_1(U_i, b_0)$ and $H = \pi_1(V, b_0)$. The canonical inclusions define a commutative diagram.

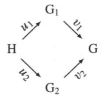

The theorem states that the homomorphism $v : G_1 \overset{H}{\star} G_2 \to G$ defined by v_1 and v_2 is an isomorphism.

Proof Let U_i' be the connected component of b_0 in U_i and B' the connected component of b_0 in B. Then $U_1' \cup U_2' = B'$ and $U_1' \cap U_2' = V$ and the fundamental groups remain the same when U_i is replaced by U_i' and B by B'. Hence B and both U_i may be assumed to be connected.

There is a commutative diagram of functors

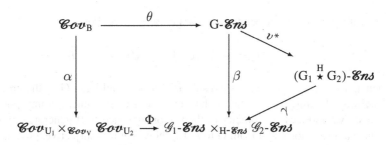

where $\theta(X) = X(b_0)$, $\Phi(X_1, X_2, f) = (X_1(b_0), X_2(b_0), f_{b_0})$, $\alpha(X) = (X|_{U_1},$ $X|_{U_2}, I)$, $\beta(F) = (F, F, I)$, $\gamma(F) = (F, F, I)$. By Theorem (4.6.8), in this diagram, θ is an equivalence of categories, and so is Φ. Clearly, this is also the case of γ. By Proposition 4.7.3, α is an equivalence of categories. Hence so is v^*, and by Proposition 4.7.1 v is an isomorphism. $\square$

Exercises 4.7. (Van Kampen's theorem)

1.—Let G_1, G_2, H be groups, $u_1 : H \to G_1$ and $u_2 : H \to G_2$ homomorphisms. Show that the amalgamated sum $G = G_1 \overset{H}{\star} G_2$ is generated by the union of the images of G_1 and G_2.

2.—Let S^n be the unit sphere in $\mathbb{R}^{n+1}$, S^n_+ the upper hemisphere, i.e.

$$S^n_+ = \{x = (x_1, \ldots, x_{n+1}) \in S^n \mid x_{n+1} \geqslant 0\},$$

and S^n_- the lower one. Let X be a fibre bundle over S^n. Assume that the fibres X_+ and X_- induced on S^n_+ and S^n_- are trivial (this condition can be shown to be necessarily satisfied). Set $s_0 = (1, 0, \ldots, 0)$, $F = X(s_0)$ and let $x_0 \in F$.

The aim is to find $\pi_1(X, x_0)$ assuming known $\pi_1(F, x_0)$.

(a) Find $\pi_1(X_+, x_0)$ and $\pi_1(X_-, x_0)$ as well as $\pi_1(X_0, x_0)$, where

$$X_0 = X_+ \cap X_-$$

(consider the cases $n = 2$ and $n \geqslant 3$ separately).

(b) Find $\pi_1(X, x_0)$ when $n \geqslant 3$.

c) Show that, if $n = 2$, then the group $\pi_1(X, x_0)$ can be identified with the quotient of $\pi_1(F, x_0)$ by a cyclic subgroup contained in the centre.

3.—Let SO_n be the group of direct linear isometries of $\mathbb{R}^n$. The aim is to find $\pi_1(SO_n)$.

(a) Show that SO_2 is homeomorphic to S^1.

(b) Let **H** denote the field of *quaternions*: it is a 4-dimensional vector space over $\mathbb{R}$ with basis $(1, i, j, k)$, and bilinear multiplication defined by

$$i^2 = j^2 = k^2 = -1, \; ij = -ji = k, \; jk = -kj = i, \; ki = -ik = j.$$

This multiplication is associative, but not commutative. For

$$x = r + ai + bj + ck, \text{ set } \|x\| = \sqrt{r^2 + a^2 + b^2 + c^2} \;;$$

Then $\|xy\| = \|x\| \cdot \|y\|$, and as a result, **H** is a skew field. Let G be the group of the quaternions having norm 1. Show that for all $x \in G$, the inner automorphism of **H** defined by x induces an isometry of the vector subspace of **H** generated by (i, j, k). Deduce a homomorphism from G to SO_3. Show that it is surjective. Give its kernel.

(c) Show that $\pi_1(SO_3) = \mathbb{Z}/(2)$.

(d) Show that, for all $n \geqslant 3$, $\pi_1(SO_n) = \mathbb{Z}/(2)$ (use Exercise 2). Deduce that for all $n \geqslant 2$, there is a topological group Spin_n, unique up to isomorphism, which is a connected 2-fold covering of SO_n.

4.—Show that in Theorem 4.7.7, the assumption that there is a universal covering of (B, b_0) is a consequence of the other assumptions.

5.—With the notation of 4.7.6, assume that u_1 and u_2 are injective. Show that j_1 and j_2 are injective.

6.—Let G and H be two groups.

(a) Show that, if the categories G-$\mathscr{E}ns$ and H-$\mathscr{E}ns$ are equivalent, then the groups G and H are isomorphic.

(b) Are all equivalences from H-$\mathscr{E}ns$ to G-$\mathscr{E}ns$ of the form f^*, where $f : G \to H$ is a homomorphism?

7.—Let E be the set of pairs (x, y) of orthonormal vectors of $\mathbb{R}^3$, i.e. with $\|x\| = \|y\| = 1$, $(x \cdot y) = 0$. Equip E with the projection $\pi : (x, y) \mapsto x$ from E onto S^2.

(a) Show that π is a fibration with fibre S^1.

(b) Show that E is homeomorphic to SO_3.

(c) Assuming there is a continuous section $\sigma : S^2 \to E$ of π, show that E is homeomorphic to $S^2 \times S^1$. With this assumption, determine $\pi_1(E)$. Compare with Exercise 3, (c). Conclude.

(d) A continuous tangent vector field on S^2 is a continuous map $\xi : S^2 \to \mathbb{R}^3$ such that, for all $x \in S^2$, $\xi(x)$ is tangent to S^2 at x, i.e. orthogonal to x. Show that there is a point of S^2 at which the continuous tangent vector field vanishes.

(e) A continuous normal tangent vector field on S^2 is a continuous map $\nu : S^2 \to \mathbb{R}^3$ such that, for all $x \in S^2$, $\nu(x)$ is normal to S^2 at x, i.e. proportional to x. Consider the space $\mathscr{T}$ of continuous tangent vector fields and the space $\mathscr{N}$ of continuous normal tangent vector fields as modules over the ring A of continuous functions $S^2 \to \mathbb{R}$. Show that $\mathscr{N}$ is free of rank 1 and that $\mathscr{N} \oplus \mathscr{T}$ is free of rank 3. Deduce that $\mathscr{T}$ is projective. Show that $\mathscr{T}$ is not free.

4.8 Graphs. Subgroups of Free Groups

4.8.1 Graphs

A *simple graph* is a pair (S, A), where S is a set and A a subset of $S \times S$ defining a reflexive and symmetric relation. In this chapter, we will simply use the term *graph*. In chapter 7 we will give a broader definition. The points of S are called *vertices* of the graph. Two vertices s and s' are said to be *connected* if $(s, s') \in A$. If (S, A) is a graph, then its *geometric realization* is defined as follows: consider the vector space $\mathbb{R}^{(S)}$ equipped with the norm $x \mapsto \sum |x_s|$, and embed S in $\mathbb{R}^{(S)}$ by identifying s with the element of the canonical basis having index s. For $s, t \in S$, let $[s, t]$ be the segment with endpoints s and t, i.e. the set of $(1 - \theta)s + \theta t$ for $\theta \in [0, 1]$. The geometric realization of (S, A) is the space

$$|S, A| = \bigcup_{(s,t) \in A} [s, t],$$

equipped with the distance induced by the norm.

The segments $[s, t]$ with $(s, t) \in A$ are called the *edges*. If s, t, s', t' are distinct points,

$$[s, t] = \{s\} \text{ if } t = s, \ [s, t] = [t, s], \ [s, t] \cap [s, t'] = \{s\} \text{ if } t \neq t',$$

and

$$[s, t] \cap [s', t'] = \varnothing.$$

More precisely, if

$$[s, t] \cap [s', t'] = \varnothing,$$

then

$$(\forall x \in [s, t]) \ (\forall x' \in [s', t']) \ \|x - x'\| = 2.$$

For $s \in S$, the union of the edges of the graph having s as an endpoint is called the *star* of s. It is a contractible neighbourhood of s in $|S, A|$, hence simply connected (4.3.17, Corollary 4.6).

Let (S, A) and (S', A') be two graphs. A morphism from (S, A) to (S', A') is a map f from S to S' such that $f \times f : S \times S \to S' \times S'$ maps A to A'. Then there is continuous map f_* from $|S, A|$ to $|S', A'|$ corresponding to f. For $S \subset S'$ and $A \subset A'$, (S, A) is said to be a *subgraph* of (S', A'); then $|S, A|$ is a subspace of $|S', A'|$ with the induced metric. Let (S, A) and (T, B) be two graphs, (S', A') a subgraph of (S, A) and $f : (T, B) \to (S, A)$ a morphism. Set $(T', B') = f^{-1}(S', A')$; then

$$|T', B'| = f_*^{-1}(|S', A'|).$$

If G is a graph, then its geometric realization $|G|$ is a locally connected space, and every connected component of $|G|$ is the geometric realization of a subgraph of G.

4.8.2

Proposition and Definition *Let* (S, A) *and* (T, B) *be two graphs and* $\rho : (T, B) \to$ (S, A) *a morphism.* $|T, B|$ *together with* ρ_* *is a covering of* $|S, A|$ *if and only if*

$$(\forall (x, x') \in A) \, (\forall y \in T, \rho(y) = x) \, (\exists ! \, y' \in T) \, \rho(y') = x' \; et \; (y, y') \in B.$$

If these conditions hold, then (T, B) *together with* ρ *is said to be a* **covering** *of* (S, A).

Proof Suppose it is a covering. The connected component of y in $\rho_*^{-1}([x, x'])$ is of the form $|T', B'|$, where (T', B') is a subgraph of (T, B), and ρ_* induces a homeomorphism from $|T', B'|$ onto $[x, x']$. In particular, ρ injects T' into $\{x, x'\}$, and so T' has at most two points. If $x \neq x'$, T' has two points since $|T', B'|$ is homeomorphic to $[0, 1]$; these are y and a point y' with the desired property. If $x = x'$, then $y' = y$.

Conversely, let T_x be the inverse image of x in T. Two distinct vertices y_1 and y_2 of T_x are never connected, for otherwise (y_1, y_1) and (y_2, y_2) would be two coverings distinct from (x, x).

The union of all $[x, x'[$ with x' is connected x is called the *open star* of x and written U_x. Then $\rho^{-1}(U_x) = \bigcup_{y \in T_x} U_y$. Indeed, let $z \in |T, B|$ be such that $\rho(z) \in [x, x'[$. If z is a vertex, then $\rho(z) = x$; if $z \in \,]y, y'[$, one of the points y, y', say y, projects onto x, and $z \in U_y$ and $y \in T_x$.

For all $y \in T_x$, the assumption implies that ρ induces an isomorphism from the star of y onto the star of x. The same holds for open stars.

Let y_1 and y_2 be two distinct points of T_x. Then the stars of y_1 and y_2 are disjoint. Indeed, y_1 and y_2 are not connected, and if there was a vertex z connected to y_1 and to y_2, then (z, y_1) and (z, y_2) would be two coverings of $(\rho(z), x)$. Moreover, the distance of these stars is 2.

Hence $\rho^{-1}(U_x)$ is a trivial covering of U_x.   □

4.8.3

Proposition *Let* (S, A) *be a graph. The functor* $((T, B), \rho) \mapsto (|T, B|, \rho_*)$ *defines an equivalence from the category of coverings of* (S, A) *onto the category of coverings of* $|S, A|$.

Proof

(a) *The functor is essentially surjective.* Set $X = |S, A|$ and let Y be a covering of X. Denote by p the projection of Y onto X, and set $S' = p^{-1}(S)$. Define $A' \subset S' \times S'$ by $(s', t') \in A' \Leftrightarrow (s, t) \in A$ and s' and t' are in the same connected component of $p^{-1}[s, t[$, where $s = p(s')$ and $t = p(t')$. Set $X' = |S', A'|$. The map p induces a map from S' to S. It is a morphism from (S', A') to (S, A), hence there is a map $p_* : X' \to X$.

Define $\varphi : X' \to Y$ as follows: for $(s', t') \in A'$, set $s = p(s')$ and $t = p(t')$, then $Y|_{[s,t]}$ is a trivial covering since $[s, t]$ is simply connected, and so there is a unique continuous section σ of Y over $[s, t]$ such that $\sigma(s) = s'$. Then $\sigma(t) = t'$ by definition of A'. Set $\varphi_{s',t'} = \sigma \circ p_*|_{[s',t']}$. Gluing together all $\varphi_{s',t'}$ gives a map φ commuting with the projections onto X.

Show that φ is a homeomorphism. As the interiors of the stars covering X, it suffices to show that for every star V in X, $\varphi_V : X'|_V \to Y|_V$ is a homeomorphism. As V is contractible, hence simply connected, $Y|_V$ is a trivial covering. Thus the question reduces to the case of a trivial covering.

It may then be assumed that $Y = X \times F$ for some discrete set F. Equip Y with the distance defined by

$$d((x, u), (y, u)) = d(x, y) \text{ and } d((x, u), (y, v)) = 2 \text{ if } u \neq v.$$

As can be checked, φ is an isometry, and Y is isomorphic to $|S', A'|$.

(b) *The functor is fully faithful.* Let (T, B) and (T', B') be coverings of (S, A) and $f : |T, B| \to |T', B'|$ a covering morphism of |S, A|. We show that there a unique morphism $g : (T, B) \to (T', B')$ inducing f. For $s \in S$, two distinct points of $p^{-1}(s)$ are never connected since $p_*^{-1}(s)$ is discrete. Hence $p_*^{-1}(S) = T$. Similarly $p_*'^{-1}(S) = T'$, and $f(T) \subset T'$. Let $g : T \to T'$ be the map induced by f.

Let t_1 and t_2 be two vertices of T; set $s_1 = p(t_1)$ and $s_2 = p(t_2)$. Then, t_1 and t_2 are connected if and only if s_1 and s_2 are connected and t_1 and t_2 are in the same connected component in $p_*^{-1}([s_1, s_2])$. Hence, if t_1 and t_2 are connected, so are $f(t_1)$ and $f(t_2)$; hence g is a morphism from (T, B) to (T', B'). The covering (topological) morphisms g_* and f agree on the vertices, and hence at least at one point in each connected component of |T, B|; hence they are equal. Uniqueness is obvious. □

4.8.4

Definition A **tree** is a graph whose geometric realization is simply connected.

Proposition *Let (S, A) be a graph and $(S_i, A_i)_{i \in I}$ a nonempty increasing directed family of subgraphs. Assume that (S, A) is the union of (S_i, A_i), i.e. $S = \bigcup S_i$ and $A = \bigcup A_i$, and that (S_i, A_i) is a tree for all i. Then (S, A) is a tree.*

Proof Let $s \in S$. If necessary replacing I by a cofinal set, we may assume that $s \in S_i$ for all i. Then |S, A| is the union of the connected $|S_i, A_i|$ containing s, and so is connected.

Let Y be a covering of |S, A| and $t \in Y(s)$. We may assume $Y = |T, B|$ for some covering (T, B) of (S, A). Then $t \in T$. For all i, there is a unique continuous section of |T, B| over $|S_i, A_i|$ passing through t. By (4.8.3), this section stems from a unique morphism $\varphi_i : (S_i, A_i) \to (T, B)$. By uniqueness, gluing together these morphisms gives a morphism $\varphi : (S, A) \to (T, B)$ and $\varphi_* : |S, A| \to |T, B|$ is a continuous section passing through t. By (4.3.11), Y is a trivial covering. $\square$

4.8.5

A graph is connected if so is its geometric realization.

Proposition *Every connected graph* (S, A) *has a spanning tree, i.e. a subgraph* (T, B) *with* $T = S$.

Proof By the previous proposition, the set of trees in (S, A) is inductive. Hence, by Zorn's theorem, there is a maximal tree (T, B). We show that $T = S$. If $S - T \neq \varnothing$, there exists $(t_0, t_0') \in A$ with $t_0 \in T$ and $t_0' \in S - T$; otherwise (S, A) would not be connected. Set

$$T' = T \cup \{t_0'\} \quad \text{and} \quad B' = B \cup \{(t_0, t_0'), (t_0', t_0), (t_0', t_0')\}.$$

We show that (T', B') is a tree, which will contradict maximality.
For $\theta \in [0, 1]$ and $t \in T'$, set

$$h(\theta, t) = t \text{ if } t \neq t_0' \text{ and } h(\theta, t_0') = (1 - \theta)t_0 + \theta t_0'.$$

For each value of θ, extend h to a continuous map from $[0, 1] \times \mathbb{R}^{(T')}$ to $\mathbb{R}^{(T')}$ by linearity. This map can be checked to be continuous and to induce a homotopy (4.3.16) between a retraction from $|T', B'|$ onto $|T, B|$ and the identity of $|T', B'|$. Therefore $|T, B|$ is a deformation retract of $|T', B'|$, of the same homotopy type (4.3.16) as $|T, B|$, and so is simply connected. $\square$

4.8.6

Theorem *Let* (S, A) *be a graph and* $s_0 \in S$. *Then* $\pi_1(|S, A|, s_0)$ *is a free group.*

Replacing (S, A) by the connected component of s_0 if necessary, which leaves π_1 unchanged, (S, A) can be assumed to be connected. Then, by the previous proposition, there is a tree (S, B). As B contains all (s, s), $A - B = C \sqcup C'$, where C and C' are mutually symmetric sets, i.e. $C' = \{(x', x) \mid (x, x') \in C\}$. The theorem then follows from the more precise following proposition:

Proposition *The group* $\pi_1(|S, A|, s_0)$ *has a basis indexed by* C.

Proof Set

$$U_1 = |S, A| - \bigcup_{(x,x')\in C} \left]\frac{x+x'}{2}, x'\right[= |S, B| \cup \bigcup_{(x,x')\in C} \left[x, \frac{x+x'}{2}\right]$$

and define U_2 likewise by replacing C by C'. The sets U_1 and U_2 are closed in $|S, A|$ since $\left]\frac{x+x'}{2}, x'\right[$ is open for all (x, x'), and their union is $|S, A|$.

The space U_1 is of the same homotopy as $|S, B|$. Indeed, define a retraction $\rho : U_1 \to |S, B|$ by $\rho(x) = x$ for $x \in |S, B|$ and $\rho(x) = s$ for

$$x \in \left[s, \frac{s+s'}{2}\right], \quad (s, s') \in C.$$

This retraction is continuous at each vertex since $\|\rho(x) - s\| \leqslant \|x - s\|$ for all vertices s and all $x \in U_1$ such that $\|x - s\| < 1$; it is continuous at each point that is not a vertex since in the neighbourhood of such a point it agrees with a constant map, the identity say. The map $(\theta, x) \mapsto (1 - \theta)\rho(x) + \theta x$ from $[0, 1] \times U_1$ to U_1 is a homotopy between ρ and 1_{U_1}, and so U_1 deformation retracts onto $|S, B|$. In particular, U_1 is simply connected. So is U_2.

The space $U_1 \cap U_2$ consists of the connected subset $|S, B|$ and of the isolated points $\frac{x+x'}{2}$ for $(x, x') \in C$. By (4.7.5, Theorem) the group $\pi_1(|S, A|, s_0)$ has a basis indexed by C.   □

4.8.7

Proposition *Every free group is isomorphic to a group of the form* $\pi_1(|S, A|, s_0)$, *where* (S, A) *is a connected graph and* $s_0 \in S$.

Proof Let Λ be a set and L the free group on Λ (4.7.4). Define a graph (S, A) by setting $S = \{0\} \cup \{1, 2\} \times \Lambda$, and connecting 0 to all points and $(1, \lambda)$ to $(2, \lambda)$ for all $\lambda \in \Lambda$. Connecting only 0 to all points gives a tree (S, B). In the construction of 4.8.6, we may take $C = \{((1, \lambda), (2, \lambda))\}_{\lambda \in \Lambda}$, which can be identified with Λ, and so $\pi_1(|S, A|, 0)$ is isomorphic to L.   □

4.8.8

Lemma *Let* (S, A) *be a connected graph. Then the space* $|S, A|$ *has a simply connected covering.*

Proof The space $|S, A|$ is locally contractible. Indeed, if s is a vertex, then every ball centered at s with radius <2 is contractible, and a point which is not a vertex has a fundamental system of neighbourhoods consisting of segments. Hence $|S, A|$ must necessarily be locally connected, and the lemma follows by (4.4.5).   □

4.8.9

> **Theorem** *All subgroups of a free group are free.*

Proof Let L be a free group and H a subgroup of L. By (4.8.7), there is a connected graph (S, A) and $s_0 \in S$ such that $\pi_1(|S, A|, s_0)$ is isomorphic to L. Identify this fundamental group with L by an isomorphism. By (4.6.11, (a)), there is connected pointed covering (X, x_0) of $(|S, A|, s_0)$ such that $\pi_1(X, x_0)$ is isomorphic to H. By (4.8.3), the space X is homeomorphic to the geometric realization of a graph, and so $H = \pi_1(X, x_0)$ is free by (4.8.6). $\qquad\qquad\qquad\qquad\square$

Remark We have also proved (3.5.1) that a subgroup of a commutative group is free, i.e. of a free $\mathbb{Z}$-module is a free $\mathbb{Z}$-module. These two results only appear to be similar. For example, a submodule of a free $\mathbb{Z}$-module of rank n is free of rank $\leqslant n$, whereas a free group on a set of two elements contains subgroups that are not finitely generated (4.8, Exercises 3 and 4).

Exercises 4.8. (Graphs)
1.—Let (S, A) be a graph. Consider the topology $\mathscr{T}_0$ on $|S, A|$ defined as follows: a subset U in $|S, A|$ is open with respect to $\mathscr{T}_0$ if and only if $U \cap [s, t]$ is open in $[s, t]$ for all $(s, t) \in A$.

(a) Show that the topology $\mathscr{T}_0$ is finer than the topology $\mathscr{T}$ defined in (4.8.1), and that these topologies agree if and only if (S, A) is *locally finite*, i.e. if each vertex is only connected to finitely many vertices.

(b) Let $s_0 \in S$. Show that the identity map

$$i : (|S, A|, \mathscr{T}_0) \to (|S, A|, \mathscr{T})$$

induces an isomorphism on the fundamental group with basepoint s_0. Is this map a homotopy equivalence?

2.—Let G be a free group. Show that any two bases of G are equipotent (consider the Abelianization of G, i.e. the quotient de G by the normal subgroup generated by the commutators).

3.—Consider the lemniscate $B = \{z \mid |1 - z^2| = 1\}$ in $\mathbb{C}$.

(a) Draw B. Show that it is homeomorphic to the geometric realization of a graph and give the latter's Cartesian diagram.

(b) Consider the covering X over B defined by the function $\log(1 - z)$, i.e. $X = \{(z, x) \in \mathbb{C}^2 \mid z \in B, e^x = 1 - z\}$ equipped with the projection $(z, x) \mapsto z$. Describe X.

(c) Describe the pointed covering Q of B whose fundamental group can be identified with the commutator group of $\pi_1(B, 0)$.

(d) Consider the polynomials $P_1 = \frac{3X-X^3}{2}$ and $P_2 = 1 - 4X^2 + 2X^4$. Show that
$P_1^{-1}(B)$ and $P_2^{-1}(B)$ are coverings of B. Describe these coverings.

(e) Which of the above coverings are Galois?

4.—Let $L(a, b)$ be a free group on a set $\{a, b\}$ with two elements.

(a) Let $f : L(a, b) \to \mathbb{Z}$ be the homomorphism defined by $f(a) = 1$, $f(b) = 0$.
Show that $(a^n \cdot b \cdot a^{-n})_{n \in \mathbb{Z}}$ is a basis for the kernel of f.

(b) Let $g : L(a, b) \to \mathbb{Z}^2$ be defined by $g(a) = (1, 0)$, $g(b) = (0, 1)$. Show that
the kernel of g is the commutator group of $L(a, b)$ with basis $(a^p \cdot b^q \cdot a \cdot b \cdot a^{-1} \cdot
b^{-1} \cdot b^{-q} \cdot a^{-p})_{p,q \in \mathbb{Z}}$, or $(a^p \cdot b^q \cdot a \cdot b^{-q} \cdot a^{-p-1})_{p,q \in \mathbb{Z}}$.

5.—Show that any countable graph is of homotopy equivalent to a graph all of whose
vertices are connected to at most three others.

6.—Show that all trees are contractible.

7.—Let $U \subset \mathbb{R}^2$ be an open set $\mathbb{R}^2 - K$ for some compact not necessarily locally
connected set K, and let $x_0 \in U$. The aims is to show that $\pi_1(U, x_0)$ is a free group
with countable basis.

(a) Let $P = [a, b] \times [c, d]$ be a rectangle in $\mathbb{R}^2$ and K a compact subset in the
interior of P. Show that there is a sequence (Q_n) of rectangles such that

(i) $P - K = \bigcup Q_n$;
(ii) $\overset{\circ}{Q}_i \cap \overset{\circ}{Q}_j = \varnothing$ for $j \neq i$;
(iii) each $a \in P - K$ has a neighbourhood meeting only finitely many Q_i.

Show that, if $K \neq \varnothing$, then the rectangles Q_i can be ordered so that

$$(\forall n) \quad \partial Q_n \not\subset \bigcup_0^{n-1} Q_i \,,$$

where ∂Q_n denotes the boundary of Q_n.

(b) Let (Q_n) be a sequence of rectangles satisfying the conditions of (a), and for
all n, let $k(n)$ be the smallest k such that $Q_k \cap Q_n \neq \varnothing$. Show that $k(n)$ tends to
infinity as n tends to infinity.

(c) Let $A \subset \mathbb{R}^2$ be a finite union of rectangles Q_i satisfying condition (ii) of (a),
and Q a rectangle such that $Q \cap A \underset{\neq}{\subset} \partial Q$. Set $A' = A \cup Q$. Let $\Gamma = (I, T)$ be a graph,
where I is the set of indices i, and

$$T \subset \{(i, j) \mid Q_i \cap Q_j \neq \varnothing\}.$$

Assume that there is a homeomorphism embedding the geometric realization of Γ in A under a such that the image of each vertex i is contained in the interior of the rectangle Q_i, and the image of each edge meets Q_i along a line segment. Assume that A deformation retracts onto $|\Gamma|$; let ρ be a retraction and h a homotopy from 1_A to $\iota \circ \rho$. Show that there is a graph Γ' containing Γ and such that the same situation holds for A'. Show that ρ' and h' can then be chosen in such a way that h' agrees with h except on Q_i such that $Q_i \cap Q \neq \varnothing$.

(d) Keeping the notation of (a), show that $P - K$ deformation retracts onto the geometric realization of a countable graph. Deduce that the same holds for $\mathbb{R}^2 - K$. Conclude.

(e) Show that, if U is connected and if K only has finitely many connected components, then $\pi_1(U, x_0)$ has a basis with the same number of elements as the number of connected components of K. Does this still hold if U is not connected? What if K has infinitely many connected components? (Take K to be the Cantor set in $\mathbb{R} \subset \mathbb{R}^2$.)

(f) If U is an arbitrary open subset of $\mathbb{R}^2$ and $x_0 \in U$, is the group $\pi_1(U, x_0)$ free?

8.—Let K be a compact subset of $\mathbb{R}^2$.

(a) Show that the category $\mathcal{C}ov_K$ can be identified with $\varinjlim \mathcal{C}ov_U$, the inductive limit being that of neighbourhoods U of K.

(b) Show that K is simply connected if and only if its complement is connected.

4.9 Loops

4.9.1 Paths

Let B be a topological space, and $b, b' \in B$. A *path* from b to b' in B is a continuous map c from an interval $[0, a] \subset \mathbb{R}_+$ to B such that $c(0) = b$ and $c(a) = b'$. The *length* of the path is defined to be a. Let c_0 and c_1 be two paths from b to b'. A *homotopy* from c_0 to c_1 is a family $(h_s)_{s \in [0,1]}$ of paths from b to b', such that $h_0 = c_0$, $h_1 = c_1$, such that the length a_s of h_s depends continuously on s and that the map $(s, t) \mapsto h_s(t)$ from $\{(s, t) \mid s \in [0, 1]$ and $t \in [0, a_s]\}$ to B is continuous. If there is a homotopy from c_0 to c_1, they are said to be *homotopic* and we write $c_0 \simeq c_1$. The homotopy relation between paths from b to b' is an equivalence relation. The set of homotopy classes of paths from b to b' is written $S(b, b')$.

4.9.2 Standard Paths

A path of length 1 is said to be *standard*.

If c_0 and c_1 are two homotopic standard paths, then there is a standard homotopy from c_0 to c_1, i.e. a homotopy (h_s) such that h_s is standard for all s, or equivalently a continuous map $h : [0, 1] \times [0, 1] \to B$ such that $h(s, 0) = b$, $h(s, 1) = b'$, $h(0, t) = c_0(t)$ and $h(1, t) = c_1(t)$ for $s, t \in [0, 1]$. Indeed, if (h_s) is a homotopy, $\bar{h}_s(t) = h_s(a_s t)$ defines a standard homotopy $(\bar{h}_s)$.

Moreover, every path is homotopic to a standard path. Indeed, if c is a path of length a, define a homotopy (h_s) from c to a standard path by $h_s(t) = c\left(\frac{at}{s+(1-s)a}\right)$; to check continuity, note that, for $a \neq 0$, the denominator does not vanish, and for $a = 0$, the function h is constant.

Hence, $S(b, b')$ can be defined as the set of standard homotopy classes of standard paths from b to b'.

4.9.3 Juxtaposition

Let $b, b', b'' \in B$, c a path from b to b' of length a and c' a path from b' to b'' of length a'. Define a path c'' from b to b'' of length $a + a'$ by

$$c''(t) = c(t) \text{ for } t \in [0, a] \text{ and } c''(t) = c'(t - a) \text{ for } t \in [a, a + a'].$$

We say that c'' is the juxtaposition of c and c', and is written $c' \cdot c$.

Juxtaposition is associative in the following sense: Let $b, b', b'', b''' \in B$, c a path from b to b', c' from b' to b'' and c'' from b'' to b'''; then

$$(c'' \cdot c') \cdot c = c'' \cdot (c' \cdot c).$$

For all $b \in B$, the path e_b of zero length defined by $e_b(0) = b$ is the identity in the following sense: $c \cdot e_b = c$ for all paths c with initial point b and $e_b \cdot c = c$ for all paths c with endpoint b.

Juxtaposition is compatible with the homotopy relation: if c_0 and c_1 are two homotopic paths from b to b', c_0' and c_1' two homotopic paths from b' to b'', then the paths $c_0' \cdot c_0$ and $c_1' \cdot c_1$ are homotopic. Thus there is a well-defined map $(\gamma, \gamma') \mapsto \gamma' \cdot \gamma$, also called *juxtaposition* from $S(b, b') \times S(b', b'')$ to $S(b, b'')$. This law is associative is the above sense, and for all b the class ε_b of e_b is the identity. Moreover, for all $\gamma \in S(b, b')$, there is an inverse $\gamma^{-1} \in S(b', b)$: let c be a representative of γ of length a, define c^* by $c^*(t) = c(a - t)$, c^* is a path from b' to b; setting $h_s = (c|_{[0,sa]})^* \cdot c|_{[0,sa]}$, this gives a homotopy $(h_s)_{s \in [0,1]}$ from e_b to $c^* \cdot c$ and a similar homotopy from $e_{b'}$ to $c \cdot c^*$; so $\gamma^{-1} \cdot \gamma = \varepsilon_b$ and $\gamma \cdot \gamma^{-1} = \varepsilon_{b'}$.

4.9.4 Lifting of Paths

Proposition *Let* X *be a covering of* B, $b, b' \in$ B, $x \in$ X(b) *and* c *a path from* b *to* b'. *Then there is a unique path* $\tilde{c}$ *in* X *with initial point* x *and such that* $\pi \circ \tilde{c} = c$.

Proof Let a be the length of c. The path c is a map from $[0, a]$ to B and c^*X is a covering of $[0, a]$, which is trivial since $[0, a]$ is simply connected. Hence there is a unique section of c^*X passing through $(0, x)$, and a unique lifting $\tilde{c}$ corresponding to it with the desired property.   □

4.9.5

Proposition and Definition *Let* X *be a covering of* B, $\gamma \in$ S(b, b') *and* $x \in$ X(b). *Choose a representative* c *of* γ *and let* $\tilde{c}$ *be the unique lifting of* c *with initial point* x. *Then the endpoint* x' *of* $\tilde{c}$ *in* X(b') *does not depend on the choice of* c. *It is denoted by* $\gamma \cdot_X x$, *or simply* $\gamma \cdot x$.

Proof Let c_0 and c_1 be two homotopic paths from b to b' with respective lengths a_0 and a_1, and let (h_s) be a homotopy from c_0 to c_1. This homotopy defines a continuous map $h : \Delta \to$ B, where $\Delta = \{(s, t) \mid s \in [0, 1]$ and $t \in [0, a_s]\}$. The space Δ deformation retracts onto $[0, 1]$, so is simply connected, and h^*X is a trivial lifting of Δ. Hence there is a continuous section of h^*X passing through $(0, 0, x)$, and thus a continuous map

$$\tilde{h} : \Delta \to \text{X such that } \tilde{h}(0, 0) = x \text{ and } \pi \circ \tilde{h} = h \,.$$

Since $t \mapsto \tilde{h}(0, t)$ is a lifting of c_0 with initial point x, $\tilde{h}(0, t) = \tilde{c}_0(t)$; and $\tilde{h}(s, 0) = x$ for $s \in [0, 1]$ since $s \mapsto \tilde{h}(s, 0)$ is a continuous map from the connected interval $[0, 1]$ to the discrete set X(b). In particular $\tilde{h}(1, 0) = x$. Since $t \mapsto \tilde{h}(1, t)$ is a lifting of c_1 with initial point x, $\tilde{h}(1, t) = \tilde{c}_1(t)$. The map $s \mapsto \tilde{h}(s, a_s)$ is continuous from the connected interval $[0, 1]$ to the discrete set X(b'), and so is constant and

$$\tilde{c}_0(a_0) = \tilde{h}(0, a_0) = \tilde{h}(1, a_1) = \tilde{c}_1(a_1) \,.$$

□

Remark Let $x \in$ X(b), $\gamma \in$ S(b, b') and $\gamma' \in$ S(b', b''). Then

$$(\gamma' \cdot \gamma) \cdot x = \gamma' \cdot (\gamma \cdot x) \,.$$

This follows readily from the definitions.

4.9.6 Poincaré Group

Let (B, b_0) be a pointed space. A *loop* in (B, b_0) is a path in B from b_0 to b_0. The set $S(b_0, b_0)$ of homotopy classes of loops is a group with respect to the juxtaposition law called the singular fundamental group or the *Poincaré group* of (B, b_0). Il will be denoted it by $\widetilde{\pi}_1(B, b_0)$.

If $b_0, b_1 \in B$ are in the same arc-connected component, then the groups $\widetilde{\pi}_1(B, b_0)$ and $\widetilde{\pi}_1(B, b_1)$ are isomorphic. Indeed, for all $\beta \in S(b_0, b_1)$, the map $\gamma \mapsto \beta \cdot \gamma \cdot \beta^{-1}$ is an isomorphism from $\widetilde{\pi}_1(B, b_0)$ onto $\widetilde{\pi}_1(B, b_1)$.

B is said to be *simply arc-connected* if it is arc-connected and there exists $b_0 \in B$ such that $\widetilde{\pi}_1(B, b_0) = \{e\}$; if this is the case then, for all $b \in B$, $\widetilde{\pi}_1(B, b) = \{e\}$. B is said to be *locally simply arc-connected* if all points of B have a fundamental system of simply arc-connected neighbourhoods.

4.9.7 Functoriality

Let $f : (B, b_0) \to (B', b_0')$ be a pointed space morphism. For all loops c in (B, b_0), $f \circ c$ is a loop in (B', b_0') and the map $c \mapsto f \circ c$ is compatible with the homotopy relation and with juxtaposition. Hence $f_* : \widetilde{\pi}_1(B, b_0) \to \widetilde{\pi}_1(B', b_0')$ is a group homomorphism.

Together with $f \mapsto f_*$, the map $\widetilde{\pi}_1$ is a functor from the category of pointed est spaces to the category of groups. This functor commutes with products.

4.9.8 Poincaré Group of a Product

Proposition *Let (B, b_0) and (B', b_0') be pointed topological spaces. The homomorphism*

$$\widetilde{\pi}_1(B \times B', (b_0, b_0')) \to \widetilde{\pi}_1(B, b_0) \times \widetilde{\pi}_1(B', b_0')$$

defined by the projections is an isomorphism.

Proof A standard loop in $(B \times B', (b_0, b_0'))$ is given by a pair (γ, γ'), where γ (resp. γ') is a standard loop in (B, b_0) (resp. (B', b_0')); similarly a standard homotopy in $(B \times B', (b_0, b_0'))$ is given by a pair of standard homotopies. The proposition follows.   □

4.9.9   The Poincaré Group and the Fundamental Group

For any covering X of B, the map $(\gamma, x) \mapsto \gamma \cdot x$ is an action of the group $\widetilde{\pi}_1(\mathrm{B}, b_0)$ on $\mathrm{X}(b_0)$. In particular, for all $\gamma \in \widetilde{\pi}_1(\mathrm{B}, b_0)$, the map $\gamma_\mathrm{X} : x \mapsto \gamma \cdot x$ is a permutation of $\mathrm{X}(b_0)$. If $f : \mathrm{X} \to \mathrm{X}'$ is a covering morphism of B, then the map $f_{b_0} : \mathrm{X}(b_0) \to \mathrm{X}'(b_0)$ is compatible with the actions of $\widetilde{\pi}_1(\mathrm{B}, b_0)$, i.e. $f(\gamma \cdot x) = \gamma \cdot f(x)$ for $x \in \mathrm{X}(b_0)$ and $\gamma \in \widetilde{\pi}_1(\mathrm{B}, b_0)$. Hence the maps γ_X form an automorphism of the functor $\mathrm{X} \mapsto \mathrm{X}(b_0)$, and so is an element $\gamma. \in \pi_1(\mathrm{B}, b_0)$. The map $\varepsilon : \gamma \mapsto \gamma.$ from $\widetilde{\pi}_1(\mathrm{B}, b_0)$ to $\pi_1(\mathrm{B}, b_0)$ is a group homomorphism.
    We aim to prove the following result:

Theorem *Let* B *be a locally simply arc-connected space and* $b_0 \in \mathrm{B}$. *Then* $\varepsilon :$ $\widetilde{\pi}_1(\mathrm{B}, b_0) \to \pi_1(\mathrm{B}, b_0)$ *is an isomorphism.*

    This theorem will be proved in 4.9.14. Surjectivity is easy (4.9.10). To prove injectivity, the existence of a covering X of (B, b_0) such that γ acts non trivially on $\mathrm{X}(b_0)$, for nonzero elements $\gamma \in \widetilde{\pi}_1(\mathrm{B}, b_0)$ needs to be shown. The universal covering satisfies this property, but this will be more clearly seen by considering a different construction of the universal covering from that given in section 4.1.

4.9.10   Surjectivity

Proposition *Let* (B, b_0) *be a locally simply arc-connected pointed space having a universal covering. Then* ε *is surjective.*

Proof Let $\gamma \in \pi_1(\mathrm{B}, b_0)$ and (E, t) be a universal pointed covering of (B, b_0); set $u = \gamma_\mathrm{E}(t)$. Since B is locally simply arc-connected, so is E, and as E is connected, it is arc-connected. Hence there is a path c from t to u in E, and $p \circ c$ is a loop in (B, b_0), where p is the projection $\mathrm{E} \to \mathrm{B}$. Writing α for the class of $p \circ c$ in $\widetilde{\pi}_1(\mathrm{B}, b_0)$, $\alpha \cdot t = u$, and so $\varepsilon(\alpha)(t) = u = \gamma(t)$; thus $\varepsilon(\alpha) = \gamma$ since $\pi_1(\mathrm{B}, b_0)$ acts simply transitively on $\mathrm{E}(b_0)$ by (4.6.2).   □

4.9.11

Proposition *All locally arc-connected, simply arc-connected spaces are simply connected.*

Proof Since all coverings of B are sums of connected coverings, it suffices to shows that every connected covering is trivial. Let X be a connected covering of B, $b \in \mathrm{B}$, $x, x' \in \mathrm{X}(b)$. Since X is locally arc-connected and connected, it is arc-connected, and so there is a path c from x to x', and $p \circ c$ is a loop in (B, b). Writing α for the class of $p \circ c$ in $\widetilde{\pi}_1(B, b)$, $\alpha \cdot x = x'$; however $\alpha = e$, so $x = x'$. Hence X has degree 1, and so is trivial.   □

Corollary *All connected, locally simply arc-connected spaces have a simply connected covering.*

Indeed by the previous proposition, such a space is locally simply connected, and so the result follows from Theorem 4.4.5.

4.9.12 Construction of Universal Coverings as Spaces of Paths

Let (B, b_0) be a locally simply arc-connected pointed space. By the previous corollary it has a universal covering. To prove injectivity in Theorem 4.9.9, we give a new construction of this universal covering.

For all $b \in B$, let $S(b) = S(b_0, b)$ be the set of homotopy classes of paths from b_0 to b. Set $S = \bigcup_{b \in B} S(b)$, and let p be the map from S to B sending $S(b)$ to b.

For all simply arc-connected open subset U of B, define an equivalence relation $\sim_U$ on $S|_U$ as follows: let $\gamma, \gamma' \in S|_U$, $b = p(\gamma)$ and $b' = p(\gamma')$. All paths in U from b to b' are homotopic since U is simply arc-connected; let $\beta \in S(b, b')$ be the unique class containing the paths from b to b' in U. We write $\gamma \sim_U \gamma'$ if and only if $\beta \cdot \gamma = \gamma'$. Set $F_U = S|_U / \sim_U$ and let χ denote the canonical map $S|_U \to F_U$.

Lemma (a) *The map* $\tau_U : S|_U \to U \times F_U$ *defined by* $\tau_U(\gamma) = (p(\gamma), \chi(\gamma))$ *is bijective.*

(b) *If* $V \subset U$ *is simply arc-connected and open, then the map* $\iota : F_V \to F_U$ *induced by the canonical injection* $S_V \to S_U$ *is bijective.*

Proof
(a) *Injectivity*: let $\gamma, \gamma' \in S|_U$ such that $\tau_U(\gamma) = \tau_U(\gamma')$. Then $p(\gamma) = p(\gamma')$ and $\gamma' = \beta \cdot \gamma$ with $\beta \in \widetilde{\pi}_1(U, p(\gamma))$. Since U is simply connected, $\beta = e$ and $\gamma = \gamma'$.

Surjectivity: let $(b, f) \in U \times F_U$, and γ be a representative of f. Set $b' = p(\gamma)$. Let $\beta' \in S(b, b')$ be the class containing the paths from b to b' in U. Then $\tau_U(\beta' \cdot \gamma) = (b, f)$.

(b) Let $b \in V$. There is a commutative diagram

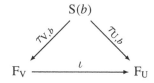

By (a), τ_U and τ_V are bijective, and hence so is ι.   □

Topology on S For every simply arc-connected U, equip F_U with the discrete topology and $U \times F_U$ with the product topology. Let $\mathcal{T}_U$ be the topology on $S|_U$ obtained by transferring that of $U \times F_U$ by τ_U. If V is an open simply arc-connected subset of U, the lemma implies that $\mathcal{T}_V$ is induced by $\mathcal{T}_U$. If U and U' are two simply arc-connected sets, then $U \cap U'$ can be covered by simply arc-connected V_λ, and for all λ, $\mathcal{T}_U$ and $\mathcal{T}_{U'}$ induce the same topology $\mathcal{T}_{V_\lambda}$ on $S|_{V_\lambda}$. Hence, by the topology gluing lemma (4.4.6, Gluing lemma) $\mathcal{T}_U$ and $\mathcal{T}_{U'}$ induce the same topology on $S|_{U \cap U'}$. This lemma also tells us that there is a unique topology $\mathcal{T}$ on S inducing $\mathcal{T}_U$ on $S|_U$ for any simply arc-connected cover U.

Equipped with the topology $\mathcal{T}$ and the projection p, S is a covering of B. We will see in subsection 4.9.15 that this covering is universal.

4.9.13

Proposition *With the notation of 4.9.12, the action of $\tilde{\pi}_1(B, b_0)$ on $S(b_0)$ is simply transitive.*

Proof As $S(b_0) = \tilde{\pi}_1(B, b_0)$, we show that $\tilde{\pi}_1(B, b_0)$ acts on itself by left translations. More precisely, for $\gamma \in \tilde{\pi}_1(B, b_0)$ and $\alpha \in S(b_0)$, we show that $\gamma \cdot_S \alpha$ defined in (4.9.5) is equal to $\gamma \cdot \alpha$, defined by juxtaposition. More generally, we show that for $b \in B$, $b' \in B$, $\alpha \in S(b)$ and $\gamma \in S(b, b')$, $\gamma \cdot_S \alpha = \gamma \cdot \alpha$.

Suppose first that γ is represented by a path c of length a contained in a simply arc-connected open subset U of B. Define $\tilde{c} : [0, a] \to S$ by $\tilde{c}(t) = \gamma_t \cdot \alpha$, where γ_t is the homotopy class of $c|_{[0,t]}$. Then $p \circ \tilde{c} = c$. We check that $\tilde{c}$ is continuous. For this, it suffices to check that $\tau_U \circ \tilde{c}$ is continuous. Now, $\tau_U = (p, \chi)$, $p \circ \tilde{c} = c$ is continuous, and $\chi \circ \tilde{c}$ is constant. Indeed $\tilde{c}(t) \underset{U}{\sim} \alpha$ since γ_t is the class in $S(b, c(t))$ of paths contained in U. Hence, $\tilde{c}$ is the lifting of c in S with initial point α, and $\gamma \cdot_S \alpha = c(\alpha) = \gamma \cdot \alpha$.

By the subdivision lemma (Proposition 4.1.10), all paths can be obtained by juxtaposition of paths contained in a simply arc-connected open set. So the result holds for all γ. □

4.9.14 Proof of Theorem 4.9.9

Let (B, b_0) be a locally simply arc-connected pointed space. By (4.9.11, Corollary), the connected component of b_0 has a universal covering; so (B, b_0) has a universal pointed cover. Moreover B is locally arc-connected, and so ε is surjective by (4.9.10, Proposition).

Let γ be an identity element in $\tilde{\pi}_1(B, b_0)$. By (4.9.13, Proposition), γ acts non trivially on $S(b_0)$; so $\varepsilon(\gamma) \neq e$, and ε is injective. □

4.9.15

Corollary *With the notation of (4.9.12),* S *is a universal covering of* B.

Proof If necessary replacing B by the connected component of b_0, B can be assumed to be connected. By (4.9.11, Corollary), B has a universal covering E. The fundamental group $\pi_1(B, b_0)$ acts simply transitively on $E(b_0)$. By Theorem 4.9.9 and Proposition 4.9.13, $\pi_1(B, b_0)$ also acts simply transitively on $S(b_0)$. The $\pi_1(B, b_0)$-sets $S(b_0)$ and $E(b_0)$ are therefore isomorphic. By Theorem 4.6.8, it follows that the coverings S are E isomorphic. $\square$

Exercises 4.9. (Loops)
1.—Let C_n be the circle in $\mathbb{R}^2$ centered at $\left(\frac{1}{2^n}, 0\right)$ and passing through O. Set

$$X = \bigcup C_n, \qquad X_n = C_0 \cup \cdots \cup C_n .$$

(a) For all n, let $f_n : X_{n+1} \to X_n$ be the map inducing the identity on X_n and sending C_{n+1} onto O. Show that X can be identified with the projective limit of the system defined by all X_n and f_n. Show that the category $\mathscr{C}ov_X$ can be identified with $\varinjlim \mathscr{C}ov_{X_n}$. Deduce that $\pi_1(X, 0) = \varprojlim \pi_1(X_n, 0)$.

(b) Show that the map $\varepsilon : \widetilde{\pi}_1(X, 0) \to \pi_1(X, 0)$ is injective, but not surjective.

(c) Let Y be the cone in $\mathbb{R}^3$, with basis X and vertex $(0, 0, 1)$. Set Z to be the union of Y and of its symmetric Y' with respect to 0 (cf. Zisman [1], 3.3.5.3, p. 129). Show that Z is simply connected but not simply connected by loops, i.e. $\widetilde{\pi}_1(Z, 0) \neq 0$. In particular ε_Z is surjective, but not injective.

(d) In a topological space, two compact subsets A and B whose intersection is arc-connected form a Van Kampen pair for π_1 (resp. for $\widetilde{\pi}_1$) if $\pi_1(A \cup B, x_0)$ is the sum of $\pi_1(A, x_0)$ and $\pi_1(B, x_0)$ amalgamated by $\pi_1(A \cap B, x_0)$ for $x_0 \in A \cap B$ (resp. same for $\widetilde{\pi}_1$). Among the following pairs, which are Van Kampen for π_1? for $\widetilde{\pi}_1$?

- (X, X'), where X' is the symmetric of X with respect to 0;
- (Y, Y');
- (X_1, X_1'), where X_1 is the union of the segment $[0, (1, 0)]$ and of the translation of X by the vector $(1, 0)$, and where X_1' is the symmetric of X_1 with respect to 0.

2.—Let X be a topological space, $\mathscr{U} = (U_i)_{i \in I}$ a simply arc-connected open cover of X, and $\mathscr{U}' = (U_i')_{i \in I}$ a shrinking of $\mathscr{U}$ (i.e. an open cover such that $\overline{U}_i' \subset U_i$ for all i). For every 2-element subset $\{i, j\}$ of I, let C_{ij} be the set of connected components of $U_i \cap U_j$ containing at least one point of $U_i' \cap U_j'$.

Let Γ be a graph with $I \cup \bigcup C_{ij}$ as its set of vertices in which a vertex $c \in C_{ij}$ is connected to the vertices i and j.

(a) Show that Γ is connected if and only if X is connected.

(b) For all $i \in I$ let $x_i \in U_i$ for all $c \in C_{ij}$; let $x_c \in c \cap U_i' \cap U_j$, $\gamma_{c,i}$ and $\gamma_{c,j}$ paths respectively from x_i and x_j to x_c. Define a continuous map $\varphi : |\Gamma| \to X$. Show that

$$\varphi_* : \pi_1(\Gamma, e) \to \pi_1(X, \varphi(e))$$

is surjective for $e \in |\Gamma|$.

(c) Show that for all locally simply arc-connected compact spaces X and all $x \in X$, the group $\pi_1(X, x)$ is finitely generated.

(d) Show that the result continues to hold when X is compact, locally simply connected (not arc-connected) [there is no map φ, but $\pi_1(X, x)$ is isomorphic to a quotient of $\pi_1(\Gamma, e)$].

3. *(Partial blowup of* $\mathbb{R}^2$ *at all points of the axis* Oy *.)*—Set $U = \mathbb{R}^* \times \mathbb{R} = \mathbb{R}^2 - (\{0\} \times \mathbb{R})$ and $E = \mathbb{R} \times \;]-1, 1[$. Denote by X the disjoint union $U \sqcup E$.
For $(y_0, p) \in E$ and $\varepsilon > 0$, set

$$V_\varepsilon(y_0, p) = \left\{ (x, y) \; \middle| \; 0 < |x| < \varepsilon, \left| \frac{y - y_0}{x} - p \right| < \varepsilon \right\} \sqcup \left(\{y_0\} \times \;]p - \varepsilon, p + \varepsilon[\right).$$

(a) Show that there is a unique topology on X inducing the topology of $\mathbb{R}^2$ on U, that of $\mathbb{R}_{\text{discr}} \times \;]-1, 1[$ on E, where $\mathbb{R}_{\text{discr}}$ is the space obtained by equipping $\mathbb{R}$ with the discrete topology, such that U is open and $V_\varepsilon(y_0, p)$ form a fundamental system of neigbourhoods of $(y_0, p) \in E$. Equip X with this topology.

(b) Show that X is a topological surface. Define an $\mathbb{R}$-analytic manifold structure on X.

(c) Show that the space X is Hausdorff, arc-connected, locally compact, admitting a countable dense set but that it is not the countable union of compact sets. Show that X is neither metrizable, nor normal.

(d) Show that for $b_0 \in X$, the Poincaré group $\tilde{\pi}_1(X, b_0)$ can be identified with the fundamental group $\pi_1(X, b_0)$ and is not countable.

(e) Can X be embedded in the transfinite line (chap. 1, §6, Exercise 1)?

Reference

1. M. Zisman, *Topologie algébrique éléémentaire* (Collection U, Armand Colin, 1972)

Chapter 5
Galois Theory

Unless stated otherwise, all algebras considered in this chapter are assumed to be unital associative commutative, and the homomorphisms to be unital.

The letter K will denote a field.

5.1 Extensions

5.1.1 Finite Algebras

Let A be a K-algebra. Denote by ι_A the unique homomorphism from K to A, defined by $\iota(x) = x \cdot 1$. If $A \neq 0$, then ι_A is an injection, called canonical.

A is said to be a **finite** algebra on K if A is a finite dimensional K-vector space. The **degree** of A, written $(A : K)$ or $\deg_K A$, is the dimension of A as a K-vector space.

5.1.2

Let A be a K-algebra and $x \in A$.

Proposition and Definition *The following conditions are equivalent:*

(i) the subalgebra $K[x]_A$ of A generated by x is finite over K;
(ii) there is a nonzero polynomial $P \in K[X]$ such that $P(x) = 0$;
(iii) the powers of x are linearly dependent.

If these conditions hold, then x is said to be **algebraic** *over K.*

Proof (ii) is clearly equivalent to (iii). If the powers of x are linearly independent, then the subalgebra $K[x]_A$ of A is not finite, and so (i) implies (iii). (ii) $\Rightarrow$ (i) since

© Springer Nature Switzerland AG 2020
R. Douady and A. Douady, *Algebra and Galois Theories*,
https://doi.org/10.1007/978-3-030-32796-5_5

the algebra homomorphism $\chi : K[X] \to A$ defined by $\chi(X) = x$ vanishes at P and so factorizes through the finite dimensional K-vector space $K[X]/(P)$.   □

The polynomials $P \in K[X]$ such that $P(x) = 0$ are sometimes called the equations of x. They form an ideal with a single monic generator (i.e. whose leading coefficient is 1), called the **minimal polynomial** of x in A over K. Its degree is the **degree** of x over K, as well as that of the subalgebra of A generated by x.

5.1.3

Proposition *Let A be a K-algebra. The algebraic elements of A over K form a subalgebra of A.*

Proof Let $x, y \in A$ be algebraic over K. The subalgebra generated by $(x + y)$ (resp. xy) is contained in the algebra $K[x, y]_A$ isomorphic to a quotient of $K[x]_A \otimes K[y]_A$ (3.8.16). Thus $x + y$ and xy are algebraic over K.   □

Remark This proposition can be proved directly by taking condition (ii) de 5.1.2. as the definition of algebraic elements, but this is harder. Note that the above proof does enable us to find an equation of $x + y$ (resp. xy) from those of x and y.

5.1.4

Let A be a K-algebra.

Definition We say that A is an **extension** of K if A is a field with respect to its ring structure.

A is said to be a **trivial extension** of K if ι_A is an isomorphism. All K-algebras of degree 1 are trivial extensions of K.

If A is a *extension field* of K, i.e. a field containing K as a subfield, then A equipped with a natural structure of K-algebra is an extension of K; all extensions are isomorphic to an extension of this type.

If A is of the form $K[X]/(P)$ then A is a finite extension of K if and only if P is irreducible (3.2.10, Lemma).

Proposition *If A is finite and integral, then A is an extension of K.*

Proof Let $a(\neq 0) \in A$. The map $x \mapsto ax$ from A to A is injective and hence surjective since A is a finite dimensional vector space. Hence there exists $x \in A$ such that $ax = 1$, and so A is a field.   □

5.1.5

Chinese Remainder Theorem *Let* A *be a ring,* $(B_i)_{i \in I}$ *a finite family of rings and for all* $i \in I$, $\varphi_i : A \to B_i$ *a ring homomorphism. Assume that, for each* $i \in I$, φ_i *is surjective, and that for all* $(i, j) \in I \times I$ *with* $i \neq j$, $\mathrm{Ker}\varphi_i + \mathrm{Ker}\varphi_j = A$. *Then the map* $\Phi : A \to \prod_i B_i$ *defined by* $\Phi(x) = (\varphi_i(x))_{i \in I}$ *is surjective.*

Proof For all $(i, j) \in I \times I$ with $i \neq j$, there exists $u_{ij} \in A$ such that $\varphi_i(u_{ij}) = 1$ and $\varphi_j(u_{ij}) = 0$. Indeed, for $u \in \mathrm{Ker}\varphi_j$ and $v \in \mathrm{Ker}\varphi_i$ such that $u + v = 1$, $\varphi_j(u) = 0$ and $\varphi_i(u) = \varphi_i(1 - v) = 1$. Take $u_{ij} = u$.

For all $i \in I$, set $u_i = \prod_{j \neq i} u_{ij}$; then $\varphi_i(u_i) = 1$ and $\varphi_j(u_i) = 0$ for all $j \neq i$.

Let $b = (b_i)_{i \in I} \in \prod_i B_i$ and for all $i \in I$, let $x_i \in A$ be such that $\varphi_i(x_i) = b_i$. Setting $x = \sum_{i \in I} u_i x_i$, $\Phi(x) = b$, $\qquad \square$

Remark The condition $\mathrm{Ker}\varphi_i + \mathrm{Ker}\varphi_j = A$ holds in particular if $\mathrm{Ker}\varphi_i$ and $\mathrm{Ker}\varphi_j$ are distinct maximal ideals. See also (3.5.9).

5.1.6

Corollary 5.1 *Let* A *be a finite* K-*algebra.*

(1) *The set* X *of maximal ideals of* A *is finite.*

(2) *The map* $\Phi : A \to \prod_{\mathfrak{m} \in X} A/\mathfrak{m}$ *is surjective and its kernel consists of the nilpotent ideals of* A.

Proof (1) Let Y be a finite subset of X. By the Chinese remainder theorem, $A \to \prod_{\mathfrak{m} \in Y} A/\mathfrak{m}$ is surjective, and so card $Y \leqslant (A : K)$. Hence card $X \leqslant (A : K)$.

(2) By the Chinese remainder theorem and by (1), Φ is surjective; by 5.1.4, all prime ideals are maximal and by 3.1.6, the intersection of all prime ideals of A is the set of nilpotent elements of A. $\qquad \square$

An algebra with no nonzero nilpotent elements is said to be **reduced** (see 3.1.7).

Corollary 5.2 *Every reduced finite* K-*algebra is isomorphic to a finite product of finite extensions of* K.

5.1.7

Dedekind Theorem *Let* A *be a* K-*algebra and* L *an extension of* K. *Consider the* L-*vector space* $\mathscr{L}_K(A, L)$ *of* K-*linear maps from* A *to* L. *In this vector space, the* K-*algebra homomorphisms from* A *to* L *are linearly independent.*

Proof By (3.8.17), $\text{Hom}_{K\text{-}\mathscr{a}\mathscr{l}\mathscr{g}}(A, L)$ can be identified with the set $\text{Hom}_{L\text{-}\mathscr{a}\mathscr{l}\mathscr{g}}(L \otimes_K A, L)$. Let $(\varphi_i)_{i \in I}$ be a finite family of elements of $\text{Hom}_{K\text{-}\mathscr{a}\mathscr{l}\mathscr{g}}(A, L)$ and $(\tilde{\varphi}_i)_{i \in I}$ the family of corresponding elements in $\text{Hom}_{L\text{-}\mathscr{a}\mathscr{l}\mathscr{g}}(L \otimes_K A, L)$. For each $i \in I$, $\tilde{\varphi}_i$ is surjective; as L is a field, the kernel $\mathfrak{m}_i$ of $\tilde{\varphi}_i$ is a maximal ideal and $L \otimes_K A = \mathfrak{m}_i \oplus L \cdot 1$. If $\varphi_i \neq \varphi_j$ for $i \neq j$, then moreover $\mathfrak{m}_i \neq \mathfrak{m}_j$ since $\tilde{\varphi}_i(1) = \tilde{\varphi}_j(1) = 1$.

By the Chinese remainder theorem, the map $\Phi : x \mapsto (\tilde{\varphi}_i(x))_{i \in I}$ from $L \otimes_K A$ to L^I is surjective.

Assuming there is a relation $\sum a_i \varphi_i = 0$, $\sum a_i \tilde{\varphi}_i = 0$ and the image of Φ is contained in the hyperplane $\{(t_i) \in L^I \mid \sum a_i t_i = 0\}$, which contradicts the surjectivity of Φ.   □

5.1.8 Transcendental Elements

Let A be an extension of K. An element $x \in A$ is said to be **transcendental** over K if it is not algebraic, in other words if the homomorphism $\varphi : K[X] \to A$ defined by $\varphi(X) = x$ is injective.

A family $(x_i)_{i \in I}$ of elements of A is said to be **algebraically free** if the homomorphism $\phi : K[X_i]_{i \in I} \to A$ defined by $\phi(X_i) = x_i$ is injective, in other words if there is a homomorphism $\Phi : K(X_i)_{i \in I} \to A$ such that $\Phi(X_i) = x_i$ (where $K(X_i)$ denotes the field of fractions of $K[X_i]$). A maximal algebraically free family is a **transcendental basis** of A over K. If $(x_i)_{i=1,...,k}$ is a transcendental basis of A, then the field A equipped with Φ is an algebraic extension of $K(X_1, ..., X_k)$.

Proposition and Definition *All transcendental bases of A over K have the same cardinality. This cardinal is called the **transcendental degree** of A over K.*

Lemma (Exchange Lemma) *Let $(x_1, ..., x_k)$ be a transcendental basis of A over K and $y \in A$ a transcendental element over K. Then there exists $i_0 \in \{1, ..., k\}$ such that the family (x_i') defined by $x_{i_0}' = y$, $x_i' = x_i$ for $i \neq i_0$ is a transcendental basis of A over K.*

Proof By maximality of $(x_1, ..., x_k)$, there is a nonzero polynomial $P_0 \in K[X_1, .., X_k, Y]$ such that $P(x_1, ..., x_k, y) = 0$, and $P \notin K[Y]$ since y is transcendental. Choose i_0 such that the degree of P with respect to X_{i_0} is nonzero, and define x_i' as in the lemma. Then x_{i_0} is algebraic over $K(x_1', ..., x_k')$, and so are all elements of A. The family $(x_i)_{i \neq i_0}$ is algebraically free, and y is transcendental over $K(x_i)_{i \neq i_0}$; otherwise x_{i_0} would be algebraic over this field. Hence (x_i') is algebraically free, and thus a transcendental basis.   □

Proof of the Proposition The proof is by induction on k.

(E_k) *For any extension A of a field K, if A has a transcendental basis with k elements, all other transcendental bases of A also have k elements.*

Statement (E_0) is immediate: if the empty family is a transcendental basis, then A is algebraic over K, and every transcendental basis is empty. Let $k > 0$, and assumed that (E_{k-1}) holds. We prove (E_k). Let A be an extension of K with two transcendental bases $(x_i)_{i \in I}$ and $(y_j)_{j \in J}$ with card $I = k$. Then $J \neq \varnothing$; choose $j_0 \in J$. The element y_{j_0} is transcendental over K, and it follows from the lemma that there exists i_0 such that $(x_i)_{i \neq i_0}$ is a transcendental basis of A over $K(y_{j_0})$. But $(y_j)_{j \neq j_0}$ is also a transcendental basis of A over $K(y_{j_0})$. By the induction hypothesis, card $(J - \{j_0\}) = k - 1$, and so card $J = k$.

This proves the proposition when A has a finite transcendental basis. Assume that A has two transcendental bases $(x_i)_{i \in I}$ and $(y_j)_{j \in J}$ with I and J infinite. For each $j \in J$, it is possible to find a finite subset I_j of I such that y_j is algebraic over $K(x_i)_{i \in I_j}$. The extension A is algebraic over $K(x_i)_{i \in I'}$, where $I' = \bigcup I_j$, and so $I' = I$. Let $S \subset I \times J$ be the set of (i, j) such that $i \in I_j$. As J is an infinite set, card $S =$ card J. The projection of S onto I is I. So card $S \geqslant$ card I, and card $J \geqslant$ card I. Similarly card $I \geqslant$ card J, and thusù card $J =$ card I, $\qquad \square$

Exercises 5.1 (Extensions)
1.—Find equations over $\mathbb{Q}$ for the following numbers:

$$\sqrt{2} + \sqrt{3}, \ \sqrt{2} + \sqrt[3]{2}, \ \left(1 + \frac{\sqrt{5}}{2}\right)\left(1 + \frac{i\sqrt{3}}{2}\right).$$

$$x + \sqrt{2} \quad \text{where} \quad x^3 + x + 1 = 0.$$

2.—Show that, if x has equation P, then

$$\sum \frac{(-1)^r}{r!s!} d^{\frac{r+s}{2}} P^{(r)} P^{(s)}$$

is an equation for $x + \sqrt{d}$, the sum being over all pairs (r, s) of integers $0 \leqslant r \leqslant$ d°(P), $0 \leqslant s \leqslant$ d°(P) with $r + s$ even, and $P^{(r)}$ denoting the r-th derivative of P. Interpret this formula when the field K has nonzero characteristic.

3.—(a) Let B be an integral ring, P and Q polynomials in B[X] of respective degrees p and q, and Ω the closure of the field of fractions of B. Show that the following conditions are equivalent:

(i) P and Q have a common root in Ω;
(ii) there are nonzero polynomials u with degree $\leqslant q - 1$ and v with degree $\leqslant p - 1$ such that $uP + vQ = 0$.

Let $B[X]_m$ be the B-module of polynomials of degrees $\leqslant m$ (isomorphic to B^{m+1}), and Res(P, Q) the resultant of P and Q (3.7.11).

(b) For P, Q $\in$ K[X, Y, Z], define the resultant $\mathrm{Res}_Z(P, Q) \in$ K[X, Y] of P and Q as elements of (K[X, Y])[Z].

Let A be a K-algebra, $x, y \in$ A algebraic elements over K having respectively P (of degree p) and Q (of degree q) as equations. Show that

$$R_1 = \mathrm{Res}_X(P(X), Q(Z - X)) \; ;$$
$$R_2 = \mathrm{Res}_X\left(P(X), X^q Q\left(\frac{Z}{X}\right)\right)$$

are respectively equations for $x + y$ and xy over K.

4.—(a) Draw the curve in $\mathbb{R}^2$ with parametric representation:

$$\begin{cases} x = \cos 2t \\ y = \cos 3t \, . \end{cases}$$

Show that this curve is a subset of an algebraic curve and give its equation.

(b) Let $\mathbb{R}(\cos 2T, \cos 3T)$ be the field of fractions of the ring of functions generated by the real constants and the functions $t \mapsto \cos 2t$ and $t \mapsto \cos 3t$. Show that this field is a finite extension of $\mathbb{R}(\cos 2T)$. Give an equation for $\cos 3T$.

5.—Let K be a field and M a commutative monoid. Consider the vector space K^M of maps from M to K. Show that in this vector space the homomorphisms from M to the multiplicative monoid K are linearly independent (consider the algebra $K^{(M)}$ of M). Does this result generalize to non-commutative monoids?

6.—Let A be a K-algebra, B and C subalgebras of A. If the canonical map $B \otimes_K C \to A$ is injective, then B and C are said to be *linearly disjoint* (over K). Let L be an extension of K in B. Show that B and C are linearly disjoint over K if and only if this is also the case of L and C over B as well as that of B and the algebra $L \cdot C$ generated by L and C over L.

7.—A non-trivial ring is said to be **local** if it has a unique maximal ideal. The aim is to show that every finite K-algebra is a unique product of local K-algebras. Let A be a finite K-algebra and $(\mathfrak{m}_i)_{i \in I}$ the family of maximal ideals. Set $\mathfrak{N} = \bigcap \mathfrak{m}_i$.

(a) Show that $\mathfrak{N}^r = 0$ for some r. Deduce that the product ideal of $\mathfrak{m}_i^r$ is 0.

(b) Show that the annihilator of $x \in \mathfrak{m}_i^r$ is not contained in $\mathfrak{m}_i$. Deduce that $\bigcap \mathfrak{m}_i^r = 0$.

(c) Show that $\mathfrak{m}_i^r + \mathfrak{m}_j^r = A$ for $j \neq i$. Deduce that the canonical map $A \to \prod A/\mathfrak{m}_i^r$ is an isomorphism. Show that $A/\mathfrak{m}_i^r$ is local.

(d) Let $f : A \to \prod_{j \in J} B_j$ be an isomorphism, where B_j is a local algebra for all j. Show that there is a bijection $s : I \to J$ and an isomorphism from $A/\mathfrak{m}_i^r$ onto $B_{s(i)}$ for all i.

(e) Let (k_i) be integers. Show that the product ideal of $\mathfrak{m}_i^{k_i}$ is equal to their intersection.

8.—Let A and B be finite algebras over K, $f : A \to B$ an injective homomorphism, and $(\mathfrak{m}_i)_{i \in I}$ and $(\mathfrak{n}_j)_{j \in J}$ the respective families of maximal ideals of A and B. Show that there is a surjective map $f^* : J \to I$ and, for all j, a homomorphism

$$f_j : A/\mathfrak{m}_{f^*(j)} \to B/\mathfrak{n}_j$$

making the diagram

$$
\begin{array}{ccc}
A & \longrightarrow & \prod A/\mathfrak{m}_i \\
\downarrow{\scriptstyle f} & & \downarrow{\scriptstyle f_*} \\
B & \longrightarrow & \prod B/\mathfrak{n}_j
\end{array}
$$

where f_* is defined by all f_j, commutative. Show that f^* and all f_j are uniquely determined.

9.—Let $P = \pm 1 + \prod_{i=0}^{k}(X - a_i)$ be a polynomial in $\mathbb{Z}[X]$, where $a_i \in \mathbb{Z}$ take at least three distinct values for $i \neq 0$, $a_0 = 0$, and $|a_i| \geqslant 2$.

(a) Show that P has a root in $\mathbb{C}$ with absolute value < 1, while the other roots have absolute value > 1.

(b) Show that P is irreducible in $\mathbb{Q}[X]$.

(c) Can $\mathbb{Z}$ be replaced by $\mathbb{Z}[i]$? By what other subring of $\mathbb{C}$ can it replaced?

10. *(Liouville numbers)*—(a) Let P be a polynomial of degree d with integral co-efficients and $x \in \mathbb{R}$ a root of P. Let $\left(\frac{a_n}{b_n}\right)$ ($a_n \in \mathbb{Z}$, $b_n \in \mathbb{N}^*$) a sequence of rational numbers tending to x while remaining distinct from x. Show the existence of a constant c such that

$$\left| \frac{a_n}{b_n} - x \right| \geq c \left(\frac{1}{b_n} \right)^d$$

(note that $b_n^d \, P\left(\frac{a_n}{b_n}\right) \in \mathbb{Z}$).

(b) Show that if (u_n) is a sequence of integers equal to 1 or 2, then the real number $\sum \frac{u_n}{10^{n!}}$ is transcendental, i.e. is not algebraic over $\mathbb{Q}$. Show that such numbers form a set having the cardinality of the continuum, and homeomorphic to the Cantor set (2.8, Exercise 2).

(c) Is the criterion obtained in (a), applied to partial sums of the series $\sum \frac{1}{n!}$, sufficient to show that e is transcendental?

11. *(Transcendence of e)*—(a) For $P \in \mathbb{R}[X]$, set

$$L(P) = \int_0^{\infty} e^{-x} P(x) \, dx .$$

Show that if P is of degree $\leqslant n$, then $L(P) = P(0) + P'(0) + \cdots + P^{(n)}(0)$, where $P^{(k)}$ denotes the k-th derivative of P.

(b) Let n be a fixed integer and for $p \in \mathbb{N}$, set

$$P_p(x) = \frac{1}{(p-1)!} x^{p-1}(x-1)^p \cdots (x-n)^p .$$

Show that $M_p = L(P_p)$ is an integer congruent to $\pm(n!)^p \pmod{p}$.

(c) Let $\tau_{-k} P$ denote the polynomial $x \mapsto P(x+k)$. Set

$$M_{p,k} = L(\tau_{-k} P_p) = e^k \int_k^\infty e^{-x} P_p(x)\, dx .$$

Show that for $p \leqslant k \leqslant n$, $M_{p,k}$ is an integral multiple of p.

(d) Suppose that $a_n e^n + a_{n-1} e^{n-1} + \cdots + a_0 = 0$. Then

$$\sum a_k (M_{p,k} + \epsilon_{p,k}) = 0 ,$$

where $\epsilon_{p,k} = \dfrac{e^k}{(p-1)!} \displaystyle\int_0^k P_p(x)\, dx$. Show that $\epsilon_{p,k}$ tends to 0 as p tends to infinity.

(e) Suppose that $a_n e^n + \cdots + a_0 = 0$ with $a_i \in \mathbb{Z}$. Show that $\sum a_k M_{p,k} = 0$ for sufficiently large p.

Assuming $a_0 \neq 0$, find a contradiction by taking p to be a sufficiently large prime.

5.2 Algebraic Extensions

5.2.1

Definition An extension A of K is **algebraic** if every element of A is algebraic over K or, equivalently, if as an algebra, A is generated by algebraic elements over K.

Every finite extension is algebraic.

Proposition *Every subalgebra of an algebraic extension is an extension.*

Proof Let A be an algebraic extension of K and B a subalgebra of A. The extension A is the union of finite sub-extensions (A_i), $i \in I$. For all $i \in I$, $B \cap A_i$ is a finite integral algebra and hence an extension (5.1.4, Proposition) and $B = \bigcup (B \cap A_i)$ is an extension. $\square$

5.2.2

Proposition *Let A be an algebraic extension of* K. *Every endomorphism of A is an automorphism.*

Proof Let $f : A \to A$ be an extension endomorphism. Then f is injective since A is a field. For all $P \in K[X]$, let R_P be the set of roots of P in A. The endomorphism f induces an injective map from R_P in R_P, which is therefore surjective since R_P is finite. As $A = \bigcup_{P \in K[X]} R_P$, f is surjective. □

5.2.3

Proposition and Definition *The following conditions are equivalent:*

(i) all polynomials $P \in K[X]$ of degree > 0 have at least one root in K;
(ii) all polynomials can be uniquely written as $\alpha(X - a_1) \ldots (X - a_d)$ with α, a_1, ..., $a_d \in K$;
(iii) every algebraic extension over K is trivial.

If these conditions hold, then K is said to be **algebraically closed.**

Proof (ii) ⇒ (i) is immediate. We show (i) ⇒ (iii).
Let A be an extension of K, $x \in A$. Embed K in A by ι_A and let P be the minimal polynomial of x, a a root of P. Then $P = (X - a)Q$ with $d°Q < d°P$, $P(x) = 0$ and $Q(x) \neq 0$. Hence $x - a = 0$ since A is integral and $x = a \in K$. Thus $A = K$.

(iii) ⇒ (ii). As K[X] is factorial, it suffices to show that all irreducible polynomials have degree 1. Let P be an irreducible polynomial. The K-algebra K[X]/(P) is finite, and integral since K[X] is factorial. So it is an algebraic extension of K (5.1.4, Proposition). It is trivial by assumption, so that P has degree 1, □

5.2.4 Example. d'Alembert's Theorem

The field $\mathbb{C}$ of complex numbers is algebraically closed.

Proof Let $P \in \mathbb{C}[Z]$ be a polynomial of degree $\geqslant 1$ and suppose that P has no roots in $\mathbb{C}$. Then the function $f = \frac{1}{P}$ is holomorphic on $\mathbb{C}$, and Cauchy's formula applies:

$$f(x) = \frac{1}{2\pi} \int_0^{2\pi} f\left(x + re^{i\theta}\right) d\theta.$$

As r tends to infinity, the second term tends to 0; so $f(x) = 0$, and f vanishes everywhere, which is absurd. □

For other proofs, see 4.3, Exercise 10, 4.4, Exercise 5, 4.6, Exercise 4 and 6.1, Exercise 3 (b).

5.2.5

Definition An **algebraic closure** of K is an algebraic extension of K that is an algebraically closed field.

Theorem *Every field has an algebraic closure.*

Cardinality Lemma *For any algebraic extension A of K,*

$$\text{Card}A \leqslant \text{Card}A(K[X] \times \mathbb{N}) .$$

Indeed, set $\Gamma = \{(P, x) \mid P \in K[X] - \{0\}, x \in A \text{ and } P(x) = 0\}$. The map $(P, x) \mapsto P$ from Γ to $K[X]$ has finite fibres since every polynomial has only finitely many roots. So

$$\text{card } \Gamma \leqslant \text{card } (K[X] \times \mathbb{N}) .$$

The map $(P, x) \mapsto x$ from Γ to A is surjective since A is an algebraic extension of K. Hence card $A \leqslant$ card Γ, proving the lemma.

Proof of the Theorem Let U be a set containing K and such that

$$\text{card } U > \text{card } (K[X] \times \mathbb{N}) .$$

Let $\mathscr{E}$ be the set of algebraic extensions of K whose underlying set is in U. Define the following order relation on $\mathscr{E}$: "A $\leqslant$ B" if A is a sub-extension of B, i.e. the underlying set of A is in the underlying set of B and the laws of A are induced by those of B.

The ordered set $\mathscr{E}$ is inductive. Let Ω be a maximal element. We show by contradiction that Ω is algebraically closed.

Let W be a non trivial algebraic extension of Ω. By the cardinality lemma, card $W \leqslant$ card$(K[X] \times \mathbb{N})$. Writing ι for the canonical injection from Ω to W,

$$\text{card } (W - \iota(\Omega)) \leqslant \text{card } W \leqslant \text{card } (K[X] \times \mathbb{N}) < \text{card } U = \text{card } (U - \Omega) .$$

Let φ be a bijection from $W - \iota(\Omega)$ onto a subset F of $U - \Omega$, and $\Omega' = \Omega \cup F$. Define $\Phi : W \to \Omega'$ by $\Phi(x) = \iota^{-1}(x)$ if $x \in \iota(\Omega)$ and $\Phi(x) = \varphi(x)$ if $x \in W - \iota(\Omega)$. The map Φ is a bijection. Use Φ to transfer the extension structure of W to Ω'. Then Ω' becomes a non trivial algebraic extension of Ω containing Ω, which contradicts the maximality of Ω. Hence Ω is algebraically closed.   □

5.2.6

Theorem *Let* A *be an algebraic extension of* K, B *a sub-extension of* A *and* Ω *an algebraic closure of* K. *Every homomorphism from* B *to* Ω *extends to a homomorphism from* A *to* Ω.

Proof Let $\mathscr{E}$ be the set of pairs (C, h) where C is a subalgebra of A and $h : C \to \Omega$ a homomorphism. Equip $\mathscr{E}$ with the order defined by: "$(C, h) \leqslant (C', h')$" if and only if $C \subset C'$ and $h'|_C = h$. The ordered set $\mathscr{E}$ is inductive. Let $f : B \to \Omega$ be a homomorphism and $(\overline{C}, \overline{h})$ a maximal element in $\mathscr{E}$ and an upper bound of (B, f).

We show by contradiction that $\overline{C} = A$. Let $x \in A - \overline{C}$ and $P = \sum a_i X^i$ be the minimal polynomial of x over $\overline{C}$. The polynomial $\sum \overline{h}(a_i) X^i \in \Omega[X]$ has at least one root ξ since Ω is algebraically closed. The homomorphism $g : \overline{C}[x]_A \to \Omega$ defined by $g(x) = \xi$ extends h and $(\overline{C}[x]_A, g)$ is a strict upper bound of the maximal element $(\overline{C}, \overline{h})$, giving a contradiction. $\qquad\qquad\square$

5.2.7

Corollary 5.3 *Let* A *be an algebraic extension of* K *and* Ω *an algebraic closure of* K. *Then there is an embedding of* A *into* Ω.

This follows from the above Theorem 5.2.6 applied to (K, ι), where $\iota : K \to \Omega$ is the canonical injection.

Corollary 5.4 *Let* Ω *be an algebraic closure of* K, A *and* B *sub-extensions of* Ω. *Every homomorphism from* A *to* B *extends to an automorphism of* Ω. *In particular, every automorphism of* A *extends to an automorphism of* Ω.

Indeed, by Theorem 5.2.6, every homomorphism from A to B extends to an endomorphism of Ω that is bijective by 5.2.2.

Corollary 5.3 and Definition *Let* Ω *be an algebraic closure of* K, *and* $x, x' \in \Omega$. *The following conditions are equivalent:*

(i) x and x' have the same minimal polynomial;
(ii) there is an automorphism $g : \Omega \to \Omega$ such that $g(x) = x'$.

If these conditions hold, then x and x' are said to be **conjugate**.

Proof (i) $\Rightarrow$ (ii). Sub-extensions $K[x]_\Omega$ and $K[x']_\Omega$ of Ω are isomorphic to $K[X]/(P)$, where P is the minimal polynomial of both x and x'.

The isomorphism $f : x \mapsto x'$ from $K[x]_\Omega$ to $K[x']_\Omega$ extends to an automorphism of Ω (5.2.7. Corollary 5.2).

(ii) $\Rightarrow$ (i). Every equation of x is an equation of x', proving (i).

Corollary 5.4 *All algebraic closures of* K *are isomorphic.*

Indeed, let Ω and Ω' be two algebraic closures of K, $f : \Omega \to \Omega'$ and $g : \Omega' \to \Omega$ homomorphisms (these exist by Corollary 1). The endomorphisms $g \circ f$ and $f \circ g$ of Ω and Ω' respectively are automorphisms by 5.2.2, and so f and g are isomorphisms.

Remark Given two algebraic closures of a field, there is no method to choose an isomorphism from the one onto the other. Hence algebraic closures of K should not be identified; the notion of "the algebraic closure" of K is not well-defined. This is similar to universal coverings in the context of topology (5.2, Exercise 2).

5.2.8 *Characteristic of a Field and the Frobenius Endomorphism*

We remind the reader that the *characteristic* of a field K is the positive generator p of the kernel of the homomorphism $\epsilon : n \mapsto n \cdot 1$ from $\mathbb{Z}$ to K. If ϵ injective, then $p = 0$ and the subfield of K generated by 1 is isomorphic to $\mathbb{Q}$. If ϵ is not injective, the ring $\mathbb{Z}/(p)$, isomorphic to a subring of K, is integral and so p is prime and $\mathbb{Z}/(p)$ is isomorphic to a subfield of K; if $p \neq 0$, p is also the order of 1 in the additive group K.

For p prime, the field $\mathbb{Z}/(p)$ is written $\mathbb{F}_p$.

Theorem and Definition *Let K be a field of characteristic $p \neq 0$. Then the map $F : x \mapsto x^p$ is an endomorphism of K, called the* **Frobenius endomorphism** *of K.*

The theorem follows from the binomial formula and from the fact that the binomial coefficients $\binom{p}{k} = \frac{p!}{k!(p-k)!}$ are multiples of p for $0 < k < p$.

5.2.9 *Finite Fields*

If K is a finite field, its characteristic is a prime $p \neq 0$ and K is a finite extension of $\mathbb{F}_p$. If r is the degree of K over $\mathbb{F}_p$, then

$$\text{card } K = p^r .$$

Proposition *Let p be a prime, and $q = p^r$ $(r \geqslant 1)$ a power of p. There is a field with q elements, and all such fields are isomorphic.*

Proof Let Ω be an algebraic closure of $\mathbb{F}_p$. The endomorphism $F^r : x \mapsto x^q$ de Ω is an automorphism. Hence the elements $x \in \Omega$ left invariant by F^r form a subfield $\mathbb{F}_q$ of Ω whose elements are the roots of the polynomial $P = X^q - X$. The derivative of P is

$$qX^{q-1} - 1 = pp^{r-1}X^{q-1} - 1 = -1 \neq 0 .$$

The polynomial P has q simple roots which are therefore distinct and $\mathbb{F}_q$ has q elements.

Let K be a field with q elements, the multiplicative group K^* has $q - 1$ elements and for all $x \in K^*$, $x^{q-1} = 1$; so $x^q = x$ for all $x \in K$. As K is a finite extension of $\mathbb{F}_p$, it may be embedded in Ω, and thus is isomorphic to a subfield of $\mathbb{F}_q$, and is isomorphic to $\mathbb{F}_q$ since card $K =$ card $\mathbb{F}_q$. $\qquad \square$

Remark The field $\mathbb{F}_q$ constructed here depends on the choice of the algebraic closure Ω of $\mathbb{F}_p$. We do not give any intrinsic definition of $\mathbb{F}_q$ for $q = p^\nu$, $\nu > 1$. There is no natural isomorphism between two fields with q elements.

5.2.10

Theorem *If* K *is a finite field, then the multiplicative group* K^* *is cyclic.*

We prove the following more general result:

Proposition *Let* K *be a field. Every finite subgroup* G *of the multiplicative group* K^* *is cyclic.*

Proof The multiplicative group G is isomorphic to an additive group $\bigoplus \mathbb{Z}/p_i^{r_i}$, where all p_i are primes (3.5.10, Proposition). We show that all p_i are distinct. Suppose that for $i \neq j$ $p_i = p_j = p$, then the group G would contain a subgroup of the form $(\mathbb{Z}/(p))^2$, and hence at least p^2 elements x such that $x^p = 1$. This is impossible since the polynomial $X^p - 1$ has at most p roots. By (3.5.9, Corollary), G is isomorphic to $\mathbb{Z}/\prod p_i^{r_i}$, and so is cyclic. $\qquad \square$

Remark When $K = \mathbb{C}$, the finite subgroups can be explicitly described. The map $t \mapsto e^{2i\pi t}$ defines an isomorphism from $\mathbb{Q}/\mathbb{Z}$ onto the group of elements of finite order of $\mathbb{C}^*$. Every subgroup of $\mathbb{Q}/\mathbb{Z}$ is generated by the image of a number $\frac{1}{q}$, and so every finite subgroup of $\mathbb{C}^*$ is of the form $\left\{ e^{2i\pi \frac{p}{q}} \right\}_{0 \leqslant p \leqslant q-1}$.

Exercises 5.2. (Algebraic extensions)
1.—Find a generator of the group $(\mathbb{Z}/p\mathbb{Z})^*$ for each prime $p < 100$. For which primes is 2 a generator?

2. *(Non-functoriality of the algebraic closure)*—(a) Consider the complex fields $K = \mathbb{Q}(i)$, $L = K(\sqrt{2})$, $M = K(\sqrt[4]{2})$ and the field Ω of algebraic numbers over $\mathbb{Q}$. Let $f \in \mathrm{Aut}_K(L)$ be defined by $f(\sqrt{2}) = -\sqrt{2}$. Determine $\mathrm{Aut}_K(M)$. Show that there is an automorphism g of M extending f, but none satisfying $g \circ g = 1_M$.

(b) Show that every automorphism Ω preserves M. Show that there is no automorphism h of Ω inducing f and satisfying $h \circ h = 1_\Omega$.

(c) Let $\mathscr{C}$ be the category of fields. Show that there is no functor $S : \mathscr{C} \to \mathscr{C}$ and no functorial morphism $\iota : 1_{\mathscr{C}} \to S$ such that, for all $K \in \mathscr{C}$, the field $S(K)$, considered an extension of K by $\iota_K : K \to S(K)$, is an algebraic closure of K. (Note the similarity with 4.4, Exercise 1.)

3.—Let p be a prime and Ω an algebraic closure of the field $\mathbb{F}_p = \mathbb{Z}/(p)$.

(a) Show that if p does not divide n, then the equation $x^n - 1 = 0$ has n distinct roots in Ω.

(b) Let r be an integer > 0. Show that the set of fixed points of F^r, where F is the Frobenius automorphism of Ω, is a subfield $\mathbb{F}_{p^r}$ of Ω with p^r elements.

(c) Show that $\mathbb{F}_{p^r}$ is the unique subfield of Ω with p^r elements. Deduce that every finite subfield of Ω is of the form $\mathbb{F}_{p^r}$, and that every finite field of characteristic p is isomorphic to some $\mathbb{F}_{p^r}$.

(d) Show that the group of automorphisms from $\mathbb{F}_{p^r}$ onto $\mathbb{F}_p$ is generated by the Frobenius automorphism, and is isomorphic to $\mathbb{Z}/(r)$.

(e) Show that $\mathbb{F}_{p^r}$ is contained in $\mathbb{F}_{p^s}$ if and only if r divides s. Are there infinite subfields in Ω other than Ω? Show that the group of automorphisms from Ω onto $\mathbb{F}_p$ is isomorphic to the profinite completion $\widehat{\mathbb{Z}}$ of $\mathbb{Z}$ (see 2.9.5).

(f) Show that the multiplicative group of Ω is isomorphic to the direct sum of the l-primary components M_l of $\mathbb{Q}/\mathbb{Z}$ for some prime $l \neq p$ (M_l is the subgroup of $\mathbb{Q}/\mathbb{Z}$ consisting of the elements whose order is a power of l).

(g) Show that the ring of endomorphism of Ω^* is isomorphic to $\prod_{l \neq p} \widehat{\mathbb{Z}}_l$ (3.3, Exercise 9) and that this gives an injective homomorphism from the additive group $\widehat{\mathbb{Z}}$ to the multiplicative group of invertible elements of this ring.

4. *(Another proof of the existence of an algebraic closure)*—If $(A_i)_{i \in I}$ is an infinite family of algebras, set $\bigotimes_{i \in I} A_i = \varinjlim \bigotimes_{j \in J} A_j$, the direct limit being taken over the finite subsets J of I. Let $(A_i)_{i \in I}$ be the family of extensions $K[X_1, \ldots, X_n]/\mathfrak{p}$ of K, where $\mathfrak{p}$ is a maximal ideal of $K[X_1, \ldots, X_n]$, and let $\mathfrak{M}$ be a maximal ideal of $\bigotimes_{i \in I} A_i$. Show that $\bigotimes_{i \in I} A_i/\mathfrak{M}$ is an algebraic closure of K.

5.—Let A be a finite K-algebra, B a subalgebra of A and Ω an algebraic closure of K. Show that every homomorphism from B to Ω extends to a homomorphism from A to Ω (use 5.1, Exercise 8). Does this result generalize to infinite algebras A, but all of whose elements are algebraic over K?

6.—(a) Let A and B be two integral rings and $f : A \to B$ an injective homomorphism, M the field of fractions of A. Assume that B, considered an A-algebra by using f, is a finitely generated A-module. Show that $M \otimes_A B$ can be identified with the field of fractions of B.

(b) Let $P \in K[X, Y]$ be an irreducible polynomial not in $K[X]$. Show that the field of fractions of $K[X, Y]/(P)$ is a natural finite extension of $K(X)$.

5.3 Diagonal Algebras

5.3.1 Algebra of Functions on a Finite Set

Let X be a a finite set with d elements. The algebra K^X of K-valued functions defined on X is a finite K-algebra over K of degree d isomorphic to K^d equipped with the multiplication

$$((x_1, \ldots, x_d), (y_1, \ldots, y_d)) \mapsto (x_1 y_1, \ldots, x_d y_d).$$

For all $x \in X$, the map $\delta_x : f \mapsto f(x)$ from K^X to K is an algebra homomorphism, and its kernel $\mathfrak{m}_x$ is a maximal ideal.

For all subsets Y of X, the elements $f \in K^X$ such that $f|_Y = 0$ form an ideal $\mathscr{I}_Y$ and the algebra $K^X / \mathscr{I}_Y$ can be identified with K^Y. If Y and Z are subsets of X, then

$$\mathscr{I}_{Y \cap Z} = \mathscr{I}_Y + \mathscr{I}_Z$$
$$\mathscr{I}_{Y \cup Z} = \mathscr{I}_Y \cap \mathscr{I}_Z = \mathscr{I}_Y \cdot \mathscr{I}_Z$$
$$Y \subset Z \Longleftrightarrow \mathscr{I}_Y \supset \mathscr{I}_Z.$$

Proposition *With the above notations, the map* $Y \mapsto \mathscr{I}_Y$ *from* $\mathfrak{P}(X)$ *to the set of ideals of* K^X *is a bijection.*

Proof The map is clearly injective; we show that it is surjective. Let $\mathscr{I}$ be an ideal of K^X and Y the set of $x \in X$ such that $f(x) = 0$ for all $f \in \mathscr{I}$. For all $x \notin Y$, there exists $u_x \in \mathscr{I}$ such that $u_x(x) \neq 0$. Let e_x be the element of K^X defined by $e_x(x) = 1$ and $e_x(x') = 0$ for all $x' \neq x$. Then

$$e_x = \frac{e_x}{u_x(x)} u_x \in \mathscr{I}.$$

Any $f \in \mathscr{I}_Y$ can be written

$$f = \sum_{x \in X} f(x) e_x = \sum_{x \in X - Y} f(x) e_x \in \mathscr{I}.$$

Hence $\mathscr{I}_Y \subset \mathscr{I}$; as $\mathscr{I} \subset \mathscr{I}_Y$, $\mathscr{I}_Y = \mathscr{I}$. $\qquad\square$

Corollary *The map* $x \mapsto \mathfrak{m}_x$ *is a bijection from X onto the set of maximal ideals of* K^X *and* $x \mapsto \delta_x$ *is a bijection from X onto* $\mathrm{Hom}_{K\text{-}\mathscr{alg}}(K^X, K)$.

Indeed, the first map is bijective because singletons are the minimal nonempty subsets. Clearly, $x \mapsto \delta_x$ is injective; let $f : K^X \to K$ be a homomorphism, $\mathrm{Ker} f$ a maximal ideal and so of the form $\mathfrak{m}_x$, the maps f and δ_x agree on $\mathfrak{m}_x$ and on $K \cdot 1$, hence on K^X.

5.3.2   The Gelfand Transform

Let A be a K-algebra, and $X = \text{Hom}_{K\text{-}\mathit{alg}}(A, K)$. For all $a \in A$, let $\widehat{a}$ be the map $\xi \mapsto \xi(a)$ from X to K. For $a \in A$, $b \in A$, $\lambda \in K$,

$$\widehat{a+b} = \widehat{a} + \widehat{b}, \quad \widehat{ab} = \widehat{a}\,\widehat{b}, \quad \widehat{\lambda a} = \lambda\widehat{a}, \quad \text{and} \quad \widehat{1} = 1 \, .$$

In other words the map $a \mapsto \widehat{a}$ from A to K^X is an algebra homomorphism. This homomorphism is called the **Gelfand transform** of A (see 2.6.7).

If A is a finite algebra, then by the Chinese remainder theorem, the Gelfand transform of A is surjective (5.1.5, Theorem and Remark).

Remark Associating to every element a of a set A a function $\widehat{a}$ on a set X of functions on A by $\widehat{a}(\xi) = \xi(a)$ often gives interesting mathematical results, for instance the Gelfand transform, biduality, Čech compactification, Dirac measures.

5.3.3

Proposition and Definition *Let* A *be a finite* K-*algebra of degree* d. *Set* $X = \text{Hom}_{K\text{-}\mathit{alg}}(A, K)$. *The following conditions are equivalent:*

   (i) A *is isomorphic to an algebra of functions on a finite set;*
   (ii) *the Gelfand transform* $\gamma : A \to K^X$ *is an algebra isomorphism;*
   (iii) *for all* $a \in A$, $a \neq 0$, *there exists* $\xi \in X$ *such that* $\xi(a) \neq 0$;
   (iv) $\text{card } X = d$;
   (v) $\text{card } X \geqslant d$.

If these conditions hold, then A *is* **diagonal.**

Proof (ii) $\Rightarrow$ (iv) and (i) is obvious.
   (i) $\Rightarrow$ (iii) since if $A = K^Y$, then for all $a \neq 0$ there exists $x \in Y$ such that $a(x) \neq 0$; take $\xi = \delta_x$.
   (iii) $\Rightarrow$ (v): γ is injective, proving (v).
   (v) $\Rightarrow$ (ii) and (iv) since γ is surjective by the Chinese remainder theorem.
   (iv) $\Rightarrow$ (v) is obvious. $\square$

5.3.4   Example. (The Fourier Transform for $\mathbb{Z}/(n)$.)

Let G be the group $\mathbb{Z}/(n)$. Equip the $\mathbb{C}$-vector space $\mathbb{C}^G$ with the convolution law defined on the canonical basis by

$$e_x * e_y = e_{x+y}, \text{ which gives } f * g = h \text{ with } h(x) = \sum_{y \in G} f(y)g(x - y)$$

for $f, g \in \mathbb{C}^G$. Let A denote the algebra $(\mathbb{C}^G, *)$. We show that A is diagonal.

For all maps φ from G to $\mathbb{C}$, there is a unique $\mathbb{C}$-linear map $\xi : A \to \mathbb{C}$ such that $\xi(e_x) = \varphi(x)$. It is an algebra homomorphism if and only if $\xi(e_0) = 1$ and $\xi(e_x * e_y) = \xi(e_x)\xi(e_y)$, i.e. $\varphi(0) = 1$ and $\varphi(x + y) = \varphi(x)\varphi(y)$ for $x, y \in G$. The set $X = \mathrm{Hom}_{\mathbb{C}\text{-}\mathit{alg}}(A, \mathbb{C})$ can thus be identified with $\mathrm{Hom}_{G_c}(G, \mathbb{C} - \{0\})$. For any n-th root of unity θ, the map $x \mapsto \theta^x$ is a group homomorphism from G to $\mathbb{C} - \{0\}$. This give n distinct homomorphisms from G to $\mathbb{C} - \{0\}$. Hence condition (v) of 5.3.3 holds and A is diagonal.

We express the Gelfand transform of A explicitly. For all $k \in G$, set $\theta_k = e^{\frac{2i\pi k}{n}}$ and let φ_k be the homomorphism $p \mapsto \theta_k{}^p$ from G to $\mathbb{C}^* - \{0\}$. The homomorphism ξ_k from A to $\mathbb{C}$ corresponding to φ_k is defined by

$$\xi_k(f) = \sum_{p \in G} f(p)\varphi_k(p) = \sum_{p \in G} f(p)e^{\frac{2i\pi kp}{n}} .$$

The set X contains $\xi_0, \ldots, \xi_{n-1}$, and so $X = \{\xi_0, \ldots, \xi_{n-1}\}$ since card $X = n$. Hence X can be identified with G. Thus for all $f \in A$,

$$\hat{f}(k) = \sum_{p \in G} f(p)e^{\frac{2i\pi kp}{n}} ,$$

and $f \mapsto \hat{f}$ is an isomorphism from the algebra $(\mathbb{C}^G, *)$ onto the algebra $(\mathbb{C}^G, \cdot)$, where $\cdot$ denotes function multiplication.

5.3.5

Proposition *(a) All subalgebras of a diagonal algebra are diagonal.*

(b) All quotient algebras of a diagonal algebra are diagonal.

(c) The product algebra of two diagonal algebras is diagonal.

(d) The tensor product algebra of two diagonal algebras is diagonal.

(e) Let A be an algebra, B and C diagonal subalgebras of A. Then the subalgebra B · C of A generated by B and C is diagonal.

Proof (a) Condition (iii) of (5.3.3) holds for subalgebras. (b) follows from (5.3.1), (c) and (d) from $K^X \times K^Y = K^{X \cup Y}$ and $K^X \otimes K^Y = K^{X \times Y}$. Finally (e) follows as B · C is isomorphic to a quotient of $B \otimes C$. $\square$

5.3.6

Proposition *Let $\mathscr{Setf}$ be the category of finite sets, and $\mathscr{D}$ the category of diagonal algebras over K. Then $\mathscr{D}$ is anti-equivalent to $\mathscr{Setf}$.*

Proof Let F be the contravariant functor $X \mapsto K^X$ from $\mathscr{Setf}$ to $\mathscr{D}$ and G the contravariant functor $A \mapsto \mathrm{Hom}_{K\text{-}\mathscr{alg}}(A, K)$ from $\mathscr{D}$ to $\mathscr{Setf}$. For all $A \in \mathscr{D}$, the Gelfand transform $\gamma_A : A \to K^{G(A)} = F(G(A))$ is an isomorphism, and these γ_A form an isomorphism from the identity functor $1_{\mathscr{D}}$ to $F \circ G$. For all finite sets X, the map $\beta_X : x \mapsto \delta_x$ from X to $\mathrm{Hom}_{K\text{-}\mathscr{alg}}(K^X, K) = G(F(X))$ is bijective, and these β_X form an isomorphism from the identity functor of $\mathscr{Setf}$ to $G \circ F$. The functors F and G are therefore anti-equivalent.   □

ADDENDA. Let X and Y be objects of $\mathscr{Setf}$ and f a map from X to Y. The homomorphism $f^* : K^Y \to K^X$ is surjective (resp. injective) if and only if f is injective (resp. surjective). The subsets (resp. quotient sets) of X are in bijective correspondence with the quotient algebras (resp. with subalgebras) of K^X.

In particular a diagonal algebra only has finitely many subalgebras.

Exercises 5.3. (Diagonal algebras)

1.—A *Boolean ring* is a ring A all of whose elements are idempotents (i.e. $x^2 = x$ for all x).

(a) Show that all finite Boolean rings are $\mathbb{Z}/(2)$-diagonal algebras.

(b) Let X be a set. Show that $\mathfrak{P}(X)$, equipped with the laws $\triangle$ and $\cap$, where

$$Y \triangle Z = (Y \cup Z) - (Y \cap Z),$$

is a Boolean ring. Show that every Boolean ring is embedded in a ring of this type, and that every finite Boolean ring is of this type with X finite.

2.—(a) Show that all reduced algebras over an algebraically closed field are diagonal.

(b) Show that an algebra over $\mathbb{F}_{p^r}$, where $x^{p^r} = x$ for all x, is diagonal.

3.—Let G be a finite commutative group, and A the convolution $\mathbb{C}^G$.

(a) Show that the algebra A is diagonal.

(b) Set $\widehat{G} = \mathrm{Hom}_{\mathscr{alg}}(A, \mathbb{C})$. Show that $\widehat{G}$ can be identified with $\mathrm{Hom}_{\mathscr{Gr}}(G, \mathbb{C}^*)$ or with $\mathrm{Hom}_{\mathscr{Gr}}(G, \mathbb{Q}/\mathbb{Z})$.

(c) The diagonal map $G \to G \times G$ defines a homomorphism $\Delta : A \to A \otimes A$, giving a composition law on $\widehat{G}$. Show that this law agrees with the law given by the identification with $\mathrm{Hom}_{\mathscr{Gr}}(G, \mathbb{C}^*)$.

(d) Show that $\widehat{\widehat{G}}$ can be identified with G.

(e) Show that, in the previous questions, $\mathbb{C}$ can be replaced by an algebraically closed field of characteristic p as long as p does not divide the order of G. For what values of r can $\mathbb{F}_{p^r}$ be taken?

(f) For $f, g \in \mathbb{C}^G$, $\widehat{(f \cdot g)} = \frac{1}{n}(\hat{f} * \hat{g})$, where n is the order of G. Is there a $\mathbb{C}$-linear bijection $f \mapsto f'$ from $\mathbb{C}^G$ onto some $\mathbb{C}^{G'}$, where G′ is a group such that $(f * g)' = f' \cdot g'$ and $(f \cdot g)' = f' * g'$ for $f, g \in \mathbb{C}^G$?

5.4   Etale Algebras

5.4.1

Definition Let A be a finite K-algebra. An extension L of K is said to **diagonalize** A if $L \otimes_K A$ is a diagonal L-algebra.

L diagonalizes A if and only if

$$\operatorname{card} \operatorname{Hom}_{K\text{-}\mathscr{alg}}(A, L) = \deg_K A .$$

5.4.2

Proposition *Let* $P \in K[X]$ *be a polynomial of degree d, and* L *an extension of* K. L *diagonalizes* $K[X]/(P)$, *if and only if* P *has d distinct roots in* L.

Proof The algebra $L \otimes_K K[X]/(P)$ is isomorphic to $L[X]/(P)$ and the degree of $L[X]/(P)$ over L is d. The map $\varphi \mapsto \varphi(x)$ is a bijection from $\operatorname{Hom}_{L\text{-}\mathscr{alg}}(L[X]/(P), L)$ onto the set of roots of P in L.

There are d distinct homomorphisms from $L[X]/(P)$ to L if and only if P has d distinct roots. $\qquad\square$

5.4.3

Proposition and Definition *Let* Ω *be an algebraic closure of* K *and* A *a finite* K-algebra. *The following conditions are equivalent:*

(i)   *there is an extension* L *of* K *diagonalizing* A;
(ii) *there is a finite extension* L *of* K *diagonalizing* A;
(iii) Ω *diagonalizes* A.

If these conditions hold, then A *is said to be an* **etale** *algebra.*

Proof (ii) $\Rightarrow$ (iii). Indeed, L embeds in Ω and $\Omega \otimes_K A = \Omega \otimes_L (L \otimes_K A)$.

(iii) $\Rightarrow$ (i) is obvious.

(i) $\Rightarrow$ (ii). Let d be the degree of A on K. There are d distinct homomorphisms $\xi_1, \ldots, \xi_d$ from A to L. The subalgebra L' of L generated by $\xi_1(A), \ldots, \xi_d(A)$ is a finite extension of K contained in L, and $\xi_1, \ldots, \xi_d$ are d homomorphisms from A to L'. So L' diagonalizes A. $\square$

5.4.4

Proposition and Definition *Let* $P \in K[X]$. *The following conditions are equivalent:*

  (i) *the algebra* $K[X]/(P)$ *is etale;*
 (ii) *the roots of* P *in* Ω *are distinct;*
(iii) g.c.d.$(P, P') = 1$;
 (iv) *The discriminant of* P *is nonzero.*

If these conditions hold, then the polynomial P *called* **separable.**

(i) $\Leftrightarrow$ (ii) is an immediate consequence of (5.4.2).

(ii) $\Leftrightarrow$ (iii) follows from the characterization of simple roots.

(iii) $\Leftrightarrow$ (iv) follows from (3.7.12) and (3.7.11, Proposition).

Let A be a K-algebra. An algebraic element $x \in A$ over K is said to be **separable** if its minimal polynomial is separable, in other words if the subalgebra $K[x]_A$ generated by x is etale. If A is algebraic over K, then A is said to be **separable** if all elements of A are separable.

5.4.5

Proposition *(a) All subalgebras of an etale algebra are etale.*

(b) All quotient algebras of an etale algebra are etale.

(c) The product algebra of two etale algebras is etale.

(d) The tensor product algebra of two etale algebras is etale.

(e) Let A *be an algebra,* B *and* C *etale subalgebras of* A, *then the subalgebra* $B \cdot C$ *of* A *generated by* B *and* C *is etale.*

Proof Let L be an extension of K. By (5.3.5), if L diagonalizes an algebra A, then L diagonalizes every subalgebra of A, and every quotient algebra of A. If L diagonalizes algebras B and C, then it diagonalizes the product algebra $B \times C$ and the tensor product algebra $B \otimes C$.

If L diagonalizes the subalgebras B and C of an algebra A, then it also diagonalizes $B \cdot C$. $\square$

Corollary 5.5 *An algebra is etale if and only if it is finite and separable.*

Corollary 5.2 and Definition *Let* A *be a* K-*algebra. The algebraically separable elements over* K *of* A *form a subalgebra of* A, *written* A_{sep} *and called the* **separable closure** *of* K *in* A.

5.4.6

Proposition *Let* L *be an etale extension of* K. *An etale algebra* A *over* L *is etale over* K.

Proof Let Ω be an algebraic closure of L. Then $\Omega \otimes_K L \approx \Omega^d$ and $\Omega \otimes_L A \approx \Omega^e$ with $d \in \mathbb{N}^*$ and $e \in \mathbb{N}$, and so

$$\Omega \otimes_K A \approx (\Omega \otimes_K L) \otimes_L A \approx \Omega^d \otimes_L A \approx (\Omega \otimes_L A)^d \approx \Omega^{ed}. \qquad \square$$

5.4.7

Let A be a finite cyclic extension of K, then there is an irreducible polynomial P such that $A = K[X]/(P)$.

Proposition *The following conditions are equivalent:*

(i) A *is not etale;*
(ii) $P' = 0$;
(iii) K *has characteristic* $p \neq 0$ *and* $P \in K[X^p]$.

Proof (ii) $\Leftrightarrow$ (i): The g.c.d. of (P, P') is a divisor of P and so is 1 or P. If $P' \neq 0$, then $\deg P' < \deg P$. Hence P does not divide P' and g.c.d.$(P, P') = 1$.

If $P' = 0$, g.c.d.$(P, P') = P \neq 1$ since $A \neq 0$, and (i) $\Leftrightarrow$ (ii) thanks to (5.4.4).

(ii) $\Leftrightarrow$ (iii). If the characteristic of K is 0 then the derivative of a polynomial of degree $d > 0$ is a nonzero polynomial of degree $d - 1$.

If the characteristic of K is $p \neq 0$ then, for $P = \sum a_k X^k$ $P' = \sum k a_k X^{k-1}$. Hence $P' = 0$ if and only if $a_k = 0$ for all k that is not a multiple of p, in other words if $P \in K[X^p]$, $\qquad \square$

5.4.8

Theorem and Definition *Let* p *be the characteristic of* K. *The following conditions are equivalent:*

(i) Every finite extension of K *is etale.*

(ii) Either $p = 0$, *or* $p \neq 0$ *and the Frobenius endomorphism of* K *is surjective.*

If these conditions hold, then K *is said to be a* **perfect** *field.*

Proof Condition (i) is equivalent to

(i$'$) *Every finite cyclic extension of* K *is etale.*

Indeed (i) $\Rightarrow$ (i$'$) is obvious, and (i$'$) $\Rightarrow$ (i) thanks to (5.4.5, (e) since every finite extension is generated by cyclic extensions.

(ii) $\Rightarrow$ (i$'$). Let E $=$ K[X]/(P) be a finite cyclic extension. Since P is irreducible, if K is of characteristic 0 then E is etale (5.4.7, (iii) $\Rightarrow$ (i)).

Suppose that K is of characteristic $p \neq 0$. Then P $\notin$ K[X^p]. Indeed, if P $\in$ K[X^p], then P $= \sum a_k X^{pk}$, where for all k, $a_k = b_k^p$ and $b_k \in$ K, and so P $=$ Q^p for some Q $= \sum b_k X^k$. Hence P is reducible.

(i$'$) $\Rightarrow$ (ii). Suppose that $p \neq 0$ and that F is not surjective. Let $a \in$ K, $a \notin$ F(K), and Ω be an algebraic closure of K and $\alpha \in \Omega$ such that $\alpha^p = a$. Then $X^p - a = (X - \alpha)^p$. The minimal polynomial P of α is monic and divides $X^p - a$, and so P $= (X - \alpha)^d$, where $1 < d \leqslant p$, since $\alpha \notin$ K. Therefore α is a multiple root of P and the algebra K[X]/(P) is a non-etale finite cyclic extension (in fact, $d = p$ necessarily holds). $\qquad\square$

Example Every finite field is perfect. Indeed, as F : K $\to$ K is injective, it is necessarily surjective if K is finite.

Counterexample If the characteristic of K is $p \neq 0$, then the field K(X) is not perfect. Indeed, there is no $f \in$ K(X) such that $f^p = X$.

5.4.9

Proposition *Let* A *be a finite* K-*algebra.*

(a) If K *is perfect, then* A *is etale if and only if* A *is reduced.*

(b) Let A *be generated by* $x_1, \ldots, x_r$ *with respectively* P$_1$, ..., P$_r$ *as minimal polynomials. Then* A *is etale if and only if* ($\forall i$) g.c.d.(P$_i$, P$_i'$) $= 1$.

(c) With the notation of (b), *if* K *is of characteristic* $p \neq 0$ *and moreover* A *is an extension, then* A *is etale if and only if* ($\forall i$) P$_i \notin$ K[X^p].

Proof (a) Let L be an extension of K diagonalizing A; then A $=$ K $\otimes_K$ A embeds in L $\otimes_K$ A $=$ L^d for some d. The algebra L^d being a product of fields is reduced and hence so is A.

Conversely, A being finite and reduced is a finite product of finite extensions over a perfect field, and so is etale (5.4.5, c).

(b) and (c): A is etale if and only if ($\forall i$) K[X]/(P$_i$) is etale (5.4.5, a, e), in other words, if and only if ($\forall i$) g.c.d.(P$_i$, P$_i'$) $= 1$ (5.4.4).

If A is an extension, all P_i are irreducible, $K[X]/(P_i)$ are finite cyclic extensions, and by (5.4.7) A is etale if and only if $(\forall i)\ P_i \notin K[X]^p$. □

Corollary *The tensor product of two reduced finite algebras over a perfect field is a reduced finite algebra. In particular, the tensor product of two finite extensions is a reduced algebra.*

Proof This corollary follows from (5.4.9, a) and (5.4.5, d).

Remark (1) In general, the tensor product of two extensions is not an extension. For example $\mathbb{C} \otimes_{\mathbb{R}} \mathbb{C} \approx \mathbb{C} \times \mathbb{C}$.

(2) The previous corollary has a converse. Over any non-perfect field K, there are finite extensions A and B such that $A \otimes_K B$ is not reduced (5.4, Exercise 3).

5.4.10 Example

Let K be a field of characteristic p, G a finite commutative group of order n indivisible by p. Then K^G equipped with convolution is an etale algebra (Maschke's theorem).

Indeed, for $x \in G$, let $e_x \in K^G$ be defined by $e_x(y) = 0$ if $x \neq y$ and $e_x(x) = 1$. If x has order k, then e_x is a root of $P(X) = X^k - 1$. As p does not divide n, neither does it divide k; k is invertible in K and the ideal of $K[X]$ generated by $(X^k - 1)$ and kX^{k-1} is the whole ring, and so $K[X]/(X^k - 1)$ is etale. Since the elements e_x generate K^G, the algebra $(K^G, *)$ must be etale by (5.4.9, b).

5.4.11 Primitive Element Theorem

Theorem *If* K *is infinite, every etale algebra over* K *is cyclic.*

Lemma 5.1 *An etale algebra has only finitely many subalgebras.*

Let A be an etale algebra and L an extension of K such that $L \otimes_K A$ is diagonal. For any subalgebra B of A, $L \otimes_K B$ is a subalgebra of $L \otimes_K A$ and $B = A \cap (L \otimes_K B)$. So $B \mapsto L \otimes_K B$ is an injection from the set of subalgebras of A to the set of subalgebras of $L \otimes_K A$, which is finite. □

Lemma 5.2 *Let* K *be an infinite field,* E *a finite dimensional vector space over* K *and let* F_i *be finitely many strict subspaces. Then* $\bigcup F_i \neq E$.

Indeed, each F_i can be embedded in a hyperplane H_i defined by an equation $h_i = 0$. We may assume that $E = K^n$. Every $h_i \in K[x_1, \ldots, x_n]$ is a homogeneous polynomial of degree 1. So $h = \prod h_i$ is $\neq 0$. As K is infinite, the polynomial function $K^n \to K$ represented by h is nonzero. There exists $x \in K^n$ such that $h(x) \neq 0$ and hence $x \notin \prod H_i$. □

Proof of the Theorem Let A be an etale K-algebra. It is finite and has finitely many strict subalgebras A_i (Lemma 1). So $\bigcup A_i \neq A$. Let $x \in A - \bigcup A_i$. The subalgebra generated by x is not strict, hence must be A. $\qquad\square$

Definition A generator of A is called a **primitive element** of A.

Corollary (Primitive element Theorem) *All etale extensions* L *of* K *are cyclic.*

If K is infinite, this amounts to the theorem.

If K is finite, then L^* is cyclic and any of its generators generates L.

Exercises 5.4. (Etale algebras)

1.—Let E be an algebraic extension of K and Ω an algebraic closure of K. Show that the following conditions are equivalent:

(i) E is separable;
(ii) All finite sub-extensions of E are etale;
(iii) All Ω-algebra homomorphisms from $\Omega \otimes_K E$ to Ω separate the elements of $\Omega \otimes_K E$;
(iv) For all finite dimensional vector subspaces V of E, $\text{Hom}_K(V, \Omega)$ is generated as a vector space over Ω by $f|_V$, $f \in \text{Hom}_{K\text{-}alg}(E, \Omega)$.

2.—(a) Let A and B be K-algebras and $f : A \to B$ a surjective homomorphism such that $I = \text{Ker} f$ is a nilpotent ideal. Let $P \in K[X]$ be a separable polynomial and $y \in B$ such that $P(y) = 0$. Show that there is a unique $x \in A$ such that $f(x) = y$ and $P(x) = 0$ (x may be constructed through successive approximations by noting that for all $x' \in f^{-1}(y)$, $P'(x')$ is invertible in A).

Deduce that f induces a bijection from A_{sep} onto B_{sep}.

(b) Let A and B be finite K-algebras and $f : A \to B$ a surjective homomorphism. Show that f induces a surjective homomorphism

$$f_* : A_{\text{sep}} \to B_{\text{sep}}$$

(see [1], Exercise 7).

(c) Show that if K is perfect then, for all K-algebras A, $A = A_{\text{sep}} \oplus \mathfrak{N}$, where $\mathfrak{N}$ is the nilradical of A.

3.—Let K be a non-perfect field of characteristic p, and $a \in K$ an element without any p-th root in K. Set $L = K[X]/(X^p - a)$. Show that L is an extension of K, but that the algebra $L \otimes_K L$ is not reduced.

4.—Let k be an infinite field of characteristic p; set $K = k(X, Y)$ and $L = k(X^{1/p}, Y^{1/p})$ (or, if preferred, $L = k(X', Y')$ and $K = k(X'^p, Y'^p)$).

(a) Show that L is a non cyclic extension of K of degree p^2.

(b) For all $a \in k$, let M_a be the subfield of L generated by K and $(Y - aX)^{1/p}$. Show that these subfields are mutually distinct.

5.—Let K be a field of characteristic $p \neq 0$ and $G = \mathbb{Z}/(p^r)$. Show that the algebra of the group G is isomorphic to $K[T]/(T^{p^r})$.

6.—Let k be a field, and A $= k[X]$, B $= k[X, Y]/(P)$, where P $\in k[X, Y]$ is a polynomial of degree d in Y; set K $= k(X)$ and E $= K \otimes_A B$.

Show that E is etale over K if and only if there exists x in an algebraic closure Ω of k such that $P(x, Y) \in \Omega[Y]$ has degree $\geqslant d - 1$ and distinct roots in Ω.

5.5 Purely Inseparable Extensions

In this section, K is assumed to be of characteristic $p \neq 0$.

5.5.1

Proposition *Let Ω be an algebraic closure of* K *and* P $\in$ K[X] *an irreducible monic polynomial. The following conditions are equivalent:*

(i) the polynomial P *has only one (possibly multiple) root in Ω;*
(ii) there exists $r \in \mathbb{N}$ and $a \in$ K such that P $= X^{p^r} - a$.

Proof (ii) $\Rightarrow$ (i). The Frobenius endomorphism F $: x \mapsto x^p$ of Ω is injective since Ω is a field. Hence the same holds for F^r.

(i) $\Rightarrow$ (ii). Let r be the largest integer for which $P(X)$ can be written as $Q(X^{p^r})$. Then $Q(Y) \notin K[Y^p]$, for otherwise $P(X)$ could be written as $Q_1(X^{p^{r+1}})$. Moreover Q is irreducible since so is P. By (5.4.7) and (5.4.4), the roots of Q in Ω are distinct; as P has only one root, so does Q and thus Q is of degree 1.   □

5.5.2

Definition Let L be an extension of K. An element $x \in$ L is said to be **radical** over K if there is an integer r such that $x^{p^r} \in$ K. If all elements of L are radical over K, L is a **purely inseparable extension** (or radical) of K.

5.5.3

Proposition *Let* L *be an algebraic extension of* K *and Ω an algebraic closure of* K. *The following conditions are equivalent:*

(i) L *is purely inseparable;*
(ii) there is no K*-algebra homomorphism from* L *to* Ω.

Proof (i) $\Rightarrow$ (ii). Let $x \in L$. There exist $r \in \mathbb{N}$ and $a \in K$ such that $x^{p^r} = a$. If $f : L \to \Omega$ is a homomorphism, then $f(x)$ is necessarily the unique root of $X^{p^r} - a$ in Ω.

not (i) $\Rightarrow$ not (ii). Suppose $x \in L$ is not radical. The minimal polynomial of x has at least two distinct roots in Ω, and thus there are two distinct homomorphisms from $K[x]_L$ to Ω. Embedding in Ω, which is possible by (5.2.6), gives two distinct homomorphisms from L to Ω. $\square$

5.5.4

Proposition *Every algebraic extension* L *of* K *is a purely inseparable extension of* L_{sep}.

Proof Let $x \in L$, P its minimal polynomial, r the largest integer for which $P(X)$ can be written as $Q(X^{p^r})$. Then $Q \notin K[X^p]$ and Q is irreducible, so that Q is separable, and $x^{p^r} \in L_{sep}$ since its minimal polynomial is separable. $\square$

Remark Let L be an extension of K. Then the set L_{rad} of the radicals over K in L is a sub-extension of L. Indeed L_{rad} is the increasing union of $(F^r)^{-1}(K)$, and so is a subfield, and $L_{rad} \supset K$. In general, even when the extension L is algebraic, it is not separable over L_{rad}. In other words, an algebraic extension of K is always a purely inseparable extension of a separable extension of K, but not necessarily a separable extension of a purely inseparable extension of K (5.5, Exercise 3).

However, if Ω is an algebraic closure of K, then Ω is a separable extension of Ω_{rad} since Ω_{rad} is a perfect field.

5.5.5

Theorem (MacLane's criterion) *Let* A *be an algebraic* K-*algebra over* K, *and* Ω *an algebraic closure of* K.

(a) The following conditions are equivalent:

(i) A *is separable;*
(ii) $\Omega_{rad} \otimes_K A$ *is reduced;*
(iii) $\Omega \otimes_K A$ *is reduced.*

(b) A *is a separable extension of* K *if and only if* $\Omega_{rad} \otimes_K A$ *is an extension of* K.

Proof All the properties considered continue to hold for the direct limit. Hence A may be assumed to be finite over K.

(a) (i) $\Rightarrow$ A is etale $\Rightarrow$ $\Omega \otimes_K A$ is diagonal $\Rightarrow$ (iii) $\Rightarrow$ (ii). We show that (ii) $\Rightarrow$ (i): if (ii) holds, then A is reduced and hence is a product of extensions. So it suffices to prove that these extensions are separable. We do so by showing that if L is a not a separable extension, then $\Omega_{rad} \otimes_K L$ is not reduced: let $x \in L - L_{sep}$ and P its minimal polynomial. Then $P(X) = Q(X^p)$ for some polynomial Q; let $Q_1 \in \Omega_{rad}[X]$ be the polynomial all of whose coefficients are the roots of the p-th coefficients of Q. Then $Q_1^p = P$, and so $(Q_1(x))^p = 0$. But $Q_1(x) \neq 0$. Indeed, $K[x]_L = K[X]/(P)$; hence the subalgebra $\Omega_{rad}[x]$ of $\Omega_{rad} \otimes_K L$ can be identified with $\Omega_{rad}[X]/(P)$, and the minimal polynomial of x in $\Omega_{rad} \otimes_K L$ is P. However, Q_1 is not a multiple of P since its degree is strictly smaller. Therefore $\Omega_{rad} \otimes_K L$ is not reduced.

(b) If $\Omega_{rad} \otimes_K A$ is an extension of K, then the algebra A is an extension of K; it is separable by (a). Assume that A is a separable extension. Then $\Omega_{rad} \otimes_K A$ is a finite reduced algebra over Ω_{rad} and can be written $\prod_{i \in I} E_i$, where all E_i are extensions of Ω_{rad} (5.1.6, Corollary 1). These E_i can also be considered extensions of A. Embed A in Ω. By (5.5.3), the set $\mathrm{Hom}_{A\text{-}\mathcal{alg}}(\Omega_{rad} \otimes_K A, \Omega) = \mathrm{Hom}_{K\text{-}\mathcal{alg}}(\Omega_{rad}, \Omega)$ consists of a single element. If there were distinct elements i and j in I, projecting respectively the product onto E_i and E_j and embedding E_i and E_j in Ω would give two distinct A-homomorphisms from $\Omega_{rad} \otimes_K A$ to Ω. Hence, I has a unique element, and $\Omega_{rad} \otimes_K A$ is an extension of K. $\qquad\square$

5.5.6 Characteristic 0 Conventions

If K is a field of characteristic 0, the identity map of K is called the *Frobenius endomorphism*. Un element of an extension of K belonging to K is said to be *radical*. Only the trivial extension is *purely inseparable*. With these conventions, the results of 5.5.3, 5.5.4 and 5.5.5 hold in characteristic 0.

Exercises 5.5. (Purely inseparable extensions)
1.—(a) Let Ω and Ω' be two algebraic closures of K. Show that there is a unique isomorphism from Ω_{rad} to Ω'_{rad}. (Thus Ω_{rad} and Ω'_{rad} can be identified, and *the* purely inseparable closure of K is well-defined.)

(b) Show that every purely inseparable and perfect extension of K can be uniquely identified with Ω_{rad}.

(c) Consider the direct system

$$K \xrightarrow{F} K \xrightarrow{F} K \xrightarrow{F} \cdots \xrightarrow{F} K \xrightarrow{F} \cdots$$

indexed by $\mathbb{N}$, where F is the Frobenius endomorphism. Let E be the direct limit of this system. Show that, equipped with $f_0^\infty : K \to E$, E is an extension of K isomorphic to Ω_{rad}.

(d) State a functorial property of the purely inseparable closure.

2.—(a) Show that if K is perfect, then every finite extension of K is a perfect field $(\deg_K(F(L)) = \deg_{F(K)}(F(L)) = \deg_K(L))$. Deduce that all algebraic extensions of K are perfect fields.

(b) Let K be a field and Ω an algebraic closure of K. Show that $\Omega = \Omega_{rad} \otimes_K \Omega_{sep}$ (the term on the right can be identified with a perfect sub-extension of Ω containing Ω_{sep}).

3.—Let k be a field of characteristic $p \neq 0, 2$. Consider the subfields $M = k(X, Y^p)$, $L = k(X^p, Y^p)$ and

$$K = \{f \in L \mid f(Y, X) = f(X, Y)\}$$

in $E = k(X, Y)$. (a) Show that L is a separable extension of K and that M is a purely inseparable extension of L. Determine M_{sep} relative to K.

(b) Let $f \in M$ be such that $f(Y, X) = f(X, Y)$. Show that $f \in K$ (reduce to the case $f \in k[X, Y]$).

(c) Show that the purely inseparable closure of K in M is K. Deduce that M is not a separable extension of a purely inseparable extension of K.

4.—Let L an algebraic extension of the field K. Consider the canonical homomorphism $\varphi : L_{rad} \otimes_K L_{sep} \to L$ inducing the canonical injection on L_{rad} and L_{sep}.

(a) Show that φ is injective (see 5.5.5).

(b) Show that the image of φ is the separable closure M of L_{rad} in L (reduce to the case of finite L and prove the equalities

$$(M : L_{rad}) = \text{card Hom}_{L_{rad}\text{-}\mathscr{alg}}(M, \Omega) = \text{card Hom}_{K\text{-}\mathscr{alg}}(M, \Omega)$$
$$= \text{card Hom}_{K\text{-}\mathscr{alg}}(L_{sep}, \Omega) = (L_{sep} : K) = (L_{rad} \otimes_K L_{sep} : L_{rad})).$$

In particular $L = L_{rad} \otimes_K L_{sep}$ if and only if L is a separable extension of L_{rad}.

(c) Let Ω be an algebraic closure of K, $x \in \Omega$, P its minimal polynomial, and L the sub-extension of Ω generated by the roots of P. Show that $L = L_{rad} \otimes_K L_{sep}$ (consider the symmetric functions of the roots of P, each counted only once).

5.—Let A be a K-algebra and B a subalgebra of A. We will say that A is **radical** over B if $\Omega \otimes_K A$ is generated by $\Omega \otimes_K B$ and by nilpotent elements.[1]

(a) Show that for an extension A of K, the definition is equivalent to that of (5.5.2).

(b) Show that every algebraic algebra A over K is radical over A_{sep} (use 5.4, Exercise 2).

(c) Generalize to the algebras of Proposition 5.5.3.

---

[1] This convention is not usual.

6.—Let A be a finite K-algebra.

(a) Let Ω be an algebraic closure of K. Show that all K-algebra homomorphisms from A_{sep} to Ω extend uniquely to A. Deduce that the cardinality of $\mathrm{Hom}_{K\text{-}\mathcal{A}lg}(A, \Omega)$ is equal to $\deg_K(A_{sep})$.

(b) Let L be an extension of K. Show that the separable closure of L in $L \otimes_K A$ is $L \otimes_K A_{sep}$.

(c) If A and B are two finite K-algebras, show that

$$(A \otimes_K B)_{sep} = A_{sep} \otimes_K B_{sep} .$$

Extend this result to algebraic K-algebras.

7.—Let Ω be an algebraic closure of K and $\Omega' \subset \Omega$ a sub-extension. Assume that all $P \in K[X]$ have at least one root in Ω'.

(a) Show that every finite separable extension K embeds in Ω' (use the primitive element theorem). Show that $\Omega'_{sep} = \Omega_{sep}$.

(b) Prove that, for all n, every separable polynomial in K[X] has at least one root in $\Omega'_n = F^n(\Omega') \cap \Omega_{sep}$. Show that $\Omega' = \Omega$.

(c) Show that in 5.2, Exercise 4, the family (A_i) of extensions K can be taken to be of the form $K[X]/(P)$, where P is an irreducible polynomial.

5.6 Finite Galois Extensions

5.6.1

Definition A finite extension L of K is said to be **Galois** if L diagonalizes itself (5.4.1).

If L is a Galois extension of K, the group $G = \mathrm{Aut}_K L$ is called the **Galois group** of L over K.

5.6.2

Proposition *Let* L *be a Galois extension of* K *and* A *a sub-extension of* L. *Then* L *diagonalizes* A *over* K, *and as an* A-*algebra,* L *is a Galois extension of* A.

Proof $L \otimes_K A$ is a subalgebra of $L \otimes_K L$, and $L \otimes_A L$ is a quotient L-algebra of $L \otimes_K L$ since $(\lambda, \mu) \mapsto \lambda \otimes_A \mu$ is A-bilinear and so necessarily K-bilinear, and thus defines a map $\epsilon : \lambda \otimes_K \mu \mapsto \lambda \otimes_A \mu$ whose image contains the simple tensors; hence

ϵ is surjective. Besides, ϵ is a L-algebra homomorphism, and so $L \otimes_K A$ and $L \otimes_K L$ are diagonal L-algebras.   □

Remark A Galois extension of a Galois extension is not necessarily Galois (5.6, Exercise 3).

5.6.3

Let L be a finite extension of K and Ω an algebraic closure of K containing L. Set

$$d = \deg_K L, \ G = \text{Aut}_K L, \ \text{Fix}_G L = \{x \in L \mid (\forall g \in G) \ g(x) = x\}.$$

Theorem *The following conditions are equivalent:*

(i) L is Galois;
(ii) card G = d;
(ii') card G $\geqslant$ d;
(iii) L is etale and every automorphism of Ω preserves L;
(iv) Fix$_G$L = K;
(v) for all $x \in L$, the roots of the minimal polynomial of x in Ω are simple and belong to L.

Proof By 5.3.3, (i) $\Leftrightarrow$ (ii) $\Leftrightarrow$ (ii') since

$$G = \text{Aut}_K L = \text{End}_K L = \text{Hom}_L (L \otimes_K L, L).$$

(i) $\Rightarrow$ (iii). As L is etale, card $\text{Hom}_{K\text{-}\mathpzc{alg}}(L, \Omega) = d$. As $G \subset \text{Hom}_{K\text{-}\mathpzc{alg}}(L, \Omega)$ and (i) $\Rightarrow$ (ii), $G = \text{Hom}_{K\text{-}\mathpzc{alg}}(L, \Omega)$, proving (iii).

(iii) $\Rightarrow$ (ii). As L is etale, there are d distinct homomorphisms from L to Ω. By (5.2.6) and (5.2.2), these homomorphisms are induced by automorphisms of Ω and so are automorphisms of L, proving (ii).

(iii) $\Rightarrow$ (iv). All points of K are fixed under the action of G. Assume $K \neq \text{Fix}_G L$; let $x \in \text{Fix}_G L - K$ and P its minimal polynomial. As $x \notin K$, $d°P \geqslant 2$; and as L is etale, x is a simple root of P. Let y be another root of P in Ω. The homomorphism $\varphi : K[x] \to \Omega$ defined by $\varphi(x) = y$ extends to an automorphism of Ω which induces an automorphism of L not fixing x, giving a contradiction.

(iv) $\Rightarrow$ (v). Let $\alpha \in L$ and $(a_i)_{i \in I}$ such that $i \mapsto a_i$ is a bijection from I onto the set of conjugates of α in L. Set $P(X) = \prod(X - a_i)$. For $g \in G$, define

$$g_* : L[X] \to L[X] \text{ by } g_*(\lambda) = g(\lambda) \text{ for } \lambda \in L \text{ and } g_*(X) = X.$$

Then

$$g_*(P) = \prod(X - g(a_i)) = \prod(X - a_i)$$

since g induces a permutation of these a_i. Hence the coefficients of P are invariant under G, and P $\in$ K[X]. The roots of P are simple and in L. Hence the minimal polynomial of α, which divides P, has the same property (in fact this polynomial may be shown to be P).

(v) $\Rightarrow$ (iii). Clearly, L is etale. For all automorphisms φ of Ω and all $x \in$ L, $\varphi(x)$ is a root of the minimal polynomial of x and so $\varphi(x) \in$ L by assumption. Hence φ preserves L. $\square$

5.6.4

Corollary 5.6 *Let* L *and* L' *be Galois extensions of* K *contained in* Ω. *Then the product extension* L $\cdot$ L' *is Galois.*

This follows from (i) $\Leftrightarrow$ (iii) in Theorem 5.6.3.

Corollary 5.7 *Let* L *and* L' *be Galois extensions of* K *with* K $\subset$ L' $\subset$ L $\subset$ Ω. *Set* G $=$ Aut$_K$L, G' $=$ Aut$_K$L', *and* H $=$ Aut$_{L'}$L. *Then* H *is a normal subgroup of* G *and* G' *identified with* G/H.

Proof Every element of G leaves L' invariant, and so there is a restriction homomorphism $\rho :$ G $\rightarrow$ G'. Every element of G' extends to an automorphism of Ω, leaving L invariant. Hence ρ is surjective. The kernel of ρ is obviously H. $\square$

5.6.5 Decomposition Field

Let P $\in$ K[X] be a separable polynomial (5.4.4). The K-sub-extension E of Ω generated by the roots of P is Galois (Theorem 5.6.3, (iii) $\Rightarrow$ (i)). It is called the *decomposition field* of P.

Proposition *With this notation, the group* G $=$ Aut$_K$E *acts transitively on the set* X *of roots of* P *in* Ω *if and only if* P *is irreducible.*

Proof Suppose that P is not irreducible. Then P $=$ P$_1 \cdot$ P$_2$ for some P$_1$ and P$_2$ of degree > 0. Then X $=$ X$_1 \cup$ X$_2$, where X$_i =$ P$_i^{-1}(0)$. These two sets are disjoint since the roots of P are simple, and they are invariant under G. Hence G does not act transitively on X.

Conversely, let α and β be two roots of P. If P is irreducible, it is the minimal polynomial of α as well as of β. Hence there is an automorphism σ from K[α]$_\Omega$ onto K[β]$_\Omega$ such that $\sigma(\alpha) = \beta$. Then σ extends to an automorphism of Ω inducing an automorphism of E. $\square$

5.6.6

Proposition *Let* A *be an etale* K*-algebra. There is a finite Galois extension diagonalizing* A.

Proof Let Ω be an algebraic closure of K and $x_1, \ldots, x_r$ generators of A. For each i, let P_i be the minimal polynomial of x_i and L_i the sub-extension of Ω generated by the roots of P_i. For each i, L_i diagonalizes the subalgebra A_i of A generated by x_i, and the sub-extension L of Ω generated by all L_i diagonalizes all A_i, and hence diagonalizes A. The roots of P_i are simple since A is etale, and so L is etale; every automorphism of Ω permutes the roots of P_i for each i, and so preserves L, and L is Galois. $\square$

Corollary *Let* Ω *be an algebraic closure of* K *and* L *an etale extension of* K *contained in* Ω. *Then there is a finite Galois extension of* K *in* Ω *and containing* L.

Proof Let $E \subset \Omega$ be a finite Galois extension diagonalizing L. Then $L \subset E$ since the image of any homomorphism, in particular the canonical injection, from L to Ω is in E.

5.6.7

Proposition *Let* L *be a field,* H *a finite group of automorphisms of* L, *and set* $F = \text{Fix}_H L$. *Then* L *is a finite Galois extension of* F *and* $\text{Aut}_F(L) = H$.

Proof (a) *Assume that* L *is finite over* F. Set $H' = \text{Aut}_F L$. Then $H' \supset H$, and $F \subset \text{Fix}_{H'} L \subset \text{Fix}_H L = F$; so $F = \text{Fix}_{H'} L$, and by (5.6.3, (iv) $\Rightarrow$ (i)), L is Galois over F. In particular, L is etale, and hence cyclic by the primitive element Theorem (5.4.11, Corollary). Let $a \in L$ such that $L = F[a]$. The polynomial

$$P = \prod_{g \in H} (X - g(a))$$

is in F[X] since its coefficients are fixed by H, and the minimal polynomial Q of a over F divides P (in fact $Q = P$). For all $f \in H'$, $f(a)$ is another root of Q, hence of P, and there exists $g \in H$ such that $f(a) = g(a)$. The automorphisms f of g agree on F and at a, and so $f = g$ and $H' = H$.

(b) *We now prove that* L *is finite over* F. First, L is algebraic over F. Indeed, for all $a \in L$, the polynomial $P = \prod_{g \in H} (X - g(a))$ is in F[X] and $P(a) = 0$. The extension L is the directed union of finite sub-extensions L_i over F, each preserved by H. For each i, let H_i be the image of H in $\text{Aut}_F L_i$. Then $\text{Fix}_{H_i} L_i = \text{Fix}_H L_i = F$, and so $\text{Aut}_F L_i = H_i$ by (a). As L_i is Galois, $\deg_F L_i = \text{card } H_i \leqslant \text{card } H$; hence finally $\deg_F L \leqslant \text{card } H < \infty$. $\square$

5.7 Finite Galois Theory

5.7.1

Let L be a finite Galois extension of K, A a finite K-algebra diagonalized by L, $G = \mathrm{Aut}_K L$, and $X = \mathrm{Hom}_{K\text{-}\mathcal{alg}} (A, L)$. The set X can be identified with $\mathrm{Hom}_{L\text{-}\mathcal{alg}} (L \otimes_K A, L)$. Set $g \cdot \xi = g \circ \xi$ for all $g \in G$ and $\xi \in X$. Thus G acts on X.

The Gelfand transform of the L-algebra $L \otimes_K A$ is the map $\gamma : t \mapsto \hat{t}$ from $L \otimes_K A$ to L^X with $X = \mathrm{Hom}_{L\text{-}\mathcal{alg}} (L \otimes_K A, L)$ defined by $\hat{t}(\xi) = \xi(t)$ for all $\xi \in X$.

For all t of the form $\lambda \otimes a$, $(\widehat{\lambda \otimes a})(\xi) = \xi(\lambda \otimes a) = \lambda \xi(a)$. As G acts on L, G can be made to act on $L \otimes_K A$ by $g_*(\lambda \otimes a) = g(\lambda) \otimes a$. The Gelfand transform is an isomorphism since $L \otimes_K A$ is diagonal, and the action of G on $L \otimes_K A$ can be transferred on L^X giving an action of G on L^X which will be written $(g, f) \mapsto g \perp f$. This action is defined by the commutativity of the diagram:

$$
\begin{array}{ccc}
L \otimes_K A & \xrightarrow{\ \gamma\ } & L^X \\
{\scriptstyle g_* = g \otimes 1_A} \Big\downarrow & & \Big\downarrow {\scriptstyle f \mapsto g \perp f} \\
L \otimes_K A & \xrightarrow{\ \gamma\ } & L^X
\end{array}
\tag{5.1}
$$

5.7.2

Proposition *The law* $\perp : G \times L^X \to L^X$ *is given by the formula*

$$(g \perp f)(\xi) = g(f(g^{-1} \circ \xi)).$$

Proof Let $\xi \in X$ and $g \in G$. The map $\varphi : f \mapsto (g \perp f)(\xi)$ from L^X to L is a K-algebra homomorphism. Its kernel is a maximal ideal of L^X, and is of the form $\mathfrak{m}_\eta$, with $\eta \in X$. The map g_* in diagram (1) is g-linear; as γ is L-linear, $f \mapsto g \perp f$ is g-linear, in other words, for $\lambda \in L$ and $f \in L^X$, $\varphi(\lambda f) = g(\lambda)\varphi(f)$. For all $f \in L^X$, there exists $f_1 \in \mathfrak{m}_\eta$ such that $f = f(\eta) \cdot 1 + f_1$. Hence $\varphi(f) = g(f(\eta))$. η remains to be determined. For $\lambda \in L$ and $a \in A$, by definition of $\perp$:

$$(g \perp (\widehat{\lambda \otimes a}))(\xi) = (\widehat{g(\lambda) \otimes a})(\xi) = g(\lambda) \cdot \xi(a).$$

Besides,

$$(g \perp (\widehat{\lambda \otimes a}))(\xi) = \varphi(\widehat{\lambda \otimes a}) = g(\widehat{\lambda \otimes a}(\eta)) = g(\lambda \cdot \eta(a)) = g(\lambda) \cdot g(\eta(a)),$$

and so taking $\lambda = 1$, $\xi(a) = g(\eta(a))$. Therefore $\eta = g^{-1} \circ \xi$. $\qquad\square$

5.7.3

Let L be a Galois extension of K and G $= \mathrm{Aut_K} \mathrm{L}$ its Galois group. Let G-$\mathcal{Setf}$ denote the category of finite sets on which G acts, and for X and Y objects of G-$\mathcal{Setf}$, let $\mathrm{Hom_G(X, Y)}$ be the set of morphisms from X to Y, i.e. of maps φ from X to Y such that, for all $x \in \mathrm{X}$ and $g \in \mathrm{G}$, $\varphi(g \cdot x) = g \cdot \varphi(x)$. Let $\mathcal{D}$ be the category of finite K-algebras diagonalized by L. For $\mathrm{A} \in \mathcal{D}$, let S(A) be the set $\mathrm{Hom_{K\text{-}\mathcal{alg}}(A, L)}$, where G acts by $(g, f) \mapsto g \circ f$. This defines a contravariant functor S from $\mathcal{D}$ to G-$\mathcal{Setf}$. For $f : \mathrm{A} \to \mathrm{B}$, the map $f^* : \mathrm{S(B)} \to \mathrm{S(A)}$ is defined by

$$f^*(\eta) = \eta \circ f.$$

The aim of this section is to prove the following result:

Theorem *The functor* S : $\mathcal{D}$ $\to$ G-$\mathcal{Setf}$ *is an anti-equivalence of categories.*

This theorem combines Propositions 5.7.4 and 5.7.5.

5.7.4

Proposition *The functor* S *is fully faithful.*

In other words, for all algebras A and B of $\mathcal{D}$, the map $f \mapsto f^*$ from $\mathrm{Hom_{K\text{-}\mathcal{alg}}}$ (A, B) to $\mathrm{Hom_G(S(B), S(A))}$ is bijective.

Lemma *For all objects* A *of* $\mathcal{D}$, $\mathrm{Fix_G(L \otimes_K A)} = \mathrm{A}$.

Proof of the Lemma The statement does not involve the algebra structure of A, only its vector space structure. Hence A can be identified with K^d; the lemma is then a consequence of Theorem 5.6.3, (i) $\Rightarrow$ (iv).

Proof of the Proposition For any morphism $\varphi : \mathrm{S(B)} \to \mathrm{S(A)}$ of G-$\mathcal{Setf}$, there is an L-algebra morphism $\varphi^* : \mathrm{L^{S(A)}} \to \mathrm{L^{S(B)}}$ defined by $\varphi^*(h) = h \circ \varphi$ compatible with the action $\perp$ of G. In other words, $\varphi^*(g \perp h) = g \perp \varphi^*(h)$. Indeed, for $\xi \in \mathrm{S(A)}$,

$$\varphi^*(g \perp h)(\xi) = (g \perp h)(\varphi(\xi)) = g(h(g^{-1} \circ \varphi(\xi))) = g(h(\varphi(g^{-1} \circ \xi)))$$
$$= (g \perp (h \circ \varphi))(\xi) = (g \perp \varphi^*(h))(\xi).$$

As the Gelfand transforms

$$\gamma_\mathrm{A} : \mathrm{L} \otimes_\mathrm{K} \mathrm{A} \to \mathrm{L^{S(A)}} \text{ and } \gamma_\mathrm{B} : \mathrm{L} \otimes_\mathrm{K} \mathrm{B} \to \mathrm{L^{S(B)}}$$

are isomorphisms, there is a unique homomorphism

$$\varphi^\otimes : L \otimes_K A \to L \otimes_K B$$

making the diagram

$$
\begin{array}{ccc}
L \otimes_K A & \xrightarrow{\gamma_A} & L^{S(A)} \\
\varphi^\otimes \downarrow & & \downarrow \varphi^* \\
L \otimes_K B & \xrightarrow{\gamma_B} & L^{S(B)}
\end{array}
$$

commutative. This homomorphism is compatible with the actions of G, and so $\varphi^\otimes$ induces a K-algebra homomorphism f from

$$A = \mathrm{Fix}_G(L \otimes_K A) \text{ to } B = \mathrm{Fix}_G(L \otimes_K B).$$

We show that $f^* : S(B) \to S(A) = \varphi$, i.e. that

$$(\forall \eta \in \mathrm{Hom}_{K\text{-}\mathscr{alg}}(B, L)) \quad \eta \circ f = \varphi(\eta).$$

Let $\delta_\eta : L^{S(B)} \to L$ be the homomorphism defined by $\delta_\eta(h) = h(\eta)$. By definition of the Gelfand transform, $\eta = \delta_\eta \circ \gamma_B \circ \iota_B$, where $\iota_B : B \to L \otimes_K B$ is the canonical injection. Hence

$$\eta \circ f = \delta_\eta \circ \gamma_B \circ \iota_B \circ f = \delta_\eta \circ \gamma_B \circ \varphi^\otimes \circ \iota_A = \delta_\eta \circ \varphi^* \circ \gamma_A \circ \iota_A$$

$$= \delta_{\varphi(\eta)} \circ \gamma_A \circ \iota_A = \varphi(\eta).$$

So indeed $f^* = \varphi$, which shows that the map $f \mapsto f^*$ is surjective.

We show that if $\varphi = u^*$ for some homomorphism u from A to B, then $f = u$. It will follow that $u \mapsto u^*$ is injective. For all $\eta \in S(B)$,

$$\delta_\eta \circ \gamma_B \circ \iota_B \circ u = \eta \circ u = u^*(\eta) = \varphi(\eta) = \delta_{\varphi(\eta)} \circ \gamma_A \circ \iota_A$$

$$= \delta_\eta \circ \varphi^* \circ \gamma_A \circ \iota_A = \delta_\eta \circ \gamma_B \circ \iota_B \circ f.$$

As $\bigcap_{\eta \in S(B)} \mathrm{Ker} \delta_\eta = 0$ and $\gamma_B \circ \iota_B$ is injective, it follows that $u = f$. □

5.7.5

Proposition *The functor S is essentially surjective.*

In other words, any finite G-set is isomorphic to a G-set of the form S(A), where $A \in \mathscr{D}$.

Proof (a) Let H be a subgroup of G, and set $F = \text{Fix}_H(L)$. The restriction $\rho : G \to \text{Hom}_{K\text{-}\mathscr{alg}}(F, L)$ is surjective. Indeed, embed L in an algebraic closure Ω of K; then any K-homomorphism of F to L extends to a homomorphism from L to Ω (Theorem 5.2.6), inducing an automorphism of L (Theorem 5.6.3, (i) $\Rightarrow$ (iii)). Let $f, g \in G = \text{Aut}_K(L)$; then, $\rho(f) = \rho(g)$ if and only if $h \in G$ defined by $g = f \circ h$ is in $\text{Aut}_F(L)$. Now, $\text{Aut}_F(L) = H$ by (5.6.7), and so $S(F) = \text{Hom}_{K\text{-}\mathscr{alg}}(F, L)$ can be identified with the set $G/H = \{fH\}_{f \in G}$, and this identification is compatible with the left actions of G.

(b) Let A and B be objects of $\mathscr{D}$. Then, $S(A \times B) = S(A) \sqcup S(B)$. Indeed, $L \otimes_K A = L^X$ with $X = S(A)$, and $L \otimes_K B = L^Y$ with $Y = S(B)$, and

$$L \otimes_K (A \times B) = (L \otimes A) \times (L \otimes B) = L^{X \sqcup Y}.$$

So

$$S(A \times B) = \text{Hom}_{L\text{-}\mathscr{alg}}(L^{X \sqcup Y}, L) = X \sqcup Y.$$

(c) The proposition then follows from the fact that every finite G-set is of the form $\bigsqcup_{i \in I} X_i$, where I is finite, and where, for all $i \in I$, X_i is isomorphic to a G-set of the form G/H_i, where H_i is a subgroup of G. $\qquad\square$

Exercises 5.6 (Finite Galois extensions) and 5.7 (Finite Galois theory)
1.—Let K be a field of characteristic $\neq 2$. Show that every extension of degree 2 of K is Galois and give its Galois group. What happens in characteristic 2?

2.—Let $K = \mathbb{C}(Z)$ and $L = K[X]/(P)$, where $P(X) = X^3 - 3X + 2Z$.

(a) Show that L is an algebraic extension of K. Give its degree.

(b) Find the group of K-automorphisms of L. Is L a Galois extension of K?

(c) Let $\widetilde{L}$ be a decomposition field of P (i.e. the field generated in an algebraic closure of K by the roots of P). Show that $\widetilde{L}$ is a Galois extension of K. Give the degree of $\widetilde{L}$ over K and over L. Find the number of intermediate fields between K and $\widetilde{L}$.

3.—Let Ω be the algebraic closure of $\mathbb{Q}$ in $\mathbb{C}$. Set

$$\varphi = \frac{1 + \sqrt{5}}{2} \quad \text{and} \quad \varphi' = \frac{1 - \sqrt{5}}{2},$$

and consider the field
$$L = \mathbb{Q}(\varphi) = \mathbb{Q}(\varphi') = \mathbb{Q}(\sqrt{5}).$$

(a) Show that fields $L_1 = L(\sqrt{\varphi})$ and $L_2 = L(\sqrt{\varphi'})$ of Ω are conjugate over $\mathbb{Q}$, but not over L. Show that L_1 is a Galois extension of L, which is a Galois extension of $\mathbb{Q}$, but that L_1 is not a Galois extension of $\mathbb{Q}$. Is this also true of L_2?

(b) Let M be the smallest Galois sub-extension of Ω containing L_1 and L_2. Give the degree of M over K? Describe the Galois group $G = \text{Aut}_K(M)$. Is a normal subgroup of a normal subgroup of G necessarily a normal subgroup of G?

4.—Let K be of characteristic $p \geqslant 0$ and Ω an algebraic closure of K. Let the integer n not be a multiple of p.

(a) Show that, for all $a \neq 0$, the polynomial $X^n - a$ is separable.

(b) Show that the group μ_n of n-th roots of unity in Ω is isomorphic to $\mathbb{Z}/(n)$. In the following, assume that $\mu_n \subset K$.

(c) Let $a \in K$ be nonzero, and α a n-th root of a in Ω. Set $L = K(\alpha)$. Show that any conjugate of α can be written $\zeta\alpha$, where $\zeta \in \mu_n$. Deduce that L is Galois and that the group $G = \text{Aut}_K(L)$ can be identified with a subgroup of μ_n.

(d) Show that $X^n - a \in K[X]$ is irreducible if and only if there is no m-th root of a of K, where $m \neq 1$ is a divisor of n. Show that then $G = \mu_n$ ($\prod_{g \in G} g(\alpha)$ may be considered).

5.—Let E be the subfield of $\mathbb{C}$ generated by the roots of $X^4 - 5$. Give the Galois groups of E over

$$\mathbb{Q}, \quad \mathbb{Q}(\sqrt{5}), \quad \mathbb{Q}(\sqrt{-5}), \quad \mathbb{Q}(i).$$

6.—Let $P \in K[X]$ a monic polynomial of degree d, Ω an algebraic closure of the field K and $x_1, \ldots, x_d$ the roots of P in Ω. Set

$$\Delta = \prod_{i<j}(x_i - x_j)^2.$$

(a) Show that, up to sign, Δ is the discriminant of P (resulting from P and its derivative), and that in particular $\Delta \in K$. Give the sign.

(b) Let E be the sub-extension of Ω generated by $x_1, \ldots, x_d$. Every $g \in \text{Aut}_K(E)$ induces a permutation of $\{x_1, \ldots, x_d\}$, hence, assuming these elements are all distinct, i.e. $\Delta \neq 0$, an embedding of $G = \text{Aut}_K E$ in $\mathfrak{S}_d$. Show that Δ is a square in K if and only if G is contained in the alternating group $\mathfrak{A}_d$ (see 5.8.7). More generally, show that $G \cap \mathfrak{A}_d = \text{Aut}_L E$, where $L = K(\sqrt{\Delta})$.

(c) Suppose that $G = \mathfrak{S}_d$ and $d \geqslant 5$. Show that L is the only strictly intermediate Galois extension between K and E. Study the cases $d = 4, d = 3$.

7.—Let k be a non-perfect field of characteristic p, $a \in k$, $a \notin k^p$ and $K = k(T)$, where T is an indeterminate. Consider the algebraic extension $L = K(x)$ of K generated by a root of

$$f(X) = X^p - T^{p-1}X - a.$$

(a) Show that L is a Galois extension of K.

(b) Show that K is not perfect.

(c) Let $b \in k$ ($b \neq a$ and $b \notin k^p$), and $Q = L(y)$ the extension of L generated by a root y of

$$g(X) = X^{p-1} - T^{p-1}X - b.$$

Study the Galois extension Q of K: Galois group, intermediate extensions.

8.—(A) Let K be an infinite field, E a Galois extension of K of degree d, G the Galois group $\text{Aut}_K E$ and $g_1, \ldots, g_d \in G$.

(a) Let $(e_1, \ldots, e_d)$ be a basis for the vector space E over K. Show that

$$x \mapsto \det_{(e_1,\ldots,e_d)}(g_1(x), \ldots, g_d(x))$$

is a K-polynomial map that does not vanish everywhere (use 5.1.7). Deduce that there exists $x \in E$ whose conjugates form a basis for the K-vector space E (such a basis is called a *normal basis*).

(b) Show that the elements $g_i(x)$ form a normal basis if and only if the determinant of the matrix

$$(g_i \circ g_j(x))_{\substack{1 \leqslant i \leqslant d \\ 1 \leqslant j \leqslant d}}$$

with entries in E is nonzero.

(B) Let K be a finite field with $q = p^r$ elements, where p is prime.

(a) Show that $\text{Aut}_{\mathbb{F}_p} K$ is cyclic of order r, generated by the Frobenius automorphism.

(b) Show that, for any r' dividing r, there is a unique subfield K' of K with $p^{r'}$ elements. Show that $\text{Aut}_{K'} K$ is cyclic and give a generator.

(C) (a) Let E be a vector space of finite dimension d over a field k and φ an endomorphism of E. The minimal polynomial of φ is assumed to be of degree d. Show that there exists $x \in E$ such that the smallest vector subspace of E preserved by φ containing x is E (the results on principal ring modules may be used).
Show that then $(x, \varphi(x), \ldots, \varphi^{d-1}(x))$ is a basis for E.

(b) Let K be a field and E a finite Galois extension of E such that the Galois group $G = \text{Aut}_K E$ is cyclic of order d. If φ is a generator of G, show that the minimal polynomial of φ is $X^d - 1$ (5.1.7 may be used).

(c) Prove that all finite fields have a normal basis.

9.—Let K be a field, K' a not necessarily algebraic extension of K, and $P \in K[X]$ an irreducible separable polynomial. Its coefficients may be taken to be in K', but it is not necessarily irreducible in $K'[X]$. Let Ω and Ω' be respective algebraic closures of K and K', E the sub-extension of Ω generated by the roots of P and E' a sub-extension of Ω' generated by a root x' of P and its conjugates.

(a) Set $L = K(x)$, where x is a root of P, and $L' = K'(x')$. Show that L' is isomorphic to a quotient of $K' \otimes_K L$.

(b) Show that $G = \text{Aut}_K E$ can be identified with the group of permutations of $S(L) = \text{Hom}_{K\text{-}\mathscr{alg}}(L, \Omega)$ induced by the automorphisms of Ω.

(c) Show that $G' = \text{Aut}_{K'}(E')$ is isomorphic to a subgroup of G.

5.8 Solvability

In this section, K is a field of characteristic p with $p = 0$ or prime, and Ω an algebraic closure of K.

5.8.1

Proposition *Let the integer n not be a multiple of p.*
(a) The multiplicative group $\mu_n = \{u \in \Omega \mid u^n = 1\}$ is isomorphic to the additive group $\mathbb{Z}/(n)$.
(b) The K-sub-extension M_n of Ω generated by μ_n is Galois.
(c) The Galois group of $\text{Aut}_K M_n$ can be identified with the subgroup of the multiplicative group $\mathbb{Z}/(n)^*$ of invertible elements in the ring $\mathbb{Z}/(n)$. In particular, $\text{Aut}_K(M_n)$ is commutative.

Proof (a) By Proposition 5.2.10, the group μ_n is cyclic. The roots of $Z^n - 1$ are simple, the derivative nZ^{n-1} only vanishes at 0 (because of the assumption that p does not divide n), and hence there are n of them in Ω and $\text{Card}(\mu_n) = n$.

(b) The extension M_n is etale (Proposition 5.4.9) and invariant under all automorphisms of Ω. Hence it is Galois (Theorem 5.6.3).

(c) The map sending $m \in \mathbb{Z}$ to the endomorphism $x \mapsto m \cdot x$ of $\mathbb{Z}/(n)$ is a ring homomorphism from $\mathbb{Z}$ to $\text{End}(\mathbb{Z}/(n))$, with respect to multiplication in the initial set and composition in the end one. Taking quotients, this gives an isomorphism $\mathbb{Z}/(n) \to \text{End}(\mathbb{Z}/(n))$ assigning to the elements of $\mathbb{Z}/(n)^*$ the automorphisms of $\mathbb{Z}/(n)$.

Transfer using the isomorphism $\phi : \mathbb{Z}/(n) \to \mu_n$ shows that the map from $\mathbb{Z}$ to $\text{End}(\mu_n)$ sending m to the map $u \mapsto u^m$ gives a group isomorphism from $\mathbb{Z}/(n)$ onto $\text{End}(\mu_n)$, which in turn induces a group isomorphism from $\mathbb{Z}/(n)^*$ (equipped with multiplication) onto $\text{Aut}(\mu_n)$ (equipped with composition). Note that this isomorphism does not depend on the choice of ϕ.

Any automorphism of the extension M_n of K induces a group automorphism μ_n, and this action is injective since M_n is generated by μ_n. Hence $\text{Aut}_K(M_n)$ can be identified with a subgroup of $\text{Aut}(\mu_n)$, which can itself be identified with $\mathbb{Z}/(n)^*$. No choice is involved in these identifications.

Remark For $K = \mathbb{Q}$, the group $\mathrm{Aut}(M_n)$ can be shown to be identified with the whole of $\mathbb{Z}/(n)^*$ (irreducibility of cyclotomic polynomials, 5.8, Exercise 2).

5.8.2

For $a \in K^*$ and $n \in \mathbb{N}^*$, let $R_K(a, n)$, or simply $R(a, n)$, be the extension of K generated by the n-th roots of a in Ω, i.e. by the elements $x \in \Omega$ such that $x^n = a$. If n is not a multiple of p, then the extension $R(a, n)$ is Galois: Indeed it is the decomposition of the separable polynomial $X^n - a$.

If x is an n-th root of a, the other n-th roots of a are $u.x$ for $u \in \mu_n$. Hence $M_n = R_K(1, n) \subset R_K(a, n)$.

More generally, let $a_1, ..., a_k \in K$ and $n_1, ..., n_k$ integers > 0. Write $R_K(a_1, n_1 ; ... ; a_k, n_k)$ for the extension of K generated by $R_K(a_1, n_1)$, ..., $R_K(a_k, n_k)$ in Ω. This extension contains M_n for $n = \mathrm{l.c.m.}(n_1, ..., n_k)$. If no n_i is a multiple of p, this extension is Galois, being generated by Galois extensions.

Proposition *Let* $E = R_k(a_1, n_1 ; ... ; a_k, n_k)$ *and* $n = \mathrm{l.c.m.}(n_1, ..., n_k)$. *Suppose that* μ_n *is contained in* K *and that* n *is not a multiple of* p. *Then the Galois group of* $G = \mathrm{Aut}_K(E)$ *can be identified with a subgroup of* $\mu_{n_1} \times \cdots \times \mu_{n_k}$. *In particular,* G *is commutative.*

Proof Let $g \in G$. For each $i \in \{1, ..., k\}$, let x_i be a n_i-th root of a_i. Then $g(x_i) = u_i \cdot x_i$ for some $u_i \in \mu_{n_i}$ which does not depend on the choice of x_i. For if x_i' is another n_i-th root of a_i, then $x_i' = w_i \cdot x_i$ with $w_i \in \mu_{n_i}$, and $g(x_i') = w_i \cdot g(x_i)$ since $w_i \in K$. Assigning to each $g \in G$ the family $(u_1, ..., u_k)$ defines a map $\varphi : G \to \mu_{n_1} \times \cdots \times \mu_{n_k}$ which is injective since E is generated by the elements x_i.

It is also a group homomorphism. For if $\varphi(g) = (u_1, ..., u_k)$ and $\varphi(h) = (v_1, ..., v_k)$, then $g \circ h(x_i) = u_i \cdot v_i \cdot x_i$. $\square$

5.8.3

Proposition *Let the integer* n *not be a multiple of* p. *Suppose* $\mu_n \subset K$.

An extension E *of* K *contained in* Ω *is of the form* $R_K(a, n)$ *if and only if* E *is Galois and* $G = \mathrm{Aut}_K(E)$ *cyclic of order dividing* n.

Proof (a) *Necessity:* It is Proposition (5.8.2), with $k = 1$.

(b) *Sufficiency:* Suppose G is cyclic of order m dividing n; let g be a generator of G. Consider E as a vector space over K and g as a K-linear endomorphism of E.

The endomorphism g satisfies $g^m - I = 0$, and the roots of $Z^m - 1$ are simple and in K. Hence g is diagonalizable; in other words, E has a basis consisting of eigenvectors of g. The eigenvalues are in μ_m. If x and y are eigenvectors of g with respective eigenvalues u and v, then

$$g(xy) = g(x).g(y) = ux.vy = uv.xy$$

and xy are eigenvectors for the eigenvalue uv. Consequently, the eigenvalues of g form a subgroup of μ_m.

Moreover, $y \mapsto xy$ is a linear injection of the eigenspace corresponding to v into the eigenspace corresponding to uv. As a result, the eigenspaces corresponding to the various eigenvalues all have the same dimension.

But the eigenspace corresponding to the eigenvalue 1 is $\mathrm{Fix}_G(E) = K$, and so has dimension 1. As E is an m-dimensional vector space, there are m distinct eigenvalues, and all elements of μ_m are eigenvalues.

Let w be a generator of μ_m and z an eigenvector of g for w. Then z^i, $i = 0, ...,$ $m - 1$, are the eigenvectors for all the eigenvalues of g. Hence they form a basis for E. In particular E is generated by z as extension of K.

Set $b = z^m$ and $a = z^n$. The m-th roots of b (resp. the n-th roots of a) are $u \cdot z$ with $u \in \mu_m$ (resp. $u \in \mu_n$). Hence they are in E and

$$E = R(b, m) = R(a, n).$$

□

Remark If $K = \mathbb{Q}$, G has order n, and E is then called a **Kummer extension.**

5.8.4 Solvable Extensions

Definition Let $L \subset \Omega$ be a finite extension of K. Then L is said to be a **solvable extension** of K if there is a finite sequence $(L_0, ..., L_N)$ of extensions with $K = L_0 \subset L_1 \subset \cdots \subset L_N \subset \Omega$, $L \subset L_N$ and L_{i+1} is of the form $R_{L_i}(a_{i,1}, n_{i,1} ; ... ; a_{i,k_i}, n_{i,k_i})$ for $0 \leqslant i \leqslant N - 1$.

This definition implies that a sub-extension of a solvable extension of K is solvable, an extension generated by finitely many solvable extensions is solvable, and a solvable extension of a solvable extension of K is a solvable extension of K.

In this definition, imposing $k_i = 1$ for all i does not restrict the notion of solvability,

A finite radical extension, i.e. a purely inseparable one, is solvable. An extension L of K is solvable if and only if so is L_{sep}: indeed L_{sep} is a sub-extension of L and L is a radical extension of L_{sep}.

5.8.5 Separable Solvable Extensions

A separable extension L of K is solvable if and only if the Galois extension E generated by L is solvable. Indeed L is a sub-extension of E, and E is generated by L and its conjugates under the action of $\mathrm{Aut}_K(\Omega)$.

If L is a separable solvable extension of K, then there is a sequence $(L_0, ..., L_N)$ satisfying the conditions of definition 5.8.4., where no $n_{i,j}$ is a multiple of p. This follows from the next lemma.

Lemma *If $n = p^r m$ where m is not a multiple of p, then for $a \in K^*$ $R(a, n)_{sep} = R(a, m)$.*

Proof The extension $R(a, m)$ is separable, and $R(a, n)$ is a purely inseparable extension of $R(a, m)$ since it is generated by the p^r-th roots of the m-th roots of a. □

If L is a separable solvable extension of K, there is a sequence $(L_0, ..., L_N)$ satisfying conditions of definition 5.8.4, with all L_i Galois. Indeed, by the previous reduction, all the extensions L_i may be assumed to be separable. Then, each time n-th roots of an element are added, we also add those of all its conjugates over K.

5.8.6 Solvable Groups

Definition A group G is said to be **solvable** if there is a decreasing finite sequence $(G_0, ..., G_N)$ of subgroups of G with $G_0 = G$, $G_N = \{e\}$, G_{i+1} normal in G_i and G_i/G_{i+1} commutative for $0 \leqslant i \leqslant N - 1$.

A subgroup of a solvable is solvable. A quotient group of a solvable group is solvable. Given a group G and a normal subgroup H, if H and G/H are solvable, then so is G.

If G is a finite solvable group, then there is a sequence $(G_0, ..., G_N)$ satisfying the conditions of the above definition with G_i/G_{i+1} *cyclic* for all i. This follows from the fact that every finite commutative group is the direct sum of cyclic groups (Theorem 3.5.8, with $A = \mathbb{Z}$).

5.8.7 The Groups $\mathfrak{S}_n$ and $\mathfrak{A}_n$

If the set X is finite, then let $\mathfrak{S}(X)$ be the permutation group on X (*symmetric group*), and $\mathfrak{A}(X)$ the kernel of the homomorphism $\epsilon : \mathfrak{S}(X) \to \{+1, -1\}$ which assigns to each permutation its signature (3.3.18) (*alternating group*). For $Y \subset X$, embed $\mathfrak{S}(Y)$ in $\mathfrak{S}(X)$ by extending the permutations by the identity on $X - Y$. Then $\mathfrak{A}(Y) = \mathfrak{S}(Y) \cap \mathfrak{A}(X)$. If X has n elements, then $\mathfrak{S}(X)$ and $\mathfrak{A}(X)$ are respectively isomorphic to $\mathfrak{S}_n = \mathfrak{S}(\{1, ..., n\})$ and $\mathfrak{A}_n = \mathfrak{A}(\{1, ..., n\})$.

Let the set X be finite, $k \geqslant 2$ and $a_1, ..., a_k$ distinct points of X. Write $[a_1, ..., a_n]$ for the permutation ζ defined by $\zeta(a_i) = a_{i+1}$ for $1 \leqslant i \leqslant k - 1$, $\zeta(a_k) = a_1$ and $\zeta(x) = x$ for $x \in X - \{a_1, ..., a_k\}$. Such a permutation is called a *k-cycle* and

$\{a_1, ..., a_k\}$ is the *support* of ζ. Cycles with disjoint supports commute. Every permutation on X can, up to order, be uniquely written as a product of cycles with disjoint supports.

If $\sigma = \zeta_1 ... \zeta_\nu$ with disjoint supports, where ζ_i is a k_i-cycle and $k_1 \geqslant \cdots \geqslant k_\nu$, then σ will be said of *type* $(k_1, ..., k_\nu)$. Two permutations on X are conjugate in $\mathfrak{S}(X)$ if and only they are of the same type.

If σ and σ' are of the same type $(k_1, ..., k_n)$ and if some k_i is even, or any two k_i are equal, or $k_1 + \cdots + k_\nu \leqslant n - 2$, then the permutations σ and σ' are conjugate by an element of $\mathfrak{A}(X)$. In all three cases, there exists $\tau \in \mathfrak{S}(X) - \mathfrak{A}(X)$ commuting with ζ. However, if $\sigma' = \alpha\sigma\alpha^{-1}$, then $\sigma' = \alpha'\sigma\alpha'^{-1}$ when $\alpha' = \alpha\tau$ with τ commuting with σ; if moreover $\epsilon(\tau) = -1$, one of α, α' is in $\mathfrak{A}(X)$.

5.8.8 Simplicity of $\mathfrak{A}_n$ for $n \geqslant 5$

A group G is said to be *simple* if its only normal subgroups are G and $\{e\}$.

Proposition *Let* X *be an* n *element set with* $n \geqslant 5$. *Then the group* $\mathfrak{A}(X)$ *is simple.*

This proposition follows from the next two lemmas.

Lemma 5.1 *Let* H *be a nontrivial normal subgroup of* $\mathfrak{A}(X)$. *Then* H *contains a 3-cycle or an element of type* $(2, 2)$.

Proof Let $\sigma \in H$ be a nontrivial element, $(k_1, ..., k_\nu)$ the type of σ and write $\sigma = \zeta_1 \cdots \zeta_\nu$, where ζ_i is a k_i-cycle and the supports are all disjoint.

If there is some $k_i \neq 3$, then there is a 3 element set Δ such that $Y = \Delta \cup \sigma(\Delta)$ has 4 elements. Indeed if $k_1 > 3$ and $\zeta_1 = [a_1, ..., a_{k_1}]$ then we may take $\Delta = \{a_1, a_2, a_3\}$; if $k_\nu = 2$, then $\nu > 1$ for otherwise $\epsilon(\sigma) = -1$ and we could take $\Delta = \text{supp}(\zeta_\nu) \cup \{x\}$, where x is not a fixed element.

In this case, let α be a cycle of support Δ. Then $\alpha' = \sigma\alpha\sigma^{-1}$ is a cycle of support $\sigma(\Delta)$, and $\gamma = \alpha'\alpha^{-1}$ is in $\mathfrak{A}(Y)$ and is nontrivial. Hence γ is a 3-cycle or an element of type $(2, 2)$. But γ can be written as $\sigma\sigma'^{-1}$, where $\sigma' = \alpha\sigma\alpha^{-1} \in H$, and so $\gamma \in H$. This proves the lemma in this case.

It remains to prove the case where all $k_i = 3$. If $\nu = 1$, σ itself is a 3-cycle. If $\nu \geqslant 2$, write $\zeta_1 = [a_1, a_2, a_3]$, $\zeta_2 = [b_1, b_2, b_3]$, and set $\alpha = [a_1, b_1][a_2, b_2]$. Then $\sigma' = \alpha\sigma\alpha^{-1} = [b_1, b_2, a_3][a_1, a_2, b_3] \zeta_3 \cdots \zeta_\nu$, and $\gamma = \sigma\sigma'^{-1} = [a_1, b_1][a_3, b_3] \in$ H. $\qquad\square$

Lemma 5.2 *Let* H *be a normal subgroup of* $\mathfrak{A}(X)$ *with* $\text{Card}(X) \geqslant 5$. *If* H *contains a 3-cycle or an element of type* $(2, 2)$, *then* $H = \mathfrak{A}(X)$.

Proof First note that if there is an element of type $(2, 2)$, then there is also a 3-cycle. Indeed, if $\sigma = [a_1, a_2][b_1, b_2] \in$ H, choose $x \in X$ different from a_1, a_2, b_1, b_2, which is possible since $\text{Card}(X) \geqslant 5$. Set $\alpha = [a_1, a_2, x]$. Then

$$\gamma = \sigma\alpha\sigma^{-1}\alpha^{-1} = [x, a_2, a_1] \in H.$$

Hence this is the case where H contains a 3-cycle. Then all 3-cycles are in H since all of them are conjugate in $\mathfrak{A}(X)$. The proof then follows by a "jeu de taquin".

Take $X = \{1, ..., n\}$, and let $\sigma \in \mathfrak{A}_n$. We show that $\sigma \in H$. Construct by induction a sequence $(\sigma_i)_{0 \leqslant i \leqslant n-2}$ such that σ_i fixes $1, ..., i$ with $\sigma_0 = \sigma$ and $\sigma_i = h_i \sigma_{i-1}$, where $h_i \in H$ for $1 \leqslant i \leqslant n - 2$. If $\sigma_{i-1}(i) = i$, then take $h_i = I$; otherwise take $h_i = [\sigma_{i-1}(i), i, x]$, where $x \in \{i + 1, ..., n\}$ is different from $\sigma_{i-1}(i)$, which is possible. Then σ_{n-2} fixes $1, ..., n - 2$, and as $\epsilon(\sigma_{n-2}) = 1$, $\sigma_{n-2} = I$, and so $\sigma = (h_{n-2}h_{n-3}...h_1)^{-1} \in H$.   □

5.8.9

Corollary *For an n element set X with* $n \geqslant 5$*, the only normal subgroups of* $\mathfrak{S}(X)$ *are* $\{I\}$, $\mathfrak{A}(X)$ *and* $\mathfrak{S}(X)$.

Proof Let H be a normal subgroup of $\mathfrak{S}(X)$. Then $H' = H \cap \mathfrak{A}(X)$ is a normal subgroup of $\mathfrak{A}(X)$, and so is $\mathfrak{A}(X)$ or $\{I\}$.

If $H' = \mathfrak{A}(X)$, i.e. if $H \supset \mathfrak{A}(X)$, then $H = \mathfrak{A}(X)$ or $H = \mathfrak{S}(X)$, since the only subgroups of $\mathfrak{S}(X)/\mathfrak{A}(X) = \{+1, -1\}$ are $\{1\}$ and the whole group.

If $H' = \{I\}$, then H has at most 2 elements. If H is nontrivial, then the nontrivial element must be left invariant by all inner automorphisms of $\mathfrak{S}(X)$, hence is central, which is impossible.   □

5.8.10

Theorem *A finite Galois extension* E *of* K *is a solvable extension if and only if the Galois group* $G = \mathrm{Aut}_K(E)$ *is solvable and its order is not a multiple of p.*

Proof (a) *necessity:* Suppose that E is solvable and that $(L_0, ..., L_N)$ satisfies the conditions of definition 5.8.4., where for all i, L_i is a Galois extension of K and no $n_{i,j}$ is a multiple of p for $i \in \{0, ..., N\}$ and $j \in \{1, ..., k_i\}$ (5.8.5). Set $K' = M_n$, where n is the l.c.m. of $n_{i,j}$.

For $i \in \{0, ..., N\}$, let L_i' be the extension generated by K' and L_i in Ω. All L_i' are Galois extensions of K' (5.6.4, Corollary 1), and

$$L_{i+1}' = R_{L_i'}(a_{i,1}, n_{i,1} ; ... ; a_{i,k_i}, n_{i,k_i}) .$$

The Galois groups G_i' of $\mathrm{Aut}_{L_i'}(L_N')$ are normal subgroups of G_0' (5.6.4, Corollary 2), $G_N' = \{e\}$ and by Proposition 5.8.2., for $i \in \{0, ..., N - 1\}$, the quotient group $G_i'/G_{i+1}' = \mathrm{Aut}_{L_i'}(L_{i+1}')$ is commutative. So the group $G_0' = \mathrm{Aut}_{K'}(L_N')$ is solvable.

By Proposition 5.8.1, (c), the group $\Gamma = \text{Aut}_K(K')$ is commutative, and hence solvable. As $\Gamma = G''/G_0'$, where $G'' = \text{Aut}_K(L_N')$ is solvable (5.8.6). The group $G = \text{Aut}_K(E)$ can be identified with a quotient of G'' (5.6.4, Corollary 2), hence is also solvable.

(b) *Sufficiency:* Suppose that G is solvable and that its order n is not a multiple of p, and let $(G_0, \ldots, G_N)$ be a sequence of subgroups satisfying the conditions of definition 5.8.6, with G_i/G_{i+1} cyclic of order n_i. The order n of G is the product of n_i, and so no n_i is a multiple of p. Let m be the l.c.m. of the integers n_i, $K' = M_m$, E' the extension generated by K' and E in Ω, and G' the Galois group of $\text{Aut}_{K'}(E')$. Setting $G_i' = G' \cap G_i$, the group G_i'/G_{i+1}' can be identified with a subgroup of G_i/G_{i+1}. Hence it is cyclic of order n_i' dividing n_i as well as m.

Set $L_i' = \text{Fix}_{G_i'}(E')$. Then

$$K' = L_0' \subset \cdots \subset L_N' = E',$$

E' is a Galois extension of L_i' for all i, with $\text{Aut}_{L_i'}(E') = G_i'$, and for $i \leqslant N - 1, L_{i+1}'$ is a Galois extension of L_i' with $\text{Aut}_{L_i'}(L_{i+1}') = G_i'/G_{i+1}' \approx \mathbb{Z}/(n_i')$.

By Proposition 5.8.3, the extension L_{i+1}' of L_i' is of the form $R_{L_i'}(a_i, n_i')$. Hence E' is a solvable extension of K'. As K' is a solvable extension of K, so is the case of E' and hence also of its sub-extension E.

5.8.11 *Example of a Non Solvable Extension of $\mathbb{Q}$*

Let E be the decomposition field of the polynomial $f = X^5 - 3X - 1$ over $\mathbb{Q}$ in $\mathbb{C}$. We show that E is a non solvable extension of $\mathbb{Q}$.

(1) *f is irreducible in $\mathbb{Q}[X]$.* The polynomial f has a root z_1 such that $|z_1| < 1$, the others have absolute value > 1. Indeed, as z travels once round the unit circle in $\mathbb{C}$, $f(z) = -3z\left(1 - \frac{z^5-1}{3z}\right)$ travels round 0 once. Were f reducible, there would be f_1 and f_2 of degree $\geqslant 1$ in $\mathbb{Z}[X]$ such that $f = f_1.f_2$ (3.7.8, Lemma 1). The leading coefficients and the constant terms of f_1 and f_2 should be 1 or -1, contradicting the fact that one of them should have absolute value > 1.

(2) *f has 3 real roots.* Considering the sign of $f(x)$ for $x = -2, -1, 1$ and 2, shows that f has at least 3 real roots. The derivative $f' = 5X^4 - 3$ has only 2 real roots, and so f has at most 3 real roots.

(3) *$G = \text{Aut}_{\mathbb{Q}}(E)$ is isomorphic to $\mathfrak{S}_5$.* The group G acts on the set Δ of roots of f, which has 5 elements. Hence there is a homomorphism $G \to \mathfrak{S}(\Delta)$, which is injective since Δ generates E as an extension of $\mathbb{Q}$. Complex conjugation induces a transposition τ permuting the two non real roots of f. The group G acts transitively on Δ since f is irreducible (Proposition 5.6.5); if S is the stabilizer of a point of Δ, then $\text{Card}(G)/\text{Card}(S) = 5$, and so $\text{Card}(G)$ is a multiple of 5. By Sylow's theorem, G contains a subgroup of order 5, necessarily generated by an element of order 5

which can only be a 5-cycle. But as in $\mathfrak{S}_5$, a subgroup in $\mathfrak{S}(\Delta)$ containing a 2-cycle and a 5-cycle is the whole group. Hence G can be identified with $\mathfrak{S}(\Delta)$.

(4) *The group* $\mathfrak{S}_5$ *is not solvable.* This follows from Corollary 5.8.9. As a result, E is a non solvable extension of $\mathbb{Q}$ by Theorem 5.8.10.

Exercises 5.8. (Solvability)

1.—Let p be a prime. A finite group G is a p-group if the order of G is a power of p.

(a) Let G be a p-group and X a finite G-set; set $Y = \mathrm{Fix}_G X$. Show that card $Y \equiv$ card X (mod p).

(b) Making G act on itself by inner automorphisms, show that the centre of G is not trivial.

(c) Deduce that G is solvable.

2. *(Cyclotomic polynomials)*— Let n be an integer, G_n the group of invertible elements of the ring $\mathbb{Z}/(n)$, and φ_n its cardinality. Suppose G_n is cyclic, and let Ω be an algebraic closure of $\mathbb{Q}$. Set $\Phi_n(X) = \prod(X - \zeta_i)$, where ζ_i are the primitive n-th roots of unity. The aim is to show that $\Phi_n \in \mathbb{Q}[X]$ is irreducible (this property holds even when G_n is not cyclic, but the proof is different).

(a) Show that $X^n - 1 = \prod \Phi_d$, where d is a divisor of n. Deduce that $\Phi_n \in \mathbb{Z}[X]$ (use the results of 3.7.8). For any field K, let Φ_n^K be the image of Φ_n in K[X].

(b) Show that for any field K whose characteristic p does not divide n, the roots of Φ_n^K in an algebraic closure of K are the primitive n-th roots of unity.

(c) Prove that if there is a prime p such that $\Phi_n^{\mathbb{F}_p}$ is irreducible, then the irreducibility of Φ_n in $\mathbb{Q}[X]$ follows (use 3.7.8 to reduce from the case $\mathbb{Q}$ to $\mathbb{Z}$).

(d) Let p be a prime whose class modulo n generates G_n. Show that the field $\mathbb{F}_{p^r}$ has no primitive n-th root of unity for $r < \varphi_n$. Deduce that $\Phi_n^{\mathbb{F}_p}$ is irreducible.

(e) Conclude by assuming the following result: an arithmetic progression whose first term and difference are coprime contains infinitely many primes.

(f) Show that $L = \mathbb{Q}[X]/(\Phi_n)$ is a Galois extension of $\mathbb{Q}$ and that its Galois group can be identified with G_n. Each embedding σ of L in Ω defines an isomorphism of L onto the sub-extension E of Ω generated by the n-th roots of unity. Hence there is an isomorphism σ_* from G_n onto $\mathrm{Aut}_{\mathbb{Q}}(E)$. Does this isomorphism depend on σ?

3.—Let K be of characteristic different from 2 and 3 and $P \in K[X]$ an irreducible polynomial of degree 3. Let E be the extension generated in an algebraic closure of K by roots of P and the cubic roots 1, j, j^2 of 1.

(a) Show that E is a solvable extension of K. More precisely, show that $\widetilde{K} = K(j)$ has degree 2 or 1 over K and that there is an extension F of degree 2 or 1 of $\widetilde{K}$ such that $E = F(\sqrt[3]{a})$ for some $a \in F$ is an extension of degree 3 of F.

In the following, the aim is to reduce the search for the roots of P to extractions of square or cubic roots.

(b) Let $x \in E$ be a root of P and $\sigma \in \mathrm{Aut}_K(E)$ an element of order 3. The other two roots of P are $y = \sigma(x)$ and $z = \sigma^2(x)$. Let F, V and W be the eigenspaces of σ considered a $\widetilde{K}$-linear map from E to E for the eigenvalues 1, j and j^2 respectively. Let v and w be the respective projections of x onto V and W. Express v and w in terms of x, y, z and j. Show that $F = K(v^3) = K(w^3)$, and that $vw \in K$.

(c) Suppose $P = X^3 + pX + q$. Show that $x = v + w$ and $vw = -\frac{p}{3}$.

(d) Replacing x by $v + w$ in $x^3 + px + q = 0$, calculate $v^3 + w^3$. Deduce that v^3 and w^3 are the two roots of a second degree equation over K and give this equation.

(e) Express x by radicals (Cardano's formula).

4.—Let K be of characteristic different from 2 and 3, P an irreducible polynomial of degree 4 over K, and x, y, z, t the roots of P in an algebraic closure.

(a) Show that $u = xy + zt$, $v = xz + yt$ and $w = xt + yz$ are the roots of a degree 3 polynomial over K.

(b) Show that xy and zt are the roots of a second degree equation over $K(u)$. Show that so are $x + y$ and $z + t$.

(c) Show that $x + y$ and xy generated the same quadratic extension L of $K(u)$. Show that x and y are the roots of a second degree equation over L.

(d) Let $E = K(x, y, z, t)$. Assume that the Galois group $\mathrm{Aut}_K(E)$ is isomorphic to $\mathfrak{S}_4$. Give the list of subgroups of $\mathfrak{S}_4$ and for each of them, give the generators of the corresponding sub-extension of E.

(e) What can be said about the Galois group $\mathrm{Aut}_K(E)$ if $P(X) = X^4 + pX^2 + q$ (square of a quadratic)? if $P(X) = aX^4 + bX^3 + cX^2 + bX + a$ (reciprocal polynomial)?

5.—(a) Let K and L be subfields of $\mathbb{C}$ such that $K \subset L$ and $[L : K] = 2$. Show that every point of L can be obtained from points of K by a ruler and compass construction.

(b) Let E be a subfield of $\mathbb{C}$ generated by $e^{2i\pi/17}$. Show that E is a Galois extension of $\mathbb{Q}$ and that the Galois group $G = \mathrm{Aut}_{\mathbb{Q}}E$ can be identified with a group of automorphisms of $\mathbb{Z}/(17)$. Using 5.2.10, deduce that G is isomorphic to a cyclic group of order 2^k, $k \leqslant 4$ (in fact $G \approx \mathbb{Z}/(16)$). Show that there is a sequence $E_0, \ldots, E_k$ of subfields of E with $\mathbb{Q} = E_0 \subset E_1 \subset \cdots \subset E_k = E$ and $[E_i : E_{i-1}] = 2$. Are the subfields E_i Galois extensions of $\mathbb{Q}$?

(c) Show that a regular polygon with 17 sides can be constructed using ruler and compass (an explicit construction is not asked for). What properties of the number 17 have been used?

Give a ruler and compass construction of a regular pentagon.

5.9   Infinite Galois Theory

In this section, Ω will denote an algebraic closure of the field K, Ω_{sep} the separable closure of K in Ω, and G the group $\text{Aut}_K \Omega$.

5.9.1

Proposition $G = \text{Aut}_K \Omega_{sep}$.

In other words, every automorphism of Ω induces an automorphism of Ω_{sep}, and every automorphism of Ω_{sep} extends uniquely to an automorphism of Ω.

Proof The existence of an extension follows from 5.2.7, Corollary 2. As Ω is a purely inseparable extension of Ω_{sep} by 5.5.4, uniqueness is a consequence of 5.5.3 applied to Ω_{sep} and Ω.

5.9.2   *Profinite Structure of* G

Let $\mathcal{E}$ be the set of finite Galois extensions of K in Ω. Ordered by inclusion, it is a directed set: the sub-extension of Ω generated by two finite Galois extensions is etale and invariant under G, hence Galois (5.6.3). For objects E and F of $\mathcal{E}$ such that $E \subset F$, every K-automorphism of F induces a K-automorphism of E. Hence there is a restriction homomorphism $\rho_E^F : \text{Aut}_K F \to \text{Aut}_K E$. By 5.6.4, Ω_{sep} is the union of $E \in \mathcal{E}$. As a result,

$$G = \text{Aut}_K \Omega_{\text{sep}} = \varprojlim_{E \in \mathcal{E}} \text{Aut}_K E \, .$$

Equip G with the topology obtained by transferring the topology of the projective limit via this identification. The group G then becomes a profinite group.

5.9.3   *The Functor* S

Let A be an algebraic and separable algebra over K (see 5.4.4). Set $S(A) = \text{Hom}_{K\text{-}\mathscr{alg}}(A, \Omega_{\text{sep}}) = \text{Hom}_{K\text{-}\mathscr{alg}}(A, \Omega)$. Indeed, let $x \in A$, $f \in S(A)$, and P be the minimal polynomial of x; since $P(f(x)) = 0$, the minimal polynomial of $f(x)$ divides P, and $f(x)$ is separable.

For any finite subalgebra B of A, the set $\text{Hom}_{K\text{-}\mathscr{alg}}(B, \Omega)$ is finite. Then $S(A) = \varprojlim \text{Hom}_{K\text{-}\mathscr{alg}}(B, \Omega)$, the inverse limit being taken over all finite subalgebras of A. This identification turns $S(A)$ into a profinite space.

The group G acts on S(A) by $(g, f) \mapsto g \circ f$. This action is continuous. Indeed, for any etale algebra B, there is a finite Galois extension E diagonalizing B; then the action of G on S(B) factorizes through $Aut_K(E)$, hence is continuous. Therefore the inverse limit S(A) of finite G-sets S(B), for finite subalgebras B of A, is an object of G-$\mathcal{Prof}$ (2.8.6).

This defines a contravariant functor from the category $\mathcal{A}$ of separable K-algebraic algebras to the category G-$\mathcal{Prof}$ of profinite spaces on which the action of G is continuous.

5.9.4

> **Theorem** *The functor* S : $\mathcal{A}$ → G-$\mathcal{Prof}$ *is an anti-equivalence of categories.*

Proof Let $\mathcal{D}$ be the category of etale K-algebras, and for any finite Galois extension E of K in Ω, let $\mathcal{D}_E$ be the sub-category of $\mathcal{D}$ consisting of algebras diagonalized by E. Set $G_E = Aut_K E$. By Theorem 5.7.3., the functor $S_E : A \to Hom_{K-\mathcal{alg}}(A, E) = Hom_{K-\mathcal{alg}}(A, \Omega)$ is an anti-equivalence from $\mathcal{D}_E$ onto G_E-$\mathcal{Setf}$.

As $\mathcal{D} = \varinjlim \mathcal{D}_E$, and $G = \varprojlim G_E$, ù G-$\mathcal{Setf} = \varinjlim G_E$-$\mathcal{Setf}$ by 2.9.4. The functors S_E mutually induce each other, and so by passing to the direct limit, defines an anti-equivalence from $\mathcal{D}$ onto G-$\mathcal{Setf}$. This anti-equivalence is readily seen to be given by the functor S. It is also immediate that $\mathcal{A}$ can be identified with the category $\underset{\rightrightarrows}{\mathcal{D}}$ defined by $(\underset{\rightrightarrows}{\mathcal{D}})^\circ = (\underset{\rightrightarrows}{\mathcal{D}^\circ})$. Moreover, G-$\mathcal{Prof}$ = G-$\mathcal{Setf}$ by 2.8.6. As a result, S is an anti-equivalence of categories.   □

5.9.5 *Dictionary*

Let A be a separable algebra over K.

(a) If $A = A_1 \times A_2$, then $S(A) = S(A_1) \sqcup S(A_2)$. If $A = A_1 \otimes A_2$, then $S(A) = S(A_1) \times S(A_2)$. Indeed, the functor S transforms sums into products and conversely.

(b) Let A and B be separable algebras. A homomorphism $f : A \to B$ is injective if and only if $f^* : S(B) \to S(A)$ is surjective. Indeed consider the fibre product $A' = A \times_B A$. The G-set $S(A')$ is sum of two copies of S(A) amalgamated by S(B), i.e. the quotient of $S(A) \sqcup S(A) = \{1, 2\} \times S(A)$ by the equivalence relation identifying $(1, x)$ and $(2, x)$ for $x \in f^*(S(B))$, (it is the fibre product in the opposite category). Then f is injective if and only if the two projections of A' onto A agree, or equivalently the two injections of S(A) into $S(A')$ agree, which is the case if and only if f^* is surjective.

(c) Similarly, f is surjective if and only if f^* is injective. A direct consequence of the definition of S is that if f is surjective, then f^* is injective. If f is not surjective, then f factorizes in $A \xrightarrow{\tilde{f}} B' \xrightarrow{\iota} B$, where ι is injective but not an isomorphism. Then $f^* = \tilde{f}^* \circ \iota^*$, and ι^* is surjective but not an isomorphism; hence ι^* is not injective since $S(B)$ is compact, and f^* is not injective.

(d) A is an extension of K if and only if G acts transitively on $S(A)$. Indeed, $S(\Omega) = G$ and A is an extension of K if and only if there is an injective homomorphism from A to Ω.

   More generally, the orbits of the action of G on $S(A)$ are in bijective correspondence with the maximal ideals of A.

(e) Assigning to each sub-extension L of Ω_{sep} the stabilizer of the canonical injection $\iota_L \in S(L)$, i.e. $\mathrm{Aut}_L(\Omega_{sep})$, defines a bijection between the sub-extensions of Ω_{sep} and the closed subgroups of G. The inverse is given by $H \mapsto \mathrm{Fix}_H(\Omega_{sep})$.

(f) The sub-extensions L and L' of Ω_{sep} are isomorphic if and only if the sub-groups $\mathrm{Aut}_L(\Omega_{sep})$ and $\mathrm{Aut}_{L'}(\Omega_{sep})$ of G are conjugate.

5.9.6

Proposition and Definition *Let L be a sub-extension of Ω. The following conditions are equivalent:*

(i) L is separable and invariant under G;
(ii) L is the union of finite Galois sub-extensions of Ω_{sep};
(iii) L is contained in Ω_{sep} and the sub-group $\mathrm{Aut}_L(\Omega)$ of G is normal.

If these conditions hold, then L is said to be **Galois**.

Proof (ii) $\Rightarrow$ (i). Immediate from Theorem 5.6.3.

   (i) $\Rightarrow$ (ii). Let $x \in L$. The extension generated by the conjugates of x is etale and invariant under G, hence is Galois (5.6.3, Theorem).

   (i) $\Rightarrow$ (iii). For $g \in G$, $\mathrm{Aut}_{g(L)}\Omega = g \cdot \mathrm{Aut}_L\Omega \cdot g^{-1}$. If L is preserved by G, then $g \cdot \mathrm{Aut}_L\Omega \cdot g^{-1} = \mathrm{Aut}_L\Omega$ for all $g \in G$.

   (iii) $\Rightarrow$ (i). The same formula shows that $\mathrm{Aut}_L\Omega = \mathrm{Aut}_{g(L)}\Omega$. Hence, by (5.6.4, Corollary 2), $L = g(L)$. $\square$

Reference

1. J. Lelong-Ferrand, J.-M. Arnaudis, *Cours de mathématiques*, tome 1, *Algèbre* (Dunod, 1971)

Chapter 6
Riemann Surfaces

Introduction

We begin this chapter with the definition of a Riemann surface. Let B be a compact connected Riemann surface, and $\mathcal{M}(B)$ the field of meromorphic functions on B. If X is a compact connected Riemann surface over B (i.e. equipped with a non constant morphism $\pi : X \to B$), then the field $\mathcal{M}(X)$ is a finite extension of $\mathcal{M}(B)$. Moreover, there is a finite subset Δ of B such that $X' = X - \pi^{-1}(\Delta)$ is a connected finite cover of $B - \Delta$. The functors $X \mapsto \mathcal{M}(X)$ and $X \mapsto X'$ give an equivalence between the category $\mathscr{V}_B^1$ of compact connected Riemann surfaces over B, and respectively, the opposite category of the category of finite extensions of $\mathcal{M}(B)$, and the direct limit category of categories of connected covers of $B - \Delta$, where $\Delta \subset B$ is finite.

The study of Riemann surfaces can thus be undertaken either in a purely algebraic framework or a purely topological one. In each, the category $\mathscr{V}_B^1$ can be identified (Chap. 4 for the topological framework, Chap. 5 for the algebraic one) with the category of finite sets acted upon by a profinite group G which may be called G-$\mathcal{A}\ell g$ or G-$\mathcal{T}op$. These two groups may be identified, and thus G-$\mathcal{A}\ell g$ can be described topologically. In Section 6.4, we give a description of the Galois group of the algebraic closure of $\mathbb{C}(Z) = \mathcal{M}(\Sigma)$ (where Σ is the Riemann sphere), which cannot be done purely algebraically.

In Section 6.7, we study the automorphism group of a Riemann surface, and assuming it to be finite we give an upper bound for its order when its genus is $\geqslant 2$. For this homology theory is required. We have chosen the theory obtained from triangulation. These tools are developed in Sections 6.5 and 6.6. In Section 6.8, we show that the bound obtained for the order of the automorphism group is not reached in genus 2, but is in genus 3. In Section 6.10 we show the existence of a Riemann surface of genus 3 whose automorphism group has order 168. We construct this latter example in the framework of Poincaré geometry, which is introduced in Section 6.9.

© Springer Nature Switzerland AG 2020
R. Douady and A. Douady, *Algebra and Galois Theories*,
https://doi.org/10.1007/978-3-030-32796-5_6

6.1   Riemann Surfaces, Ramified Coverings

6.1.1   Riemann Surfaces

A Hausdorff topological space X is said to be a **topological surface** if every point
of X has an open neighbourhood homeomorphic to some open subset of $\mathbb{R}^2$.

Let X be a topological surface. A **complex chart** for X is a homeomorphism φ
from an open subset U of X onto an open subset of $\mathbb{C}$; U is said to be the **domain** of
the chart φ, and φ a chart **centered** at x_0 if $\varphi(x_0) = 0$ and if the image of φ is a disk
centered at 0. Let φ and ψ be two charts for X with respective domains U and V.
The map $\gamma : x \mapsto \psi(\varphi^{-1}(x))$ is a homeomorphism from the open subset $\varphi(U \cap V)$
of $\mathbb{C}$ onto the open subset $\psi(U \cap V)$ of $\mathbb{C}$, and is called the **transition map** from φ
to ψ. A $\mathbb{C}$-**analytic** atlas on X is a family of charts whose domains cover X such that
all transition maps are holomorphic, i.e. $\mathbb{C}$-analytic.

Two $\mathbb{C}$-analytic atlases $(\varphi_i)_{i \in I}$ and $(\psi_j)_{j \in J}$ are **equivalent** if for all (i, j), the
transition maps from φ_i to ψ_j and from ψ_j to φ_i are holomorphic, i.e. if the family
obtained by their union is also an atlas. This is an equivalence relation. A $\mathbb{C}$-**analytic
structure** on X is an equivalence class of analytic atlases. A topological surface with
an analytic structure is called a **Riemann surface**, or a **1-dimensional $\mathbb{C}$-analytic
manifold**, or a **complex curve**.

Let X be a Riemann surface whose structure is defined by an atlas (φ_i), and U_i
the domain of φ_i. Let Y be an open subset of X. The restrictions of φ_i to $U_i \cap Y$
form an atlas for Y. The structure of Y defined by this atlas only depends on that of
X. The structure on Y is said to be **induced** by that of X.

6.1.2   Example. The Riemann Sphere

Consider the set $\Sigma = \mathbb{C} \cup \{\infty\}$ and equip Σ with the topology inducing the ordinary
topology on $\mathbb{C}$, and for which the open neighbourhoods of ∞ are complements of
the compact subsets of $\mathbb{C}$.

The space Σ is homeomorphic to the sphere

$$S^2 = \{(x, y, z) \in \mathbb{R}^3 \mid x^2 + y^2 + z^2 = 1\}.$$

Indeed, as can be checked, the stereographic projection

$$(x, y, z) \mapsto \frac{x + iy}{1 - z}, \quad (0, 0, 1) \mapsto \infty$$

is a homeomorphism from S^2 onto Σ.

The charts $\varphi_0 = 1_{\mathbb{C}} : \Sigma - \{\infty\} \to \mathbb{C}$ and $\varphi_1 : \Sigma - \{0\} \to \mathbb{C}$, defined by $\varphi_1(z) = \frac{1}{z}$ and $\varphi_1(\infty) = 0$, form an analytic atlas on Σ, the transition map being $z \mapsto \frac{1}{z}$ from

$\mathbb{C} - \{0\}$ to itself. This atlas defines a $\mathbb{C}$-analytic structure on Σ. With this structure, Σ is a compact Riemann surface called the *Riemann sphere*.

6.1.3   Analytic Maps

Let X be a Riemann surface and φ a chart for X with domain U. If $f : X \to \mathbb{C}$ is a function, the function $f \circ (\varphi^{-1})$ defined on $\varphi(U)$ is called an **expression** of f in the chart φ, and **holomorphic** if its expression in every chart in an atlas defining the structure de X is holomorphic, this property being independent of the choice of atlas. If Y is a Riemann surface, ψ a chart with domain V and $f : X \to Y$ is a continuous map, the map $\psi \circ f \circ \varphi^{-1}$ defined on $\varphi(U \cap f^{-1}(V))$ is called an **expression** of f in the above charts, and analytic if there exist atlases $(\varphi_i)_{i \in I}$ and $(\psi_j)_{j \in J}$ of X and Y such that its expression in the charts φ_i and ψ_j is holomorphic for all $(i, j) \in I \times J$. This property is independent of the equivalence class of the given atlases.

If $D \subset X$ is a discrete closed set, then a holomorphic function $f : X - D \to \mathbb{C}$ is called **meromorphic** on X with poles in D if for all $a \in D$, there is a neighbourhood of a in V and two holomorphic functions $g, h : V \to \mathbb{C}$ such that $h^{-1}(0) \subset \{a\}$ and $f = \frac{g}{h}$ on $V - \{a\}$ (then $f := \frac{g}{h}$ on V). The ring of meromorphic functions on X with poles in D is written $\mathcal{M}(X, D)$. let

$$\mathcal{M}(X) = \lim_{\longrightarrow} \mathcal{M}(X, D)$$

be the direct limit over the set of discrete closed subsets of X; the elements of $\mathcal{M}(X)$ are called *meromorphic functions on X*.

6.1.4

Proposition and Definition *Let X and Y be Riemann surfaces, $f : X \to Y$ an analytic map, $x_0 \in X$, and $y_0 = f(x_0)$. Assume that f is not constant in the neighbourhood of x_0. Let ψ be a chart for Y centered at y_0. Then there is a chart φ for X centered at x_0 such that the expression of f in the charts φ and ψ is $z \mapsto z^d$, for some number d independent of the choice of charts. It is called the **ramification index** of f at x_0.*

Proof Let φ_0 be an analytic chart for X in the neighbourhood of x_0 such that $\varphi_0(x_0) = 0$, and $\dot{f}$ the expression of f in the charts φ_0 and ψ. Then $f(0) = 0$. We give a series expansion of $\dot{f}$ in the neighbourhood of 0. Let $\dot{f}(z) = \sum a_k z^k$, d the smallest integer for which $a_d \neq 0$, and $c \in \mathbb{C}$ such that $c^d = a_d$. Then $\dot{f}(z) = (cz)^d (1 + u(z))$ for some holomorphic u such that $u(0) = 0$. Since there is a holomorphic function $w \mapsto \sqrt[d]{w}$ in the neighbourhood of 1, the function $1 + u$ may be written as h^d for some holomorphic h in the neighbourhood of 0 with $h(0) = 1$. Hence $\dot{f}(z) = (cz \cdot h(z))^d$. Set $\varphi(x) = c\,\varphi_0(x) \cdot h(\varphi_0(x))$. The implicit function theorem implies that φ is an analytic chart in the neighbourhood of x_0, and $\psi(f(x)) = (\varphi(x))^d$ for x near x_0.

If $z \mapsto z^d$ is the expression of f, then there is a neighbourhood U_0 of x_0 such that for any neighbourhood U of x_0 in U_0, there is a neighbourhood V of y_0 such that for all $y \neq y_0 \in V$, d is the cardinality of $f^{-1}(y) \cap U$. This gives a characterization of d independent of the charts. $\Box$

6.1.5

Proposition *Let*

$$
\begin{array}{ccc}
X & \xrightarrow{\psi} & X' \\
\alpha \downarrow & & \downarrow \beta \\
B & \xrightarrow{\varphi} & B'
\end{array}
$$

be a commutative diagram, where X, X', B *and* B' *are Riemann surfaces,* α, β, φ *analytic maps, and* ψ *a continuous map. Assume that* β *is not constant on any connected component of* X'. *Then* ψ *is analytic.*

Proof Since this is a local result, X' and B' may be assumed to be open in $\mathbb{C}$, $\beta : z \mapsto z^d$, and X connected. The function $\psi^d = \beta \circ \psi = \varphi \circ \alpha$ is analytic; so the set $\Delta = \psi^{-1}(0) = \{x \mid (\psi(x))^d = 0\}$ is the whole of X, in which case ψ is either constant, or a closed discrete set. Then ψ is continuous on X, and analytic on $X - \Delta$, so also on X by a well-known result on holomorphic functions.[1] $\Box$

6.1.6

Proposition *Let* B *be a connected Riemann surface,* X *a Riemann surface equipped with a proper analytic map* $\pi : X \to B$. *The following conditions are equivalent:*

(i) *there is no connected component of* X *on which* π *is constant;*
(ii) *the fibres de* π *are finite;*
(iii) *for all connected components* X_i *of* X, $\pi(X_i) = B$;
(iv) π *is open;*
(v) *for all* $x \in X$, *there is a chart* φ *for* X *in the neighbourhood of* x *with* $\varphi(x) = 0$ *and a chart* ψ *for* B *with* $\psi(\pi(x)) = 0$ *such that the expression of* π *in these charts takes the form* $z \mapsto z^d$.

---

[1]Cartan [1], 2.7, Corollary of th. 2.7.4, p. 74; see also [2], 8.9, making sure that $a_n = 0$ for $n < 0$.

Proof (i) $\Rightarrow$ (v) follows from Proposition 6.1.4 by noting that, according to the analytic continuation principle, if π is constant in the neighbourhood of x, then it also is on the connected component of x.

(v) $\Rightarrow$ (iv) since the map $z \mapsto z^d$ is open.

(iv) $\Rightarrow$ (iii): let X_i be a connected component of X. Then X_i is clopen and nonempty. Hence $\pi(X_i)$ is clopen since π is proper and nonempty; so $\pi(X_i) = B$.

(iii) $\Rightarrow$ (i) is obvious.

(v) $\Rightarrow$ (ii): each point of X is isolated in its fibre, hence the fibres are discrete. As they are compact, they are finite.

(ii) $\Rightarrow$ (i) is obvious. $\qquad\qquad\qquad\qquad\qquad\qquad\qquad\qquad\qquad\qquad\qquad\quad$ $\square$

6.1.7 Ramified Coverings

Definition Let B be a topological surface. A **finite ramified covering** B is a topological surface X equipped with a proper continuous map $\pi : X \to B$ satisfying the following condition:

(RR) For all $x \in X$, there is a complex chart for X centered at x and a complex chart for B centered at $\pi(x)$ such that the expression of π in these charts takes the form $z \mapsto z^d$.

Examples (1) Every finite covering of B is a finite ramified covering of B. However a finite ramified covering B is not in general a covering of B.

(2) Let B be a Riemann surface and X a Riemann surface equipped with a proper analytic map $\pi : X \to B$. Assume that there is no connected component of X on which π is constant. Then X is a finite ramified covering of B and is called **a finite ramified analytic covering** of B.

Remarks (1) If X is a finite ramified covering of B, for any point x the integer d in condition (RR) is independent of the choice of charts, and is called the *local degree* or the *ramification index* of X at x. The set of points for which $d > 1$ is discrete and closed in X, its projection Δ in B is discrete and closed in B and is called the *ramification set* and $X|_{B-\Delta}$ is a finite covering of $B - \Delta$.

(2) If X is a finite ramified analytic covering of B, then the projection $\pi : X \to B$ is open with finite fibres.

(3) If B is connected then for all closed discrete $\Delta \subset B$, $B - \Delta$ is connected. Indeed, if h is a locally constant function on $B - \Delta$, it extends to a locally constant function $\bar{h}$ on B; then $\bar{h}$ is constant and so is h. Hence the *degree* of a finite ramified covering over a connected component of B is well defined.

6.1.8

Proposition *Let* B *be a topological surface,* (X, π) *a ramified covering of* B, $\Delta \subset$ B *the ramification set of* π, $b \in$ B, φ *a chart for* B *centered at* b *with domain* U, *and* V *a connected component of* $\pi^{-1}(U)$. *Set* $U^* = U - \{b\}$. *Suppose that* $U^* \cap \Delta = \varnothing$. *Then the fibre* $V(b) = V \cap \pi^{-1}(b)$ *of* V *at* b *only contains* x, *and there is a chart* ψ *of* X *centered at* x *with domain* V *such that the expression of* π *in the charts* ψ, φ *is* $z \mapsto z^d$, *where* d *is the local degree of* π *at* x.

Lemma *Any two connected* d-*fold coverings of* U^* *are isomorphic.*

Proof of the Lemma The chart φ induces a homeomorphism from U^* to a pointed disk $D^* = D - \{0\}$. For $\beta \in U^*$, the fundamental group $\pi_1(U^*, \beta)$ is isomorphic to $\mathbb{Z}$. By (4.6.9) the isomorphism classes of connected d-fold coverings of U^* are in bijective correspondence with (conjugation classes of) subgroups of index d in $\mathbb{Z}$, and it is unique. □

Proof of the Proposition The set $V(b)$ is finite, and so $V^* = V - V(b)$ is connected; equipped with the restriction of π it is a connected component of U^* of some degree d. The image of φ is a disk $D = D_r$. Set $r' = r^{1/d}$, $D' = D_{r'}$. The map $p : z \mapsto z^d$ turns D'^* into a connected component of D^* of degree d. By the lemma, the coverings V^* and $\varphi^*(D'^*, p)$ of U^* are isomorphic. Thus there is a homeomorphism $\psi : V^* \to D'^*$ such that $p \circ \psi = \varphi \circ p : V^* \to D^*$.

We show that $V(b)$ is a singleton. Let x_i be the points of $V(b)$ with $i \in I$, where I is finite. The map $\pi : V \to U$ is proper, and so there is a neighbourhood U' of b in U and a partition $(V'_i)_{i \in I}$ of $V' = V \cap \pi^{-1}(U')$ into nonempty open sets such that $x_i \in V'_i$ for all i. And we may assume that there is some $\rho > 0$ such that $U' = \varphi^{-1}(D_\rho)$. Then $\psi(V'_i \cap V^*)$ form a partition of $D^*_{\rho'}$ into nonempty open sets, where $\rho' = \rho^{1/d}$. Hence, $\mathrm{Card}\, V(b) = 1$.

Let $V(b) = \{x\}$ and extend ψ to a bijection from V onto D', also denoted ψ, by setting $\psi(x) = 0$. We show that it is a homeomorphism. If $x' \in V^*$ tends to x, then $\pi(x')$ tends to b, and $\varphi(\pi(x')) = (\psi(x'))^d$ tends to 0; so $\psi(x')$ tends to 0. Hence ψ is continuous. If $\psi(x')$ tends to 0, then $\pi(x') = \varphi^{-1}((\psi(x'))^d)$ tends to b, and so x' tends to x since $\pi; V \to U$ is proper and $V(b) = \{x\}$. Hence ψ^{-1} is continuous.

Then $\psi : V \to D'$ is a chart for X centered at x, the expression of π in this chart is $z \mapsto z^d$, and d is necessarily the local degree of π at x. □

6.1.9

Let B be a topological surface, $a, b \in$ B with $a \neq b$, and $\gamma \in \pi_1(B - \{a\}, b)$, identified with the Poincaré group (4.9.9). The element γ will be said to *go round* a in B if there is a chart $\varphi : U \to \mathbb{D}$ for B centered at a, a point $b' \in U$ and a class β of paths from b to b' such that $\gamma = \beta^{-1} \cdot \alpha \cdot \beta$, where α is the class of the loop in (U, b') whose image under φ is $t \mapsto \varphi(b') \cdot e^{2\pi i t}$.

Proposition *Let* B *be a topological surface and* X *a finite ramified covering of* B; Δ *the ramification set. Let* $b \in B - \Delta$, $a \in \Delta$ *and* $\gamma \in \pi_1(B - \Delta, b)$ *an element going round* a *in* $B - (\Delta - \{a\})$. *Set* σ_γ *be the permutation of* X(b) *defined by* γ. *Then the order of* σ_γ *in the permutation group of* X(b) *is* l.c.m.$(r_1, \ldots, r_k)$, *where* r_i *are the ramification indices of* X *at points of* X(a).

Proof The class of paths β enables the identification of X(b) with X(b') and of $\pi_1(X - \Delta, b)$ with $\pi_1(X - \Delta, b')$; moreover these identifications are compatible with all actions. We may assume that $b = b'$, $\beta = e_b$ and $\gamma = \alpha$.

Let X$(a) = \{x_1, \ldots, x_k\}$. By Proposition 6.1.8, there are charts $\psi_1, \ldots, \psi_k$ de X with domains $V_1, \ldots, V_k$, centered respectively at $x_1, \ldots, x_k$, such that the expression of the projection $\pi : X \to B$ in the charts ψ_i and φ is $z \mapsto z^{r_i}$. The loop with expression $t \mapsto \varphi(b) \cdot e^{2\pi i t}$ can be lifted in V_i to paths with expressions $t \mapsto \lambda_i e^{2\pi i \frac{t}{r_i}}$, where $\lambda_i^{r_i} = \varphi(b)$. Hence the permutation σ_α preserves each set $V_i(b)$, and its expression in $V_i(b)$, identified by ψ_i with the set of r_i-th roots of $\varphi(b)$, is multiplication by $e^{2\pi i \frac{1}{r_i}}$. Hence $(\sigma_\alpha)^\nu = 1$ if and only if ν is a common multiple to r_i. $\qquad\square$

6.1.10

Proposition *Let* X *be a finite ramified covering of* B. *For any* $\mathbb{C}$-*analytic structure on* B, *there is a unique* $\mathbb{C}$-*analytic structure on* X *such that* $\pi : X \to B$ *is analytic.*

Proof Let δ be the ramification set of π. Let $(\varphi_i)_{i \in I}$ be an atlas for B consisting of centered charts such that, for all i, φ_i is a chart with domain U_i centered at b_i with $(U_i - \{b_i\}) \cap \Delta = \varnothing$. For each i, let $V_{i,j}$, $j \in J_i$ be the connected components of $\pi^{-1}(U_i)$, and $J = \{(i, j) \mid j \in J_i\}$. For each $(i, j) \in J$, let $\psi_{i,j}$ be a chart of X such that the expression of π in the charts $\psi_{i,j}$ and φ_i is of the form $z \mapsto z^{d_{i,j}}$ (6.1.8, Proposition). The maps $\psi_{i,j}$ form an atlas for X, and as a result of (6.1.5) the transition maps are analytic. This atlas defines an analytic structure on X with the desired property. Uniqueness follows from (6.1.5). $\qquad\square$

Corollary *Let* B *be a Riemann surface. The forgetful functor identifies the category* $\mathscr{V}_B$ *of finite analytically ramified covering over* B *with the category* $\mathscr{RR}$ *of finite topological ramified coverings of* B *and continuous maps over* B.

This is a restatement of the previous proposition for objets and of Proposition 6.1.5 for morphisms.

6.1.11

Proposition *Let* B *be a topological surface and* Δ *a closed discrete set in* B. *Then the functor* $\rho : X \mapsto X|_{B-\Delta}$ *from the category* $\mathscr{RR}_\Delta$ *of finite ramified coverings* *of* B *whose ramification set is contained in* Δ *to the category* $\mathcal{C}\!\mathit{ovf}_{B-\Delta}$ *of finite coverings of* B $-$ Δ *is an equivalence of categories.*

A quasi-inverse functor is given by the "end compactification" (see 6.1, Exercise 2).

Proof (a) We show that ρ is essentially surjective. Suppose first that B is a disk D and $\Delta = \{0\}$. As ρ commutes with finite sums, it suffices to show that every connected finite covering of B $-$ $\Delta = $ D $-$ $\{0\}$ is the restriction of a ramified covering of D. Now, any two finite sets with the same cardinality on which $\mathbb{Z}$ acts transitively are isomorphic as $\mathbb{Z}$-sets. Hence by 4.6.8, every connected d-fold covering of D $-$ $\{0\}$ is isomorphic to $(\tilde{D} - \{0\}, z \mapsto z^{d})$, where $\tilde{D}$ is the disk with radius $\sqrt[d]{r}$, r being the radius of D, and this covering is the restriction of the ramified covering $\tilde{D}$ of D.

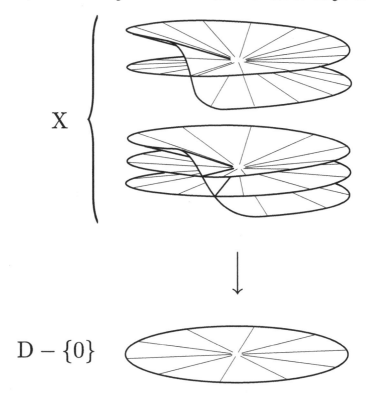

Finally, in this case, a covering X of D $-$ $\{0\}$ is the restriction of a ramified covering Y of D obtained by adding a point over 0 for each connected component of X.

(b) In the general case, let X be a finite covering of B $-$ Δ. Each $b \in \Delta$ has a neighbourhood U such that $X|_{U-b}$ is the restriction of a finite ramified covering Y_b of U. Then $Y_b = X|_{U-b} \cup F_b$ for $F_b = Y_b(b)$. By the gluing lemma for topologies (4.4.6), the set $Y = X \sqcup \bigsqcup_{b \in \Delta} F_b$ can be equipped with a topology inducing that of X and that of all Y_b. Then Y become a ramified covering of B whose restriction to B $-$ Δ is X.

(c) ρ *is fully faithful.* Let X and Y be objects of $\mathscr{R}\mathscr{R}_\Delta$ and f a covering morphism from $X|_{B-\Delta}$ to $Y|_{B-\Delta}$. Let $x_0 \in X$ be such that $b_0 = \pi(x_0) \in \Delta$, and $Y(b_0) = \{y_1, \ldots, y_k\}$. Let $W_1, \ldots, W_k$ be mutually disjoint respective neighbourhoods of $y_1, \ldots, y_k$ in Y. As Y is proper over B, there is an open neighbourhood U of b_0 in B such that $Y|_U \subseteq W_1 \cup \ldots \cup W_k$. Let V be a neighbourhood of x_0 in X homeomorphic to a disc, such that $V \cap \pi^{-1}(\Delta) = \{x_0\}$ and $\pi(V) \subset U$. Then $V - \{x_0\}$ is connected and in $X|_{B-\Delta}$. The set $f(V - \{x_0\})$ is connected and in $W_1 \cup \ldots \cup W_k$, hence in some W_i, W_{i_0} say. For any neighbourhood U' of b_0 in B the set $f^{-1}(W_{i_0}|_{U'}) = V|_{U'} - \{x_0\}$ is the trace over $X - \{x_0\}$ of a neighbourhood of x_0. However, as U' varies, $W_{i_0}|_{U'}$ form a fundamental system of neighbourhoods of y_{i_0}. Hence the map f, extended by $x_0 \mapsto y_{i_0}$, is continuous at x_0. Extending likewise at each point of $\pi^{-1}(\Delta)$ gives a continuous continuation of f to a morphism from X to Y. As $X|_{B-\Delta}$ is dense in X, this continuation is unique.   □

6.1.12

Theorem *Let* B *be a Riemann surface,* I *the ordered set of discrete closed subsets of* B *and* $\mathscr{V}_B$ *the category of finite analytically ramified coverings of* B. *Define a functor* $\omega : \mathscr{V}_B \to \varinjlim_{\Delta \in I} \mathscr{C}\!ov_{B-\Delta}$ *by associating to each object* $X \in \mathscr{V}_B$ *the underlying space of* $X_{B-\Delta}$, *where* Δ *is the ramification set of* X. *This functor is an equivalence of categories.*

Proof By Corollary 6.1.10, the category $\mathscr{V}_B$ can be identified with

$$\mathscr{R}\mathscr{R} = \bigcup_\Delta \mathscr{R}\mathscr{R}_\Delta = \varinjlim \mathscr{R}\mathscr{R}_\Delta.$$

The theorem then reduces to Proposition 6.1.11 by passing to the direct limit. □

Exercises 6.1. (Riemann surfaces)

1.—Show that the following spaces are topological surfaces:
   (a) The torus $\mathbb{T}^2 = S^1 \times S^1 = \mathbb{R}^2/\mathbb{Z}^2$ (it can be embedded in $\mathbb{R}^4$:

$$(\alpha, \theta) \mapsto (\cos \alpha, \sin \alpha, \cos \theta, \sin \theta)$$

or in $\mathbb{R}^3$:

$$(\alpha, \theta) \mapsto ((R + r \cos \theta) \cos \alpha, (R + r \cos \theta) \sin \alpha, r \sin \theta)$$

with $r < R$).

(b) The open Möbius strip

$$\left\{ \left(\left(1 + u\cos\frac{t}{2}\right)\cos t, \left(1 + u\cos\frac{t}{2}\right)\sin t, u\sin\frac{t}{2} \right) \;\middle|\; u \in \,]-1,1[,\ t \in \mathbb{R} \right\}.$$

(c) The projective plane $P^2\mathbb{R}$, i.e. the quotient of S^2 by the equivalence relation identifying two opposite points. If $a \in P^2\mathbb{R}$, then the space $P^2\mathbb{R} - \{a\}$ is homeomorphic to the open Möbius strip.

(d) The half cone $\{(t\cos\theta, t\sin\theta, t) \mid t \in \mathbb{R}_+,\ \theta \in \mathbb{R}\} \subset \mathbb{R}^3$.

(e) Plücker's conoid:

$$\{(1+u)\cos\theta, (1-u)\sin\theta, u) \mid u \in [-1,+1],\ \theta \in \mathbb{R}.\} \subset \mathbb{R}^3.$$

(f) The cone in $\mathbb{R}^4$ over a knot in $\mathbb{R}^3$:

$$\{(s(R+r\cos 3\theta)\cos 2\theta, s(R+r\cos 3\theta)\sin 2\theta, sr\sin 3\theta, s) \mid s \in \mathbb{R}_+,\ \theta \in \mathbb{R}\} \subset \mathbb{R}^4,$$

for fixed r and R with $r < R$.

2. *(End compactification)*—For a space X, let $\pi_0(X)$ denote the set of its connected components. Let $\mathscr{C}$ be the category of locally connected, locally compact spaces with finitely many connected components, and proper maps.

(a) Let $X \in \mathscr{C}$. Show that the relatively compact open subsets U such that $\pi_0(X - U)$ is finite form a cofinal set in the set of relatively compact subsets of X.

(b) Let $B(X) = \varprojlim \pi_0(X - K)$ be the inverse limit over the relatively compact subsets of X (the elements of $B(X)$ are the *ends* of X). Define a topology on $\widehat{X} = X \sqcup B(X)$ which makes it compact. Show that $X \mapsto \widehat{X}$ is a functor of $\mathscr{C}$ in the category of compact spaces.

(c) Give an example of a connected topological surface X such that $\widehat{X}$ is not a topological surface.

(d) With the notation of Proposition 6.1.11, show that if B is compact, X can be identified with the end compactification of $X|_{B-\Delta}$.

3.—(a) *(Maximum principle)* Let X be a Riemann surface, $x \in X$ and $f : X \to \mathbb{C}$ a holomorphic function. Deduce from 6.1.4 that $|f|$ has a nonzero upper or lower bound only if f is constant in the neighbourhood of x.

(b) *(Yet another proof of d'Alembert's theorem.)* Show that if a holomorphic function $f : \mathbb{C} \to \mathbb{C}$ is proper (i.e. $|f(z)|$ tends to ∞ as $|z|$ tends to ∞), then it vanishes at least at one point. In particular every non constant polynomial has at least one root in $\mathbb{C}$.

(c) Show that if a holomorphic function $f : \mathbb{C} \to \mathbb{C}$ is proper, then it is necessarily a polynomial.

6.2 Ramified Coverings and Etale Algebras

6.2.1 Separation Theorem

The next result will be assumed:

Theorem *Let* X *be a compact Riemann surface, and* $a, b \in$ X *with* $a \neq b$. *Then there is a meromorphic function* f *on* X *defined at* a *and* b *such that* $f(a) \neq f(b)$.

For a proof, see 6.2, Exercise 1, or R. Gunning, *Lectures on Riemann Surfaces* [3].

Remark The theorem still holds when X is not compact, but the proof is different. If X is not compact but is connected, then there is a better result: the *holomorphic* functions separate the points.

6.2.2 Analytic Criterion for Meromorphism

Proposition *Let* X *be a Riemann surface,* D *a discrete closed subset of* X, $f :$ X $-$ D $\to \mathbb{C}$ *a holomorphic function,* $a \in$ X, φ *a chart centered at* a *with domain* U. *Then* f *is meromorphic at* a *if and only if there is an open neighbourhood* U' *of* a *in* U *and constants* c *and* k *such that,*

$$\forall x \in U' - \{a\}, \ |f(x)| \leqslant \frac{c}{|\varphi(x)|^k} \, .$$

Proof Assume that the condition holds. Then the function $x \mapsto (\varphi(x))^k f(x)$ is holomorphic on U $-$ $\{a\}$, and bounded in the neighbourhood of a and so extends to a holomorphic function on U.

If f is meromorphic at a, then in the neighbourhood of a, $f(x) = \frac{g(x)}{h(x)}$, where $h(x) = \varphi(x)^k u(x)$, with u is holomorphic and $u(a) \neq 0$. Then u is invertible in the neighbourhood of a and $\frac{g}{u}$ is bounded by a constant c in a neighbourhood U' of a. Then

$$|f(x)| \leqslant \frac{c}{|\varphi(x)|^k} \ \text{ for } \ u \in U' - \{a\} \, .$$

$\square$

6.2.3 The Functor $\mathcal{M}$

If X is a Riemann surface, let $\mathcal{M}(X)$ be the $\mathbb{C}$-algebra of meromorphic functions on X. If X is connected, then $\mathcal{M}(X)$ is a field. Let X and Y be two Riemann surfaces and $f :$ X $\to$ Y an analytic map which is not constant on any connected component of X. Define $f^* : \mathscr{M}(Y) \to \mathscr{M}(X)$ by $f^*(h) = h \circ f$. If D is the set of poles of h, the function $f^*(h)$ is defined on $X - f^{-1}(D)$. Then $(g \circ f)^* = f^* \circ g^*$.

Theorem *Let* B *be a compact connected Riemann surface and* X *an analytically ramified d-fold covering d of* B. *Then* $\mathscr{M}(X)$ *is an etale algebra of degree d over* $\mathscr{M}(B)$.

Proof (a) Let $f \in \mathscr{M}(X)$. We show that f is algebraic of degree $\leqslant d$ over $\mathscr{M}(B)$. This will follow if there is a polynomial $P \in \mathscr{M}(B)[Z]$ such that $P(f) = 0$. Let Δ_r (resp. Δ_p) be the ramification set of X (resp. the projection of the set of poles of f), and $\Delta = \Delta_r \cup \Delta_p$. Let $b \in B - \Delta$, $X(b) = \{x_1, \ldots, x_d\}$ and $a_i(b)$ the values of the elementary symmetric functions $f(x_1), \ldots, f(x_d)$, i.e.,

$$a_1(b) = \sum_i f(x_i), \ a_2(b) = \sum_{i<j} f(x_i) f(x_j), \ldots, \ a_d(b) = \prod_i f(x_i).$$

We show that for all i, $a_i \in \mathscr{M}(B)$. All a_i are meromorphic on $B - \Delta_r$. Indeed, let $b \in B - \Delta_r$ and U a neighbourhood of b such that $X|_U$ is a trivial covering of U, i.e. $X|_U = X_1 \cup \ldots \cup X_d$, π inducing an isomorphism from each X_i onto U. Define the meromorphic functions f_i on U by $f_i \circ \pi|_{X_i} = f|_{X_i}$. Then

$$a_1|_U = f_1 + \cdots + f_d, \ a_2|_U = \sum_{i<j} f_i f_j, \ldots, \ a_d|_U = f_1 \ldots f_d.$$

Hence all a_i are meromorphic on U. Let φ be a chart centered at $b \in \Delta_r$ with domain U. For $x_i \in X(b)$, let ψ_i a chart of X centered at x_i with domain V_i and such that the expression of π in the charts ψ_i and φ_i is of the form $z \mapsto z^{d_i}$; then by 6.2.2 there is an open neighbourhood V_i of x_i, $V_i' \subset V_i$, and constants c_i and k_i such that, $\forall x_i' \in V_i'$,

$$|f(x_i')| \leqslant \frac{c_i}{|\psi_i(x_i')|^{k_i}} = \frac{c_i}{|\varphi(b')|^{k_i/d_i}} \quad \text{where} \quad b' = \pi(x_i').$$

Set $c = \sup_i c_i$ and $k = \sup_i \frac{k_i}{d_i}$. Then $|a_j(b')| \leqslant \binom{d}{j} \frac{c^j}{|\varphi(b')|^{jk}}$ and $a_i \in \mathscr{M}(B)$.

Set $P(Z) = Z^d + \sum_i (-1)^i a_i Z^{d-i} \in \mathscr{M}(B)[Z]$. We show that $P(f) \in \mathscr{M}(X)$ is null. Let $b \in B - \Delta$ and $x \in X(b)$. The polynomial $P_d(Z) = Z^d + \sum_{i=1}^d (-1)^i a_i(b) Z^{d-i}$ has roots $f(x_1), \ldots, f(x_d)$, where $X(b) = \{x_1, \ldots, x_d\}$. In particular,

$$f(x)^d + \sum_{i=1}^d (-1)^i a_i(b) f(x)^{d-i} = 0.$$

In other words, $P(f)(x) = 0$. As this holds for all $x \in X|_{B-\Delta}$, $P(f) = 0$.

(b) We show that $\mathscr{M}(X)$ is etale of degree $\leqslant d$ over $\mathscr{M}(B)$. For this, X may be assumed to be connected. Then $\mathscr{M}(X)$ is an extension of $\mathscr{M}(B)$. It is the directed union of finitely generated sub-extensions E_λ of $\mathscr{M}(B)$. Each E_λ is finitely generated.

It is generated by algebraic elements and so is finite and hence etale since $\mathcal{M}(B)$ is of characteristic 0, and so cyclic by the primitive element theorem, thus of degree $\leqslant d$. Therefore the degree of $\mathcal{M}(X)$ is at most d.

(c) We show that $\mathcal{M}(X)$ is of degree $\geqslant d$.

The separation theorem has the following corollary:

Lemma *Let X be a Riemann surface and $a_1, \ldots, a_d$ distinct points of X. Then there is a meromorphic function f on X, defined at each point $a_1, \ldots, a_d$ with distinct values $f(a_1), \ldots, f(a_d)$.*

Proof of the Lemma We show that for all i, j with $i \neq j$, there is a function f_{ij} defined at $a_1, \ldots, a_d$ and such that $f(a_i) \neq f(a_j)$. Let $i \neq j$. By the separation theorem, there is a function g defined at a_i and a_j with $g(a_i) \neq g(a_j)$. If necessary adding a constant, we may assume that $g(a_k) \neq 0$ for all $k \in \{1, \ldots, d\}$. Then $f_{ij} = \frac{1}{g}$ has the required property. Let A be the vector subspace of $\mathcal{M}(X)$ consisting of the functions defined at $a_1, \ldots, a_d$. For each (i, j) the set V_{ij} of $f \in A$ such that $f(a_i) = f(a_j)$ is a strict vector subspace of A. Hence $\bigcup V_{ij} \neq A$ and any function $f \in A - (\bigcup V_{ij})$ has the required property.   □

End of the Proof of the Theorem Let $b_0 \in B - \Delta_r, X(b_0) = \{x_1, \ldots, x_d\}$, and $f \in \mathcal{M}(X)$ a function defined at $x_1, \ldots, x_d$ with distinct values. We show that f has degree $\geqslant d$ over $\mathcal{M}(B)$, i.e. that the minimal polynomial of f over $\mathcal{M}(B)$ has degree $\geqslant d$. Let $P = \sum_0^k c_i Z^i$ be a polynomial with coefficients $c_i \in \mathcal{M}(B)$ and such that $P(f) = 0$, i.e., for all $x \in X$ and $b = \pi(x)$ for which $f(x)$ and $c_i(b)$ are defined, $\sum_0^k c_i(b)(f(x))^i = 0$. We show that $k \geqslant d$. Note that for b sufficiently near b_0, f is defined on $X(b)$ with distinct values at its points. If b is sufficiently near b_0 and is not a pole for any c_i, then $P_b = \sum c_i(b)Z^i \in \mathbb{C}[Z]$ has at least d distinct roots: the values of f at the points of $X(b)$. Hence $k \geqslant d$. This result in particular applies to the minimal polynomial of f. Consequently, $\deg_{\mathcal{M}(B)} f \geqslant d$ and $\deg_{\mathcal{M}(B)} \mathcal{M}(X) \geqslant d$.   □

Remark Let X be a finite analytically ramified covering of B, $b \in B - \Delta$, and $f \in \mathcal{M}(X)$ a function defined on $X(b)$ with distinct values at its points. The above proof shows that f generates $\mathcal{M}(X)$ as an algebra over $\mathcal{M}(B)$.

6.2.4

The main aim of this section is to prove the next result:

> **Theorem** *Let $\mathcal{V}_B$ be the category of finite analytically ramified coverings of B and $\mathcal{E}_B$ the category of etale algebras over $K = \mathcal{M}(B)$. Then, the functor $\mathcal{M} : X \mapsto \mathcal{M}(X)$ from $\mathcal{V}_B$ to $\mathcal{E}_B$ is an anti-equivalence of categories.*

The proof will be given in 6.2.9.

Corollary 6.1 *Let* X *and* Y *be finite analytically ramified coverings of* B, *and* $\Delta \subset B$ *a finite set such that* X *and* Y *are unramified over* B $-$ Δ. *Then* $\mathrm{Hom}_{\mathscr{E}_B}(\mathscr{M}(X), \mathscr{M}(Y))$ *can be identified with* $\mathrm{Hom}_{\mathscr{C}ov\!f(B-\Delta)}(Y|_{B-\Delta}, X|_{B-\Delta})$.

Corollary 6.2 *Let* X *be a connected finite analytically ramified covering of* B, *and* $\Delta \subset B$ *a finite set containing the ramification set of* X. *Then* $\mathscr{M}(X)$ *is Galois extension of* $\mathscr{M}(B)$ *if and only if* $X|_{B-\Delta}$ *is a Galois covering of* B $-$ Δ. *In this case, the algebraic Galois group* $\mathrm{Aut}_{\mathscr{M}(B)}\mathscr{M}(X)$ *can be identified with the opposite group to the topological Galois group* $\mathrm{Aut}_{B-\Delta}(X|_{B-\Delta})$. *In particular these Galois groups are isomorphic.*

6.2.5 Construction of a Covering $\mathscr{S}(E, \zeta)$

Let E be an etale algebra over $\mathscr{M}(B)$ and $\zeta \in E$ a primitive element, i.e. generating E as algebra (5.4.11). Let

$$P(Z) = Z^d + a_1 Z^{d-1} + \cdots + a_d$$

be the minimal polynomial of ζ in E. Denote by $\Delta_{E,\zeta}$ the set of elements $b \in B$ where either the coefficients of P are not defined or $P_b = Z^d + a_1(b)Z^{d-1} + \cdots + a_d(b) \in \mathbb{C}[Z]$ has multiple roots.

Lemma *The set* $\Delta_{E,\zeta}$ *is finite.*

Proof Let $P = Z^d + a_1 Z^{d-1} + \cdots + a_d$ be a monic polynomial over $\mathscr{M}(B)$. Consider the discriminant σ of P (3.7.12). It is in $\mathscr{M}(B)$. Denote by Δ' the set of points of B where some a_i is not defined, and let $b \in B - \Delta'$. All the roots of $P_b \in \mathbb{C}[Z]$ are distinct if and only if its discriminant $\sigma(b)$ is nonzero.

If $\mathscr{M}(B)[Z]/(P)$ is etale, then by 5.4.4, σ is a nonzero element of $\mathscr{M}(B)$. Then the set Δ'' of the zeros of σ is finite, and $\Delta_{E,\zeta} = \Delta' \cup \Delta''$ is finite $\square$

Denote by $\mathscr{S}(E, \zeta)$ the subspace of $(B - \Delta) \times \mathbb{C}$ consisting of pairs (b, z) such that $P(b, z) = 0$. By 4.3.4, Example 6, $\mathscr{S}(E, \zeta)$ is a finite covering of $B - \Delta_{E,\zeta}$.

6.2.6 Functoriality

Let E and F be two etale algebras over $\mathscr{M}(B)$, ζ and θ primitive elements of E and F respectively, P and Q their minimal polynomials over $\mathscr{M}(B)$ and $\alpha : E \to F$ an algebra homomorphism. We do not assume that $\alpha(\zeta) = \theta$. We are going to define a morphism $\alpha^* : \mathscr{S}(F, \theta) \to \mathscr{S}(E, \zeta)$ in the category $\varprojlim \mathscr{C}ov\!f(B - \Delta)$.

The element $\alpha(\zeta) \in F$ takes the form $R(\theta)$, where $R \in \mathscr{M}(B)[Z]$. Let Δ'_R be the set of points of B where the coefficients of R are not all defined. Define a map $\widetilde{R}$ from $(B - \Delta'_R) \times \mathbb{C}$ to itself by $\widetilde{R}(b, z) = (b, R(b, z))$.

Lemma and Definition *The map* $\widetilde{R}$ *induces a map from* $\mathscr{S}(F, \theta)|_{B-\Delta'_R}$ *to* $\mathscr{S}(E, \zeta)$, *which does not depend on the choice of* R. *It is written* α^*.

Proof Since $P(R(\theta)) = P(\alpha(\zeta)) = \alpha(P(\zeta)) = 0 \in F$, $P \circ R = Q \cdot V$ for some $V \in \mathscr{M}(B)[Z]$. In other words, $P(b, R(b, z)) = Q(b, z)V(b, z)$ for $b \in B$ such that the coefficients of P, Q, R, V are defined and $z \in \mathbb{C}$. If $(b, z) \in \mathscr{S}(F, \theta)$, then $Q(b, z) = 0$; so $P(b, R(b, z)) = 0$, and $\widetilde{R}(b, z) = (b, R(b, z)) \in \mathscr{S}(E, \zeta)$.

Let $R_1 \in \mathscr{M}(B)[Z]$ be another polynomial such that $\alpha(\zeta) = R_1(\theta)$. Then $R - R_1$ vanishes at θ, and so is of the form $Q \cdot W$ with $W \in \mathscr{M}(B)[Z]$. If $Q(b, z) = 0$, $R(b, z) = R_1(b, z)$; so $\widetilde{R}$ and $\widetilde{R}_1$ agree on $\mathscr{S}(F, \theta)$.   □

Let E, F and G be etale algebras over $\mathscr{M}(B)$, ζ, θ and λ primitive elements of E, F and G respectively, $\alpha : E \to F$ and $\beta : F \to G$ homomorphisms. Then $(\beta \circ \alpha)^* = \alpha^* \circ \beta^*$. Indeed, let R_1 and $R_2 \in \mathscr{M}(B)[Z]$ be such that $\alpha(\zeta) = R_1(\theta)$ and $\beta(\theta) = R_2(\lambda)$. So $\beta(\alpha(\zeta)) = \beta(R_1(\theta)) = R_1(\beta(\theta)) = R_1(R_2(\lambda))$. Hence for R_3 such that $\beta \circ \alpha(\zeta) = R_3(\lambda)$, we may choose $R_3 = R_1 \circ R_2$. Then $(\beta \circ \alpha)^*$ is induced by $\widetilde{R}_3 = \widetilde{R}_1 \circ \widetilde{R}_2$, and so $(\beta \circ \alpha)^* = \alpha^* \circ \beta^*$ since α^* and β^* are respectively induced by $\widetilde{R}_1$ and $\widetilde{R}_2$.

6.2.7

Let X be a finite analytically ramified covering of B and Δ_X its ramification set. Let $f \in \mathscr{M}(X)$ and Δ'_f be the projection of the set of points where f is not defined. Assume that there exists $b \in B - \Delta_X$ such that f is defined on $X(b)$ with distinct values at its points. Then we know (6.2.3, Remark) that f is a primitive element of $\mathscr{M}(X)$ and the covering $\mathscr{S}(\mathscr{M}(X), f) \to B - \Delta_f$ may be considered. Set $\Delta = \Delta_X \cup \Delta'_f \cup \Delta_f$.

Proposition *The map* $(\pi, f) : X|_{B-\Delta} \to (B - \Delta) \times \mathbb{C}$ *induces a covering isomorphism from* $X|_{B-\Delta}$ *onto* $\mathscr{S}(\mathscr{M}(X), f)|_{B-\Delta}$.

Proof Let P be the minimal polynomial of f in $\mathscr{M}(X)$. For $x \in X$ such that f is defined at x and the coefficients of P are defined at $b = \pi(x)$, $P(b, f(x)) = 0$. Hence (π, f) is indeed a continuous map from $X|_{B-\Delta}$ to $\mathscr{S}(\mathscr{M}(X), f)$; since it commutes with the projections, it is a covering morphism. When $b \in \Delta$ (which can be shown never to occur), take $b' \in B - \Delta$ sufficiently near b so that the assumptions made still hold for b. Then the covering morphism (π, f) is bijective on the fibres at b'. As a result of (4.3.6. Corollary 4.2) and (6.1.7 remarque 3), it is an isomorphism. □

A Riemann surface of X is said to be *algebraic* if there is a polynomial $Q \in \mathbb{C}[T, Z]$ (where T and Z are indeterminates), and finite sets $\Theta \subset X$ and $\Lambda \subset V = Q^{-1}(0)$ such that $X - \Theta$ and $V - \Lambda$ are isomorphic Riemann surfaces.

Corollary *Every compact Riemann surface is algebraic.*

Proof Let X be a compact Riemann surface and $\pi \in \mathcal{M}^{**}(X)$ (i.e. π is a meromorphic function on X which is not constant on any connected component). Then (X, π) is an analytically ramified covering of Σ, and $\mathcal{M}(X)$ equipped with π^* is an etale algebra over $\mathcal{M}(\Sigma) = \mathbb{C}(T)$. Choose a primitive element $f \in \mathcal{M}(X)$, and let $P \in \mathbb{C}(T)[Z]$ be its minimal polynomial. Then, there are $Q \in \mathbb{C}[T, Z]$ and $D \in \mathbb{C}[T]$ such that $P(T, Z) = Q(T, Z)/D(T)$. Define Δ as above. Then $X|_{B-\Delta}$ is isomorphic to $\mathcal{S}(\mathcal{M}(X), f)|_{B-\Delta} = \{(x, z) \in (\mathbb{C} - \Delta) \times \mathbb{C} \mid Q(x, z) = 0\}$. $\qquad\square$

6.2.8

Let E be an etale algebra over $\mathcal{M}(B)$ and $\zeta \in E$ a primitive element. Consider the covering $\mathcal{S}(E, \zeta)$ of $B - \Delta$. By (6.1.12), there is an analytically ramified covering $\widehat{\mathcal{S}}(E, \zeta)$ of B such that $\widehat{\mathcal{S}}(E, \zeta)|_{B-\Delta} = \mathcal{S}(E, \zeta)$. Denote by Z the map from $\mathcal{S}(E, \zeta)$ to $\mathbb{C}$ induced by the second projection $B \times \mathbb{C} \to \mathbb{C}$.

Proposition *(a) The function Z is meromorphic on $\widehat{\mathcal{S}}(E, \zeta)$.*
(b) There is a unique isomorphism $\varphi : E \to \mathcal{M}(\widehat{\mathcal{S}}(E, \zeta))$ such that $\varphi(\zeta) = Z$.

To prove (a), we use the criterion for meromorphism (6.2.2). To do so, we will need an upper bound.

Lemma (Upper bound of the roots of a polynomial) *Let*

$$P = Z^d + a_1 Z^{d-1} + \cdots + a_d$$

be a polynomial over $\mathbb{C}$ and z a root of P. Then

$$|z| \leqslant \sup(1, |a_1| + \cdots + |a_d|).$$

Proof If $|z| \geqslant 1$, then $z = -a_1 - \frac{a_2}{z} - \cdots - \frac{a_d}{z^{d-1}}$, and so $|z| \leqslant |a_1| + \cdots + |a_d|$. $\qquad\square$

Proof of the Proposition (a) The function Z is clearly holomorphic over $B - \Delta$. Let $b \in \Delta$ and $x \in X(b)$, where $X = \widehat{\mathcal{S}}(E, \zeta)$. Let φ be a chart for B centered at b and ψ a chart of X centered at x such that the expression of π in the charts ψ and φ is $z \mapsto z^r$. The coefficients $a_1, \ldots, a_d$ of the minimal polynomial P of ζ are meromorphic functions on B. Let $\dot{a}_1, \ldots, \dot{a}_d$ be their expression in the chart φ. For x' near x, the number $Z(x')$ is a root of $Z^d + a_1(b')Z^{d-1} + \cdots + a_d(b')$, where $b' = \pi(x')$, and so

$$|Z(x')| \leqslant \sup(1, |a_1(b')| + \cdots + |a_d(b')|).$$

However, $a_i(b') = \dot{a}_i(\varphi(b')) = \dot{a}_i(\psi(x')^r)$. Being meromorphic, the functions a_i can be bounded above as described in (6.2.2). Hence the same holds for Z, and Z is meromorphic.

(b) For $(b, z) \in \mathscr{S}(E, \zeta)$, $P(b, z) = 0$. In other words, $P(Z) = 0$ in $\mathscr{M}(X)$. We show that P is the minimal polynomial of Z in $\mathscr{M}(X)$. Let Q be a polynomial over $\mathscr{M}(B)$ such that $Q(Z) = 0$. We show that Q is a multiple of P. If necessary dividing Q by P, the degree of Q can be assumed to be strictly less than the degree d of P. For all $b \in B - \Delta$ where the coefficients of Q are defined, the points of $X(b)$ are the pairs (b, z), z being a root of P_b, and so the roots of P_b are roots of Q_b. As there are d distinct ones, $Q_b = 0$. Hence $Q = 0$.

So there is a unique homomorphism of $\mathscr{M}(B)$-algebras $\varphi : E \to \mathscr{M}(X)$ such that $\varphi(\zeta) = Z$, and φ is injective. By assumption, the algebra E has degree d, and X is a ramified d-fold covering, and so by (6.2.3, Theorem) $\mathscr{M}(X)$ is an algebra of degree d. Hence φ is an isomorphism. $\qquad\Box$

6.2.9 Proof of Theorem 6.2.4

(a) *The functor $\mathscr{M}$ is essentially surjective.* Let E be an etale algebra over $\mathscr{M}(B)$, and ζ a primitive element of E. By Proposition 6.2.8, E is isomorphic to $\mathscr{M}(\widehat{\mathscr{S}}(E, \zeta))$.

(b) *The functor $\mathscr{M}$ is fully faithful.* Let X and Y be finite analytically ramified covering of B and $u : \mathscr{M}(X) \to \mathscr{M}(Y)$ a $\mathscr{M}(B)$-algebra homomorphism. We show that there is a morphism $h : Y \to X$ such that $h^* = u$. Let f and g be primitive elements of $\mathscr{M}(X)$ and $\mathscr{M}(Y)$ respectively, satisfying assumptions 6.2.7. The isomorphism $(\pi, f) : X|_{B-\Delta} \to \mathscr{S}(\mathscr{M}(X), f)$ extends to an isomorphism $\tilde{f} : X \to \widehat{\mathscr{S}}(\mathscr{M}(X), f)$. Define $\tilde{g} : Y \to \widehat{\mathscr{S}}(\mathscr{M}(Y), g)$ likewise. By 6.1.12 the morphism $u^* : \mathscr{S}(\mathscr{M}(Y), g) \to \mathscr{S}(\mathscr{M}(X), f)$ extends to a morphism which is also written $u^* : \widehat{\mathscr{S}}(\mathscr{M}(Y), g) \to \widehat{\mathscr{S}}(\mathscr{M}(X), f)$. Define $h : Y \to X$ by the commutativity of the diagram

$$
\begin{array}{ccc}
Y & \xrightarrow{\ \ h\ \ } & X \\
{\scriptstyle \tilde{g}}\downarrow & & \downarrow{\scriptstyle \tilde{f}} \\
\widehat{\mathscr{S}}(\mathscr{M}(Y), g) & \xrightarrow{\ u^*\ } & \widehat{\mathscr{S}}(\mathscr{M}(X), f)
\end{array}
$$

We show that $h^* = u$. As f is primitive, it suffices to show that $u(f) = h^*(f) = f \circ h : Y \to \mathbb{C}$. Now, $f \circ h = Z \circ \tilde{f} \circ h = Z \circ u^* \circ \tilde{g}$. For $y \in Y(b)$, $u^*(\tilde{g}(y)) = u^*(b, g(y)) = \widetilde{R}(b, g(y)) = (b, R(b, g(y)))$, where R is a polynomial over $\mathscr{M}(B)$ such that $u(f) = R(g)$. Hence $Z \circ u^* \circ \tilde{g} = R(g) = u(f)$, and so $h^* = u$.

We next show uniqueness. Let h_1 be such that $h_1^* = u$. Then $h_1^*(f) = h^*(f)$, in other words $f \circ h_1 = f \circ h$, where $\tilde{f} \circ h_1 = \tilde{f} \circ h$, and so $h_1 = h$ since $\tilde{f}$ is an isomorphism. $\qquad\Box$

Remark Consider the category $\mathcal{V}_B$ of finite analytically ramified coverings of B, the category $\mathcal{E}_B$ of etale algebras over $\mathcal{M}(B)$ and the category $\mathcal{R} = \varinjlim \mathcal{C}ovf_{B-\Delta}$. For each $E \in \mathcal{E}_B$ choose a primitive element. This defines a functor $\vec{\delta} : \mathcal{E}_B \to \mathcal{R}$. Then up to functorial isomorphism, there is a commutative diagram:

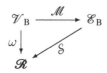

where ω is the forgetful functor. In this diagram, the three arrows are either equivalences or anti-equivalences of categories.

6.2.10   Dictionary

Let X and Y be finite analytically ramified coverings of S.

(a) The degree of the algebra $\mathcal{M}(X)$ over $\mathcal{M}(B)$ is equal to the degree of X over B, i.e. to the degree of the covering $X|_{B-\Delta}$ of $B - \Delta$ (see 6.2.3).

(b) $\mathcal{M}(X \sqcup Y) = \mathcal{M}(X) \times \mathcal{M}(Y)$. The algebra $\mathcal{M}(X)$ is an extension of $\mathcal{M}(B)$ if and only if X is connected. More generally, the connected components of X are in bijective correspondence with the maximal ideals of $\mathcal{M}(X)$.

(c) In general, the fibre product $X \times_B Y$ is not a ramified covering of B, but a finite analytically ramified covering Z of B satisfying $Z|_{B-\Delta} = (X \times_B Y)|_{B-\Delta}$, where $\Delta = \Delta_X \cup \Delta_Y$, may be constructed. Then $\mathcal{M}(Z) = \mathcal{M}(X) \otimes_{\mathcal{M}(B)} \mathcal{M}(Y)$.

(d) Suppose that X is connected. Set $\Delta = \Delta_X \cup \Delta_Y$. Then $X|_{B-\Delta}$ trivializes the covering $Y|_{B-\Delta}$ is and only if $\mathcal{M}(X)$ diagonalizes $\mathcal{M}(Y)$. In particular $X|_{B-\Delta_X}$ is a Galois covering if and only if $\mathcal{M}(X)$ is a Galois extension of $\mathcal{M}(B)$. In this case, $\mathrm{Aut}_{\mathcal{M}(B)}(\mathcal{M}(X)) = \mathrm{Aut}_B(X) = \mathrm{Aut}_{B-\Delta}(X|_{B-\Delta})$ since the functor $\mathcal{M}$ is fully faithful.

(e) Let $f : Y \to X$ be morphism. Then $f^* : \mathcal{M}(X) \to \mathcal{M}(Y)$ is injective (resp. surjective) if and only f is surjective (resp. injective).

(f) Let $f \in \mathcal{M}(X)$. Then f is a primitive element of $\mathcal{M}(X)$ if and only if there exists $b \in B - \Delta_X$ such that f defines an injective map from $X(b)$ to $\mathbb{C}$. This condition has been seen to be sufficient (6.2.3, Remark). We show that it is necessary: suppose that f is primitive and consider the morphism $(\pi, f) : X|_{B-\Delta} \to S(\mathcal{M}(X), f)$. Its image is a clopen subset S_1 of $S(\mathcal{M}(X), f)$, which can be embedded in an analytic covering $\widehat{S_1}$. The morphism $\tilde{f} : X \to \widehat{S_1}$ extending (π, f) defines a homomorphism $\tilde{f}^* : \mathcal{M}(\widehat{S_1}) \to \mathcal{M}(X)$ which is surjective since f is in the image. Hence $\deg_B(\widehat{S_1}) \geqslant \deg_B(X) = \deg_{B-\Delta}(S(\mathcal{M}(X), f))$, and so $S_1 = S(\mathcal{M}(X), f)$ and f is an isomorphism. Thus, for $b \in B - \Delta$, the map f induces an injection from $X(b)$ to $\mathbb{C}$.

6.2.11

Let E be an etale algebra over $\mathcal{M}(B)$. Then, up to isomorphism, there is a unique analytically ramified covering X of B such that E is isomorphic to $\mathcal{M}(X)$. Let Δ_E be the ramification set of X. It is said to be the ramification set of E.

If F is a subalgebra of E, then $\Delta_F \subset \Delta_E$. Indeed, if $F \approx \mathcal{M}(Y)$ is a surjective morphism $f : X \to Y$, and if $b \notin \Delta_E$, then there is a local section of X passing through all points of $X(b)$. Hence composing with f gives a local section of Y passing through all points of $Y(b)$ and $b \notin \Delta_F$.

Similarly, if F is a quotient of E, then $\Delta_F \subset \Delta_E$.

If E and F are etale algebras over $\mathcal{M}(B)$, then $\Delta_{E \times F} = \Delta_E \cup \Delta_F$ and $\Delta_{E \otimes F} \subset \Delta_E \cup \Delta_F$ by 6.2.10, (b) and (c). If E and F are nonzero, then $\Delta_{E \otimes F} = \Delta_E \cup \Delta_F$ since E and F can then be identified with subalgebras of $E \otimes F$.

If F and G are subalgebras of an etale algebra E over $\mathcal{M}(B)$, then $\Delta_{F \cdot G} = \Delta_F \cup \Delta_G$. Indeed $\Delta_{F \cdot G} \subset \Delta_F \cup \Delta_G$ since $F \cdot G$ is a quotient de $F \otimes G$, and the opposite inclusion holds since F and G are subalgebras of $F \cdot G$.

6.2.12

Proposition *With the notation of (6.2.5) and (6.2.11), the set Δ_E is the intersection of all $\Delta_{E,\zeta}$, where ζ is a primitive element of the $\mathcal{M}(B)$-algebra E.*

Proof For a primitive element ζ, $S(E, \zeta)$ is an (unramified) covering of $B - \Delta_{E,\zeta}$, and so $\Delta_E \subset \Delta_{E,\zeta}$. Hence $\Delta_E \subset \bigcap \Delta_{E,\zeta}$.

Let $b \in B - \Delta_{E,\zeta}$, and $X(b) = \{x_1, ..., x_d\}$. By the separation theorem (6.2.1), there is a function $f \in \mathcal{M}(X)$ defined at $x_1,...,x_d$, with distinct values at these points. Let $P \in \mathcal{M}(B)[Z]$ be the minimal polynomial of f on $\mathcal{M}(B)$. For all $b' \in B - \Delta_P$, $P_{b'} \in \mathbb{C}[Z]$ vanishes at $f(x')$ for $x' \in X(b')$, and so the coefficients of $P_{b'}$ are the elementary symmetric functions of these numbers (with sign $(-1)^{d-i}$). Hence these coefficients are defined at b and the roots of P_b are $f(x_1),...,f(x_d)$. As these are distinct, the discriminant of P_b is nonzero and $b \notin \Delta_{E,f}$. Since the degree of P is d, the function f is a primitive element of E and $b \notin \bigcap \Delta_{E,\zeta}$. Thus $\bigcap \Delta_{E,\zeta} \subset \Delta_E$. $\square$

Exercises 6.2. (Ramified coverings and etale algebras)

1. *(Proof of the separation theorem)*—Let X be a Riemann surface. A *differential form*[2] α *of degree* 1 on X assigns to each point $x \in X$ a $\mathbb{R}$-linear map α_x from the tangent space $T_x X$, a 1-dimensional $\mathbb{C}$-vector space, to $\mathbb{C}$. For a C^1 function f on X, the *differential* of f is the 1-form df, where $df_x(u) = u.f$ denotes the derivative of f along the vector $u \in T_x X$. If X is an open subset of $\mathbb{C}$, every 1-form can be

---

[2]See Cartan [4] for the notion of basis for differential forms on manifolds, and Weil [5] for notions specific to complex analytic manifolds.

written $\alpha = f_1(z)\,dx_1 + f_2(z)\,dx_2$ in a chart for X with coordinate $z = x_1 + ix_2$. In particular

$$df = \frac{\partial f}{\partial x_1}dx_1 + \frac{\partial f}{\partial x_2}dx_2 \,.$$

If for all x, the map α_x is $\mathbb{C}$-linear (resp. $\mathbb{C}$-antilinear), α is said to be of *type* $(1, 0)$ (resp. $(0, 1)$). The forms of type $(1, 0)$ (resp. $(0, 1)$) with respect to a chart are $g(z)\,dz$ (resp. $h(z)\,d\bar{z}$).

A *differential 2-form* η assigns to each point $x \in X$ an alternating $\mathbb{R}$-bilinear map $\eta_x : T_xX \times T_xX \to \mathbb{C}$. It is said to be *real* (resp. *positive*, resp. *strictly positive*) if

$$(\forall x \in X)\ (\forall u \in T_xX - \{0\})\quad \eta_x(u, iu) \in \mathbb{R}\ (\text{resp.} \geqslant 0, \ \text{resp.} > 0)\,.$$

If η is a 2-form, then define $\int_X \eta$ as in Bourbaki [6], §10.

If α and β are 1-forms, define the 2-form $\alpha \wedge \beta$ by

$$(\alpha \wedge \beta)_x(u, v) = \alpha_x(u)\beta_x(v) - \alpha_x(v)\beta_x(u)\,.$$

Finally the *exterior derivative* of a C^1 1-form α is a 2-form written $d\alpha$. whose definition can be found in [4]. For practical calculations, in particular for question b,γ), it is sufficient to know the formulas $d(f\,dg) = df \wedge dg$ (for C^1 functions f, g) and $\int_X d\alpha = 0$ for a C^1 1-form α with compact support.

A. *Hilbert norm*

[For the exercise, the results in this part may be assumed.]

Let $\mathbb{D} \subset \mathbb{C}$ be the open unit disc. If $f : \mathbb{D} \to \mathbb{C}$ is a continuous function then for $A \subset \mathbb{D}$ set

$$\|f\|_A = \left(\int_A |f(x + iy)|^2\,dxdy\right)^{\frac{1}{2}}\,.$$

If $\alpha = u\,dx + v\,dy$ is a differential form then set

$$\|\alpha\|_A = \left(\|u\|_A^2 + \|v\|_A^2\right)^{\frac{1}{2}}\,.$$

(a) Show that if A and B are compact subsets in $\mathbb{D}$ such that $A \subset \overset{\circ}{B}$, then there is a constant q such that for any C^1 function f on $\mathbb{D}$ (i.e. continuously $\mathbb{R}$-differentiable), there is a constant M satisfying

$$\|f - \mathrm{M}\|_A \leqslant q\,\|df\|_B\,.$$

Does this inequality continue to hold with $B = A$ when A is a disc?

(b) For a C^1 function f on $\mathbb{D}$, define

$$\frac{\partial f}{\partial z} \ \text{and} \ \frac{\partial f}{\partial \bar{z}} \ \text{by} \ df = \frac{\partial f}{\partial z}dz + \frac{\partial f}{\partial \bar{z}}d\bar{z}\,.$$

Let h be a square-integrable function on $\mathbb{D}$. Show that the following conditions are equivalent:

(i) h agrees nearly everywhere with a holomorphic function;
(ii) for any C^1 function f with compact support in $\mathbb{D}$,

$$\int_{\mathbb{D}} h \frac{\partial f}{\partial \bar{z}}\, dxdy = 0 .$$

B. *Forms of type $(1, 0)$ or $(0, 1)$*

Let X be a connected compact Riemann surface.

(a) Show that every 1-form on X is the sum of a unique form of type $(1, 0)$ and a unique form of type $(0, 1)$.

(b) For every 1-form α, define a form $*\alpha$ by

$$(*\alpha)_x = -\tau \circ \alpha_x \circ J ,$$

where $\tau : \mathbb{C} \to \mathbb{C}$ is conjugation $z \mapsto \bar{z}$, and J multiplication by i. Show that, if α is of type $(1, 0)$ (resp. $(0, 1)$), then $*\alpha$ is of type $(0, 1)$ (resp. $(1, 0)$). Show that, for every form α, $\alpha \wedge *\alpha$ is real positive, and strictly positive at all points where α does not vanish.

Show that if α are β two 1-forms, then

$$\alpha \wedge *\beta = \overline{\beta \wedge *\alpha} .$$

Set $(\alpha|\beta) = \int_X \alpha \wedge *\beta$. With this scalar product, the space of continuous 1-forms is a pre-Hilbert space. Let E^1 be its completion. Show that $E^1 = E^{1,0} \oplus E^{0,1}$, where $E^{1,0}$ is the closure of the space of continuous forms of type $(1, 0)$ and $E^{0,1}$ is defined likewise, and that these subspaces are orthogonal.

(c) For a C^1 function f, let $d'f$ and $d''f$ be the parts of type $(1, 0)$ and $(0, 1)$ of df. Show that $*d'f = -id''\bar{f}$ and $*d''f = id'\bar{f}$. Assuming that f is C^2, calculate $d(f\, d\bar{f})$, and show that $\|d'f\| = \|d''f\|$. Does this inequality hold when f is C^1?

C. *Hilbert Complex*

Let ω be a continuous 2-form strictly positive on X. Let E^0 be the completion of the space of C^1 functions with respect to the norm

$$f \mapsto \left(\int |f|^2 \omega + \int df \wedge *df \right)^{\frac{1}{2}} .$$

The maps $f \mapsto df$, $f \mapsto d'f$ and $f \mapsto d''f$ extend to maps $d : E^0 \to E^1$, $d' : E^0 \to E^{1,0}$, and $d'' : E^0 \to E^{0,1}$. Using A, (a) show that $d : E^0 \to E^1$ is a strict morphism (i.e. that the algebraic isomorphism $E^0/\operatorname{Ker} d \to \operatorname{Im} d$ is a homeomorphism). Generalize the equality proved in B, (c) to $f \in E^0$. Deduce that $d'' : E^0 \to E^{0,1}$ is a strict morphism. What is its kernel?

D. *Finiteness theorem*

(a) For 1-forms α and β, set

$$\langle \alpha, \beta \rangle = \int \alpha \wedge \beta \,.$$

Show that $(\alpha, \beta) \mapsto \langle \alpha, \beta \rangle$ extends to a bilinear form on $E^1 \times E^1 \to \mathbb{C}$, which defines an isomorphism between $E^{1,0}$ and the dual of $E^{0,1}$.

(b) Define E^2 to be the dual of E^0. Show that $\alpha \mapsto d\alpha$ extends to a continuous linear map $d : E^1 \to E^2$ characterized by

$$\langle f, d\alpha \rangle = -\langle df, \alpha \rangle \,.$$

(c) Let $d'' : E^{1,0} \to E^2$ be the map induced by d. Show that its kernel is the space Ω of holomorphic differential forms on X, i.e. whose local expressions are $f(z)dz$ for f holomorphic (use A, (b)).

d) Let H denote the cokernel of $d'' : E^0 \to E^{0,1}$. Show that the scalar product $(\alpha, \beta) \mapsto \langle \alpha, \beta \rangle$ defines an isomorphism between H and the dual of Ω (a particular case of the Serre duality theorem).

(e) Using (4.6, Exercise 13) and (4.9, Exercise 2), show that Ω is finite dimensional. Deduce that so is H (finiteness theorem).

(e') Using Montel's theorem for normal families, show that Ω is locally compact. Deduce another proof of the finiteness of the dimension of Ω, hence of H.

E. *Construction of meromorphic functions*

Let $x \in X$ and (U, Z) an analytic chart for X centered at x.

Let η be a C^1 function with compact support in U, equal to 1 in the neighbourhood of x. Set

$$\alpha_n = d'' \left(\eta \cdot \frac{1}{Z^n} \right).$$

(a) Show that the form α_n is continuous of type (0, 1) with support in U. Extend it by 0 to a form on X.

(b) Show that there exist nonzero numbers c_k such that the form $\sum_1^{g+1} c_k \alpha_k$, where $g = \dim H = \dim \Omega$ (analytic genus), can be written as $d'' h$ with $h \in E^0$.

(c) Show that h is holomorphic on $X - U$ and in the neighbourhood of x. Set $f = h - \eta$. Show that f is holomorphic on $U - \{x\}$, meromorphic on X and has a pole of order $\geqslant 1$ and $\leqslant g + 1$ at x.

(d) Let $y \neq x \in X$. Show that there is a meromorphic function f on X, such that $f(y) = 1$ with a pole at x. Conclude by considering $\frac{1}{f}$.

6.3 Extensions of $\mathbb{C}$ with Transcendence Degree 1

6.3.1

Consider the Riemann sphere Σ (6.1.2).

Proposition *Let* X *be a Riemann surface. Extend every meromorphic function* f *on* X *to a map* $\tilde{f} : X \to \Sigma$ *by assigning the value* ∞ *to the poles of* f. *This gives a bijection* $f \mapsto \tilde{f}$ *from* $\mathscr{M}(X)$ *onto the set of analytic maps from* X *to* Σ, *non constant and distinct from* ∞ *on every connected component of* X.

Proof Let $f \in \mathscr{M}(X)$ and $x_0 \in X$. In the neighbourhood of x_0, $f = \frac{g}{h}$ for some holomorphic g and h. If necessary simplifying the fraction, g and h can be assumed not to both vanish at x_0. If $h(x_0) \neq 0$, then the function $\varphi_0 \circ \tilde{f} = f$ is holomorphic in the neighbourhood of x_0. If $g(x_0) \neq 0$, $\varphi_1 \circ \tilde{f} = \frac{h}{g}$ is holomorphic in the neighbourhood of x_0. Hence f is analytic.

Let $h : X \to \Sigma$ be an analytic map and set $D = h^{-1}(\infty)$. Assume that there is no connected component of X in D. Then $f = \varphi_0 \circ h : X - D \to \mathbb{C}$ is meromorphic on X; indeed, f is holomorphic on $X - D$. Let $x_0 \in D$; the function $g = \varphi_1 \circ h$ is holomorphic and is not identically zero in the neighbourhood of x_0, and f agrees with the meromorphic function $\frac{1}{g}$. $\qquad\square$

Remark Let $f = \frac{g}{h}$ be a meromorphic function and $x \in X$. If g has a zero of order $k > 0$ at x and $h(x) \neq 0$, then $f(x) = 0$; so $\tilde{f}(x) = 0$. The ramification index of X over Σ at x for $\tilde{f}$ is k. Indeed $f = \varphi^k u$ for some chart φ centered at x and $u(x) \neq 0$. Then $\frac{u}{g} = v^k$ in the neighbourhood of x for some holomorphic v such that $v(x) \neq 0$. Then $f = \psi^k$, where $\psi = \varphi v$, and ψ is another chart centered at x. Similarly, if $g(x) \neq 0$ and h has a zero of order $k > 0$, then $\tilde{f}(x) = \infty$ and the ramification index of X at x is k.

6.3.2 Meromorphic Functions on the Riemann Sphere

Let $P \in \mathbb{C}[Z]$. The polynomial P defines a function $\mathbb{C} \to \mathbb{C}$, meromorphic on Σ. Indeed, it is holomorphic on $\mathbb{C}$, and in the neighbourhood of ∞, $P \circ \varphi_1 = \left(z \mapsto P(\frac{1}{z})\right)$ is a rational fraction. More generally, any rational fraction defines a meromorphic function on Σ.

Theorem *Every meromorphic function on the Riemann sphere* Σ *is a rational fraction.*

Proof Let U be an open subset of Σ and f a meromorphic function on U. Let $a \in U$ be a pole of f. Define the *polar part* $P_a f$ of f at a as follows: if $a \neq \infty$, then f

can be uniquely written as $f(z) = \sum_{k=1}^{N} \frac{c_k}{(z-a)^k} + h(z)$ with h holomorphic in the neighbourhood of a; then the polar part is $\sum_{k=1}^{N} \frac{c_k}{(z-a)^k}$.

If $a = \infty$, f can be uniquely written as

$$f(z) = \sum_{k=1}^{N} c_k z^k + h(z),$$

where h extends to a holomorphic function in the neighbourhood of ∞ in Σ; then the polar part is $\sum_{k=1}^{N} c_k z^k$. In both cases, a is the unique pole of the rational fraction $P_a f$, but is not a pole of $f - P_a f$.

Let f be a meromorphic function on Σ and $a_1, \ldots, a_n$ the poles of f. Then $h = f - \sum_{1}^{N} P_{a_i} f$ is holomorphic on Σ, and necessarily constant. Indeed, $|h|$ reaches its maximum at some point x. By the maximum principle, h is constant in its neighbourhood, and hence, by analytic continuation, constant on Σ. (Liouville's theorem could also be used: every bounded holomorphic function on $\mathbb{C}$ is constant). Thus, f is a rational fraction. $\qquad\square$

6.3.3 Example: Homographies

A rational fraction $f = x \mapsto \frac{ax+b}{cx+d}$, where $ad - bc \neq 0$, defines an analytic map from Σ to Σ called a **homography.** All homographies are automorphisms of Σ. The inverse of f is given by $y \mapsto \frac{-dy+b}{cy-a}$.

Proposition *All automorphisms of Σ are homographies.*

Proof Let f be an automorphism of Σ. Write f as $\frac{P}{Q}$, where P and Q are coprime polynomials in $\mathbb{C}[X]$. The zeros of P being the zeros of f, P has at most one zero, which must be simple by Remark 6.3.1. Hence the degree of P is at most 1. Similarly f can only have one pole, and it is simple. Hence the degree of Q is at most 1. Thus $f = x \mapsto \frac{ax+b}{cx+d}$ for some a, b, c, d such that $ad - bc \neq 0$, for otherwise f would be constant. $\qquad\square$

Corollary *The automorphism group of Σ is generated by the translations $x \mapsto x + b$, the homotheties $x \mapsto ax$ and the map $x \mapsto \frac{1}{x}$.*

6.3.4 Transcendence degree of $\mathcal{M}(X)$

Proposition *Let X be a nonempty connected compact Riemann surface. Then $\mathcal{M}(X)$ is a finitely generated extension of $\mathbb{C}$ with transcendence degree 1.*

Proof Let f be a non-constant meromorphic function on X (such functions exist by 6.2.1). Consider X as a finite analytically ramified covering of Σ by f (6.3.1). Then $\mathscr{M}(X)$ is a finitely generated extension of $\mathscr{M}(\Sigma) = \mathbb{C}(Z)$ by (6.2.3).   □

6.3.5

Let $\mathscr{U}$ denote the category of connected compact Riemann surfaces and non-constant analytic (hence surjective) maps, and $\mathscr{K}$ the category of finitely generated extensions of $\mathbb{C}$ with transcendence degree 1.

Theorem *The functor* $\mathscr{M} : \mathscr{U} \to \mathscr{K}$ *is an anti-equivalence of categories.*

Proof (a) *The functor* $\mathscr{M}$ *is essentially surjective.* Let $E \in \mathscr{K}$, then E is isomorphic to a finite extension of $\mathbb{C}(Z) = \mathscr{M}(\Sigma)$ (6.3.4). It is necessarily etale since $\mathbb{C}(Z)$ is of characteristic 0. So E is isomorphic to an algebra $\mathscr{M}(X)$, where X is a finite analytically ramified covering of Σ (6.2.4), which is necessarily connected since E is a field.

(b) *The map* $f \mapsto f^*$ *from* $\mathrm{Hom}_{\mathscr{U}}(X, \Sigma)$ *to* $\mathrm{Hom}_{\mathscr{K}}(\mathscr{M}(\Sigma), \mathscr{M}(X))$ *is bijective.* The set $\mathrm{Hom}_{\mathscr{U}}(X, \Sigma)$ can be identified with the set of non constant meromorphic functions on X, as well as with the set of transcendent functions $f \in \mathscr{M}(X)$ over $\mathbb{C}$ since $\mathbb{C}$ is algebraically closed. The set $\mathrm{Hom}_{\mathscr{K}}(\mathscr{M}(\Sigma), \mathscr{M}(X)) = \mathrm{Hom}_{\mathbb{C}\text{-}\mathscr{alg}}(\mathbb{C}(Z), \mathscr{M}(X))$ can also be identified with the set of transcendental elements of $\mathscr{M}(X)$ over $\mathbb{C}$. The identification of $\mathrm{Hom}_{\mathscr{U}}(X, \Sigma)$ with $\mathrm{Hom}_{\mathscr{K}}(\mathscr{M}(\Sigma), \mathscr{M}(X))$ thus obtained can be checked to be given by the map $f \mapsto f^*$.

(c) *The functor* $\mathscr{M}$ *is fully faithful.* Let X and Y be objects of $\mathscr{U}$ and $h : \mathscr{M}(X) \to \mathscr{M}(Y)$ a $\mathbb{C}$-algebra homomorphism. Let $u : X \to \Sigma$ be a non constant analytic map. The homomorphism $h \circ u^* : \mathscr{M}(\Sigma) \to \mathscr{M}(Y)$ is of the form v^*, where $v \in \mathrm{Hom}_{\mathscr{U}}(Y, \Sigma)$ by (b). If $f : Y \to X$ is a morphism such that $h = f^*$, then necessarily $u \circ f = v$. Indeed, $(u \circ f)^* = f^* \circ u^* = h \circ u^* = v^*$, and so $u \circ f = v$ by (b).

Consider $\mathscr{M}(X)$ and $\mathscr{M}(Y)$ as algebras over $\mathscr{M}(\Sigma)$ using u^* and v^*. Then h is an algebra homomorphism. By Theorem 6.2.4, there is a unique Σ-morphism $f : Y \to X$ such that $f^* = h$.   □

Corollary *The category of compact Riemann surfaces and analytic maps non constant on the connected components is anti-equivalent by* $\mathscr{M}$ *to the category of* $\mathbb{C}$-*algebras that are finite products of finitely generated extensions of* $\mathbb{C}$ *with transcendence degree 1.*

Exercises 6.3. (Extensions of $\mathbb{C}$ with transcendence degree 1)
1.—Embed the space X of 4.5, Exercise 3 in $\mathbb{C}$ in such a way that A and B go round the points 1 and -1 respectively.

Give the inverse image of X under the map $x \mapsto \frac{-x^3+3x}{2}$.

Interpret the analogy between 4.5, Exercise 3 and 5.6 and 5.7, Exercise 2.

2.—Changing 5.6 and 5.7, Exercise 3, accordingly, construct an example of a Galois covering Z of a Galois covering Y of a space X with Galois covering Z.

3. *(Algebraic definition of ramification)*— Let E be an extension of $\mathbb{C}$ with transcendence degree 1. Any map $\varphi : E \to \Sigma = \mathbb{C} \cup \{\infty\}$ such that

$$\varphi(x + y) = \varphi(x) + \varphi(y) \text{ if } (\varphi(x), \varphi(y)) \neq (\infty, \infty)$$

and

$$\varphi(xy) = \varphi(x).\varphi(y) \text{ if } (\varphi(x), \varphi(y)) \neq (0, \infty) \text{ and } \neq (\infty, 0)$$

is called a *place* of E.

(a) Show that if $E = \mathcal{M}(X)$ then, for all $x \in X$ the map $\delta_x : f \mapsto f(x)$ is a place of E, and that this gives a bijection from X onto the set of places of E. (Consider first the case $E = \mathbb{C}(\mathbb{Z}) = \mathcal{M}(\Sigma)$.)

(b) Show that if E is finitely generated, then the set X of places of E can be equipped with the structure of a compact Riemann surface identifying E with $\mathcal{M}(X)$.

(c) Let φ be a place of E, and $\mathfrak{D}_\varphi$ the subring of E consisting of $f \in E$ such that $\varphi(f) \neq \infty$. Set $\mathfrak{m}_\varphi = \varphi^{-1}(0)$. Show that $\mathfrak{m}_\varphi$ is the unique maximal ideal of $\mathfrak{D}_\varphi$. Show that, if E is finitely generated, then $\mathfrak{D}_\varphi$ is a principal ring. Describe the ideals of $\mathfrak{D}_\varphi$.

(d) Let X and Y be connected compact Riemann surfaces, $h : Y \to X$ a surjective analytic map, $y \in Y$ and $x = h(y)$. Set $E = \mathcal{M}(X)$, $F = \mathcal{M}(Y)$, so that E can be identified with a subextension of F, $\mathfrak{D}_x = \mathfrak{D}_{\delta_x} \subset E$; define $\mathfrak{D}_y$, $\mathfrak{m}_x$ and $\mathfrak{m}_y$ similarly. Show that $\mathfrak{m}_x \mathfrak{D}_y = \mathfrak{m}_y{}^k$, where k is the ramification index of Y over X at y.

6.4 Determination of Some Galois Groups

6.4.1 Free Profinite Groups

For a finite set J, let L(J) be the free group on J (4.7.4) and $(X_i)_{i \in J}$ the canonical basis for L(J). Let $\widehat{L}(J)$ be the profinite completion of L(J) (2.9.5). For $J \subset J'$, define $\rho_J^{J'} :$ $L(J') \to L(J)$ by $\rho_J^{J'}(X_i) = X_i$ if $i \in J$ and $\rho_J^{J'}(X_i) = e$ if $i \notin J$. Passing to profinite completions gives a continuous homomorphism $\widehat{\rho}_J^{J'} : \widehat{L}(J') \to \widehat{L}(J)$.

For an arbitrary set I, the inverse limit $\widehat{L}(I)$ of $\widehat{L}(J)$ for finite J in I is called the *free profinite group* on I.

For finite $J \subset I$, define the elements $(X_i^J) \in \widehat{L}(J)$, $i \in I$, by taking X_i^J to be the image of $X_i \in L(J)$ if $i \in J$ and e if $i \notin J$. Passing to the inverse limit gives elements $(\widehat{X}_i) \in \widehat{L}(I)$, $i \in I$.

Let G be a profinite group. A family $(x_i)_{i \in I}$ of elements of G tends to e if for all neighbourhoods U of e in G, the set $\{i \mid x_i \notin U\}$ is finite. Equivalently, if $G = \varprojlim G_\lambda$, where all G_λ are finite groups, then the family $(\pi_\lambda(x_i))_{i\in I}$ has finite support for all λ, $\pi_\lambda : G \to G_\lambda$ being the canonical map.

Proposition *Let* I *be a set.*

(a) The covariant factor $\Phi : \overleftarrow{\mathscr{Grp}} \to \mathscr{Ens}$ *which assigns to every profinite group* G *the set of families* $(x_i)_{i\in I}$ *of elements of* G *tending to* e *is represented by* $\widehat{L}(I)$ *equipped with* $(\widehat{X}_i)$.

(b) Let L′ *be a profinite group and* (X'_i) *a family of elements of* L′ *tending to* e. *Assume that for all finite groups* G *and families* $(x_i)_{i\in I}$ *of elements of* G *with finite support, there is a unique continuous homomorphism* $f : L′ \to G$ *such that* $f(X'_i) = x_i$ *for all* i. *Then there is a unique isomorphism from* $\widehat{L}(I)$ *onto* L′ *transforming* $\widehat{X}_i$ *into* X'_i.

Proof (a) *Under the assumptions of* (b), *the profinite group* L′ *equipped of* (X'_i) *represents* Φ. Indeed, if $G = \varprojlim G_\lambda$, where all G_λ are finite groups and the family $(x_i)_{i\in I}$ tends to e in G, then for all λ the family $(\pi_\lambda(x_i))_{i\in I}$ has finite support in G_λ; hence there is a unique $f_\lambda : L′ \to G_\lambda$ such that $f_\lambda(X'_i) = \pi_\lambda(x_i)$. Passing to the inverse limit gives a unique $f : L′ \to G$ such that $f(X'_i) = x_i$.

The profinite group $\widehat{L}(I)$ *equipped with* $(\widehat{X}_i)_{i\in I}$ *satisfies the assumptions of* (b). First, the family $(\widehat{X}_i)$ is readily seen to tend to e in $\widehat{L}(I)$. Let G be a finite group and $(x_i)_{i\in I}$ a family of elements of G with finite support J. The homomorphism $u : L(J) \to G$ defined by $u(X_i) = x_i$ for $i \in J$ extends to profinite completions in $\hat{u} : \widehat{L}(J) \to \widehat{G} = G$, and $f = \hat{u} \circ \widehat{\rho}_J : \widehat{L}(I) \to G$ satisfies $f(\widehat{X}_i) = x_i$.

(b) If $f′ : \widehat{L}(I) \to G$ satisfies $f′(\widehat{X}_i) = x_i$ for $i \in I$, then $f′$ factorizes through some $\widehat{L}(J′)$. Indeed Ker $f′$ is a neighbourhood of e in $\widehat{L}(I)$, and so contains a set $\widehat{\rho}_J^{-1}(U)$, where U is a neighbourhood of e in $\widehat{L}(J′)$, and hence necessarily Ker $\widehat{\rho}_{J′}$. For $i \notin J′$, $X'^{J′}_i = e$. So $x_i = e$ and $i \notin J$; thus $J \subset J′$. It follows that $f′ = f$. ☐

The essential point of this section is to prove that if Ω is an algebraic closure of $\mathbb{C}(Z)$, then the Galois group $\mathrm{Aut}_{\mathbb{C}(Z)}(\Omega)$ is isomorphic to the free profinite group on the set $\mathbb{C}$.

6.4.2   The Homomorphism θ_x

Let B be a connected compact Riemann surface, Δ a finite subset of B and $b \in B - \Delta$. Let E be a finite Galois extension of $\mathscr{M}(B)$ such that $\Delta_E \subset \Delta$ and X is an analytically ramified covering of B equipped with an isomorphism $\iota : \mathscr{M}(X) \to E$ identifying these two algebras. Let $x \in X(b)$.

Define a homomorphism $\theta_x : \pi_1(B - \Delta, b) \to \mathrm{Aut}_{\mathscr{M}(B)}E$ as follows: the fundamental group $\pi_1(B - \Delta, b)$ acts on $X(b)$ since $X|_{B-\Delta}$ is a covering. Let $\gamma \in \pi_1(B - \Delta, b)$; as $\mathrm{Aut}_B(X)$ acts simply transitively on $X(b)$, there is a unique automorphism g of X such that $g(x) = \gamma \cdot x$. Let $\theta_x(\gamma)$ be the automorphism g^* of $E = \mathscr{M}(X)$. For $\gamma′ \in \pi_1(B - \Delta, b)$, let $g′$ be the automorphism of X corresponding to $\gamma′$ and $g″$

the one corresponding to $\gamma \cdot \gamma'$. Then $g''(x) = \gamma \cdot \gamma' \cdot x = \gamma \cdot g'(x) = g'(\gamma \cdot x) = g'(g(x))$. So $g'' = g' \circ g$ and $\theta_x(\gamma \cdot \gamma') = g''^* = g^* \circ g'^* = \theta_x(\gamma) \circ \theta_x(\gamma')$.

Remarks (1) The homomorphism θ_x is surjective and its kernel is the stabilizer of x in $\pi_1(B - \Delta, b)$; it is normal and of finite index.

(2) The homomorphism θ_x depends on the choice of x.

Functoriality

Let E and F be finite Galois extensions of $\mathscr{M}(B)$ such that $\Delta_E \subset \Delta$ and $\Delta_F \subset \Delta$. Let X and Y be analytically ramified coverings of B equipped with isomorphisms identifying $\mathscr{M}(X)$ with E and $\mathscr{M}(Y)$ with F. Let $f : E \to F$ be a homomorphism. There is a unique B-morphism $u : Y \to X$ such that $u^* = f$.

Proposition *Choose* $y \in Y(b)$ *and set* $x = u(y)$. *Then* f *is compatible with the actions defined by* θ_x *and* θ_y; *in other words, for* $t \in E$ *and* $\gamma \in \pi_1(B - \Delta, b)$, $f(\theta_x(\gamma, t)) = \theta_y(\gamma, f(t))$.

Proof Let $\gamma \in \pi_1(B - \Delta, b)$, and take $g \in \mathrm{Aut}_B(X)$, $g_1 \in \mathrm{Aut}_B(Y)$ such that $g(x) = \gamma \cdot x$ and $g_1(y) = \gamma \cdot y$. Then

$$g(u(y)) = g(x) = \gamma \cdot x = \gamma \cdot u(y) = u(\gamma \cdot y) = u(g_1(y)).$$

As F in an extension, $Y|_{B-\Delta}$ is a connected covering, and by (4.3.5) the diagram

$$
\begin{array}{ccc}
Y & \xrightarrow{\;\;g_1\;\;} & Y \\
u \downarrow & & u \downarrow \\
X & \xrightarrow{\;\;g\;\;} & X
\end{array}
$$

is commutative. Applying the functor $\mathscr{M}$ gives a commutative diagram

$$
\begin{array}{ccc}
F & \xleftarrow{\;\;\theta_y(\gamma)\;\;} & F \\
f \uparrow & & f \uparrow \\
E & \xleftarrow{\;\;\theta_x(\gamma)\;\;} & E
\end{array}
$$

$\square$

6.4.3 *Galois Group of* Ω_Δ

Let B be a connected Riemann surface, Δ a finite subset of B, $b \in B - \Delta$ and Ω an algebraic closure of $\mathscr{M}(B)$. Denote by Ω_Δ the union of finite sub-extensions E of Ω such that $\Delta_E \subset \Delta$. By 6.2.11, it is a directed union, and so Ω_Δ is a sub-extension of Ω. It is obviously preserved by automorphisms of Ω, and separable since the characteristic is 0. Hence Ω_Δ is a Galois extension, usually infinite, of $\mathcal{M}(B)$. By 6.2.11, for any finite extension F of $\mathcal{M}(B)$ in Ω_Δ, $\Delta_F \subset \Delta$.

Let (E_i) be a directed family of finite Galois extensions of $\mathcal{M}(B)$ such that $\bigcup E_i = \Omega_\Delta$. For each i, identify E_i with an algebra $\mathcal{M}(X_i)$, where X_i is a finite analytically ramified covering of B. For $i \leqslant j$, $E_i \subset E_j$; so there is a B-morphism $p_i^j : X_j \to X_i$ such that $(p_i^j)^*$ is a canonical injection of E_i in E_j. These X_i and p_i^j form a projective system of ramified coverings of B.

By Tychonoff's theorem, $\varprojlim X_i(b) \neq \emptyset$. Let $x \in \varprojlim X_i(b)$, and for all i let $p_i(x)$ be the image of x in $X_i(b)$. Passing to the inverse limit starting from $\theta_{p_i(x)}$ gives a homomorphism $\theta_x : \pi_1(B - \Delta, b) \to \mathrm{Aut}_{\mathcal{M}(B)}\Omega_\Delta = \varprojlim \mathrm{Aut}_{\mathcal{M}(B)}E_i$. As the Galois group $\mathrm{Aut}_{\mathcal{M}(B)}\Omega_\Delta$ is profinite, the homomorphism θ_x induces a continuous homomorphism $\widehat{\theta_x}$ of the profinite completion of $\pi_1(B - \Delta, b)$ in $\mathrm{Aut}_{\mathcal{M}(B)}\Omega_\Delta$.

Proposition *The homomorphism* $\widehat{\theta_x} : \widehat{\pi}_1(B - \Delta, b) \to \mathrm{Aut}_{\mathcal{M}(B)}\Omega_\Delta$ *is an isomorphism.*

Proof The family (E_i) may be assumed to consist of all the finite Galois extensions in Ω_Δ. For all i, $\theta_{p_i(x)}$ gives an isomorphism $h_i : G/N_i \to \mathrm{Aut}_{\mathcal{M}(B)}E_i$, where $G = \pi_1(B - \Delta, b)$ and N_i is the stabilizer of $p_i(x)$ in G. Passing to the projective limit, we get an isomorphism $h : \varprojlim G/N_i \to \mathrm{Aut}\,\Omega_\Delta$.

Let N be a normal subgroup of finite index in G. Then there is a finite Galois covering Y of $B - \Delta$ and $y \in Y(b)$ with stabilizer N (4.6.11 and 4.6.10). Y can be completed to an analytically ramified covering $\widehat{Y}$ of B. Then $\mathcal{M}(\widehat{Y})$ is a Galois extension of $\mathcal{M}(B)$, which can be embedded in Ω. It then is in Ω_Δ, and thus can be identified with some E_i, and $N = N_i$.

Therefore $\varprojlim G/N_i$ is a profinite completion of G, and $h = \widehat{\theta_x}$; so h is an isomorphism. $\qquad\square$

6.4.4 Fundamental Groups with Respect to Germs of Paths

We now give a description of

$$\mathrm{Aut}_{\mathcal{M}(B)}\Omega = \varprojlim \mathrm{Aut}_{\mathcal{M}(B)}\Omega_\Delta$$

the inverse limit being taken over all finite Δ in B. The problem lies in the choice of the basepoint. Indeed, fixing $b \in B$ prevents us from considering the sets Δ containing b. To overcome this problem we replace the basepoint by a germ of path B.

Let $\beta : [0, 1] \to B$ be an injective continuous map. For all finite $\Delta \subset B$, let c_Δ be the largest c such that $\beta(]0, c[) \subset B - \Delta$. Denote by $\pi_1(B - \Delta, \beta)$ the quotient of the set of pairs (t, γ) where $t \in]0, c_\Delta[$ and $\gamma \in \pi_1(B - \Delta, \beta(t))$ by the equivalence relation identifying (t_1, γ_1) to (t_2, γ_2) if γ_2 is the image of γ_1 by the isomorphism from $\pi_1(B - \Delta, \beta(t_1))$ onto $\pi_1(B - \Delta, \beta(t_2))$ defined by $\beta|_{[t_1, t_2]}$ (4.6.6). It has a natural group structure, and for all $t \in]0, c_\Delta[$ there is a canonical isomorphism χ_t

from $\pi_1(B - \Delta, \beta(t))$ onto $\pi_1(B - \Delta, \beta)$. Besides the group $\pi_1(B - \Delta, \beta)$ is the (trivial) direct limit of $\pi_1(B - \Delta, \beta(t))$.

Let X be a finite analytically ramified covering of B and for Δ take its ramification set. Denote by $\Gamma(\beta, X)$ the set of continuous coverings of $\beta|_{]0,c_\Delta[}$ in X. If X is of degree d, then it is a set with d elements since $\beta^* X|_{]0,c_\Delta[}$ is a trivial d-fold covering.

Proposition *Assume* X *is Galois. Let* $\xi \in \Gamma(\beta, X)$. *There is a unique homomorphism* $\theta_\xi : \pi_1(B - \Delta, \beta) \to \text{Aut}_{\mathcal{M}(B)} \mathcal{M}(X)$ *such that for all* $t \in]0, c_\Delta[$, $\theta_\xi \circ \chi_t = \theta_{\xi(t)}$.

Proof It suffices to show that, for t and t' in $]0, c_\Delta[$ such that $t < t'$, there is a commutative diagram

$$
\begin{array}{ccc}
\pi_1(B - \Delta, \beta(t)) & \!\!\!\!\!\!\searrow^{\theta_{\xi(t)}} & \\
\beta_* \downarrow & & \text{Aut}_{\mathcal{M}(B)} \mathcal{M}(X) \\
\pi_1(B - \Delta, \beta(t')) & \!\!\!\!\!\!\nearrow_{\theta_{\xi(t')}} &
\end{array}
$$

where $\beta_*(\gamma) = \beta|_{[t,t']} \cdot \gamma \cdot \beta|_{[t',t]}$, denoting by $\beta|_{[t',t]}$ the inverse path of $\beta|_{[t,t']}$. Let $\gamma \in \pi_1(B - \Delta, \beta(t))$ and $\gamma' = \beta_*(\gamma)$. Let η be a path with initial point $\xi(t)$ lifting a loop from the class γ. The endpoint of η is $\gamma \cdot \xi(t)$. Let $g \in \text{Aut}_B(X)$ be such that $g(\xi(t)) = \gamma \cdot \xi(t)$. Then the path $\eta' = g(\xi|_{[t,t']}) \cdot \eta \cdot \xi|_{[t',t]}$ is a path lifting a path from the class γ', with initial point $\xi(t')$ and endpoint $g(\xi(t'))$; so $\gamma' \cdot \xi(t') = g(\xi(t'))$. Hence the same $g \in \text{Aut}_B(X)$ maps $\xi(t)$ onto $\gamma \cdot \xi(t)$ and $\xi(t')$ onto $\gamma' \cdot \xi(t')$. Therefore $\theta_{\xi(t)}(\gamma) = \theta_{\xi(t')}(\gamma')$. $\qquad\qquad\square$

Functoriality

Let X and X' be finite analytically ramified Galois coverings, Δ and Δ' their ramification sets, and $f : X' \to X$ a B-morphism. The morphism f is necessarily surjective, $\Delta \subset \Delta'$ and $c_{\Delta'} \leqslant c_\Delta$. For all $\xi' \in \Gamma(\beta, X')$ there is a unique $\xi \in \Gamma(\beta, X)$ such that $\xi|_{]0,c_{\Delta'}[} = f \circ \xi'$. This element ξ is written $f_*(\xi')$.

If $\xi = f_*(\xi')$, then the homomorphism $f^* : \mathcal{M}(X) \to \mathcal{M}(X')$ is compatible with the actions of $\pi_1(B - \Delta, \beta)$ defined by θ_ξ and $\theta_{\xi'}$. This is the functoriality Proposition 6.4.2 applied to $\xi(t)$ and $\xi'(t)$ for $t \in]0, c_\Delta[$.

6.4.5 Galois Group of the Algebraic Closure of $\mathcal{M}(B)$

If Δ and Δ' are finite subsets of B such that $\Delta \subset \Delta'$, then there is a unique homomorphism $r_\Delta^{\Delta'} : \pi_1(B - \Delta', \beta) \to \pi_1(B - \Delta, \beta)$, making the diagram

$$
\begin{array}{ccc}
\pi_1(B - \Delta', \beta(t)) & \xrightarrow{\iota_*} & \pi_1(B - \Delta, \beta(t)) \\
\chi_t \downarrow & & \chi_t \downarrow \\
\pi_1(B - \Delta', \beta) & \xrightarrow{r_\Delta^{\Delta'}} & \pi_1(B - \Delta, \beta)
\end{array}
$$

commutative for all $t \in \,]0, c_\Delta[$. As Δ runs through the directed set of finite subsets of B, $\pi_1(B - \Delta, \beta)$ form an inverse system. This is also the case of their profinite completions $\widehat{\pi}_1(B - \Delta, \beta)$.

Let Ω be an algebraic closure of $\mathscr{M}(B)$ and (E_i) an increasing family including all finite Galois sub-extensions of Ω. Set $\Delta_i = \Delta_{E_i}$. Identify each E_i with an extension $\mathscr{M}(X_i)$, where X_i is a finite analytically ramified covering of B. By Tychonoff's theorem (1.7.6 and 2.5.14), $\Gamma = \varprojlim \Gamma(\beta, X_i) \neq \varnothing$. Let $\xi = (\xi_i) \in \Gamma$. For each i, the homomorphism

$$\theta_{\xi_i} : \pi_1(B - \Delta_i, \beta) \to \mathrm{Aut}_{\mathscr{M}(B)}E_i$$

induces

$$\widehat{\theta_{\xi_i}} : \widehat{\pi}_1(B - \Delta_i, \beta) \to \mathrm{Aut}_{\mathscr{M}(B)}E_i .$$

For $i \leqslant j$, the diagram

$$
\begin{array}{ccc}
\widehat{\pi}_1(B - \Delta_j, \beta) & \xrightarrow{\widehat{\theta_{\xi_j}}} & \mathrm{Aut}_{\mathscr{M}(B)}E_j \\
{\scriptstyle r^{\Delta_j}_{\Delta_i}}\big\downarrow & & \big\downarrow{\scriptstyle \rho} \\
\widehat{\pi}_1(B - \Delta_i, \beta) & \xrightarrow{\widehat{\theta_{\xi_i}}} & \mathrm{Aut}_{\mathscr{M}(B)}E_i
\end{array}
$$

where ρ is the restriction, is commutative by the functoriality seen in 6.4.4,.

Theorem *The homomorphism $\widehat{\theta_\xi} : \varprojlim \widehat{\pi}_1(B - \Delta, \beta) \to \mathrm{Aut}_{\mathscr{M}(B)}\Omega$ induced by θ_{ξ_i} by passing to the inverse limit is an isomorphism.*

Proof For each Δ, let ξ_Δ be the image of ξ in the inverse limit of $\Gamma(\beta, X_i)$ for all i such that $\Delta_i \subset \Delta$. There is an isomorphism $\widehat{\theta_{\xi_\Delta}} : \widehat{\pi}_1(B - \Delta, \beta) \to \mathrm{Aut}_{\mathscr{M}(B)}\Omega_\Delta$. This can be seen by identifying $\pi_1(B - \Delta, \beta)$ with $\pi_1(B - \Delta, \beta(t))$ for some $t \in \,]0, c_\Delta[$, and by applying Proposition 6.4.3. The theorem follows by passing to the inverse limit. $\qquad\square$

6.4.6 Galois Group of the Algebraic Closure of $\mathbb{C}(\mathbf{Z})$

We now apply the previous theorem to the case when B is the Riemann sphere Σ. For all $\Delta \subset \mathbb{C}$, set $\overline{\Delta} = \Delta \cup \{\infty\} \subset \Sigma$. These $\overline{\Delta}$ form a cofinal system of the finite subsets of Σ. Hence

$$\mathrm{Aut}_{\mathbb{C}(\mathbf{Z})}\Omega = \mathrm{Aut}_{\mathscr{M}(\Sigma)}\Omega \approx \varprojlim \widehat{\pi}_1(\Sigma - \overline{\Delta}, \beta) = \varprojlim \widehat{\pi}_1(\mathbb{C} - \Delta, \beta)$$

where β is a continuous injective map from $[0, 1]$ to $\mathbb{C}$.

We may assume that the image of β only meets each real line of $\mathbb{C}$ at finitely many points. This is for example the case when β describes an arc of a circle.

Proposition *The profinite group* $\varprojlim \widehat{\pi_1}(\mathbb{C} - \Delta, \beta)$ *is isomorphic to the free profinite group* $\widehat{L}(\mathbb{C})$ *on the set* $\mathbb{C}$.

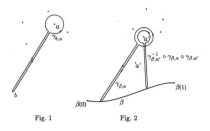

Fig. 1   Fig. 2

Proof For finite $\Delta \subset \mathbb{C}$ and $b \in \mathbb{C} - \Delta$, the group $\pi_1(\mathbb{C} - \Delta, b)$ is isomorphic to the free group $L(\Delta)$ on Δ. Likewise for $\pi_1(\mathbb{C} - \Delta, \beta)$. We describe such an isomorphism, i.e. we choose a basis for $\pi_1(\mathbb{C} - \Delta, \beta)$. Let $b \in \mathbb{C}$ be such that b, a_1, a_2 are never colinear for $a_1 \in \Delta$, $a_2 \in \Delta$, $a_1 \neq a_2$. Then define $\gamma_{b,a} \in \pi_1(\mathbb{C} - \Delta, b)$ for all $a \in \Delta$ as follows: $\gamma_{b,a}$ is the class of a loop consisting of going along a straight line from b to a, stopping at a distance ε from a, then describing an anticlockwise circle of radius ε around a and returning to b along a straight line, ε being chosen to be strictly smaller than $|a' - a|$ for all $a' \in \Delta - \{a\}$ (Fig. 1). Then $(\gamma_{b,a})_{a \in \Delta}$ form a basis for $\pi_1(\mathbb{C} - \Delta, b)$.

Next, or all Δ, define a basis $(\gamma_{\beta,a})_{a \in \Delta}$ for $\pi_1(\mathbb{C} - \Delta, \beta)$. Given the assumption on β, there exists nonzero $c' \leqslant c_\Delta$ such that $\beta(]0, c'[)$ never meets any line connecting two points of Δ. Let c'_Δ to be the greatest among all c'. As long as t is in $]0, c'_\Delta[$, $\chi_t(\gamma_{\beta(t),a}) \in \pi_1(\mathbb{C} - \Delta, \beta)$ is independent of the choice of t; we write it $\gamma_{\beta,a}$ (Fig. 2).

For $\Delta' \supset \Delta$, the homomorphism $r_\Delta^{\Delta'} : \pi_1(\mathbb{C} - \Delta', \beta) \to \pi_1(\mathbb{C} - \Delta, \beta)$ maps $\gamma_{\beta,a} \in \pi_1(\mathbb{C} - \Delta', \beta)$ to $\gamma_{\beta,a} \in \pi_1(\mathbb{C} - \Delta, \beta)$ if $a \in \Delta$ and to e if $a \notin \Delta$. In other words, there is a commutative diagram

$$
\begin{array}{ccc}
L(\Delta') & \xrightarrow{\ \alpha_{\Delta'}\ } & \pi_1(\mathbb{C} - \Delta', \beta) \\
{\scriptstyle\rho_\Delta^{\Delta'}}\Big\downarrow & & \Big\downarrow{\scriptstyle r_\Delta^{\Delta'}} \\
L(\Delta) & \xrightarrow{\ \alpha_\Delta\ } & \pi_1(\mathbb{C} - \Delta, \beta)
\end{array}
$$

where $\rho_\Delta^{\Delta'}$ is the homomorphism defined in 6.4.1 and α_Δ the isomorphism defined by the basis $(\gamma_{\beta,a})_{a \in \Delta}$. Passing to profinite limits and then to the projective limit over Δ gives an isomorphism $\hat{\alpha} : \widehat{L}(\mathbb{C}) \to \varprojlim \widehat{\pi_1}(\mathbb{C} - \Delta, \beta)$. $\square$

Corollary 6.3 *The Galois group* $\mathrm{Aut}_{\mathbb{C}(Z)}\Omega$ *is isomorphic to* $\widehat{L}(\mathbb{C})$.

Corollary 6.4 *Every finite groups is isomorphic to the automorphism group of a finite extension of* $\mathbb{C}(Z)$.

Exercises 6.4. (Some Galois groups)
1.—(a) Embed $k_0 = \mathbb{C}(Z)$ in $k = \mathbb{C}(X)$ (which is isomorphic to it) by $Z = X^3 - 3X$.

Show that k is not a Galois extension. Consider the set Γ of $(z, y) \in \mathbb{C}^2$ such that the equation $X^3 - 3X = z$ has roots x' and x'' with $y = 2x' - x''$, equipped with the projection $\pi : (z, y) \mapsto z$. Show that Γ can be completed to a ramified covering $\widehat{\Gamma}$ of Σ and that $\mathscr{M}(\widehat{\Gamma})$ is the Galois closure of k over k_0.

(b) Show that there is an involutive automorphism τ of $\widehat{\Gamma}$ whose fixed points are $a_1 = (2, 5)$ and $a_2 = (-2, -5)$.

(c) Let $e \in \Gamma$ be a point over 0, α_1 and α_2 paths from e to a_1 and a_2 respectively, and avoiding the other ramification points. Let γ_1 be a loop in $\mathbb{C} - \{2, -2\}$ consisting of a path from 0 to a point near 2 given by $\pi \circ \alpha_1$, then a circular path around 2 and finally the former path back to 0. Similarly, define γ_2 as a loop around the point -2. How do γ_1 and γ_2 act on $\Gamma(0)$?

(d) Show that, although the loops γ_1 and γ_2 go around the points 2 and -2 respectively, they do not generate the group $\pi_1(\mathbb{C} - \{2, -2\}, 0)$. Represent these loops in $\mathbb{C} - \{2, -2\}$.

2.—Let K be an algebraically closed field of characteristic 0 (not necessarily equal to $\mathbb{C}$), and Ω an algebraically closure of $K(Z)$. The aim is to show that the Galois group $\mathrm{Aut}_{K(Z)}\Omega$ is (not naturally) isomorphic to the free profinite group on the set K.

(a) Show that if a profinite group G has a basis with n elements, then every family of n elements generating G is a basis.

(b) Let X be a set, $((G_E), (\rho_F^E)_{F \subset E})$ an inverse system of profinite groups, indexed by the finite subsets E of X. Set $G_X = \varprojlim G_E$, and define the maps $\rho_E^X : G_X \to G_E$ by passing to the limit. Let Y be either a finite subset of X or $Y = X$. A family $(g_x)_{x \in Y}$ of elements in G_Y will be said to be *adapted* if $\rho_E^Y(g_x) = e$ for every finite subset E of Y such that $x \notin E$.

Show that if the maps ρ_F^E are surjective and each G_E has an adapted basis, then so does the profinite group G_X (use Tychonoff's theorem and a).

Show that every adapted basis for G_E stems from an adapted basis for G_X.

(c) If k is a finite extension of $K(Z)$, let $R(k)$ be its ramification set (6.3, Exercise 3). It is a finite set in $K \cup \{\infty\}$.

Let $E \subset K$ be a finite set, Ω_E the union of the finite extensions of $K(Z)$ in Ω such that $R(k) \subset E \cup \{\infty\}$. Set
$$G_E = \mathrm{Aut}_{K(Z)}\Omega_E .$$

Show that, if $K = \mathbb{C}$, the profinite group G_E has an adapted basis.

(d) Let K and K' be algebraically closed of characteristic 0 and $\iota : K' \to K$ an embedding enabling the identification of K' with a subfield of K. Set
$$k_0 = K(Z) \quad \text{and} \quad k_0' = K'(Z) ;$$

the field k_0' is a subfield of k_0, and any algebraic closure Ω of k_0 contains an algebraic closure Ω' of k_0'. Let E be a subset of K'. Define Ω_E and Ω_E' as in (c). Show that $\Omega_E' = \Omega' \cap \Omega_E$, and that the restriction morphism $\iota^* : G_E \to G_E'$ is an isomorphism.

Deduce that (c) continues to hold for subfields K of $\mathbb{C}$, and for arbitrary algebraically closed fields K of characteristic 0.

(e) Conclude.

3.—Which profinite groups are isomorphic to the Galois group of an extension of $\mathbb{C}(Z)$?

6.5 Triangulation of Riemann Surfaces

6.5.1 Definition of a Triangulation

The *reference segment* is $[0, 1]$. The *reference triangle* T is the triangle in $\mathbb{C}$ with vertices $\alpha_0 = 0$, $\alpha_1 = 1$ and $\alpha_2 = j + 1 = e^{i\pi/3}$ i.e. the convex envelope of $\{\alpha_0, \alpha_1, \alpha_2\}$. We parametrize T by the maps ι_0, ι_1, ι_2 defined by $\iota_0(t) = t$, $\iota_1(t) = 1 + jt$, $\iota_2(t) = j + 1 + j^2 t$. Let $\overset{\circ}{T}$ denote the interior of T.

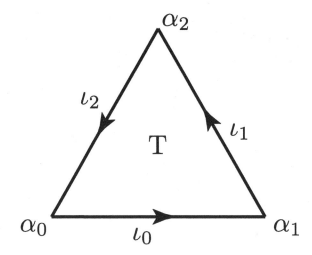

Let X be a compact topological surface. A *triangulation* τ of X is given by

- three finite sets I_0, I_1, I_2;
- a family $(s_i)_{i \in I_0}$ of points of X, called the *vertices* of τ;
- a family $(a_j)_{j \in I_1}$ of injective continuous maps from $[0, 1]$ to X, called the *edges* of τ;
- a family $(f_k)_{k \in I_2}$ of injective continuous maps from T to X, called the *faces* of τ.

For an edge a, the point $a(0)$ (resp. $a(1)$) is called the *origin* (resp. *endpoint*) of a. The *sides* of a face f are $f \circ \iota_\lambda$. The subsets $K_0(\tau) = \{s_i\}$, $K_1(\tau) = \bigcup a_j([0, 1])$ and $K_2(\tau) = \bigcup f_k(T)$ are called the *skeletons* of τ.

The above are subject to the following conditions:

(T$_1$) The origin and endpoint of an edge a_j are vertices.

(T$_2$) $a_j(]0, 1[)$ are disjoint and do not meet K_0.

(T$_3$) A side of a face f_k is either an edge a_j or a flipped edge $t \mapsto a_j(1 - t)$.

(T$_4$) $f_k(\overset{\circ}{T})$ are disjoint and do not meet K_1.

(T$_5$) $K_2 = X$.

For $j \in I_1$ and $k \in I_2$, define the *incidence number* $\varepsilon(j, k)$ by

$$
\varepsilon(j, k) = \begin{cases} 1 & \text{if } a_j \text{ is a side of } f_k \\ -1 & \text{if the flipped edge } a_j \text{ is a side of } f_k \\ 0 & \text{otherwise.} \end{cases}
$$

For $i \in I_0$ and $j \in I_1$, define the incidence number $\varepsilon(i, j)$ by

$$
\varepsilon(i, j) = \begin{cases} 1 & \text{if } s_i \text{ is the endpoint of } a_j \\ -1 & \text{if } s_i \text{ is the origin of } a_j \\ 0 & \text{otherwise.} \end{cases}
$$

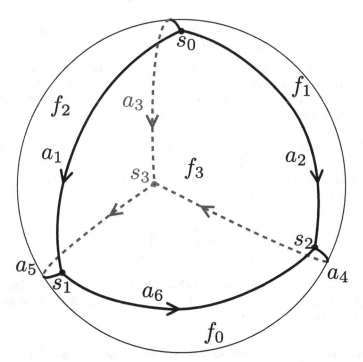

Example Tetrahedral triangulation of the sphere S^2

$$a_1(0) = s_0 \ a_1(1) = s_1 \quad f_0 \circ \iota_0 = a_4 \ f_0 \circ \iota_1 = a_5 \ f_0 \circ \iota_2 = a_6$$
$$a_2(0) = s_0 \ a_2(1) = s_2 \quad f_1 \circ \iota_0 = a_2 \ f_1 \circ \iota_1 = a_4 \ f_1 \circ \iota_2 = a_3 \circ \tau$$
$$a_3(0) = s_0 \ a_3(1) = s_3 \quad f_2 \circ \iota_0 = a_3 \ f_2 \circ \iota_1 = a_5 \ f_2 \circ \iota_2 = a_1 \circ \tau$$
$$a_4(0) = s_2 \ a_4(1) = s_3 \quad f_3 \circ \iota_0 = a_1 \ f_3 \circ \iota_1 = a_6 \ f_3 \circ \iota_2 = a_2 \circ \tau$$
$$a_5(0) = s_3 \ a_5(1) = s_1$$
$$a_6(0) = s_1 \ a_6(1) = s_2 \qquad\qquad\qquad \text{where } \tau(t) = 1 - t.$$

Incidence numbers:

On $I_0 \times I_1$,

j \ i	0	1	2	3
1	−1	1	0	0
2	−1	0	1	0
3	−1	0	0	1
4	0	0	−1	1
5	1	0	0	−1
6	−1	1	0	0

on $I_1 \times I_2$,

k \ j	1	2	3	4	5	6
0	0	0	0	1	1	1
1	0	1	−1	1	0	0
2	−1	0	1	0	1	0
3	1	−1	0	0	0	1

6.5.2 Direct $\mathbf{C}^1$ Triangulations

Let X be a Riemann surface. A triangulation τ a of X is called[3] a *direct* C^1 triangulation if the following conditions hold:

(TD$_1$) For $j \in I_1$, the map a_j is C^1 with nonzero derivative on $]0, 1[$, i.e. for $t \in]0, 1[$ there is an analytic chart φ for X centered at $a_j(t)$ such that $\varphi \circ a_j :$ $]0, 1[\to \mathbb{C}$ defined in the neighbourhood of t is continuously differentiable in the neighbourhood of t with nonzero derivative at t (this condition then holds for all charts).

(TD$_2$) For $k \in I_2$, the map f_k is C^1 on $T - \{\alpha_0, \alpha_1, \alpha_2\}$ in the real sense. Moreover, let $t \in T - \{\alpha_0, \alpha_1, \alpha_2\}$ and φ an analytic chart for X centered at $f_k(t)$, then the real functions u and v defined in the neighbourhood of t by $f(x + iy) = u(x, y) + iv(x, y)$ satisfy

$$\det \begin{pmatrix} \frac{\partial u}{\partial x} & \frac{\partial v}{\partial x} \\ \frac{\partial u}{\partial y} & \frac{\partial v}{\partial y} \end{pmatrix} > 0$$

(this condition then holds for all charts, for if $h : \mathbb{C} \to \mathbb{C}$ is a $\mathbb{C}$-linear map, then h can be written $z \cdot 1_{\mathbb{C}}$, and, as an $\mathbb{R}$-linear map from $\mathbb{R}^2$ to itself, the determinant of h is $z\bar{z} > 0$).

---

[3]The definitions given in this section are specific to the situation. Their generalization requires changes.

We next show that every compact Riemann surface has a direct C^1 triangulation.

Remark Identifying the Riemann sphere $\Sigma = \mathbb{C} \cup \{\infty\}$ with the sphere S^2, the triangulation in Example 6.5.1 is not direct (incidence table and Exercise 1).

6.5.3 Existence of Triangulations of the Riemann Sphere

Proposition *Let* Δ *be a finite subset of the Riemann sphere* Σ. *Then there is a direct* C^1 *triangulation* τ *of* Σ *such that* $K_0(\tau)$ *contains* Δ.

This proposition is an immediate consequence of the next two lemmas.

Lemma 6.1 *There is a triangulation of* Σ.

Proof The map f_0 which assigns to the point (ρ, θ) the point with polar coordinates $\left(\rho \frac{\cos(\theta - \pi/6)}{\cos \pi/6}, \frac{3}{2}\theta\right)$ is a homeomorphism from T onto the quarter circle $\{z = x + iy \mid |z| \leqslant 1, x \geqslant 0, y \geqslant 0\}$. Define $f_1, \ldots, f_7$ by

$$f_1(t) = i f_0(t), \quad f_2(t) = -f_0(t), \quad f_3(t) = -i f_0(t), \quad f_{k+4}(t) = \frac{1}{f_k(t)}.$$

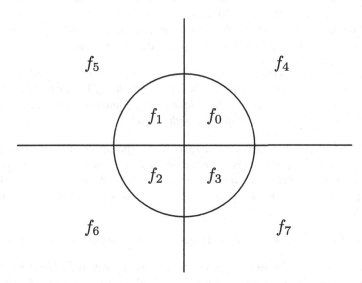

Check that $(f_k)_{k=0,\ldots,7}$ are the faces of a direct C^1 triangulation of Σ with 8 faces, 12 edges and the 6 vertices $0, 1, i, -1, -i, \infty$. $\square$

Lemma 6.2 *Let* X *be a compact Riemann surface,* τ *a direct* C^1 *triangulation of* X *and* s *a point of* X. *Then there is a direct* C^1 *triangulation* τ' *of* X *such that* $K_0(\tau') = K_0(\tau) \cup \{s\}$.

Proof If $s \in K_0$, take $\tau' = \tau$.

If $s \in K_1 - K_0$, then there exists $u \in]0, 1[$ such that $s = a_j(u)$. For all $k \in I_2$ such that $\varepsilon(j, k) \neq 0$, $v = f_k^{-1}(s)$ belongs to a side of the reference triangle T. Let α and β be the endpoints of this side, γ the opposite vertex, σ'_k and σ''_k direct affine homeomorphisms from T onto the triangles with vertices α, γ, v and β, γ, v respectively, and $\lambda_k : [0, 1] \to$ T an affine map such that $\lambda_k(0) = \gamma$, $\lambda_k(1) = v$. Replace the edge a_j by $a'_j : t \mapsto a_j(ut)$ and $a''_j : t \mapsto a_j(1 - t + ut)$; for each k such that $\varepsilon(j, k) \neq 0$, replace the face f_k by $f'_k = f_k \circ \sigma'_k$ and $f''_k = f_k \circ \sigma''_k$ and add an edge $f_k \circ \lambda_k$.

If $s \in K_2 - K_1$, then there is a unique pair $(k, t) \in I_2 \times \overset{\circ}{T}$ such that $s = f_k(t)$. Let σ (resp. σ', resp. σ'') be an affine map from T to T sending α_0, α_1, α_2 to t, α_0, α_1 (resp. t, α_1, α_2, resp. t, α_2, α_0) respectively, and λ, λ', λ'' affine maps from [0, 1] to T sending 0 to t and 1 to α_0, α_1, α_2 respectively. Replace the face f_k by faces $f_k \circ \sigma$, $f_k \circ \sigma'$, $f_k \circ \sigma''$ and add three edges $f_k \circ \lambda$, $f_k \circ \lambda'$, $f_k \circ \lambda''$.

In all cases, we obtain a triangulation τ' answering the question. $\qquad \square$

6.5.4   Lifting of a Triangulation

Let X and Y be compact topological surfaces equipped with the respective triangulations τ and τ'. A continuous map $h : X \to Y$ is *compatible with triangulations* τ and τ' if for all vertices s_i (resp. all edges a_j, resp. all faces f_k) of τ, the image $h(s_i)$ (resp. $h \circ a_j$, resp. $h \circ f_k$) is a vertex (resp. an edge, resp. a face) of τ'.

Proposition *Let* B *be a compact topological surface,* X *a finite ramified covering of* B, Δ *the ramification set of* X, *and let* τ *be a triangulation of* B *such that* $\Delta \subset K_0(\tau)$. *Then there is a triangulation* τ' *of* X, *unique up to permutation of indices, such that the projection* $\pi : X \to$ B *is compatible with* τ' *and* τ.

Proof Let f be a face (resp. an edge) of τ. The space $f^*(X)$ restricted to T $- \{\alpha_0, \alpha_1, \alpha_2\}$ (resp.]0, 1[) is a contractible covering of T $- \{\alpha_0, \alpha_1, \alpha_2\}$ (resp.]0, 1[). Hence this covering has d sections where d is the degree of X over B. Corresponding to them there are d liftings $g_\nu :$ T $- \{\alpha_0, \alpha_1, \alpha_2\} \to$ X (resp.]0, 1[$\to$ X) of f with $\nu \in \{1, \ldots, d\}$.

The next lemma shows that this liftings extend by continuity to T (resp. [0, 1]).

Lemma *Let* Z *be a topological space,* $z_0 \in$ Z. *Suppose that* z_0 *has a fundamental system of neighbourhoods* W_i *such that for all* i, $W_i - \{z_0\}$ *is connected and nonempty. Let* $\varphi : Z \to$ B *and* $\psi : Z - \{z_0\} \to$ X *be continuous maps such that* $\pi \circ \psi = \varphi|_{Z - \{z_0\}}$. *Then* ψ *can be uniquely extended to a continuous map* $\bar{\psi} : Z \to$ X *such that* $\pi \circ \bar{\psi} = \varphi$

Proof of the Lemma Let $b_0 = \varphi(z_0)$, $X(b_0) = \{x_1, \ldots, x_r\}$, and $U_1, \ldots, U_r$ disjoint open neighbourhoods of respectively $x_1, \ldots, x_r$ in X. As π is proper, there is a neighbourhood V of b_0 in B such that $\pi^{-1}(V) \subset U_1 \cup \ldots \cup U_r$. As φ is continuous, there exists W_i such that $\varphi(W_i) \subset V$. Then for $p = 1, \ldots, r$, the open sets $\psi^{-1}(U_p) \cap W_i$ form a partition of $W_i - \{z_0\}$. Hence only one of them is nonempty. Without loss of generality, assume it is $\psi^{-1}(U_1) \cap W_i$. We then show that $\psi(z)$ tends to x_1 as z tends to z_0 while remaining distinct from z_0. As V' runs through a fundamental system of neighbourhoods of b_0 in B, the set $\pi^{-1}(V') \cap U_1$ runs through a fundamental system of neighbourhoods of x_1 in X. Indeed, if U'_1 is a neighbourhood of x_1 in X, then $U'_1 \cup U_2 \cup \ldots \cup U_r$ contains some set $\pi^{-1}(V')$ and $\pi^{-1}(V') \cap U_1 = U'_1$. Then $\psi^{-1}(\pi^{-1}(V') \cap U_1) = \varphi^{-1}(V') \cap \psi^{-1}(U_1) \supset \varphi^{-1}(V') \cap (W_i - \{z_0\})$. Hence the desired extension of ψ is $\tilde{\psi}$ defined by $\tilde{\psi}(z_0) = x_1$. □

End of the Proof of the Proposition Applying the lemma to $T - \{\alpha_0, \alpha_1, \alpha_2\}$ (resp. $]0, 1[$) three times (resp. twice) lifts each face f_k (resp. each edge a_j) of τ to d faces (resp. d edges) in X. Take as vertices the points of X over the vertices of τ. These vertices, edges and faces form a triangulation τ' of X unique up to permutation of indices with the desired property. □

Remarks (1) Suppose that B is a Riemann surface and X an analytic covering of B. Then if τ is C^1 and direct, so is τ'.

Indeed, this property concerns the behaviour of τ away from vertices, and there $\pi : X \to B$ is a local isomorphism.

(2) Let $k_0(\tau), k_1(\tau), k_2(\tau)$ be the respective number of elements of I_0, I_1, I_2. Then

$$k_2(\tau') = d \cdot k_2(\tau), \quad k_1(\tau') = d \cdot k_1(\tau), \quad k_0(\tau') = d \cdot k_0(\tau) - \sum_{x \in \mathbb{R}} (e(x) - 1)$$

where R is the set of ramification points of X, $e(x)$ denoting the ramification index of x.

6.5.5

Theorem *Let X be a compact Riemann surface and $\Delta \subset X$ a finite set. There is a direct C^1 triangulation τ of X such that $K_0(\tau) \supset \Delta$.*

Proof Let $f \in \mathcal{M}(X)$ be a meromorphic function which is not constant on any of the connected components (such functions exist by the separation Theorem 6.2.1). There is an analytic map $\pi : X \to \Sigma$ corresponding to f. By Proposition 6.5.3, there is a direct C^1 triangulation τ_0 of Σ such that $K_0(\tau_0)$ contains the projection of Δ and the ramification set of X. By Proposition 6.5.4, there is a triangulation τ of X such that τ is compatible with τ and τ_0, which implies that $K_0(\tau) \supset \pi^{-1}(\pi(\Delta)) \supset \Delta$. By 6.5.4, remark 1, the triangulation τ is direct and C^1. □

6.5.6

Proposition *Let* X *be a Riemann surface, τ a direct* C^1 *triangulation. For each edge* a_j *there are exactly two faces* f_k *such that* $\varepsilon(j, k) \neq 0$. *For one of them (which will be said to be on the* **left** *of* a_j), $\varepsilon(j, k) = +1$, *for the other (which will be said to be on the* **right**), $\varepsilon(j, k) = -1$.

Lemma 6.3 *Let* J *be a neighbourhood of* 0 *in* $\mathbb{R}$ *and* $a : J \to \mathbb{R}^2$ *a* C^1 *map such that* $a(0) = 0$ *and* $a'(0) \neq 0$. *Then there is a neighbourhood* U *of* 0 *in* $\mathbb{R}^2$, *a neighbourhood* J' *of* 0 *in* J, *and a diffeomorphism* ψ *from* U *onto an open subset of* $\mathbb{R}^2$ *with Jacobian* > 0 *such that* $\psi(a(J')) = \psi(U) \cap (\mathbb{R} \times \{0\})$ *and for all* $t \in J'$, $\psi(a(t)) = (t, 0)$.

Proof Let $\lambda : \mathbb{R}^2 \to \mathbb{R}^2$ be a linear map with determinant > 0 such that $\lambda(a'(0)) = (\xi, \eta)$ for some $\xi > 0$. Then $\lambda(a(t)) = (u(t), v(t))$ for some $u'(0) = \xi > 0$. So u induces a diffeomorphism from an open interval J' of 0 in J onto a neighbourhood J'' of 0 in $\mathbb{R}$. Set $\theta(x, y) = (u^{-1}(x), y - v(u^{-1}(x)))$ for $x \in J''$, $y \in \mathbb{R}$. Then $U = \lambda^{-1}(J'' \times \mathbb{R})$ and $\psi = \theta \circ \lambda$ are as desired, the Jacobian of θ being $\det \begin{pmatrix} u'^{-1} & * \\ 0 & 1 \end{pmatrix} = u'^{-1} > 0$. □

Lemma 6.4 *Let* V *be a neighbourhood of* 0 *in* $\mathbb{R} \times \mathbb{R}_+$, *such that* $V = V \cap (\mathbb{R} \times \mathbb{R}_+^*)$ *is connected. Let* $f : V \to \mathbb{R}^2$ *be a* C^1 *injective map with Jacobian* > 0 *on* V, *such that* $f(0) = 0$. *Set* $a(t) = f(t, 0)$. *Let* U, ψ *and* J' *be as in Lemma 6.3. Assume that* $J' \times \{0\} \subset V$ *and* $f(V) \subset U$. *Then* $\psi(f(V))$ *is a neighbourhood of* 0 *in* $\mathbb{R} \times \mathbb{R}_+$.

Proof Let $D \subseteq V$ be a half-disk centered at 0 closed in $\mathbb{R} \times \mathbb{R}_+$ and let $\overset{\circ}{D}$ denote the interior of D in $\mathbb{R}^2$. The boundary $\partial D = D - \overset{\circ}{D}$ is of the form $J \cup L$, where J is the segment and L the closed half-circle. Set $g = \psi \circ f$. The set $g(L)$ is compact in $\mathbb{R}^2$ and $0 \notin g(L)$. Let D' be a closed half-circle centered at 0 in $\mathbb{R} \times \mathbb{R}_+$ such that $\overset{\circ}{D'} \cap g(L) = \varnothing$. As $g(J) \subset \mathbb{R} \times \{0\}$, $g(D) \cap \overset{\circ}{D'} = g(\overset{\circ}{D}) \cap \overset{\circ}{D'}$. Now, $g(D)$ is compact, and so closed, and $g(\overset{\circ}{D})$ is open in $\mathbb{R}^2$ by the local inversion theorem (implicit function theorem). As a result, $g(\overset{\circ}{D}) \cap \overset{\circ}{D'}$ is clopen in $\overset{\circ}{D'}$; as the latter is connected, either $g(D) \supset \overset{\circ}{D'}$ or $g(D) \cap \overset{\circ}{D'} = \varnothing$. Define $u, v \in [0, \varepsilon[\subset \mathbb{R}$ by $g(0, t) = (u(t), v(t))$. As $g(t, 0) = (t, 0)$, the Jacobian of g at 0 is $\det \begin{pmatrix} 1 & u'(0) \\ 0 & v'(0) \end{pmatrix} = v'(0)$. Thus $v'(0) > 0$.

For sufficiently small $t > 0$, $u(t)$ is small and so is $v(t) > 0$; thus $g(0, t) \in \overset{\circ}{D'}$ and $g(D) \cap \overset{\circ}{D'} \neq \varnothing$. Hence $g(D) \supset \overset{\circ}{D'}$ and $g(D) \supset D'$ since D is compact.

Indeed $g(\overset{\circ}{V})$ is connected and does not meet $\mathbb{R} \times \{0\}$ since g is injective and $(\mathbb{R} \times \{0\}) \cap U = g(J' \times \{0\})$. As a result, $g(\overset{\circ}{V}) \subseteq \mathbb{R} \times \mathbb{R}_+^*$, and $g(V) \subset \mathbb{R} \times \mathbb{R}_+$. □

Proof of the Proposition Let $j \in I_1$, and set $x = a_j\left(\frac{1}{2}\right)$. Let φ be an analytic chart for X centered at x. Define $\tilde{a} : \left]-\frac{1}{2}, \frac{1}{2}\right[\to \mathbb{R}^2$ in the neighbourhood of 0 by

$$\tilde{a}(t) = \varphi\left(a_j\left(t + \frac{1}{2}\right)\right).$$

Let J', U and ψ satisfy the conditions of Lemma 6.3 for $\tilde{a}$. Let f_k be a face of τ such that $\varepsilon(j, k) \neq 0$. Then $x_k = f_k^{-1}(x)$ is the midpoint of a side of T, i.e. $x_k = \iota_\nu\left(\frac{1}{2}\right)$ for some $\nu \in \{0, 1, 2\}$. Let $h_k : \mathbb{C} \to \mathbb{C}$ be a $\mathbb{C}$-affine map such that $h_k(t) = \iota_\nu\left(t + \frac{1}{2}\right)$. Then h_k defines a homeomorphism from a neighbourhood V_k of 0 in $\mathbb{R} \times \mathbb{R}_+$ onto a neighbourhood W of x_k in T. We may assume that $W \subset f_k^{-1}(\varphi^{-1}(U))$, $W \cap \iota_\lambda([0, 1]) = \varnothing$ for $\lambda \neq \nu$, $W \cap \overset{\circ}{T}$ connected and $V_k \supset J'$ if $\varepsilon(j, k) = 1$, while $V_k \supset -J'$ if $\varepsilon(j, k) = -1$. Set $\tilde{f}_k = \varphi \circ f_k \circ h_k : V_k \to U \subset \mathbb{R}$. If $\varepsilon(j, k) = 1$, then $\psi \circ \tilde{f}_k(t) = (t, 0)$. Then, by Lemma 6.4, $\psi \circ \tilde{f}_k(V_k)$ is a neighbourhood of 0 in $\mathbb{R} \times \mathbb{R}_+$. If $\varepsilon(j, k) = -1$, then $\psi \circ \tilde{f}_k(t) = (-t, 0)$, and by Lemma 6.4 applied to $-\psi$, $\psi \circ \tilde{f}_k(V_k)$ is a neighbourhood of 0 in $\mathbb{R} \times \mathbb{R}_-$.

Let I'_2 denote the set of k such that $\varepsilon(j, k) \neq 0$. Then

$$\Omega = \bigcup_{k \in I'_2} f_k \circ h_k(V_k)$$

is a neighbourhood of x in X. Indeed, taking $V_k = \varnothing$ for $k \notin I'_2$, $\complement\Omega \subset \bigcup_{k \in I_2} f_k(T - h_k(V_k))$, and this union is a compact set not containing x. Hence $\bigcup \psi \circ \tilde{f}_k(V_k) = \psi(\varphi(\Omega))$ is a neighbourhood of 0 in $\mathbb{R}^2$.

As all (u, v) with $v > 0$ (resp $v < 0$) belong to a neighbourhood of 0 in $\mathbb{R}^2$, there exists k such that $\varepsilon(j, k) = +1$ (resp. -1).

If $\varepsilon(j, k) = \varepsilon(j, k') = 1$ with $k \neq k'$, then the set $\psi(\tilde{f}_k(V_k)) \cap \psi(\tilde{f}_{k'}(V_{k'}))$ is a neighbourhood of 0 in $\mathbb{R} \times \mathbb{R}_+$, and so contains points (u, v) with $v > 0$, but such a point is of the form $\varphi(x')$ with $x' \in f_k(\overset{\circ}{T}) \cap f_{k'}(\overset{\circ}{T})$, which is impossible by condition (T4) of 6.5.1. Hence there is a unique k such that $\varepsilon(j, k) = 1$ and similarly a unique k such that $\varepsilon(j, k) = -1$.   □

Remark This proposition can be generalized to oriented topological surfaces, but the proof is clearly harder.

Exercises 6.5. (Triangulation of a Riemann surface)

1.—Let X be a topological surface and $\tau = (I_0, I_1, I_2, (s_i), (a_j), (f_k))$ a triangulation of X.

(a) Show that each edge is the side of two faces, i.e. that, for all $j \in I_2$, there are exactly two values of k such that $\varepsilon(j, k) \neq 0$.

(b) Suppose that X is a C^1 surface and that τ is C^1. Show that the following conditions are equivalent:

(i) $(\forall j \in I_1) (\exists k, k' \in I_2) \ \varepsilon(j, k) = 1$ and $\varepsilon(j, k') = -1$;

(ii) there is a unique orientation of X such that every face of τ is a map preserving the orientation of $\overset{\circ}{T}$;

(c) Generalize the definition of triangulations to surfaces with boundaries and give a triangulation of a closed Möbius strip. Using this triangulation, show that the Möbius strip is not orientable.

2.—Let X be a compact topological surface and τ a triangulation of X. The *combinatorial data* of τ are

- the finite sets I_0, I_1, I_2;
- the incidence functions $\varepsilon^{0,1} : I_0 \times I_1 \to \{0, 1, -1\}$ and $\varepsilon^{1,2} : I_1 \times I_2 \to \{0, 1, -1\}$;
- For $(j, k) \in I_1 \times I_2$ with $\varepsilon(j, k) \neq 0$, $\nu \in \{0, 1, 2\}$ such that the side $f_k \circ \iota_\nu$ of f_k is either the edge a_j or the flip of the edge a_j.

(a) Let T be the reference triangle (6.5.1) and $F : I_2 \times T \to X$ the map defined by $F(k, t) = f_k(t)$. Show that F is surjective, and determine from the combinatorial data the equivalence relation on $I_2 \times T$ induced by F.

(b) Show that any two compact topological surfaces admitting triangulations and with the same combinatorial data are homeomorphic.

6.6   Simplicial Homology

In this section X denotes a compact Riemann surface and $\tau = \big((s_i)_{i \in I_0}, (a_j)_{j \in I_1},$ $(f_k)_{k \in I_2}\big)$ a direct C^1 triangulation of X. The results may be generalized to an arbitrary oriented topological surface. The proofs use Proposition 6.5.6.

6.6.1   Chain Complex Associated to a Triangulation

For $\nu = 0, 1, 2$, let C_ν be the free $\mathbb{Z}$-module $\mathbb{Z}^{I_\nu}$ on I_ν. For $\nu \neq 0, 1, 2$, set $C_\nu = 0$. Denote by $[s_i]$ (resp. $[a_j]$, resp. $[f_k]$) the basis element corresponding to i (resp. j, resp. k). Define $\partial_\nu : C_\nu \to C_{\nu-1}$ by

$$\partial_2[f_k] = \sum_j \varepsilon(j, k)[a_j] \text{ and } \partial_1[a_j] = \sum_i \varepsilon(i, j)[s_i] = [a_j(1)] - [a_j(0)].$$

The ∂_ν are called *boundary operators*. The boundary of a face may be described as consisting of three edges with their incidence number, and the boundary of an edge as the "endpoint minus the origin".

Proposition and Definition $\partial_1 \circ \partial_2 = 0$.

The **chain complex associated to** τ *is* $0 \to C_2 \xrightarrow{\partial_2} C_1 \xrightarrow{\partial_1} C_0 \to 0$, *and* $H_\nu(X, \tau) =$ $\text{Ker } \partial_\nu / \text{Im } \partial_{\nu+1}$ *is said to be the* **homology of X defined by** τ.

Proof Let $k \in I_2$. Then

$$\partial_1 \circ \partial_2[f_k] = \sum_j \varepsilon(j, k)([a_j(1)] - [a_j(0)]).$$

Let $\nu \in \{0, 1, 2\} = \mathbb{Z}/(3)$, $\alpha_\nu = [f_k \circ \iota_\nu(0)]$, and define j_ν by $f_k \circ \iota_\nu = a_{j_\nu}$ or the flip of a_{j_ν}. In both cases $\varepsilon(j_\nu, k)\partial_1(a_{j_\nu}) = \alpha_{\nu+1} - \alpha_\nu$. If j is not of the form j_ν, then $\varepsilon(j, k) = 0$. Hence $\partial_1 \circ \partial_2[f_k] = \sum(\alpha_{\nu+1} - \alpha_\nu) = 0$. $\square$

Remark As may be shown, $H_\nu(X, \tau)$ can be identified with the singular homology of X, and so does not depend on τ (cf. Zisman [7], 6.3.2.7). A particular case of this result will be proved in 6.6.4; see also 6.6.2 and 6.6, Exercise 1.

6.6.2 Connected Case

Proposition *If* $X(\neq \varnothing)$ *is connected, then*
(a) $H_0(X, \tau) = \mathbb{Z}$;
(b) $H_2(X, \tau) = \mathbb{Z}$.
The class of $[s_i]$ *is the same for all* $i \in I_0$, *and this element is a generator of* $H_0(X, \tau)$. *Then* $\sum[f_k] \in \operatorname{Ker} \partial_2$, *and the class of this element is a generator of* $H_2(X, \tau)$.

Proof (a) $H_0(X, \tau) = \operatorname{Coker} \partial_1$. Define $\varepsilon : C_0 \to \mathbb{Z}$ by $\varepsilon([s_i]) = 1$ for all i. As $I_0 \neq \varnothing$, the map ε is surjective, and clearly $\varepsilon \circ \partial_1 = 0$; hence $\operatorname{Im} \partial_1 \subset \operatorname{Ker} \varepsilon$. We show the converse inclusion. The submodule $\operatorname{Ker} \varepsilon$ is generated by the $[s_i] - [s_{i'}]$. Indeed, let $i_0 \in I$ and $c = \sum \lambda_i[s_i] \in \operatorname{Ker} \varepsilon$. Then $\sum \lambda_i = 0$; so

$$\lambda_{i_0} = -\sum_{i \neq i_0} \lambda_i, \text{ and } c = \sum_{i \neq i_0} \lambda_i([s_i] - [s_{i_0}]).$$

A vertex s_i will be said to be connected to a point $x \in X$ if there is a sequence $i_0 = i, i_1, \ldots, i_N$ such that, for $n = 1, \ldots, N$, $s_{i_{n-1}}$ and s_{i_n} are not endpoints of the same edge and s_{i_N} is the vertex of a face or of an edge containing x. If two vertices s_i and $s_{i'}$ are connected, then $[s_i] - [s_{i'}] \in \operatorname{Im} \partial_1$.

For $i \in I_0$, le X_i be the set of points of X connected to s_i. These X_i form a finite partition X into nonempty closed sets, and so there is only one and any two vertices are connected. Consequently, $\operatorname{Im} \partial_1 = \operatorname{Ker} \varepsilon$, and so $H_0(X, \tau) = \operatorname{Coker} \partial_1 = \operatorname{Im} \varepsilon = \mathbb{Z}$.

(b) $H_2 = \operatorname{Ker} \partial_2$. Let $c = \sum \lambda_k[f_k] \in C_2$. If all λ_k are equal, it follows from Proposition 6.5.6 that $\partial_2 c = 0$. Conversely, suppose that $\partial_2 c = 0$. If the faces f_k and $f_{k'}$ have a common edge, then $\lambda_k = \lambda_{k'}$. For all n, the union X_n of $f_k(T)$ for all k such that $\lambda_k = n$ is a closed set. For $n \neq n'$, $X_n \cap X_{n'} \subset K_0$, and so $X'_n = X_n - K_0$ form a finite partition of $X - K_0$ into closed sets. By 6.1.7, Remark 3, the space $X - K_0$ is connected; hence there is only one $X'_n \neq \varnothing$, all λ_k are equal, and $H_2(X, \tau))$ can be identified with $\mathbb{Z}$. $\square$

Corollary *For* X *not necessarily connected,*

$$H_0(X, \tau) = H_2(X, \tau) = \mathbb{Z}^{\pi_0(X)},$$

where $\pi_0(X)$ *denotes the set of connected components of* X.

6.6.3 Barycentric Subdivision

From the triangulation τ, we define a triangulation τ' of X. Let $I_{0,1}$ (resp. $I_{1,2}$) be the set of $(i, j) \in I_0 \times I_1$ (resp. $I_1 \times I_2$) such that $\varepsilon(i, j) \neq 0$. Let $I_{0,2}$ be the set of $(i, k) \in I_0 \times I_2$ such that s_i is a vertex of f_k, i.e. $s_i = f_k(\alpha_\nu)$ for some $\nu \in \{0, 1, 2\}$.
Set $I'_0 = I_0 \sqcup I_1 \sqcup I_2$, and

$$\begin{cases} s'_i = s_i & \text{if } i \in I_0, \\ s'_j = a_j\left(\frac{1}{2}\right) & \text{if } j \in I_1, \\ s'_k = f_k(G) & \text{if } k \in I_2, \end{cases}$$

where G is the centre of mass of T.
Set $I'_1 = I_{0,1} \sqcup I_{1,2} \sqcup I_{0,2}$. For $(i, j) \in I_{0,1}$, set

$$a'_{i,j}(t) = \begin{cases} a_j\left(\frac{t}{2}\right) & \text{if } s_i = a_j(0), \\ a_j\left(1 - \frac{t}{2}\right) & \text{if } s_i = a_j(1). \end{cases}$$

For $(j, k) \in I_{1,2}$ or $I_{0,2}$, set

$$a'_{j,k} = f_k \circ h,$$

where $h : [0, 1] \to T$ is the affine map defined by $h(0) = f_k^{-1}(s'_j)$ and $h(1) = G$.
Let I'_2 be the set of $(i, j, k) \in I_0 \times I_1 \times I_2$ such that $(i, j) \in I_{0,1}$ and $(j, k) \in I_{1,2}$.
For $(i, j, k) \in I'_2$, set

$$f'_{i,j,k} = f_k \circ h,$$

where $h : T \to T$ is a direct affine homeomorphism from T onto the triangle with vertices $f_k^{-1}(s_i)$, $f_k^{-1}(s'_j)$, G.
This gives a direct C^1 triangulation $\tau' = ((s'_i)_{i \in I'_0}, (a'_j)_{j \in I'_1}, (f'_k)_{k \in I'_2})$ of X, called the *barycentric subdivision* of τ. In this triangulation, each face of τ has been replaced by 6 faces configured as:

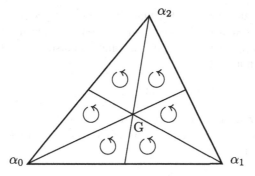

Let $C'_.$ be the chain complex associated to the triangulation τ'. Define a homo-morphism $\sigma : C. \rightarrow C'_.$ by setting $\sigma([f_k]) = \sum[f'_{i,j,k}]$, the sum being over all (i, j) such that $(i, j, k) \in I'_2$, and $\sigma([a_j]) = [a'_{i,j}] - [a'_{i',j}]$, where i and i' are defined by $s_i = a_j(0)$ and $s_{i'} = a_j(1)$, $\sigma([s_i]) = [s'_i]$. It can be checked that σ is a morphism of chain complexes.

6.6.4

Proposition *For all ν, the homomorphism σ induces an isomorphism*

$$\sigma_* : H_\nu(X, \tau) \overset{\approx}{\rightarrow} H_\nu(X, \tau').$$

Remark This proposition is obvious except for $\nu = 1$.

Proof For $C.$ and $C'_.$, define respective chain subcomplexes $F_\lambda C.$ and $F_\lambda C'_.$, where $\lambda = 0, 1, 2$, as follows.

$$F_\lambda C_\nu = C_\nu \text{ for } \nu \leqslant \lambda, \quad F_\lambda C_\nu = 0 \text{ for } \nu > \lambda ;$$

$F_\lambda C'_.$ is the submodule of $C'_.$ generated by the vertices, edges and faces of τ' contained in the λ-skeleton of τ. More explicitly

$F_0 C'_0$ is generated by $[s'_i]$ for $i \in I_0$,
$F_0 C'_1 = F_0 C'_2 = 0$,
$F_1 C'_0$ is generated by $[s'_i]$ for $i \in I_0 \sqcup I_1$,
$F_1 C'_1$ is generated by $[a'_{i,j}]$ for $(i, j) \in I_{0,1}$,
$F_1 C'_2 = 0$,
$F_2 C'_\nu = C'_\nu$.

It can be checked that $F_\lambda C.$ and $F_\lambda C'_.$ are subcomplexes, i.e. that

$$\partial(F_\lambda C_\nu) \subset F_\lambda C_{\nu-1} ;$$

likewise for C'. It can also be checked that $\sigma(F_\lambda C_\cdot) \subset F_\lambda C'_\cdot$. Set $G_\lambda C_\cdot = F_\lambda C_\cdot / F_{\lambda-1} C_\cdot$. and define $G_\lambda C'_\cdot$ similarly. The morphism σ gives a morphism of chain complexes from $G_\lambda C_\cdot$ to $G_\lambda C'_\cdot$, which in turn gives a homomorphism $\sigma_* : H_\nu G_\lambda C_\cdot \to H_\nu G_\lambda C'_\cdot$ by passing to the homology.

Lemma *The homomorphism* $\sigma_* : H_\nu G_\lambda C_\cdot \to H_\nu G_\lambda C'_\cdot$ *is an isomorphism.*

Proof of the Lemma As $G_\lambda C_\cdot = \cdots \to 0 \to C_\lambda \to 0 \to \cdots$, $H_\lambda G_\lambda C_\cdot = C_\lambda$ and $H_\nu G_\lambda C_\cdot = 0$ for $\nu \neq \lambda$. The chain complex $G_\lambda C'_\cdot$ is of the form $\bigoplus_{i \in I_\lambda} K_i$, where the chain complex K_i is as follows

$$0 \to \mathbb{Z}^6 \to \mathbb{Z}^6 \to \mathbb{Z} \to 0 \text{ if } \lambda = 2\,,$$
$$0 \to 0 \ \to \mathbb{Z}^2 \to \mathbb{Z} \to 0 \text{ if } \lambda = 1\,,$$
$$0 \ \to 0 \ \to \mathbb{Z} \to 0 \text{ if } \lambda = 0\,.$$

It may be checked that $H_\nu K_i = \mathbb{Z}$ for $\nu = \lambda$ and $H_\nu K_i = 0$ for $\nu \neq \lambda$. Hence,

$$H_\lambda G_\lambda C'_\cdot = \mathbb{Z}^{I_\lambda} = C_\lambda = H_\lambda G_\lambda C_\cdot \text{ and } H_\nu G_\lambda C'_\cdot = 0 \text{ pour } \nu \neq \lambda\,.$$

The above identification can be checked to be indeed given by σ_*, which therefore is an isomorphism.   □

End of the Proof of the Proposition We show by induction on λ that $\sigma_* : H_\nu F_\lambda C_\cdot \to H_\nu F_\lambda C'_\cdot$ is an isomorphism. For $\lambda = -1$, everything vanishes. Consider the commutative diagram of complexes

$$
\begin{array}{ccccccccc}
0 & \longrightarrow & F_{\lambda-1} C_\cdot & \to & F_\lambda C_\cdot & \to & G_\lambda C_\cdot & \longrightarrow & 0 \\
& & \downarrow & & \downarrow & & \downarrow & & \\
0 & \longrightarrow & F_{\lambda-1} C'_\cdot & \to & F_\lambda C'_\cdot & \to & G_\lambda C'_\cdot & \longrightarrow & 0
\end{array}
$$

Passing to the associated long exact sequences gives the following commutative diagram:

$$
\begin{array}{ccccccccc}
H_{\nu+1} G_\lambda C_\cdot & \to & H_\nu F_{\lambda-1} C_\cdot & \to & H_\nu F_\lambda C_\cdot & \to & H_\nu G_\lambda C_\cdot & \to & H_{\nu-1} F_{\lambda-1} C_\cdot \\
u_1 \downarrow & & u_2 \downarrow & & u_3 \downarrow & & u_4 \downarrow & & u_5 \downarrow \\
H_{\nu+1} G_\lambda C'_\cdot & \to & H_\nu F_{\lambda-1} C'_\cdot & \to & H_\nu F_\lambda C'_\cdot & \to & H_\nu G_\lambda C'_\cdot & \to & H_{\nu-1} F_{\lambda-1} C'_\cdot
\end{array}
$$

By the lemma, u_1 and u_4 are isomorphisms, by the induction hypothesis so are u_2 and u_5. Applying the five lemma[4] then shows that u_3 is an isomorphism, proving our claim. The proposition follows from taking $\lambda = 2$.   □

---

[4]See Zisman [7], 4.1.3 or Bourbaki [8], Chap. 10, § 1, cor. 3. (A X-7).

6.6.5 Dual of a Chain Complex

We construct a complex $C''_\cdot$ and a homomorphism $\sigma' : C''_\cdot \to C'_\cdot$ which is shown to give an isomorphism onto the homology. The complex $C''_\cdot$ is not associated to a triangulation, but to a "cell decomposition", a notion that has not been formally introduced. It is a triangulation where triangles are replaced by polygons. The cell decomposition considered, called the *dual* of the triangulation τ, is obtained by assembling in a cell all the faces of τ' containing a vertex of τ.

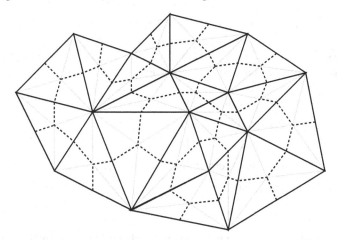

We continue with the definition of $C''_\cdot$ and σ'. Set

$$I''_0 = I_2, \quad I''_1 = I_1, \quad I''_2 = I_0, \quad \text{and} \quad C''_\nu = Z^{I''_\nu}.$$

Let $[s''_k]$ be the elements of the canonical basis C''_0, $[a''_j]$ those of C''_1 and $[f''_i]$ those of C''_2.

Define $\sigma' : C''_\cdot \to C'_\cdot$ as follows: for $i \in I_0$,

$$\sigma'([f''_i]) = \sum [f'_{i,j,k}],$$

the sum being over all (j, k) such that $(i, j, k) \in I'_2$; for $j \in I_1$,

$$\sigma'([a''_j]) = [a'_{j,k}] - [a'_{j,k'}]$$

where k is the index of the face of τ on the left of a_j and k' that of the face on the right; for $k \in I_2$, $\sigma'([s''_k]) = [s'_k]$.

Define $\partial_\nu : C''_\nu \to C''_{\nu-1}$ so that σ' is a morphism of chain complexes. This forces

$$\partial_2[f''_i] = \sum -\varepsilon(i, j)[a''_j] \quad \text{and} \quad \partial_1([a''_j]) = [s''_k] - [s''_{k'}],$$

where k (resp. k') is the index of the face on the left (resp. on the right) of a_j.

6.6.6

Proposition *For all ν, the homomorphism σ' induces an isomorphism*

$$\sigma'_* : H_\nu C''_. \to H_\nu C'_. = H_\nu(X, \tau').$$

Proof For $\lambda = 0, 1, 2$, define the respective chain subcomplexes $F'_\lambda C'_.$ and $F'_\lambda C''_.$ of $C'_.$ and $C''_.$ as follows:

$$F'_\lambda C''_\nu = C''_\nu \text{ for } \nu \leqslant \lambda \,; \; F'_\lambda C''_\nu = 0 \text{ for } \nu > \lambda \,;$$

$F'_\nu C'_.$ is the submodule of $C'_.$ generated by the vertices, edges and faces of τ' contained in the λ-skeleton of the dual cell decomposition of τ. More precisely:

$$F'_0 C'_0 \text{ is generated by } [s'_k] \text{ for } k \in I_2,$$
$$F'_0 C'_1 = F'_0 C'_2 = 0,$$
$$F'_1 C'_0 \text{ is generated by } [s'_j] \text{ for } j \in I_1 \sqcup I_2,$$
$$F'_1 C'_1 \text{ is generated by } [a'_{j,k}] \text{ for } (j, k) \in I_{1,2},$$
$$F'_1 C'_2 = 0,$$
$$F'_2 C'_\nu = C'_\nu .$$

It can be checked that $F'_\lambda C''_.$ and $F'_\lambda C'_.$ are chain subcomplexes and $\sigma'(F'_\lambda C''_.) \subset F'_\lambda C'_.$. Set $G'_\lambda C'_. = F'_\lambda C'_./F'_{\lambda-1} C'_.$ and likewise define $G'_\lambda C''_.$. Then $G'_\lambda C''_. = \cdots \to 0 \to C''_\lambda \to 0 \to \cdots$. As in Proposition 6.6.4, it suffices to show that $\sigma'_* : H_\nu G'_\lambda C''_. \to H_\nu G'_\lambda C'_.$ is an isomorphism.

This is immediate for $\lambda = 0$, and for $\lambda = 1$ the proof is similar to that of Lemma 6.6.4. The chain complex $G'_2 C'_.$ is of the form $\sum_{i \in I_0} K'_i$ where K'_i is the chain complex $0 \to \mathbb{Z}^{J_2(i)} \to \mathbb{Z}^{J_1(i)} \to \mathbb{Z} \to 0$, where $J_2(i)$ is the set of $(i, j, k) \in I'_2$ with $(j, k) \in I_1 \times I_2$, and $J_1(i)$ is the set of $(i, j) \in I_{0,1} \sqcup I_{0,2}$ with $j \in I_1 \sqcup I_2$. The proposition then follows from the following lemma:

Lemma $H_0(K'_i) = H_1(K'_i) = 0$, *and* $H_2(K'_i)$, *generated by*

$$\sum_{(i,j,k) \in J_2(i)} [f'_{i,j,k}] ,$$

is isomorphic to $\mathbb{Z}$.

Proof of the Lemma For $(i, j) \in J_1(i)$, the element $\partial_1[a'_{i,j}] = [s'_j] - [s_i]$ gives -1 in $\mathbb{Z} = (K'_i)_0$. Since s_i is not isolated, $J_1(i) \neq \varnothing$, and so

$$\partial_1 : (K'_i)_1 = \mathbb{Z}^{J_1(i)} \to \mathbb{Z} = (K'_i)_0$$

is surjective and $H_0(K'_i) = 0$.

We show that $H_1(K_i') = 0$. The submodule $\operatorname{Ker} \partial_1 : \mathbb{Z}^{J_1(i)} \to \mathbb{Z}$ is generated by all $[a_j'] - [a_{j'}']$, where $j, j' \in J_1(i)$ (see proof of 6.6.2). Two edges a_j' and $a_{j'}'$ with $j, j' \in J_1(i)$ are said to be connected if there is a finite sequence $j_0 = j, j_1, \ldots, j_N = j'$ in $J_1(i)$ such that $a_{j_{n-1}}'$ and a_{j_n}' are two sides of a same face f_{k_n}' with $k_n \in J_2(i)$ for $n = 1, \ldots, N$. Then, $\partial_2[f_{k_n}']$ having the same image in $G_2'C_.'$ as $\pm([a_{j_n}'] - [a_{j_{n-1}}'])$, $[a_j'] - [a_{j'}'] \in \operatorname{Im} \partial_2 : (K_i')_2 \to (K_i')_1$ if a_j' and $a_{j'}'$ are connected. The set $V = \bigcup_{k \in J_2(i)} f_k'(T)$ is a neighbourhood of s_i in X. The space $V - \{s_i\} = \bigcup (f_k'(T) - \{s_i\})$ is connected. Indeed, as X is a topological surface, there is a neighbourhood U of s_i such that $U - \{s_i\}$ is connected, and each $(f_k'(T) - \{s_i\})$ is connected and meets $U - \{s_i\}$.

A point $x \in V - \{s_i\}$ is said to be connected to an edge a_j' with $j \in J_1(i)$ if a_j' is connected to a side $a_{j'}'$ of a face containing x. For all $j \in J_1(i)$, let W_j be the set of all $x \in V - \{s_i\}$ connected to a_j'. The set of W_j partitions $V - \{s_i\}$ into finite sets, and hence there is only one, and any two arbitrary edges a_j' and $a_{j'}'$ with $j, j' \in J_1(i)$ are connected. Therefore $H_1(K_i') = 0$.

As each face f_k' with $k \in J_2(i)$ contains 2 edges a_j' with $j \in J_1(i)$ (the third one defining an element of F_1C_1') and as each edge a_j' with $j \in J_1(i)$ is contained in 2 faces, the sets $J_1(i)$ and $J_2(i)$ have the same number n of elements. The $\mathbb{Z}$-module $\operatorname{Im} \partial_2 = \operatorname{Ker} \partial_1$, being a submodule of a free $\mathbb{Z}$-module of rank n, is free, and $\operatorname{Im} \partial_1$ is of rank 1, while $\operatorname{Ker} \partial_1$ is of rank $n - 1$. Hence $\operatorname{Ker} \partial_2$ is free of rank 1. The element $\sum_{k \in J_2(i)} f_k'$ is in $\operatorname{Ker} \partial_1$ and generates it since it is not divisible by an integer > 1. □

This completes the proof of the proposition.

6.6.7 Remark

(1) The isomorphisms σ_* and σ_*' identify $H_\nu(X, \tau') = H_\nu(C_.')$ and $H_\nu(C_.'')$ with $H_\nu(X, \tau) = H_\nu(C_.)$.

(2) Identifying C_ν'' with the dual of $C_{2-\nu}$, i.e. with $\operatorname{Hom}_{\mathbb{Z}}(C_{2-\nu}, \mathbb{Z})$, $\partial_2'' : C_2'' \to C_1''$ becomes the opposite of the transpose of $\partial_1 : C_1 \to C_0$, and $\partial_1'' : C_1'' \to C_0''$ the transpose of $\partial_2 : C_2 \to C_1$. Indeed, the entries of the matrix representing $\partial_\nu : C_\nu \to C_{\nu-1}$ are the incidence numbers; those of the the matrix representing ∂_2'' are $-\varepsilon(i, j)$, while those of the matrix representing ∂_1'' are $\varepsilon(j, k)$.

6.6.8

Proposition *The $\mathbb{Z}$-module $H_1(X, \tau)$ is finitely generated and free.*

Lemma 6.5 *Let A be a principal ring, E and F finitely generated free A-modules and $u : E \to F$ a homomorphism. Consider the transpose of $u^\top : F^\top \to E^\top$ of u. The modules $\operatorname{Coker} u$ and $\operatorname{Coker} u^\top$ have isomorphic torsion subsets.*

Proof Let $(f_1, \ldots, f_q)$ be a basis for F adapted to Im u (3.5.2 and 3.5.7), and $a_1, \ldots, a_r \in A$ such that $(a_1 f_1, \ldots, a_r f_r)$ is a basis for Im u. Let $e_1, \ldots, e_r$ be representatives of $a_1 f_1, \ldots, a_r f_r$ in E. Completing to a basis for E by taking a basis $(e_{r+1}, \ldots, e_p)$ for Ker u, the matrix representing u is

$$
\begin{array}{c}
\quad \overbrace{}^{r} \overbrace{}^{p-r} \\
\left. r \left\{ \phantom{\begin{matrix} a_1 \\ \vdots \\ 0 \\ 0 \\ \vdots \\ 0 \end{matrix}} \right. \atop q-r \left\{ \phantom{\begin{matrix} \vdots \\ 0 \end{matrix}} \right. \right.
\begin{pmatrix}
a_1 & & 0 & 0 & & 0 \\
\vdots & \ddots & \vdots & \vdots & & \vdots \\
0 & & a_r & 0 & & 0 \\
0 & & 0 & 0 & & 0 \\
\vdots & & \vdots & \vdots & \ddots & \vdots \\
0 & & 0 & 0 & & 0
\end{pmatrix}
\end{array}
$$

and Coker u is isomorphic to $\bigoplus_{i=1}^{r} A/(a_i) \oplus A^{q-r}$. The matrix representing $u^{\top}$ is

$$
\begin{array}{c}
\quad \overbrace{}^{r} \overbrace{}^{q-r} \\
\left. r \left\{ \phantom{\begin{matrix} a_1 \\ \vdots \\ 0 \\ 0 \\ \vdots \\ 0 \end{matrix}} \right. \atop p-r \left\{ \phantom{\begin{matrix} \vdots \\ 0 \end{matrix}} \right. \right.
\begin{pmatrix}
a_1 & & 0 & 0 & & 0 \\
\vdots & \ddots & \vdots & \vdots & & \vdots \\
0 & & a_r & 0 & & 0 \\
0 & & 0 & 0 & & 0 \\
\vdots & & \vdots & \vdots & \ddots & \vdots \\
0 & & 0 & 0 & & 0
\end{pmatrix}
\end{array}
$$

Coker $u^{\top}$ is isomorphic to $\bigoplus_{i=1}^{r} A/(a_i) \oplus A^{p-r}$. In both cases, the torsion subset is $\bigoplus_{i=1}^{r} A/(a_i)$. $\qquad\square$

Lemma 6.6 *Let A be a principal ring, E. a chain complex of finitely generated free A-modules and* $E^{\top}$ *its dual. Then* $H_n E.$ *and* $H^{n+1} E^{\top}$ *have isomorphic torsion subsets.*

Proof There are exact sequences

$$0 \to H_n E. \to \operatorname{Coker} d_{n+1} \to \operatorname{Im} d_n \to 0,$$

$$0 \to H^{n+1} E^{\top} \to \operatorname{Coker} d^n \to \operatorname{Im} d^{n+1} \to 0.$$

As Im $d_n \subset E_{n-1}$ is torsion free, $H_n E.$ and Coker d_{n+1} have the same torsion subset. Likewise, $H^{n+1} E^{\top}$ and Coker d^n have the same torsion subset. As $d^n : E^n \to E^{n+1}$ is the transpose of $d_{n+1} : E_{n+1} \to E_n$, Lemma 6.6 follows from Lemma 6.5, $\square$

Remark The isomorphisms whose existence was proved in Lemmas 6.5 and 6.6 depend on the choice of an adapted basis. There are no natural ones. (see 2.3, Exercise 4, c).

Proof of the Proposition The module $H_1(X, \tau) = H_1 C.$ is finitely generated since it is a quotient of a submodule of C_1. The torsion subset of $H_1(X, \tau) = H_1 C.$ is isomorphic to that of $H^2C^\top$. Changing the sign of the differential, which does not impact the homology, enables the identification of $C^\top$ with $C''_\cdot$, and

$$H^2C^\top = H_0C''_\cdot = H_0C_\cdot = \mathbb{Z}^{\pi_0(X)} .$$

So its torsion subset is 0. Hence $H_1(X, \tau)$ is finitely generated and torsion free over $\mathbb{Z}$, and thus is free (3.5.8, Corollary 3.2).

6.6.9 Intersection Product

Define a bilinear map $(c, c'') \mapsto c \cdot c''$ from $C^\nu \times C''_{2-\nu}$ to $\mathbb{Z}$ by

$$[s_i] \cdot [f''_{i'}] = \delta^{i'}_i ,$$
$$[a_j] \cdot [a''_{j'}] = \delta^{j'}_j ,$$
$$[f_k] \cdot [s''_{k'}] = \delta^{k'}_k ,$$

where $\delta^{i'}_i = 1$ if $i = i'$ and 0 otherwise.

For $c \in C_\nu$ and $c'' \in C_{3-\nu}$, $\partial c \cdot c'' = (-1)^\nu c \cdot \partial c''$. Indeed,

$$\partial[a_j] \cdot [f''_i] = -\varepsilon(i, j) = -[a_j] \cdot \partial[f''_i]$$

and

$$\partial f_k \cdot [a''_j] = \varepsilon(j, k) = [f_k] \cdot \partial[a''_j] .$$

Hence restriction and passage to the quotient gives a map from $H_\nu C_\cdot \times H_{2-\nu}C''_\cdot$ to $\mathbb{Z}$ or $H_\nu(X, \tau) \times H_{2-\nu}(X, \tau) \to \mathbb{Z}$, called the *intersection product*. We also write it $(\gamma, \gamma') \mapsto \gamma \cdot \gamma'$, and $\gamma \cdot \gamma'$ is called the *intersection number* of γ and γ'.

Remarks 1. For $\nu = 0$ or 2, $H_0(X)$ and $H_2(X)$ can be identified with $\mathbb{Z}$ (6.6.2, proposition) if X is connected, the intersection becoming the multiplication $\mathbb{Z} \times \mathbb{Z} \to \mathbb{Z}$.

2. We have already mentioned in (6.6.1) that $H_\nu(X, \tau)$ is independent of τ. Hence $H_\nu(X)$ is well defined. For $\nu = 1$, the intersection product on $H_1(X)$ can be interpreted as follows. Let $\tau = (I_0, I_1, I_2, (s_i), (a_j), (f_k))$ and $\tilde\tau = (\tilde{I}_0, \tilde{I}_1, \tilde{I}_2, (s_{\tilde\imath}), (a_{\tilde\jmath}), (f_{\tilde k}))$ be two mutually transverse C^1 triangulations of X. This means that $\{s_i\}_{i \in I_0} \cap \{s_{\tilde\imath}\}_{\tilde\imath \in \tilde{I}_0} = \varnothing$ and that for $j \in I_1$ and $\tilde\jmath \in \tilde{I}_1$, the curves $A_j = a_j(]0, 1[)$ and $A_{\tilde\jmath} = a_{\tilde\jmath}(]0, 1[)$ intersect transversally at finitely many points. For $z = a_j(t) = a_{\tilde\jmath}(\tilde t) \in A_j \cap A_{\tilde\jmath}$, set $\theta(z) = 1$ if the basis $(a'_j(t), a'_{\tilde\jmath}(\tilde t))$ for the tangent space $T_z X$ consisting of the derived vectors is direct; set $\theta(z) = -1$ if it is inverse, and $a_j \cdot a_{\tilde\jmath} = \sum_{z \in A_j \cap A_{\tilde\jmath}} \theta(z)$. For $\xi, \eta \in H_1(X)$, consider ξ as an element in $H_1(X, \tau)$ and η as an element in $H_1(X, \tilde\tau)$. Choose respective representatives $\sum \lambda_j[a_j] \in C_1(X, \tau)$ and $\sum \mu_{\tilde\jmath}[a_{\tilde\jmath}] \in C_1(X, \tilde\tau)$ of ξ and η. The intersection product can be shown to be given by

$$\xi \cdot \eta = \sum \lambda_j \mu_{\bar{j}}([a_j] \cdot [a_{\bar{j}}]).$$

Theorem (Poincaré duality) *The intersection product defines an isomorphism from* $H_{2-\nu}(X, \tau)$ *onto the dual of* $H_\nu(X, \tau)$.

In other words, for any linear form $u : H_\nu(X, \tau) \to \mathbb{Z}$, there is a unique $\gamma' \in H_{2-\nu}(X, \tau)$ such that $(\forall \gamma \in H(X, \tau))\ u(\gamma) = \gamma \cdot \gamma'$. This theorem is in fact trivial except for $\nu = 1$.

Lemma *Let* A *be a principal ring,* E. *a chain complex of free* A-*modules such that* $H_n E.$ *are free for all* n, *and* $E^\top$ *its dual. Then* $H^n E^\top$ *can be identified with the dual of* $H_n E.$ *for all* n.

Proof There are exact sequences

$$0 \to H_n E. \to \operatorname{Coker} d_{n+1} \to \operatorname{Im} d_n \to 0$$

and

$$0 \to H_{n-1} E. \to \operatorname{Coker} d_n \to \operatorname{Im} d_{n-1} \to 0.$$

These are split since the submodules $\operatorname{Im} d_n$ of E_{n-1} and and $\operatorname{Im} d_{n-1}$ of E_{n-2} are free. Hence taking the transpose of the former gives an exact sequence $0 \to (\operatorname{Im} d_n)^\top \to \operatorname{Ker}(d_{n+1}^\top) \to (H_n E.)^\top \to 0$. As for the second sequence it shows that $\operatorname{Coker} d_n$ is free. So the exact sequence $0 \to \operatorname{Im} d_n \to E_{n-1} \to \operatorname{Coker} d_n \to 0$ is split, and the restriction $E_{n-1}^\top \to (\operatorname{Im} d_n)^\top$ is surjective. Thus $(\operatorname{Im} d_n)^\top$ has the same image in $\operatorname{Ker}(d_{n+1}^\top)$ as $E_{n-1}^\top$ under $d_n^\top$, and $(H_n E.)^\top$ can be identified with the n-th homology of $E^\top$.   □

Proof of the Theorem Renumbering and changing the sign of the differentials, which leaves the homology invariant, the complex $C''_.$ can be identified with $C^\top$. By the lemma, $H_{2-\nu}(X, \tau) = H_{2-\nu} C''_.$ can therefore be identified with the dual of $H_\nu(X, \tau)$, and this identification can be checked to be given by the intersection product.   □

6.6.10

Proposition *The intersection product* $H_1(X, \tau) \times H_1(X, \tau) \to \mathbb{Z}$ *is an alternating bilinear form.*

Proof It suffices to show that if $c \in C_1$ and $c'' \in C''_1$ are such that $\sigma c - \sigma' c'' = \partial b$ with $b \in C'_2$, then $c \cdot c'' = 0$. Let c_j be the coefficient of $[a_j]$ in c, etc. Let $j \in I_1$, and define i', i'', k', k'' by

$$\varepsilon(i', j) = -1, \ \varepsilon(i'', j) = 1, \ \varepsilon(j, k') = 1, \ \varepsilon(j, k'') = -1.$$

Then

$$c_j = (\sigma c)_{i'j} = -(\sigma c)_{i''j} = (\partial b)_{i'j} = -(\partial b)_{i''j}$$
$$= b_{i'jk'} - b_{i'jk''} = b_{i''jk'} - b_{i''jk''},$$
$$c''_j = (\sigma'c'')_{jk'} = -(\sigma'c'')_{jk''} = -(\partial b)_{jk'} = (\partial b)_{jk''}$$
$$= b_{i'jk'} - b_{i''jk'} = b_{i'jk''} - b_{i''jk''}.$$

Hence

$$2c_jc''_j = (b_{i'jk'} - b_{i'jk''})(b_{i'jk'} + b_{i'jk''} - (b_{i''jk'} + b_{i''jk''}))$$
$$= b^2_{i'jk'} - b^2_{i'jk''} - (b_{i''jk'} - b_{i''jk''})(b_{i''jk'} + b_{i''jk''})$$
$$= b^2_{i'jk'} - b^2_{i'jk''} - b^2_{i''jk'} + b^2_{i''jk''} = \sum_{(i,k)\in I_0 \times I_2} -\varepsilon(i,j)\varepsilon(j,k)b^2_{ijk}.$$

Therefore, summing over j,

$$(*) \qquad\qquad 2c \cdot c'' = \sum_{(i,j,k)} -\varepsilon(i,j)\varepsilon(j,k)b^2_{ijk}.$$

Note that, for $(i, k) \in I_{0,2}$, there are two values j' and j'' de j such that $\varepsilon(i, j)\varepsilon(j, k) \neq 0$, one with $+1$ (corresponding to face $f'_{i,j,k}$ on the left of edge a'_{ik}) and one with -1 (corresponding to the righthand face). As $(\partial b)_{ik} = 0$, $b_{ij'k} = b_{ij''k}$. Hence, the terms in the sum $(*)$ mutually cancel each other. So $2c \cdot c'' = 0 \in \mathbb{Z}$, and $c \cdot c'' = 0$. $\square$

6.6.11

Theorem and Definition *The $\mathbb{Z}$-module $H_1(X, \tau)$ is free of even rank. The number g defined by* $\dim H_1(X, \tau) = 2g$ *is called the* **genus** *of X.*

Thanks to Propositions 6.6.9 and 6.6.10, it suffices to prove the following lemma:

Lemma *Let A be a ring where $2 \neq 0$ and E a free A-module of rank n. If there is an alternating bilinear form on E defining an isomorphism from E onto its dual, then n is even.*

Proof Let (e_i) be a basis for E and let (e'_i) be the dual basis. Let u be an alternating bilinear form defining an isomorphism $\tilde{u} : E \to E^{\top}$. The homomorphism $\tilde{u}$ has matrix $M = (u_{i,j})$ where $u_{i,j} = u(e_i, e_j)$. Then $M^{\top} = -M$, where $\det M = \det M^{\top} = \det(-M) = (-1)^n \det M$. As $\det M$ is invertible, n is even. $\square$

This completes the proof of the theorem.

Remark The genus g of X can be shown not to depend on the choice of the triangulation τ (6.6, Exercise 1). In what follows, this result will be assumed.

6.6.12   The Euler-Poincaré Characteristic

Let A be an integral Noetherian ring and E. a complex of A-modules. Assume that all $H_iE.$ are finitely generated and only finitely many nonzero. Then the *Euler-Poincaré characteristic* of E. is the number

$$\chi(E.) = \sum (-1)^i \, \mathrm{rk}(H_iE.) \,.$$

Proposition *Assume that all E_i are finitely generated and only finitely many nonzero. Then*

$$\chi(E.) = \sum (-1)^i \, \mathrm{rk}(E_i) \,.$$

Proof As $\mathrm{rk}(H_iE.) = \mathrm{rk}(\mathrm{Ker}\,\partial_i) - \mathrm{rk}\,\partial_{i+1} = \mathrm{rk}\,E_i - \mathrm{rk}\,\partial_i - \mathrm{rk}\,\partial_{i+1}$,

$$\chi(E.) = \sum (-1)^i \, \mathrm{rk}(E_i) - \sum (-1)^i \, \mathrm{rk}\,\partial_i - \sum (-1)^{i-1} \, \mathrm{rk}\,\partial_i$$
$$= \sum (-1)^i \, \mathrm{rk}(E_i) \,.$$

cqfd

The *Euler-Poincaré characteristic* of X is the number

$$\chi(X) = \sum (-1)^i \, \dim H_i(X, \tau) \,.$$

By the above proposition, $\chi(X) = k_0 - k_1 + k_2$, where k_0, k_1 and k_2 are respectively the number of vertices, edges and faces of τ. If X is connected, then $\chi(X) = 2 - 2g$ by 6.6.2 and 6.6.11. If X is not connected, then $\chi(X) = \sum \chi(X_i)$, where the X_i are the connected components of X.

As seen, the Euler-Poincaré characteristic of a compact Riemann surface is always even.

6.6.13   The Riemann–Hurwitz Formula

Theorem *Let B be a compact Riemann surface and X a finite analytically ramified d-fold covering of B. Then*

$$\chi(X) = d \cdot \chi(B) - \sum_{x \in R} (e_x - 1) \,,$$

where R is the set of ramification points of X, and e_x the ramification index of x.

This theorem is an immediate consequence of 6.5.4, Remark 2.

Corollary 6.5 *(a) The number* $\sum_{x \in R} (e_x - 1)$ *is even.*
   (b) If X *and* B *are connected, then*

$$g(X) - 1 = d \cdot (g(B) - 1) + \frac{1}{2} \sum_{x \in R} (e_x - 1),$$

where $g(X)$ *(resp.* $g(B)$*) is the genus of* X *(resp. of* B*).*

Corollary 6.6 *Let* X *and* Y *be connected compact Riemann surfaces. If* $g(Y) >$
$g(X)$*, every analytic map from* X *to* Y *is constant.*

Proof Let $f : X \to Y$ be a non constant map of degree d. Then either $g(X) - 1 \geqslant$
$d \cdot (g(Y) - 1)$ or $g(Y) - 1 \geqslant 0$ since $g(Y) > g(X) \geqslant 0$;   so $g(X) - 1 \geqslant$
$d \cdot (g(Y) - 1) \geqslant g(Y) - 1$, contrary to assumption.

6.6.14 Genus, Uniformization

(1) The Riemann sphere is of genus 0. Every connected compact Riemann surface
of genus 0 is isomorphic to the Riemann sphere (6.6, Exercise 7).
   (2) Let $\Gamma \subset \mathbb{C}$ be a closed subgroup isomorphic to $\mathbb{Z}^2$; the quotient group $\mathbb{C}/\Gamma$ is
a connected compact Riemann surface of genus 1. Such a surface is called a complex
torus. Every connected compact Riemann surface of genus 1 can be shown to be
isomorphic to a complex torus.
   (3) Every connected compact Riemann surface can be shown (6.6, Exercise 6) to
be homeomorphic to a surface of the following type (a torus with g holes, where g
is the genus of the surface).

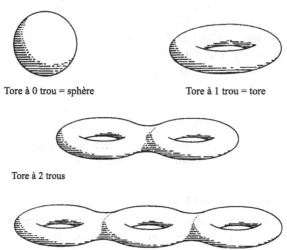

Tore à 0 trou = sphère   Tore à 1 trou = tore

Tore à 2 trous

Tore à 3 trous

(4) If X is a connected compact Riemann surface, then it is isomorphic to $\widetilde{X}/G$, where $\widetilde{X}$ is a universal covering of X (hence a simply connected Riemann surface), and G a subgroup of $\text{Aut}\widetilde{X}$ acting freely on $\widetilde{X}$.

Every simply connected Riemann surface is isomorphic to the Riemann sphere Σ, to the complex plane $\mathbb{C}$ or to the unit disk $\mathbb{D}$ (uniformization theorem, 6.6, Exercise 8). If X has genus 0, then $X \approx \widetilde{X} \approx \Sigma$ and $G = \text{Id}$. If X has genus 1, then $\widetilde{X} \approx \mathbb{C}$, the group G is isomorphic to $\mathbb{Z}^2$ and X is a complex torus. If X has genus $g \geqslant 2$, then $\widetilde{X} \approx \mathbb{D}$; the automorphisms of $\mathbb{D}$ will be studied in § 6.9.

Exercises 6.6. (Homology)

1. *(Homology and fundamental group)*—(a) Let B be a topological space and G a group. A covering of B equipped with a continuous action of G inducing a simply transitive action of G on each fibre is called a *principal G-covering* of B.

(α) Show that if (B, b_0) is a connected pointed space admitting a universal covering, then the isomorphism classes of principal pointed G-coverings of (B, b_0) are in bijective correspondence with the homomorphisms from $\pi_1(B, b_0)$ to G. Study the functoriality of this correspondence in relation to group homomorphisms.

(β) Show that the isomorphism classes of principal G-coverings correspond to the conjugation classes of homomorphisms from $\pi_1(B, b_0)$ to G. Study the case where G is commutative.

(b) α) Let X be a surface, τ a triangulation de X, and C. the chain complex associated to τ. Let G be a commutative group. Defined the chain complex $C^{\cdot}(G)$ by $C^{\nu}(G) = \text{Hom}_{\mathbb{Z}}(C_{\nu}, G)$.

Show that the isomorphism classes of principal G-coverings of X are in bijective correspondence with the elements of $H^1(C^{\cdot}(G))$ (without necessarily assuming that X is connected)

(β) Show that $H^{\nu}(C^{\cdot}(G)) = \text{Hom}_{\mathbb{Z}}(H_{\nu}(X, \tau); G)$.

(γ) Suppose that X is connected and let $x_0 \in X$. Show that $H_1(X, \tau)$ and the quotient of $\pi_1(X, x_0)$ by its commutator group are two representatives of the same covariant functor from $\mathbb{Z}\text{-}\mathcal{M}od$ to $\mathcal{E}ns$. Deduce that these groups are isomorphic. Give an isomorphism by describing its action.

(δ) Show that the genus of a Riemann surface is independent of the choice of triangulation.

2. *(Hyperelliptic curves)*—Let $P \in \mathbb{C}[X]$ be a degree d polynomial with distinct roots, $\Gamma = \{(x, y) \in \mathbb{C}^2 | y^2 = P(x)\}$ and π the projection $(x, y) \mapsto x$ from Γ to $\mathbb{C}$.

(a) Show that Γ is a ramified covering of $\mathbb{C}$ extending to a ramified covering $\widehat{\Gamma}$ of Σ. Show that $\widehat{\Gamma}$ is (resp. is not) ramified over ∞ if d is odd (resp. even).

(b) What is the genus of $\widehat{\Gamma}$?

3.—Let d be an integer $\geqslant 1$ and $\Gamma = \{(x, y) \in \mathbb{C}^2 | x^d + y^d = 1\}$.

(a) Show that Γ together with $\pi : (x, y) \mapsto x$ is a Galois ramified covering of $\mathbb{C}$. What is its degree? How many ramification points does it have? What is its ramification index?

(b) Show that Γ extends to a ramified covering $\widehat{\Gamma}$ of Σ unramified over ∞. What is the genus of $\widehat{\Gamma}$?

4.—Let $P \in \mathbb{C}[X, Y]$ be a degree d polynomial, and Γ the set of $(x, y) \in \mathbb{C}^2$ such that $P(x, y) = 0$ (algebraic curve with equation P).

(a) Suppose that the coefficient of Y^d in P is nonzero, and that the polynomial $y \mapsto P_d(1, y)$ has distinct roots, where P_d denotes the homogeneous part of P of degree d. Show that the curve Γ has d distinct non vertical asymptotes, and that the ramified covering (Γ, π) of $\mathbb{C}$, where π is the projection $(x, y) \mapsto x$, extends to a ramified covering $\widehat{\Gamma}$ of Σ, unramified over ∞.

(b) Let $\sigma \in \mathbb{C}[X]$ be the discriminant of P considered an element of $(\mathbb{C}[X])[Y]$ (3.7.12). What is the degree of the polynomial σ?

(c) Suppose that σ only has simple roots. What is the ramification of Γ over a root of σ? What is the genus of $\widehat{\Gamma}$? Is the ramified covering $\widehat{\Gamma}$ Galois?

5. *(Fundamental group of a torus with p holes.)*—Let D_0 be a disk with centre 0 and radius R in $\mathbb{R}^2$, $D_1, \ldots, D_p$ the disks of centre $C_i = (0, c_i)$ with $-R < c_1 < \cdots < c_p < R$, and sufficiently small radius r so that the disks D_i are disjoint and in the interior of D_0. Set $a_0 = (-R, 0)$, $a_i = (-r, c_i)$ and $A = D_0 - \left(\bigcup \overset{\circ}{D_i}\right)$.

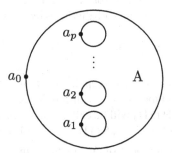

Let $f : \mathbb{R}^2 \to \mathbb{R}$ be a continuous function, > 0 on $\overset{\circ}{A}$ and < 0 on $\mathbb{R}^2 - A$, X the set of $(x, y, z) \in \mathbb{R}^3$ such that $z^2 = f(x, y)$ and π the projection $(x, y, z) \mapsto (x, y)$ of X onto A. The surface X is called a *torus with p holes*.

(a) Let u_i be the rectilinear path from a_0 to a_i, v_i the loop

$$t \mapsto (-r \cos t, c_i - r \sin t)$$

in A. Denote by Γ the union of images of all u_i and v_i, and set $U = A - \Gamma$. Show that U is contractible, and that there is a contraction

$$\varphi : [0, 1] \times U \to U$$

onto a point of ∂D_0 inducing a contraction of $\partial D_0 - \{a_0\}$. Set $\widetilde{U} = \pi^{-1}(U)$. Show that $\widetilde{U}$ is contractible.

(b) Set $X_+ = X \cap (\mathbb{R}^2 \times \mathbb{R}_+)$ and $X_- = X \cap (\mathbb{R}^2 \times \mathbb{R}_-)$. Let u_{i+} and u_{i-} be the paths lifting u_i in X_+ and X_- respectively. Set $\widetilde{\Gamma} = \pi^{-1}(\Gamma)$, and in $\pi_1(\widetilde{\Gamma}, a_0)$, let α_i be the class of the loop $u_{i-}^{-1} \cdot u_{i+}$ and β_i the class of $u_{i+}^{-1} \cdot v_i \cdot u_{i+}$. Show that $\pi_1(\widetilde{\Gamma}, a_0)$ is a free group having $(\alpha_1, \beta_1, \ldots, \alpha_p, \beta_p)$ as basis.

(c) Let $\varepsilon > 0$, and V the set of points of A at a distance $< \varepsilon$ from Γ. Show that, if ε is chosen sufficiently small, which will be henceforth assumed, then there is a deformation retraction of V onto Γ (4.3.16).

Set $\widetilde{V} = \pi^{-1}(V)$. Show that there is deformation retraction of $\widetilde{V}$ onto $\widetilde{\Gamma}$.

(d) Show that $U \cap V = V - \Gamma$ is homeomorphic to the product of a closed interval and an open interval, and that $\widetilde{U} \cap \widetilde{V}$ is homeomorphic to an annulus. Let $a_0' \in \widetilde{U} \cap \widetilde{V} \cap \partial D_0$ be a point over a_0, and $\gamma \in \pi_1(\widetilde{U} \cap \widetilde{V}, a_0')$ an element generating the group. What is the image of γ in $\pi_1(\widetilde{V}, a_0') = \pi_1(\widetilde{\Gamma}, a_0)$?

(e) Using Van Kampen's theorem, show that $\pi_1(X, a_0)$ can be identified with the quotient of a free group on $(\alpha_1, \beta_1, \ldots, \alpha_p, \beta_p)$ by the normal subgroup generated by

$$\alpha_p \beta_p^{-1} \alpha_p^{-1} \beta_p \ldots \alpha_1 \beta_1^{-1} \alpha_1^{-1} \beta_1 .$$

(f) Find a basis $\alpha_1', \beta_1', \ldots, \alpha_p', \beta_p'$ for $\pi_1(\widetilde{\Gamma}, a_0)$ such that $\pi_1(X, a_0)$ can be identified with the quotient of a free group with this basis by the normal subgroup generated by

$$\alpha_1' \beta_1' {\alpha_1'}^{-1} {\beta_1'}^{-1} \ldots \alpha_p' \beta_p' {\alpha_p'}^{-1} {\beta_p'}^{-1} .$$

6. *(Classification of surfaces.)*— We keep the notation of the previous exercise, but denote the set A by A_p and the torus X with p holes X_p.

A set of the form $S = H \cap D_0$ will be called a *segment* in A_p when H is the open half-plane such that $S \neq \varnothing$, $\overline{S} \neq D_0$, $\overline{S} \cap D_i = \varnothing$ for $1 \leqslant i \leqslant p$.

(a) Show that if S is a segment of A_p, then the inverse image $\widetilde{S}$ of S in X_p is homeomorphic to an open disc, and that $\widetilde{\overline{S}} = \pi^{-1}(\overline{S})$ is homeomorphic to a closed disc.

(b) Let $S_1, \ldots, S_q$ be segments of A_p such that $\overline{S}_1, \ldots, \overline{S}_q$ are disjoint, $A_{p,q} = A_p - (S_1 \cup \cdots \cup S_q)$, and $X_{p,q}$ the inverse image of $A_{p,q}$ in X_p. Then $X_{p,q}$ will be said to be a *torus with p holes without q disks*. The *boundary* $\partial X_{p,q}$ is the boundary of $X_{p,q}$ in X_p.

Show that $X_{p,q}$ is a surface with topological boundary, i.e. that any point has a neighbourhood homeomorphic to an open subset of $\mathbb{R}^2$ or of $\mathbb{R}_+ \times \mathbb{R}$.

Show that the homeomorphism class of $X_{p,q}$ does not depend on the integers p and q, and that if σ is a permutation of $\{1, \ldots, q\}$, there is a homeomorphism h of X_p inducing a homeomorphism of $X_{p,q}$, and such that $h(\widetilde{S}_j) = \widetilde{S}_{\sigma(j)}$ (reduce to the case where σ is a transposition, and all S_j are in $[r, R] \times \mathbb{R}$).

(c) Let p, q, p', q' be four integers, I_1 and I_2 arcs (i.e. subspaces homeomorphic to $[0, 1]$) contained in the boundary of $X_{p,q}$ and $X_{p',q'}$ respectively, D a disc, J_1 and J_2 disjoint arcs in the boundary of D, and Y the quotient of the disjoint union $X_{p,q} \sqcup$

$X_{p',q'} \sqcup D$ by an equivalence relation identifying I_1 with J_1, and I_2 with J_2, along homeomorphisms.

Show that the homeomorphism class of Y only depends on the integers p, q, p', q', and that Y is homeomorphic to $X_{p+p',q+q'-1}$.

(d) Let w_j be the rectilinear path in $A_{p,q}$ along the boundary of S_j keeping S_j to the right; w_{j+} and w_{j-} paths in $X_{p,q}$ lifting w_j to X_{p+} and X_{p-} respectively. The connected component of the boundary of $X_{p,q}$ consisting of the boundary $\tilde{S}_j$ is oriented by the loop $w_{j+} \circ w_{j-}^{-1}$. Let D be a disk with oriented boundary. Let I_1 and I_2 be two disjoint arcs in $\partial X_{p,q}$, J_1 and J_2 two disjoint arcs in ∂D, $h_1 : I_1 \to J_1$ and $h_2 : I_2 \to J_2$ homeomorphisms reversing the orientation. Let Y be the quotient of $X_{p,q} \sqcup D$ by the equivalence relation identifying I_1 with J_1 by h_1 and I_2 with J_2 by h_2. Show that Y is homeomorphic to $X_{p,q+1}$ if I_1 and I_2 are in the same connected component of $\partial X_{p,q}$, and to $X_{p+1,q-1}$ otherwise.

(e) Let X be a compact surface with an oriented triangulation τ (i.e. satisfying the conclusions of Proposition 6.5.6, which is the case if X is a Riemann surface and τ C^1 and direct). Let $f_1, \ldots, f_k$ be the faces of the dual cell decomposition of τ (see 6.6.5), and Y_ν the union of the faces $f_1, \ldots, f_\nu$. Show by induction on ν that Y_ν is the disjoint union of surfaces with boundaries of the form $X_{p,q}$. Deduce that if X is nonempty and connected, then it is homeomorphic to a torus with p holes for some value of p.

7.—Let X be a connected compact Riemann surface and $x \in X$. Suppose X has genus 0.

(a) Show that all finite unramified coverings of X are trivial (the Riemann–Hurwicz theorem may be used).

(b) Show that all homomorphisms from $\pi_1(X, x)$ to the additive group $\mathbb{C}$ are trivial (6.6, exercise 1, may be used).

(c) With the notation of (6.2, Exercise 1, D, c), show that $\Omega = 0$. Applying the result of E, (c) of this exercise, deduce that there is a meromorphic function f on X, holomorphic on $X - \{x\}$ with a simple pole at x. Show that f is necessarily an isomorphism from X onto the Riemann sphere.

8. *(Proof of the uniformization theorem)*—Let X be a non compact, connected Riemann surface. The aim is to show that if X is simply connected, then X is isomorphic to $\mathbb{C}$ or to the disk $\mathbb{D}$.

The language of singular homology is freely used: for any space Y, $H_k(Y)$ is the k-dimensional homology group of Y with coefficients in $\mathbb{Z}$. If Y is connected and locally simply path connected and if $y_0 \in Y$, then the group $H_1(Y)$ can be identified with the quotient of $\pi_1(Y, y_0)$ by its commutator group (see 2.7, Exercise 1). If Y is a non compact connected surface with boundary, then $H_2(Y) = H_2(Y, \partial Y) = 0$.

The aim is to show that X is isomorphic to $\mathbb{C}$ or $\mathbb{D}$ if $H_1(X) = 0$, a weaker assumption than $\pi_1(Y) = 0$ (hence a stronger result).

The surface X need not be assumed to be the countable union of compact spaces. This indeed is not assumed but only for the challenge. Most applications (for example if X is a universal covering of a compact surface) this is known beforehand. This mainly entails proceeding with care.

A. *Pieces*

A 2-dimensional connected compact $\mathbb{R}$-submanifold P with C^∞ boundary in X is called a *piece*. The boundary (resp. interior) of P is written ∂P (resp. $\overset{\circ}{P}$).

The piece P is said to be *full* if $X - P$ has no relatively compact connected component. Given an arbitrary piece P, the union of P and of the relatively compact connected components is said to be "P filled" and is written $\hat{P}$. Show that it is a full piece.

(a) Show that every compact subset of X is contained in a full piece.

(b) Suppose that $H_1(X) = 0$ and that P is a full piece in X. Show that $H_2(X, P) = H_2(X - \overset{\circ}{P}, \partial P) = 0$. Using the exact sequence $H_2(X, P) \to H_1(P) \to H_1(X)$, show that $H_1(P) = 0$.

B. *Harmonic functions*

If U is an open subset of X and $h : U \to \mathbb{R}$ a C^2 function, then the *Laplacian* of h is the differential 2-form Δh on U whose expression in a $\mathbb{C}$-analytic chart is $\left(\frac{\partial^2 \dot{h}}{\partial x^2} + \frac{\partial^2 \dot{h}}{\partial y^2}\right) dx \wedge dy$, where $\dot{h}$ is the expression of h (check that this form is independent of the choice of the chart). When $\Delta h = 0$, h is called *harmonic*.

Show that the real part of a holomorphic function is harmonic, and that if U is simply connected and open, then every harmonic function h on U is the real part of a holomorphic function f on U, unique up to addition of an imaginary constant. Then $g = \text{Im } f$ is said to be a *harmonic conjugate* to h. Show that all harmonic functions are $\mathbb{R}$-analytic and satisfy the maximum principle.

C. *Green's functions*

Given a piece P and a point $a \in \overset{\circ}{P}$, a function $G : P - \{a\} \to \mathbb{R}$ is a *Green's function* on P with respect to a if

(G1) G is continuous on $P - \{a\}$ and vanishes on ∂P;

(G2) G is harmonic on $\overset{\circ}{P} - \{a\}$;

(G3) If φ is a $\mathbb{C}$-analytic chart centered at a, the function G takes the form $- \log |\varphi| + h$ in the neighbourhood of a, where h is harmonic in the neighbourhood of a (check that this condition is independent of the choice of the chart φ).

Show uniqueness: there is at most one Green's function on P with respect to a.

We next aim to show existence (note that X has been assumed to be non compact, and so $\partial P \neq \varnothing$. There are two ways of proceeding. They will be described in (C′) and (C″).

C′. *Poisson's equation*

(a) Show that, for any C^∞ differential 2-form ω with compact support in $\overset{\circ}{P}$, there is a C^∞ function $h : P \to \mathbb{R}$ such that $\Delta h = \omega$ and $h = 0$ on ∂P (this requires quite powerful tools from analysis, see [9]).

(b) Let φ be a $\mathbb{C}$-analytic chart for X centered at a, defined in a neighbourhood V of a contained in $\overset{\circ}{P}$, and $\eta : V \to \mathbb{R}$ a C^∞ function with compact support in V, equal to 1 in a neighbourhood V' of a. Define the function g on $P - \{a\}$ by $g = -\eta \log |\varphi|$ on $V - \{a\}$ and $g = 0$ on $P - V$, and the differential form ω on P by $\omega = \Delta g$ on $P - \{a\}$ and $\omega = 0$ on V'. Show that there is a function $h : P - \{a\} \to \mathbb{R}$ such that $\Delta h = \omega$ and $h = 0$ on ∂P. Show that $G = g - h$ is then a Green's function on P with respect to a.

C''. *Perron families*

Show that any continuous function $u : S^1 \to \mathbb{R}$ has a unique continuous extension $h : \overline{\mathbb{D}} \to \mathbb{R}$ harmonic on $\mathbb{D}$. Show that $h(0) = \int_0^1 u(e^{2i\pi t})dt$ and that the gradient of h satisfies $|\nabla_0 h| \le \sup_{z \in S^1} |u(z)|$.

Let V be an open subset of $\mathbb{C}$ and $h : V \to \mathbb{R}$ a continuous function. Show that h is harmonic if and only if for any closed disk $\overline{D}_{c,r}$ in V, $h(c) = \int_0^1 h(c + r \cdot e^{2i\pi t})dt$. If for every closed disk $\overline{D}_{c,r}$ in V, $h(c) \le \int_0^1 h(c + r \cdot e^{2i\pi t})dt$ (resp. $h(c) \ge \int_0^1 h(c + r \cdot e^{2i\pi t})dt$), h is called *subharmonic* (resp. *superharmonic*) . Show that a locally subharmonic function is subharmonic.

Let U be an open subset of X. A continuous function $h : U \to \mathbb{R}$ is said to be subharmonic (resp. superharmonic) if its expression in every $\mathbb{C}$-analytic chart is subharmonic (resp. superharmonic).

(a) A set $\mathscr{F}$ of continuous functions $U \to \mathbb{R}$ is a *Perron family* on U if

(P1) $h \in \mathscr{F} \Rightarrow h$ is subharmonic;
(P2) $(h_1 \in \mathscr{F}$ and $h_2 \in \mathscr{F}) \Rightarrow \sup(h_1, h_2) \in \mathscr{F}$;
(P3) For $h \in \mathscr{F}$ and $\Delta \subset U$ in the domain of the chart φ, such that $\varphi(\Delta)$ is a closed disc, the continuous function $\tilde{h}$ which agrees with h on $U - \Delta$ and is harmonic on $\overset{\circ}{\Delta}$ is in $\mathscr{F}$.

Show that if U is connected and $\mathscr{F}$ is a Perron family on U, then $\sup_{h \in \mathscr{F}} h$ is either harmonic or identically $+\infty$ (Perron's theorem).

(b) Let $u : \partial P \to \mathbb{R}$ be a continuous function. Let $\mathscr{F}_u$ be the set of continuous functions $h : P \to \mathbb{R}$, subharmonic on $\overset{\circ}{P}$ and such that $h \le u$ on ∂P. Show that the restrictions of these $h \in \mathscr{F}_u$ to $\overset{\circ}{P}$ form a Perron family on $\overset{\circ}{P}$, and that $h_u = \sup_{h \in \mathscr{F}_u} h$ is a continuous function on P, equal to u on ∂P and harmonic on $\overset{\circ}{P}$ (solution to the Dirichlet problem).

In particular, if Q is a piece with non connected boundary and B a connected component of ∂Q, then there is a function h_B continuous on Q, harmonic on $\overset{\circ}{Q}$, 1 on B and 0 on all other connected components of ∂Q.

(c) Let $\varphi : V \overset{\approx}{\to} D_r$ be a chart for X centered at a with $V \subset \overset{\circ}{P}$, r' and r'' such that $0 < r' < r'' < r$. Set $V' = \varphi^{-1}(D_{r'})$ and $V'' = \varphi^{-1}(D_{r''})$, $\Gamma' = \partial V'$ and $\Gamma'' = \partial V''$. Let $h_{\Gamma'}$ be the continuous function on $P - V'$, harmonic on $\overset{\circ}{P} - \overline{V'}$, 1 on $\partial V'$ and 0 on ∂P. For $m > 0$, define $\tilde{\lambda}_m : V - \{a\} \to \mathbb{R}$ by $\tilde{\lambda}_m = -\log |\varphi| + \log r' + m$, and

$\lambda_m : P - \{a\} \to \mathbb{R}$ by $\lambda_m = \tilde{\lambda}_m$ on $\overline{V'} - \{a\}$ and $\lambda_m = m \cdot h_{\Gamma'}$ on $P - V'$, so that $\lambda_m = m$ on $\partial V'$. Show that for sufficiently large m, $m.h_{\Gamma'} \leq \tilde{\lambda}_m$ on Γ'', and that λ_m is therefore superharmonic. Choose m satisfying these properties.

Let $\mathscr{G}$ be the set of continuous functions on $P - \{a\}$, subharmonic on $\overset{\circ}{P} - \{a\}$, vanishing on ∂P and bounded above by $- \log |\varphi| + O(1)$ in the neighbourhood of a. Show that $g \in \mathscr{G} \Rightarrow g \leq \lambda_m$ and that $G = \sup_{g \in \mathscr{G}} g$ is a Green's function on P with respect to a.

D. Uniformization of a piece

Here P is assumed to be a piece in X such that $H_1(P) = 0$ and $a \in \overset{\circ}{P}$. The aim is to show the existence of a homeomorphism ϕ_P from P onto the closed disk $\overline{\mathbb{D}}$ inducing a $\mathbb{C}$-analytic isomorphism from $\overset{\circ}{P}$ onto $\mathbb{D}$, with $\phi_P(a) = 0$.

(a) Let Δ be a neighbourhood of a transformed into a closed disk by a chart centered at a. By considering the exact sequence

$$H_1(P) \to H_1(P, \Delta) \to H_0(\Delta) \to H_0(P),$$

show that $H_1(P - \{a\}, \Delta - \{a\}) = H_1(P, \Delta) = 0$. Show that the natural map $H_1(\partial \Delta) \to H_1(P - \{a\})$ is surjective. Show that if α is a closed differential 1-form on $P - \{a\}$ then, for any loop γ in $P - \{a\}$, the value of the integral $\int_\gamma \alpha$ is an integer multiple of $\int_{\partial \Delta} \alpha$.

(b) Show that Green's function $G = G_{P,a}$ has a $\mathbb{R}/2\pi\mathbb{Z}$-valued harmonic conjugate, i.e. that there is a continuous function $\Theta : P - \{a\} \to \mathbb{R}/2\pi\mathbb{Z}$ such that any lifting $\tilde{\Theta} : U \to \mathbb{R}$ of Θ to a simply connected open set is a harmonic conjugate of G.

Show that $\phi = e^{G + i\Theta}$, extended by $\phi(a) = 0$, has the desired property.

E. Erdös-Kœbe

(a) Let $f : \mathbb{D} \to \mathbb{D}$ be a holomorphic map such that $f(0) = 0$ and $|f'(0)| = \rho$ is near 1. Set $g(z) = \frac{f(z)}{z}$ and $g(0) = \rho$. Show that g is a holomorphic map from $\mathbb{D}$ to $\mathbb{D}$. Show that, for $|z| \leq r$, $|g(z) - \rho| \leq r \frac{1 - \rho^2}{1 - \rho r}$ and $|f(z) - \rho z| \leq r^2 \frac{1 - \rho^2}{1 - \rho r}$.

Let R and R' be such that $0 < R < R'$, and let $\psi : D_R \to D_{R'}$ be a holomorphic map such that $\psi(0) = 0$ and $\psi'(0) = 1$. Show that

$$|\psi(z) - z| \leq \frac{R'^2 - R^2}{R'(R - r)}.$$

(b) Let $f : D_R \to \mathbb{C}$ be an injective holomorphic map such that $f(0) = 0$ and $f'(0) = 1$. Show that $f(D_R) \supset D_{R/4}$ (Kœbe quarter theorem). Show that $f(D_{R/16}) \subset D_{R/4}$, and that the function g defined by $g(z) = \frac{f(z) - z}{z^2}$ is bounded above by $80/R$ on $D_{R/16}$, and so $|f(z) - z| \leq 80 \frac{r^2}{R}$ for $|z| \leq r \leq \frac{R}{16}$.

F. Uniformization of X

Suppose X is a non compact, connected Riemann surface such that $H_1(X) = 0$ and $a \in X$. Choose $v (\neq 0)$ in the tangent space $T_a X$. For each full piece P such that $a \in \overset{\circ}{P}$, there is a radius $R(P)$ and an isomorphism $\Phi_P : \overset{\circ}{P} \to D_{R(P)}$ such that the image of v under the tangent linear map $T_a \Phi_P$ is 1. For each P, the pair (R_P, Φ_P) is unique.

Let $R(X)$ be the upper bound of $R(P)$ for all full pieces P such that $a \in \overset{\circ}{P}$. The aim is to show that X is isomorphic to $\mathbb{D}$ if $R(X) < \infty$ and to $\mathbb{C}$ if $R(X) = \infty$.

(a) Let (P_n) be a sequence of full pieces such that $P_n \subset \overset{\circ}{P}_{n+1}$ for all n and the radii $R_n = R(P_n)$ tend to $R = R(X)$. Set X' to be the union of all $\overset{\circ}{P}_n$.

Using the inequalities proved in (E), show that, for any compact subset K of X', for sufficiently large n, Φ_n form a Cauchy sequence with respect to uniform convergence on K. Show that passing to the limit gives an isomorphism from X' onto D_R if $R < \infty$, and onto $\mathbb{C}$ if $R = \infty$.

(b) The aim is to show by contradiction that $X' = X$. Let $b \in X - X'$. Construct a sequence of full pieces $(\tilde{P}_n)$ such that, for all n, $\tilde{P}_n$ is in the interior of $\tilde{P}_{n+1}$ and $P_n \cup \{b\}$ in the interior of $\tilde{P}_n$. This gives an open subset $\tilde{X}'$ strictly containing X', and an isomorphism Ψ from D_R or from $\mathbb{C}$ onto one of its open strict subsets, tangent to the identity at 0. Deduce a contradiction.

(c) Using the transfinite line (1.6, Exercise 1) or (4.9, Exercise 3), give examples of surfaces having an $\mathbb{R}$-analytic structure, but not any $\mathbb{C}$-analytic structure.

6.7 Finite Automorphism Groups of Riemann Surfaces

6.7.1 Quotient Riemann Surface

Proposition *Let X be a connected Riemann surface and G a finite automorphism group of X. Then the quotient $Y = X/G$ is a topological surface and Y has a unique $\mathbb{C}$-analytic structure σ such that the canonical map $\chi : X \to Y$ is analytic. Let Y' be an open subset of Y and Z a Riemann surface; a continuous map $f : Y' \to Z$ is analytic with respect to σ if an only if $f \circ \chi : \chi^{-1}(Y') \to Z$ is analytic.*

Lemma *Let X be a connected Riemann surface, x_0 a point of X and H a finite automorphism group of X fixing x_0. Then H is cyclic and there is chart φ for X centered at x_0, with domain U preserved by H, such that the expressions of the elements de H are the maps $z \mapsto a \cdot z$, where a is a d-th root of unity, d denoting the order of H.*

Proof Let φ be a chart for X centered at x_0, with domain U_0. Set $U_1 = \bigcap_{g \in H} g U_0$, so that U_1 is preserved by H. For $x \in U_1$, set $p(x) = \prod_{g \in H} \varphi(g \cdot x)$. The function p is analytic on U_1 and has x_0 as a zero of order d. Hence in a neighbourhood U_2 of x_0 which may be assumed to be preserved by H, it can be written as $p = \varphi_1^d$, where φ_1 has x_0 as zero of order 1. Then the function φ_1 induces a homeomorphism from a neighbourhood U_3 of x_0, which may be assumed to be preserved by H, onto a neighbourhood of 0 in $\mathbb{C}$, containing the open disk with radius r. As $p(g \cdot x) = p(x)$ for $g \in H$, the open set U of x such that $|\varphi_1(x)| < r$, i.e. $|p(x)| < r^d$, is preserved by H, and φ_1 induces a chart for X centered at x_0, with domain U.

For $x \in U - \{x_0\}$ and $g \in H$, $\varphi_1(g \cdot x)^d = p(g \cdot x) = p(x) = \varphi_1(x)^d$, and so $\varphi_1(g \cdot x) = a_{g,x}\varphi_1(x)$, where $a_{g,x}$ is a d-th root of unity. The map $x \mapsto a_{g,x}$ is continuous, hence locally constant, hence constant since $U - \{x_0\}$ is connected, and $\varphi_1(g \cdot x) = a_g\varphi_1(x)$ for $x \in U$. The map $g \mapsto a_g$ from H to the group μ_d of d-th roots of unity is a homomorphism. The expression of g in the chart φ_1 is multiplication by a_g. If $a_g = 1$, then the map g agrees with 1_X on U, hence, by analytic continuation, also on X. The homomorphism $g \mapsto a_g$ is therefore injective and as H and μ_d are of order d, it is bijective.   □

Proof of the Proposition Let $y \in Y$, and $\chi^{-1}(y) = \{x_1, \ldots, x_r\}$. Denote by S the stabilizer of x_1; the set $\chi^{-1}(y)$ can be identified with G/S. Let φ_1 be a chart centered at x_1, with domain U_1, satisfying the properties of the lemma for S. As U_1 is preserved by S, the open set $g \cdot U_1$ only depends on the class of g in G/S. Denote it by U_k if $g(x_1) = x_k$. If necessary by shrinking U_1, all U_k may be assumed to be disjoint. The open set $W = \bigcup U_k$ is saturated and so $V = \chi(W)$ is open in Y. The map χ induces a homeomorphism from U_1/S onto V. There is a unique map $\psi : V \to \mathbb{C}$ making the diagram

$$
\begin{array}{ccc}
U_1 & \xrightarrow{\varphi_1} & \mathbb{C} \\
\chi \downarrow & & \downarrow z \mapsto z^d \\
V & \xrightarrow{\psi} & \mathbb{C}
\end{array}
$$

commutative. This map ψ is a homeomorphism from V onto an open disk of $\mathbb{C}$, i.e. a chart for Y centered at y. Besides the following property holds: let V' be an open subset of V and Z a Riemann surface; if $f : V' \to Z$ is a continuous map then, $f \circ \psi^{-1} : \psi(V') \to Z$ is analytic if and only if so is $f \circ \chi : \chi^{-1}(V') \to Z$.

The charts thereby obtained form an analytic atlas. The analytic structure σ thus defined on Y has the required properties.

If σ' is another analytic structure on Y such that $\chi : X \to (Y, \sigma')$ is analytic, then the identity map from (Y, σ) to (Y, σ') is analytic and bijective, and so is an isomorphism, and $\sigma' = \sigma$.   □

Remark The above proof shows that, for all $x \in X$, the ramification index of x for $\chi : X \to Y$ is the order d of the stabilizer of x.

6.7.2   Genus $g \geqslant 2$ Case

We show in 6.9.17 that if X is a connected compact Riemann surface of genus $\geqslant 2$, then the automorphism group of X is finite. Here, without using this result, we prove the following theorem (which, given finiteness, provides an upper bound for the number of automorphisms).

Theorem *Let* X *be a connected compact Riemann surface with genus* $g \geqslant 2$ *and* G *a finite automorphism group of* X *of order* n. *Then*

$$n \leqslant 84 \cdot (g - 1).$$

Proof Set $Y = X/G$ and let g' be the genus of Y. For each $y \in Y$, the points of $\chi^{-1}(y)$ have conjugate stabilizers, and so the same ramification index s_y, and $n = r_y s_y$, where r_y is the number of distinct points over y. Let Δ be the ramification set of X; if $\Delta = \{y_1, \ldots, y_k\}$, write s_i for s_{y_i}. As

$$g - 1 = n \cdot (g' - 1) + \frac{1}{2} \sum_{y \in \Delta} \frac{n}{s_y}(s_y - 1) = n \left((g' - 1) + \frac{1}{2} \sum_{y \in \Delta} \left(1 - \frac{1}{s_y} \right) \right),$$

$n = \frac{g-1}{A}$ with $A = g' - 1 + \frac{1}{2} \sum_1^k \left(1 - \frac{1}{s_i} \right)$, where k is the cardinality of Δ. We find a lower bound for A. Now, $A > 0$ holds. If $g' > 1$, then $A \geqslant 1$, if $g' = 1$, then $k > 0$, so that $A \geqslant \frac{1}{4}$. Suppose that $g' = 0$. Then

$$2A = k - 2 - \sum \frac{1}{s_i},$$

and so $k \geqslant 3$ since $A > 0$.
  For $k = 3$,

$$2A = 1 - \sum \frac{1}{s_i}.$$

The greatest value < 1 of $\frac{1}{s_1} + \frac{1}{s_2} + \frac{1}{s_3}$, with integers $s_i \geqslant 2$, is $1 - \frac{1}{42}$. It is reached for $(2, 3, 7)$, and so $A \geqslant \frac{1}{84}$.
  For $k = 4$,

$$2A = 2 - \sum \frac{1}{s_i}.$$

the greatest value < 2 of $\sum_1^4 \frac{1}{s_i}$, with integers $s_i \geqslant 2$, is $2 - \frac{1}{6}$. It is reached for $(2, 2, 2, 3)$, and so $A \geqslant \frac{1}{12}$.
  For $k \geqslant 5$,

$$A = -1 + \frac{1}{2} \sum_1^k \left(1 - \frac{1}{s_i} \right) \geqslant -1 + \frac{k}{4} \geqslant \frac{1}{4}.$$

In all cases, $A \geqslant \frac{1}{84}$, and so $n \leqslant 84(g - 1)$. $\qquad \square$

Remark As will be seen in section 8, this bound is not reached for $g = 2$. There is a ramified covering X of the Riemann sphere (6.8, Exercise 1) for which it is for $g = 3$; in section 10 we construct X as a quotient of the disk.

6.7.3 Genus 1 Case

Let X be a connected compact Riemann surface of genus 1 and G a finite automorphism group of X of order n. Set $Y = X/G$ and let g' be the genus of Y. As

$$n\left(g' - 1 + \frac{1}{2}\sum_k\left(1 - \frac{1}{s_i}\right)\right) = 0,$$
$$g' = 1 - \frac{1}{2}\sum_1\left(1 - \frac{1}{s_i}\right)$$

since $n \neq 0$ since it is the order of a group. Hence, either $g' = 1$ or 0.

If $g' = 1$, $k = 0$, i.e. G acts freely on X. In fact, as already mentioned, X is a complex torus, i.e. isomorphic to $\mathbb{C}/\Gamma$ with $\Gamma \approx \mathbb{Z}^2$. Then G can be identified with a subgroup of $\mathbb{C}/\Gamma$ acting by translations. The order n of G is arbitrary.

If $g' = 0$, then

$$\sum_1^k\left(1 - \frac{1}{s_i}\right) = 2,$$

and so $k = 3$ or 4 since $\frac{1}{2} \leqslant 1 - \frac{1}{s_i} < 1$.

for $k = 4$, $s_i = 2$ for all i.

For $k = 3$, $\frac{1}{s_1} + \frac{1}{s_2} + \frac{1}{s_3} = 1$. The possible solutions are $(2, 3, 6)$, $(2, 4, 4)$ and $(3, 3, 3)$. These solutions can be shown to be effectively realizable (6.7, Exercise 1).

6.7.4 Genus 0 Case

Let X be a connected compact Riemann surface of genus 0 and G a finite automorphism group of X of order n. Set $Y = X/G$; the surface Y is necessarily of genus 0. The Riemann–Hurwitz formula gives

$$\sum_1^k\left(1 - \frac{1}{s_i}\right) = 2 - \frac{2}{n}.$$

As $\frac{1}{2} \leqslant 1 - \frac{1}{s_i} < 1$, then $1 < k < 4$, and so $k = 2$ or 3 if $n > 1$.

If $k = 2$, then $\frac{1}{s_1} + \frac{1}{s_2} = \frac{2}{n}$. But as s_i is the order of a subgroup of G, $s_i \leqslant n$, and so $s_1 = s_2 = n$. There are two points x_1, x_2 in X fixed by G, the group G is cyclic of order n (6.7.1, Lemma), and G acts freely on $X - \{x_1, x_2\}$.

Example. X is the Riemann sphere, G the set of maps $z \mapsto \omega z$, where ω is an n-th root of unity. In fact, this example is typical of the general situation (6.7, Exercise 3).

If $k = 3$, then $\frac{1}{s_1} + \frac{1}{s_2} + \frac{1}{s_3} = 1 + \frac{2}{n}$. The solutions of this equation with $s_i \geqslant 2$ are:

    $(2, 2, d)$ with $n = 2d$,
    $(2, 3, 3)$ with $n = 12$,
    $(2, 3, 4)$ with $n = 24$,
    $(2, 3, 5)$ with $n = 60$.

Let us study the case $(2, 2, d), n = 2d$ (dihedral case): Let $y_3 \in Y$ correspond to $s_3 = d$. There are two points x' and x'' in the class y_3. The stabilizer H of x' is cyclic of order d and index 2 in G, and hence normal; the stabilizer of x'', being a conjugate of H, is H. Let $\sigma \in G - H$, then $\sigma(x') = x''$. The automorphism σ has at least one fixed x_1 since there is no nonzero finite group G' acting freely on X (otherwise $k(G') = 0$). Since $\chi^{-1}(y_3) = \{x', x''\}, \chi(x_1) \neq y_3$; hence it may be assumed that $\chi(x_1) = y_1$.

As $s_1 = 2$, the stabilizer of x_1 has order 2, and so σ is of order 2.

Hence there is a cyclic subgroup H of the group G of order d such that all elements of $G - H$ are of order 2; this is only possible for the *dihedral group*, i.e. the crossed product of $\mathbb{Z}/(2)$ and $\mathbb{Z}/(d)$, $\mathbb{Z}/(2)$ acting on $\mathbb{Z}/(d)$ by $a \mapsto -a$ (see 6.10.1).

Cases $(2, 3, 3)$, $(2, 3, 4)$ and $(2, 3, 5)$ are realizable as finite automorphism groups of the Riemann sphere, and uniquely so up to conjugation. The groups obtained are respectively isomorphic to $\mathfrak{A}_4$ (tetrahedral group), $\mathfrak{S}_4$ (cubic or octahedral group) $\mathfrak{A}_5$ (dodecahedral or icosahedral group) (Exercises 3, 4 and 5).

Exercises 6.7. (Finite Automorphism Groups)

1. *(Automorphisms of complex tori)*—Let X be a complex torus, i.e. a surface of the form $\mathbb{C}/\Gamma$, where Γ is a subgroup of $\mathbb{C}$ isomorphic to $\mathbb{Z}^2$.

(a) Let $f : X \to X$ be an automorphism with respect to the Riemann surface structure. Show that there is a holomorphic function $\tilde{f} : \mathbb{C} \to \mathbb{C}$ such that the diagram

$$
\begin{array}{ccc}
\mathbb{C} & \xrightarrow{\tilde{f}} & \mathbb{C} \\
\chi \downarrow & & \downarrow \chi \\
X & \xrightarrow{f} & X
\end{array}
$$

commutes. Show that $\tilde{f}'(x + \gamma) = \tilde{f}'(x)$ for $\gamma \in \Gamma$ (where $\tilde{f}'$ denotes the derivative of $\tilde{f}$). Using the maximum principle (6.1, Exercise 3), deduce that $\tilde{f}$ is of the form $x \mapsto ax + b$, with $a\Gamma = \Gamma$.

(b) Show that the number ν of nonzero elements of Γ with minimum absolute value can be 2, 4 or 6. For how many elements $a \in \mathbb{C}$, does $a\Gamma = \Gamma$?

(c) Let G_2 be the group consisting of 1 and $x \mapsto -x$. How many points in X are fixed by G_2? What is the genus of X/G_2? How many elements does the ramification set of $X \to X/G_2$ have, and what is their ramification index?

(d) Same questions for the group G_4 generated by $x \mapsto ix$ in the case $\nu = 4$, for the groups G_3 generated by $x \mapsto e^{\frac{2i\pi}{3}} x$ and G_6 generated by $x \mapsto e^{\frac{i\pi}{3}} x$ in the case $\nu = 6$.

(e) Let G be a finite automorphism group of X, and G' the subgroup consisting of the translations in G. Set $X' = X/G'$. Show that X' is a complex torus, that X is an (unramified) covering of X', and that if $G' \neq G$, then the Riemann surface of X', equipped with the equivalence relation induced by G, is isomorphic to one of the situations studied in (c) and (d).

2.—Let Y be a complex torus, a and b distinct points of Y.

(a) Show that $\pi_1(Y - \{a\}, b)$ is a free group with a 2 element basis.

(b) Construct a group G of order 8, with two elements u and v such that $uvu^{-1}v^{-1}$ is of order 2.

(c) Construct a ramified Galois covering X of Y of degree 8 whose ramification set is $\{a\}$, with ramification index 2 over a.

(d) What is the genus of Y? What is the order of the group $\mathrm{Aut}_X(Y)$?

3. *(Groups* SO_3 *and* $\mathrm{Aut}\,\Sigma$*)*—Let $SL_2\mathbb{C}$ be the group of 2×2 matrices with complex entries having determinant 1.

(a) Show that the automorphism group $\mathrm{Aut}\,\Sigma$ of the Riemann sphere can be identified with $SL_2\mathbb{C}/\{+1, -1\}$ (see 6.3.3).

(b) Let G be a finite subgroup of $\mathrm{Aut}\,\Sigma$. Show that there is a Hermitian form h on $\mathbb{C}^2$ such that all elements $g \in G$ are induced by an element $\tilde{g} \in SL_2\mathbb{C}$ preserving h.

(c) The stereographic projection (see 6.1.2) identifies Σ with the sphere $S^2 \subset \mathbb{R}^3$. Show that the group SO_3 of direct isometries of S^2 can be identified with a subgroup of $\mathrm{Aut}\,\Sigma$.

(d) Show that for any finite subgroup G of $\mathrm{Aut}\,\Sigma$, there is a subgroup G' conjugate to G contained in the group SO_3 of isometries of S^2 preserving the orientation. Show that if G is cyclic (resp. dihedral), G' can regarded as consisting of rotations about the axis Oz (resp. contained in the group generated by the rotations about the z-axis Oz and the symmetry about the xy-plane).

4. *(Subgroups of* SO_3 *and regular polyhedra)*—Let S be the Euclidean unit sphere of $\mathbb{R}^3$ and G a nontrivial finite subgroup of SO_3.

(a) Let $g \in G$ be a non-identity element, D the axis of rotation g. Show that, if D is invariant under G, then G is either cyclic or dihedral (see 6.10.1). Show that, if all non-identity element of G have order 2, then G is either cyclic of order 2 or dihedral of order 4, isomorphic to $\mathbb{Z}/(2) \times \mathbb{Z}/(2)$, consisting of 1 and of the symmetries about three orthogonal lines.

In the following, it is assumed that G is neither cyclic nor dihedral. Let $h \in G$ be an element of order $k > 2$, D the axis of rotation h, a one of the two points of $D \cap S$, H the subgroup generated by h. Let Ω be the set $G \cdot a$ of the images of a, and δ the minimum distance between two distinct points of Ω; two points of Ω are said to be *connected* if their distance is δ.

(b) Show that there is a point connected to a. Noting that the set of points of S whose distance from a is δ is a circle with radius $< \delta$, show that there are at most 5 points connected to a. Deduce that the order k of H can only be 3, 4 or 5, and that H acts transitively on the set of points connected to a.

(c) Show that G acts transitively on the set of pairs of connected points. If b and c are connected, show that the symmetry about the line generated by $b + c$ is contained in G.

(d) For connected b and c, let the arc $\gamma_{b,c}$ be the shortest path from b to c in S, and Γ the union of $\gamma_{b,c}$. Show that Γ is invariant under G and that G acts transitively on the connected components of $S - \Gamma$.

(e) Let C be a connected component of $S - \Gamma$. Show that the stabilizer H′ of C in G acts transitively on $\Phi = \overline{C} \cap \Omega$. Show that Φ is contained in a plane, and consists of the vertices of a regular polygon F. Show that the order k' of H′ must be either 3, 4 or 5.

(f) Show that the angles of F are $< \frac{2\pi}{k}$. Deduce that

$$(k - 2)(k' - 2) < 4.$$

What are the possibilities for the pair (k, k') ?

(g) Show that the convex envelope P of Ω is a polyhedron whose boundary is $\bigcup_{g \in G} gF$, and that G is the stabilizer of P in SO_3. Show that the numbers k and k' define P up to isometry and G up to conjugation. How many conjugation classes of finite subgroups that are neither cyclic nor dihedral are there in SO_3 ?

(h) Show that G is isomorphic to the quotient Λ of the free group $L(\alpha, \beta, \gamma)$ by the normal subgroup generated by α^k, $\beta^{k'}$, γ^2 and $\alpha\beta\gamma$ (use the proof given in 6.10).

5. (*Dodecahedral and icosahedral group*) — Let P be a regular dodecahedral, A the set of edges of P and G the group of direct isometries of P. Let C be a set with 5 elements, called "colours".

(a) Colour the edges of P using all the colours of C (i.e. define a map γ from A to C) in such a way that for each face f of P and each side a of f, the edge a' not contained in f terminating at the vertex of f opposite to a is of the same colour as a. Show that this colouring is unique up to permutation of C.

(b) Show that for all $g \in G$, there is a permutation $\tilde{g}$ of C making the diagram commute.

(c) Show that $g \mapsto \tilde{g}$ is an isomorphism from G onto the group $\mathfrak{A}_5$ of even permutations of C.

(d) Show that the set of faces of P can be identified with the quotient of $\mathfrak{A}_5$ by the group of circular permutations.

6. (*Construction of a Riemann surface of genus 2 with automorphism group isomorphic to* SL_2F_3)—(a) Let T be a regular tetrahedron. Show that the group H of isometries of T preserving the orientation is isomorphic to the group $\mathfrak{A}_4$ of even permutations of the vertices of T. Set $B = T/H$. Show that T is a ramified covering of B and that B is homeomorphic to the Riemann sphere.

Identify B with the Riemann sphere by a homeomorphism sending the vertices of T onto ∞.

(b) Show that there is a ramified 2-fold covering X of T whose ramification set is the set of the midpoints of the edges of T, and that it is unique up to isomorphism. Show that for any isometry f of T, there is an f-morphism $\tilde{f}$ of X. Deduce that X is a ramified Galois covering of B. Set $G = \mathrm{Aut}_B X$. Show that H is isomorphic to a quotient of G. Find the order of G.

(c) Let L be the inverse image in X of $\infty \in B$, and $V = \{O\} \sqcup L$, so that V has 9 elements.

Let F_3 be the field $\mathbb{Z}/(3)$. Show that V has a unique F_3-vector space structure such that O is the origin and the affine lines are the sets obtained

– by taking O and the 2 points of X over a vertex of T,

or

– by taking the images of 3 vertices of T under a continuous section of X over the face containing these vertices.

(d) Let GL(V) be the automorphism group of the vector space V and SL(V) the subgroup of GL(V) consisting of the elements with determinant 1. An element $g \in GL(V)$ defines a permutation of the set P(V) of lines through O. Show that $g \in SL(V)$ if and only if this permutation is even.

(e) Show that each element of G defines an automorphism of V, giving a homomorphism $\varphi : G \to GL(V)$.

Show that $\varphi(G) \subset SL(V)$ and that φ is injective, so that φ is an isomorphism from G onto SL(V).

6.8 Automorphism Groups: The Genus 2 Case

6.8.1 Homogeneity of Topological Surfaces

Proposition *Let* X *be a topological surface,* A *a closed subset of* X, x *and* y *two points in the same connected component of* X − A. *Then there is a homeomorphism* $h : X \to X$ *restricting to the identity on* A *and such that* $h(x) = y$.

Proof (a) *A particular case.* Suppose that $U = X - A$ is homeomorphic to the disk $\mathbb{D}$, and that there is a homeomorphism $\varphi : U \to \mathbb{D}$ extending to a homeomorphism from $\overline{U}$ onto $\overline{\mathbb{D}}$. It then suffices to show that if $a, b \in \mathbb{D}$, then there is a homeomorphism $h : \overline{\mathbb{D}} \to \overline{\mathbb{D}}$ inducing the identity on $S^1 = \partial\overline{\mathbb{D}}$ and such that $h(a) = b$. Define $f_a : [0, 1] \times S^1 \to \overline{\mathbb{D}}$ by $f_a(t, u) = (1 - t)a + tu$. The map f_a is surjective and, as $[0, 1] \times S^1$ is compact, it identifies $\overline{\mathbb{D}}$ with the quotient space of $[0, 1] \times S^1$ by the equivalence relation identifying the points of $\{0\} \times S^1$. Define f_b likewise. It has the same property. There is a homeomorphism h making the diagram

$$
\begin{array}{ccc}
 & [0, 1] \times S^1 & \\
{\small f_a}\swarrow & & \searrow{\small f_b} \\
\overline{\mathbb{D}} \xrightarrow{\quad h \quad} & & \overline{\mathbb{D}}
\end{array}
$$

commute and with the desired properties.

(b) *General case.* Write $x \sim y$ if there is a homeomorphism $h : X \to X$ inducing the identity on A and such that $h(x) = y$. It is an equivalence relation. The particular case implies that all $x \in X - A$ have a neighbourhood U which does not meet A and such that all $y \in U$ is equivalent to x. The classes of points of $X - A$ form a partition of $X - A$ into open subsets. Any two points in a connected component are in the same class.   □

Remark This proof applies to all dimensions.

6.8.2

Corollary 6.7 *Let* X *be a connected topological surface,* $\Delta \subset X$ *a finite set,* x *and* y *points of* X − Δ. *Then there is a homeomorphism* $h : X \to X$ *inducing the identity on* Δ *and such that* $h(x) = y$.

Indeed, $X - \Delta$ is connected (6.1.7, Remark 3).

Corollary 6.8 *Let* X *be a topological surface,* U *a connected open subset of* X, $x_1, \ldots, x_k, y_1, \ldots, y_k$ *distinct points of* U. *Then there is a homeomorphism* h *of* X *restricting to the identity on* X − U *and such that*

$$
h(x_1) = y_1, \ldots, h(x_k) = y_k.
$$

Proof For each i, let h_i be a homeomorphism of X restricting to the identity on $X - U \cup \{y_1, \ldots, y_{i-1}, x_{i+1}, \ldots, x_k\}$ and such that $h(x_i) = y_i$. Then $h = h_k \circ \cdots \circ h_1$ has the desired property.   □

Corollary 6.9 *Let* X *be a nonempty connected topological surface, and* $x_1, \ldots, x_k$ *points of* X. *Then there is an open subset* U *of* X *homeomorphic to a disk and containing* $x_1, \ldots, x_k$.

Proof The points x_i may be assumed to be distinct. Let V be an open subset of X homeomorphic to a disc, not containing any x_i, and $y_1, \ldots, y_k$ be distinct points of V. If $h : X \to X$ is a homeomorphism such that $h_i(x_i) = y_i$, then the open subset $U = h^{-1}(V)$ has the desired property.   □

6.8.3   Coalescence of Ramifications

Proposition and Definition *Let* B *be a topological surface,* X *a finite ramified covering of* B *and* Δ_X *its ramification set. Let* U *be an open subset of* B *homeomorphic to the disk* $\mathbb{D}$ *and containing* Δ_X, *and a a point of* U. *Then there is a finite ramified covering* $\widetilde{X}$ *of* B, *unique up to isomorphism, such that* $\Delta_{\widetilde{X}} \subset \{a\}$ *and* $\widetilde{X}|_W = X|_W$ *for a neighbourhood* W *of* $X - U$.

   $\widetilde{X}$ *is said to be obtained from* X *by a* **coalescence of ramifications** *at a.*

Proof Let $\varphi : U \to \mathbb{D}$ be a homeomorphism, and $Q \subset \mathbb{D}$ an annulus such that $\mathbb{D} - Q$ is a compact set containing $\varphi(\Delta_X)$ and $\varphi(a)$. Set $V = \varphi^{-1}(Q)$ and $W = (B - U) \cup V$. For $b \in Q$, $\pi_1(Q, b) = \pi_1(\mathbb{D} - \{\varphi(a)\}, b) = \mathbb{Z}$, and $\iota_* : \pi_1(Q, b) \to \pi_1(\mathbb{D} - \{\varphi(a)\}, b)$ is an isomorphism. Hence

$$\iota^* : \mathscr{C}\!ov(\mathbb{D} - \{\varphi(a)\}) \to \mathscr{C}\!ov(Q)$$

is an equivalence of categories. In particular every covering of Q extends to a covering of $\mathbb{D} - \varphi(a)$. Hence, every covering of V extends to a covering of $U - \{a\}$. The covering $X|_V$ extends to a covering Y of $U - \{a\}$. Gluing $X|_W$ and Y over V gives a finite covering of $B - \{a\}$, extending to a ramified covering $\widetilde{X}$ of B (6.1.11), with the desired property.

   If both $\widetilde{X}_1$ and $\widetilde{X}_2$ have this property, then there is a unique isomorphism from $\widetilde{X}_1|_U$ onto $\widetilde{X}_2|_U$ inducing the identity over V, hence a unique isomorphism from $\widetilde{X}_1$ onto $\widetilde{X}_2$ inducing the identity over W.   □

6.8.4

With the notation of 6.8.3, let $a_1, \ldots, a_k$ be the elements of Δ_X and $b \in W$.

Proposition *There are elements* $\gamma_1, \ldots, \gamma_k$ *in* $\pi_1(B - \Delta_X, b)$ *such that* γ_i *goes round* a_i *(see 6.1.9, Proposition) in* $B - (\Delta_X - \{a_i\})$, *and there exists* $\gamma \in \pi_1(B - \{a\}, b)$ *going round a in* B, *such that the action of* γ *on* $X(b) = \widetilde{X}(b)$ *defined by the covering* $\widetilde{X}|_{B-\{a\}}$ *agrees with the action of the product* $\gamma_k \ldots \gamma_1$ *on* $X(b)$ *defined by the covering* $X|_{B-\Delta_X}$.

Proof Keeping the notation of the former proof, we may assume that $b \in V$, and that $a_1, \ldots, a_k$ are not in $\overline{V}$. If necessary multiplying φ by a constant with absolute value 1, we may assume that $b' = \varphi(b) \in \mathbb{R}^*$, and y (6.8.2, Corollary 6.8), that $a'_\nu = \varphi(a_\nu)$ are purely imaginary and $a'_\nu = \lambda_\nu i$ with $\lambda_\nu < \lambda_{\nu+1}$.

Let $\rho > 0$ be such that $\rho \leqslant \frac{1}{2}|a'_\nu - a'_{\nu+1}|$ for all ν. For each ν, let γ'_ν be a loop in $\mathbb{D} - \{a'_1, \ldots, a'_k\}$ of the form $\beta_\nu^{-1} \cdot \alpha_\nu \cdot \beta_\nu$, where β_ν is a path from b' to $a' - \rho$ such that the real part of $\beta_\nu(t)$ is < 0 for all t, and $\alpha_\nu(t) = a' - \rho \cdot e^{2\pi i t}$. Let γ' be the loop $t \mapsto b' \cdot e^{2\pi i t}$ in Q.

Lemma *The loops* $\gamma'_k \ldots \gamma'_1$ *and* γ' *are homotopic in* $\mathbb{D} - \{a'_1, \ldots, a'_k\}$.

Proof The loop α_ν is of the form

$$\alpha_\nu^{(3)} \cdot \alpha_\nu^{(2)} \cdot \alpha_\nu^{(1)},$$

where $\alpha_\nu^{(1)}$ and $\alpha_\nu^{(3)}$ are paths in $\mathbb{R}_- + i\mathbb{R}$ and $\alpha_\nu^{(2)}$ in $\mathbb{R}_+ + i\mathbb{R}$.

Similarly, γ' is of the form $\gamma^{(3)} \cdot \gamma^{(2)} \cdot \gamma^{(1)}$. Then

$$\gamma'_k \ldots \gamma'_1 = \beta_k^{-1} \ldots \alpha_{\nu+1}^{(2)} \cdot \alpha_{\nu+1}^{(1)} \cdot \beta_{\nu+1} \cdot \beta_\nu^{-1} \cdot \alpha_\nu^{(3)} \cdot \alpha_\nu^{(2)} \ldots \beta_1.$$

Now, $\alpha_{\nu+1}^{(1)} \cdot \beta_{\nu+1} \cdot \beta_\nu^{-1} \cdot \alpha_\nu^{(3)}$ is homotopic by barycentric subdivision to the rectilinear path ε_ν de $a'_\nu + \rho i\ a'_{\nu+1} - \rho i$, and so

$$\gamma'_k \ldots \gamma'_1 \simeq \beta_k^{-1} \cdot \alpha_k^{(3)} \cdot \alpha_k^{(2)} \cdot \varepsilon_{k-1} \cdot \alpha_{k-1}^{(2)} \ldots \varepsilon_1 \cdot \alpha_1^{(2)} \cdot \alpha_1^{(1)} \cdot \beta_1.$$

In a similar way, $\beta_k^{-1} \cdot \alpha_k^{(3)}$, $\alpha_k^{(2)} \cdot \varepsilon_{k-1} \ldots \varepsilon_1 \cdot \alpha_1^{(2)}$ and $\alpha_1^{(1)} \cdot \beta_1$ are respectively homotopic to $\gamma^{(3)} \cdot \varepsilon_k$, $\varepsilon_k^{-1} \cdot \gamma^{(2)} \cdot \varepsilon_0^{-1}$ and $\varepsilon_0 \cdot \gamma^{(1)}$ where ε_0 and ε_k are rectilinear paths. Hence $\gamma'_k \ldots \gamma'_1 \simeq \gamma'$. $\square$

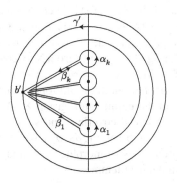

End of the Proof of the Proposition The loops $\gamma_\nu = \varphi^{-1} \circ \gamma'_\nu$ and $\gamma = \varphi^{-1} \circ \gamma'$ satisfy the required assumptions. The loops $\gamma_k \ldots \gamma_1$ and γ are homotopic in $B - \Delta_X$, and so act similarly on $X(b)$. The actions of γ on $X(b)$ defined by the coverings X and $\widetilde{X}$ agree since γ is a loop in W and $\widetilde{X}|_W = X|_W$. $\qquad\square$

6.8.5   *Coverings Ramified at a Single Point*

Proposition *Let* B *be a connected compact Riemann surface of genus* 0, $a \in B$, *and* X *be a finite ramified covering of* B *such that* $\Delta_X \subset \{a\}$. *Then* X *is a trivial covering of* B. *In particular* $\Delta_X = \varnothing$.

1st *proof* (assuming known that B is homeomorphic to S^2) The space $B - \{a\}$ is homeomorphic to $\mathbb{R}^2$, and so is simply connected. The covering $X|_{B-\{a\}}$ is trivial, and so extends to a trivial covering of B. As the extension to a ramified covering is unique up to isomorphism (6.1.11), X is a trivial covering. $\qquad\square$

2^e *proof* (without the above assumption) We may assume that $X \neq \varnothing$ and that it is connected. Let $X(a) = \{x_1, \ldots, x_k\}$, r_i the ramification index of x_i and d the degree of X. Then $d = r_1 + \cdots + r_k$. By the Riemann–Hurwitz formula (6.6.13),

$$2 - 2g(X) = 2d - \sum_{1}^{k}(r_i - 1) = 2d - \sum r_i + k = d + k .$$

As $2 - 2g(X) \leqslant 2$, $d \geqslant 1$ and $k \geqslant 1$. Therefore $d = 1$ and $k = 1$. $\qquad\square$

6.8.6

Theorem *No connected compact Riemann surface of genus* 2 *has an automorphism group of order* 84.

Proof (*By contradiction*) Let X be a Riemann surface of genus 2 and G an automorphism group of X of order 84. Set $Y = X/G$. Then X is a ramified Galois covering of Y, which implies (6.7.2) that Y has genus 0 and that X is ramified over 3 points a_1, a_2 and a_3 of Y, with ramification index 2 over a_1, 3 over a_2 and 7 over a_3.

Let $\widetilde{X}$ be a ramified covering of Y obtained from X by coalescing the ramifications at a point $a \in Y$. Let $b \in Y - \{a, a_1, a_2, a_3\}$, $\gamma_1, \gamma_2, \gamma_3 \in \pi_1(Y - \{a_1, a_2, a_3\}, b)$ and $\gamma \in \pi_1(Y - \{a\}, b)$ satisfying the conditions of Proposition 6.8.4. Let Γ be the image of $\pi_1(Y - \{a_1, a_2, a_3\}, b)$ in the permutation group $X(b)$, and σ_i the image of γ_i. By Proposition 6.8.5, X is a trivial covering, and so γ acts trivially on $X(b)$ and $\sigma_3 \circ \sigma_2 \circ \sigma_1 = 1$. By Proposition 6.1.9, $\sigma_1, \sigma_2, \sigma_3$ have respective orders 2, 3 and 7. Indeed, by (4.6.11), the group Γ is isomorphic to G, and so has 84 elements. The theorem then follows from the next lemma:

6.8.7

Lemma *There is no group* Γ *of order* 84 *having elements* u, v, w *of respective orders* 2, 3 *and* 7 *with* $w = v \cdot u$.

Proof *(By contradiction)* Set $F = \Gamma/(w)$. The set F has 12 elements and Γ acts on F. As v has order 3, the orbits of (v) can have 1 or 3 elements. However, for all $x \in F$, the stabilizer of x is a conjugate of (w), and so is of order 7 and cannot contain v. Hence there are 4 orbits of (v), each with 3 elements. For the same reason, there are 6 orbits of (u), each with 2 elements.

We next give the fixed points of $w = vu$. If x is one such point, then $vu \cdot x = x$, and so $u \cdot x = v^2 x$. Hence $x = uv^2 \cdot x$ and $vuv^2 \cdot x = v \cdot x \neq v^2 \cdot x$, since the orbit of x has 3 distinct elements. Hence $v^2 \cdot x$ is not a fixed point. Neither is $y = v \cdot x$, for otherwise $x = v^2 \cdot y$ would not be fixed. Hence the orbits of (v) contain at most one fixed point of w, which therefore fixes at most 4 fixed elements.

However the orbits of (w) have 1 or 7 elements, and there can at most be one with 7 elements since F only has 12 elements. So w fixes at least 5 points, and thus $5 \leqslant 4$. $\qquad\square$

Exercises 6.8. (Automorphism group, genus 2)
1.—Let $e, a_1, \ldots, a_k$ be distinct points on the Riemann sphere Σ.

(a) Show that $\pi_1(\Sigma - \{a_1, \ldots, a_k\}, e)$ is a free group with $k - 1$ generators. More precisely, there are $\gamma_1, \ldots, \gamma_k \in \pi_1(\Sigma - \{a_1, \ldots, a_k\}, e)$ such that $(\gamma_1, \ldots, \gamma_{k-1})$ is a basis for this group, and γ_i go round a_i in $(\Sigma - \{a_1, \ldots, a_k\}) \cup \{a_i\}$ and $\gamma_k \cdot \gamma_{k-1} \cdots \gamma_1 = e$.

(b) Let G be a finite group generated by $\alpha_1, \ldots, \alpha_{k-1}$ and set $\alpha_k = \alpha_1^{-1} \ldots \alpha_{k-1}^{-1}$. Show that there is a ramified Galois covering X of Σ with group G, whose ramification set is in $\{a_1, \ldots, a_k\}$ with ramification over a_i having index the order s_i of α_i in G.

(c) Give the genus of X in terms of the order n of G and of the indices s_i.

6.9 Poincaré Geometry

The universal covering of a Riemann surface X of genus $\geqslant 2$ can be shown to be isomorphic to a disk, so that X becomes a quotient of the disk by an automorphism group. In this section, we study the automorphisms of the disk. There is a metric on $\mathbb{D}$ with respect to which every automorphism of $\mathbb{D}$ is an isometry (Poincaré metric).

6.9.1 Homographic Transformations of Circles

A *circle* on the Riemann sphere $\Sigma = \mathbb{C} \cup \{\infty\}$ is a set which is a circle in $\mathbb{C}$, and so of the form $L \cup \{\infty\}$, where L is a line in $\mathbb{C}$. The complement of a circle on Σ has two connected components, called the *disks* limited by this circle. Hence a disk on Σ is either an open disk in $\mathbb{C}$, or an open half-plane in $\mathbb{C}$, or the complement in Σ of a closed disk in $\mathbb{C}$.

Proposition *The homographic image of a circle (resp. a disk) on Σ is a circle (resp. a disk).*

Proof The homography group is generated by the maps $x \mapsto ax + b$ and the map $h : x \mapsto \frac{1}{x}$. The proposition obviously holds for the former ones. We show that h sends a circle C on Σ to a circle. First suppose that C is a circle in $\mathbb{C}$ which not passing through 0, and that a and b are the endpoints of the diameter of C through 0 (an arbitrary diameter if C is centered at 0). Then $x \in$ C if and only if $\frac{x-a}{x-b} \in i\mathbb{R} \cup \{\infty\}$. As $\frac{1/x-1/a}{1/x-1/b} = \frac{b}{a}\frac{x-a}{x-b}$ and $\frac{b}{a} \in \mathbb{R}^*$, $x \in$ C if and only if $\frac{1}{x}$ is in the circle of diameter $[\frac{1}{a}, \frac{1}{b}]$.

If C is a circle in $\mathbb{C}$ passing through 0, then let a be such that $(0, a)$ is a diameter. So $x \in$ C if and only if $\frac{x-a}{x} \in i\mathbb{R} \cup \{\infty\}$, i.e. if $\frac{1}{x}$ is in the inverse image of $i\mathbb{R} \cup \{\infty\}$ under $y \mapsto 1 - ay$. This method gives all the circles $L \cup \{\infty\}$, where L is a line in $\mathbb{C}$ not passing through 0. As h is involutive, the image of such a circle must be a circle of $\mathbb{C}$ passing through 0.

If $C = L \cup \{\infty\}$, where $L = \lambda\mathbb{R}$ is a line passing through 0, then the image of C is $\frac{1}{\lambda}\mathbb{R} \cup \{\infty\}$.

Finally, as h is a homeomorphism of Σ, the images of the connected components of $\Sigma -$ C are the connected components of $\Sigma - h(\text{C})$, and so are disks. $\qquad\square$

6.9.2

Let $\mathbb{D}$ be the unit disk of $\mathbb{C}$ and S^1 the unit circle of $\mathbb{C}$, i.e. $\mathbb{D} = \{z \mid |z| < 1\}$ and $S^1 = \{z \mid |z| = 1\}$.

Proposition *Let $h : \Sigma \to \Sigma$ be a homography. Then, $h(S^1) = S^1$ if and only if h is of the form*

$$x \mapsto \lambda\frac{x-a}{1-\bar{a}x} \text{ with } a \in \mathbb{C} - S^1 \text{ and } \lambda \in S^1, \text{ or } x \mapsto \frac{\lambda}{x} \text{ with } \lambda \in S^1.$$

Proof Suppose that h is a homography. Let $b = h^{-1}(\infty)$. Set $h_0(x) = \frac{x-a}{1-\bar{a}x}$ with $a = \frac{1}{b}$ if $b \neq 0$ and $h_0(x) = \frac{1}{x}$ if $b = 0$. Now, $h_0(b) = \infty$ and $h_0(S^1) = S^1$. Set $u = h \circ h_0^{-1}$.

The map u is a homography such that $u(\infty) = \infty$, and hence of the form $z \mapsto \lambda z + \beta$, and $u(S^1) = S^1$, which implies that $|\lambda| = 1$ and $\beta = 0$; so $h = \lambda h_0$.

Conversely, for $x \in S^1$, $1 - \bar{a}x = x(\bar{x} - \bar{a})$. Thus $|1 - \bar{a}x| = |x - a|$ and $|h(x)| = 1$. The case $h = \left(x \mapsto \frac{\lambda}{x}\right)$ is immediate. $\qquad\square$

Corollary 6.10 *Let h be a homography. Then $h(\mathbb{D}) = \mathbb{D}$ if and only if h is of the form*

$$h_{a,\lambda} : x \mapsto \lambda \frac{x-a}{1-\bar{a}x} \text{ with } a \in \mathbb{D} \text{ and } \lambda \in S^1.$$

Proof If $h(S^1) = S^1$, the image of $\mathbb{D}$ is either $\mathbb{D}$ or $\Sigma - \bar{\mathbb{D}}$. If $h = h_{a,\lambda}$, then $h(0) = -\lambda a$ and $h(\mathbb{D}) = \mathbb{D}$ if $a \in \mathbb{D}$.

Conversely, if $h(\mathbb{D}) = \mathbb{D}$, then $h(S^1) = S^1$ since S^1 is the boundary of $\mathbb{D}$; h cannot be of the form $\frac{\lambda}{x}$, and so is of the form $h_{a,\lambda}$, and $a = h^{-1}(0) \in \mathbb{D}$. $\qquad\square$

Corollary 6.11 *The homographies h such that $h(\mathbb{D}) = \mathbb{D}$ form a group acting transitively on $\mathbb{D}$.*

Indeed the orbit of 0 is $\mathbb{D}$ since $h_{a,\lambda}^{-1}(0) = a$ for $a \in \mathbb{D}$.

6.9.3 Automorphisms of $\mathbb{D}$

An *automorphism* of $\mathbb{D}$ is an automorphism of D as a Riemann surface, i.e. a biholomorphic bijection (it follows from 6.1.6 that any holomorphic bijection is biholomorphic).

Theorem *Every automorphism of $\mathbb{D}$ is induced by a homography of Σ preserving $\mathbb{D}$.*

Proof Let f be an automorphism of $\mathbb{D}$. We first suppose that $f(0) = 0$. Then there is a holomorphic u with $|u(z)|$ tending to 1 as $|z|$ tends to 1 such that $f(z) = z \cdot u(z)$. By the maximum principle, $|u(0)| \leqslant 1$, and so $|f'(0)| \leqslant 1$ since $f'(0) = u(0)$. Set $g = f^{-1}$. Similarly, $|g'(0)| \leqslant 1$. As $g'(0) = \frac{1}{f'(0)}$, $|f'(0)| = 1$ and $|u(0)| = 1$. Hence u is a constant λ with absolute value 1, and $f(z) = \lambda z$.

In the general case, set $a = f^{-1}(0)$ and let f_0 be an automorphism of $\mathbb{D}$ induced by a homography and such that $f(a) = 0$ (see 6.9.2, Corollary 6.11). Set $g = f \circ f_0^{-1}$. Then g is an automorphism of $\mathbb{D}$ such that $g(0) = 0$, and so is $g : z \mapsto \lambda z$ with $\lambda \in S^1$, and $f = \lambda f_0$. $\qquad\square$

6.9.4 Riemannian Metrics

Let U be an open subset of $\mathbb{C}$. A *Riemannian metric*[5] on U is given by a continuous function $\mu : \mathrm{U} \to \mathbb{R}_+^*$. If $\gamma : [a,b] \to \mathrm{U}$ is a continuous differentiable path, then the *length* of γ with respect to the Riemannian metric defined by μ is $L_\mu(\gamma) = \int_a^b |\gamma'(t)| \mu(\gamma(t)) \, dt$. For x, $y \in \mathrm{U}$, the *distance* from x to y with respect to

---

[5]The Riemannian metrics considered are those that are compatible with the complex structure.

the Riemannian metric defined by μ is the lower bound $d_\mu(x, y)$ of the lengths of paths continuously differentiable from x to y in U.

Let U_1 and U_2 be open subsets of $\mathbb{C}$, μ_1 and μ_2 functions defining Riemannian metrics on U_1 and U_2 respectively. Let $f : U_1 \to U_2$ be a biholomorphic bijection. Consider the following conditions:

(i) $(\forall x \in U_1)$ $\mu_1(x) = |f'(x)|\mu_2(f(x))$;

(ii) For any continuously differentiable path γ in U_1,

$$L_{\mu_1}(\gamma) = L_{\mu_2}(f \circ \gamma) \;;$$

(iii) for $x, y \in U_1$, $d_{\mu_1}(x, y) = d_{\mu_2}(f(x), f(y))$.

(i) $\Rightarrow$ (ii) $\Rightarrow$ (iii) are immediate.[6] These conditions can be shown to be equivalent (6.9, Exercise 3). Given the function μ_2, there is a unique function μ_1 satisfying (i); the Riemannian metric defined by μ_1 is the inverse image under f of the Riemannian metric defined by μ_2 and we write $\mu_1 = f^*\mu_2$.

6.9.5   Poincaré Metric

Theorem and Definition *There is a unique Riemannian metric on* $\mathbb{D}$ *defined by a function* μ *such that* $\mu(0) = 1$, *and such that, for any automorphism* f *of* $\mathbb{D}$, $f^*\mu = \mu$.
This Riemannian metric is called the **Poincaré metric** *on* $\mathbb{D}$. *It is given by* $\mu(x) = \frac{1}{1-x\bar{x}}$.

Proof (1) *Uniqueness.* Let μ be another Riemannian metric with the required property and $x \in \mathbb{D}$. As $h_{x,\lambda}$ is an automorphism $\mathbb{D}$ sending x to 0,

$$\mu(x) = |h'_{x,\lambda}(x)| \, \mu(0) = |h'_{x,\lambda}(x)| = \frac{1}{|1 - x\bar{x}|} = \frac{1}{1 - x\bar{x}} \;.$$

(2) *Existence.* The function $\mu : x \mapsto \frac{1}{1-x\bar{x}}$ is a strictly positive continuous function on $\mathbb{D}$. Consider the Riemannian metric defined by μ. Let f be an automorphism of $\mathbb{D}$ and $x \in \mathbb{D}$. There exists $\lambda \in \mathbb{C}$ with absolute value 1 such that $h_{y,1} \circ f = h_{x,\lambda}$, where $y = f(x)$. Hence,

$$|h'_{y,1}(y)|.|f'(x)| = |h'_{x,\lambda}(x)| \;,$$

and so $|f'(x)| = \dfrac{|h'_{x,\lambda}(x)|}{|h'_{y,1}(y)|} = \dfrac{\mu(x)}{\mu(y)}$ and $f^*\mu = \mu$. $\square$

---

[6] A not necessarily holomorphic map f is an isometry with respect to the distances defined by μ_1 and μ_2 if it satisfies condition (iii). It may be shown that, as a result, f is either holomorphic, or antiholomorphic. We will prove this in the particular case of the Poincaré distance.

6.9.6

Proposition *Let x and y be two points of* $\mathbb{R} \cap \mathbb{D} =\,]-1, 1[$ *with* $x \leqslant y$, *and* γ_0 *the path* $t \mapsto t$ *from* $[x, y]$ *to* $\mathbb{D}$. *For the Poincaré metric* μ *on* $\mathbb{D}$, $d_\mu(x, y) = L_\mu(\gamma_0) =$ Argth $y -$ Argth x.

Proof Now, $L_\mu(\gamma_0) = \int_x^y \frac{1}{1-t^2}\, dt =$ Argth $y -$ Argth x.
Let γ be a continuously differentiable path from x to y. Write

$$\gamma(t) = \gamma_1(t) + i\gamma_2(t) \text{ with } \gamma_1(t) \text{ and } \gamma_2(t) \text{ reals.}$$

As $\gamma_1'(t) = \operatorname{Re}\gamma'(t)$, $|\gamma_1'(t)| \leqslant |\gamma'(t)|$, and $|\gamma_1(t)| \leqslant |\gamma(t)|$; so

$$\frac{1}{1 - (\gamma_1(t))^2} \leqslant \frac{1}{1 - |\gamma(t)|^2}$$

and

$$L_\mu(\gamma) = \int_x^y \frac{|\gamma'(t)|}{1 - |\gamma(t)|^2}\, dt \geqslant L_\mu(\gamma_1) = \int_x^y \frac{|\gamma_1'(t)|}{1 - (\gamma_1(t))^2}\, dt$$

$$\geqslant \int_x^y \frac{\gamma_1'(t)}{1 - (\gamma_1(t))^2}\, dt = \int_x^y \frac{du}{1 - u^2} = L_\mu(\gamma_0)\,,$$

$\square$

The distance d_μ is called the *Poincaré distance*.

Remark The above proof shows that if γ is a continuously differentiable path from x to y whose image is not contained in $\mathbb{R}$, then $L_\mu(\gamma) > d_\mu(x, y)$.

Corollary 6.12 *Let* $x \in \mathbb{D}$. *For the Poincaré metric* μ,

$$d_\mu(0, x) = \text{Argth } |x|\,.$$

Indeed, $x = \lambda|x|$ with $\lambda \in S^1$. As $z \mapsto \lambda z$ is an isometry with respect to the Poincaré distance, $d_\mu(0, x) = d_\mu(0, |x|) =$ Argth $|x|$.

Corollary 6.13 *The topology defined on* $\mathbb{D}$ *by the Poincaré distance is the Euclidean topology.*

Indeed, the neighbourhoods of 0 are the same. For all $x \in \mathbb{D}$, the map $h_{x,1}$ is a homeomorphism with respect to the Euclidean topology and an isometry with respect to the Poincaré distance, transforming x into 0. Hence the neighbourhoods of x are the same.

Corollary 6.14 *Every closed subset of* $\mathbb{D}$ *bounded with respect to the Poincaré distance is compact.*

6.9.7 Poincaré Circles

The *Poincaré circle* $C_{a,r}$ with centre $a \in \mathbb{D}$ and radius r is the set of $x \in \mathbb{D}$ such that $d_\mu(a, x) = r$.

Proposition *Every Poincaré circle is an Euclidean circle contained in* $\mathbb{D}$.

Proof By 6.9.6, Corollary 6.12, the Poincaré circle $C_{0,r}$ is the Euclidean circle with centre 0 and radius th r. With the notation of 6.9.2, Corollary 6.10, $C_{a,r} = h_{a,1}^{-1}(C_{0,r})$, and so, by Proposition 6.9.1, $C_{a,r}$ is a circle in $\mathbb{D}$. $\square$

Remarks (1) If $a \neq 0$, then it is not the Euclidean centre of the circle $C_{a,r}$.
(2) The Poincaré circle $C_{a,r}$ is in the Euclidean annulus consisting of u such that

$$(1 - |a|^2)\frac{\rho}{1 + \rho} < |u - a| < (1 - |a|^2)\frac{\rho}{1 - \rho} \,,$$

where $\rho = $ th r. Indeed, if $u \in C_{a,r}$, then $u = h_{a,1}^{-1}(z) = \frac{a+z}{1+\bar{a}z}$, with $z \in C_{0,r}$. Hence $|z| = \rho$, and $u - a = \frac{(1-a\bar{a})z}{1+\bar{a}z}$; so $|u - a| = \frac{1-|a|^2}{|1+\bar{a}z|}|z|$, with $1 - \rho < |1 + \bar{a}z| < 1 + \rho$. Therefore, the Euclidean radius r_e of $C_{a,r}$ satisfies

$$(1 - |a|^2)\frac{\rho}{1 + \rho} < r_e < (1 - |a|^2)\frac{\rho}{1 - \rho} \,.$$

In particular the Euclidean diameter of $C_{a,r}$ tends to 0 as $|a|$ tends to 1, the radius r staying bounded (see M.C. Escher's engraving "*Circle Limit III*", opposite 6.10.18, where all the fish have the same Poincaré diameter).

6.9.8 Geodesic Paths

Let U be an open subset of $\mathbb{C}$ and μ a Riemannian metric on U. A continuously differentiable map $\gamma : [a, b] \to$ U is a *geodesic path* if for all intervals $[a', b'] \subset [a, b]$,

$$d_\mu(\gamma(a'), \gamma(b')) = L_\mu(\gamma|_{[a',b']}) = |b' - a'| \,.$$

Lemma *For any* $x, y \in \mathbb{D}$, *there is a unique geodesic path* $\gamma : [0, d_\mu(x, y)] \to \mathbb{D}$ *from* x *to* y *with respect to the Poincaré metric. Its image is an arc of circle of* Σ *orthogonal to* S^1 *passing through* x *and* y.

Proof Suppose first that $x = 0$ and $y \in \mathbb{R}_+$. Then the map $s \mapsto$ th s from $[0, \text{Argth } t]$ to $\mathbb{D}$ is a geodesic path from 0 to y and by (6.9.6, Remarque) it is unique. Its image is an interval of $\mathbb{R}$.

In the general case, there is an automorphism h of $\mathbb{D}$ such that $h(x) = 0$ and $h(y) \in \mathbb{R}_+$. If γ_0 is the geodesic path from 0 to $h(y)$, then $\gamma = h^{-1} \circ \gamma_0$ is a geodesic path from x to y and it is the only one. The automorphism h extends to a homography, which we also write $h : \Sigma \to \Sigma$, and the image of γ is an arc of circle $h^{-1}(\mathbb{R} \cup \{\infty\})$. This circle is orthogonal to S^1 since $\mathbb{R}$ is orthogonal to S^1 and h preserves angles, its tangent linear map being $\mathbb{C}$-linear, and so is a similitude. □

6.9.9 Geodesics

Definition A **geodesic** in $\mathbb{D}$ is any set of the form $\gamma(\mathbb{R})$, where $\gamma : \mathbb{R} \to \mathbb{D}$ is a map such that, for all $[a, b] \subset \mathbb{R}$, $\gamma|_{[a,b]}$ is a geodesic path with respect to the Poincaré metric.

Proposition *The geodesics of $\mathbb{D}$ are the sets of the form $\Gamma \cap \mathbb{D}$, where Γ is a circle orthogonal to S^1.*

Proof (a) Any geodesic is of the form $\Gamma \cap \mathbb{D}$: let $\gamma : \mathbb{R} \to \mathbb{D}$ be as in the above definition. For each $[a, b]$, $\gamma([a, b])$ is in a circle $\Gamma_{a,b}$. For $[a', b'] \subset [a, b]$, $\Gamma_{a,b} \cap \Gamma_{a',b'} \supset \gamma[a', b']$ containing more than 3 points, and so $\Gamma_{a,b} = \Gamma_{a',b'}$. Hence $\Gamma_{a,b}$ does not depend on the choice of $[a, b]$. We write it Γ. Then $\gamma(\mathbb{R}) \subset \Gamma$ and γ is an isometry from Euclidean $\mathbb{R}$ onto a subset of $\Gamma \cap \mathbb{D}$ equipped with the Poincaré distance. This subset is homeomorphic to $\mathbb{R}$, and so is an open arc. It is complete and hence closed in $\Gamma \cap \mathbb{D}$. As $\Gamma \cap \mathbb{D}$ is connected and $\gamma(\mathbb{R}) \neq \varnothing$, $\gamma(\mathbb{R}) = \Gamma \cap \mathbb{D}$.

(b) Any set $\Gamma \cap \mathbb{D}$ is a geodesic: indeed, $\mathbb{R} \cap \mathbb{D}$ is a geodesic since it is the image of $t \mapsto \text{th}\, t$. Let Γ be a circle orthogonal to S^1; choose $x \in \Gamma \cap \mathbb{D}$. Let h be a homography preserving $\mathbb{D}$ and such that $h(x) = 0$. Then $h(\Gamma)$ is a circle on Σ passing through 0 and orthogonal to S^1. Hence $h(\Gamma)$ must be of the form $L \cup \{\infty\}$, where L is a line passing through 0, and we may assume that $L = \mathbb{R}$. Then $\Gamma \cap \mathbb{D} = h^{-1}(\mathbb{R} \cap \mathbb{D})$ is a geodesic. □

6.9.10

Corollary 6.15 *The geodesics passing through 0 are the sets of the form $L \cap \mathbb{D}$, where L is a line passing through 0.*

Corollary 6.16 *There is a a unique geodesic through two distinct points of $\mathbb{D}$.*

Corollary 6.17 *There is a unique geodesic passing through a tangent point to a given differentiable curve. In particular, two distinct geodesics are never tangent.*

Corollary 6.18 *The automorphism group of $\mathbb{D}$ acts transitively on the set of geodesics of $\mathbb{D}$.*

6.9.11   Isometries of $\mathbb{D}$

Let τ be the bijection $x \mapsto \bar{x}$ from $\mathbb{D}$ onto $\mathbb{D}$. For any continuously differentiable path γ in $\mathbb{D}$, $L_\mu(\tau \circ \gamma) = L_\mu(\gamma)$. Hence τ is an isometry with respect to the Poincaré distance.

Proposition *Any isometry $f : \mathbb{D} \to \mathbb{D}$ with respect to the Poincaré distance is either an automorphism of $\mathbb{D}$ or of the form $\tau \circ h$, where h is an automorphism of $\mathbb{D}$.*

Lemma 6.7 *Let $a, b \in \mathbb{D}$, $a \neq b$, and $r, r' \in \mathbb{Z}$, $r, r' > 0$. Then $C_{a,r} \neq C_{b,r'}$.*

Proof Let $\Delta = \gamma(\mathbb{R})$ be the geodesic passing through a and b, with γ satisfying the conditions of definition 6.9.9; define α and β by $\gamma(\alpha) = a$ and $\gamma(\beta) = b$. The circle $C_{a,r}$ meets Δ at two points $\gamma(\alpha_1)$ and $\gamma(\alpha_2)$ with

$$\alpha_1 = \alpha - r \quad \text{and} \quad \alpha_2 = \alpha + r \, ;$$

so $\alpha = \frac{1}{2}(\alpha_1 + \alpha_2)$. Similarly,

$$\Delta \cap C_{b,r'} = \{\gamma(\beta_1), \gamma(\beta_2)\} \text{ with } \beta = \frac{1}{2}(\beta_1 + \beta_2) \, .$$

If $C_{a,r} = C_{b,r'}$, then $\alpha = \beta$, and thus $a = b$. $\square$

Lemma 6.8 *Let $f : \mathbb{D} \to \mathbb{D}$ be an isometry fixing two distinct points a and b of $\mathbb{R} \cap \mathbb{D}$. Then $f = 1_\mathbb{D}$ or $f = \tau$.*

Let $x \in \mathbb{D} - \mathbb{R}$ and $r = d_\mu(a, x)$, $r' = d_\mu(b, x)$. The point $f(x)$ is in $C_{a,r} \cap C_{b,r'}$. Now, by Lemma 6.7, $C_{a,r}$ and $C_{b,r'}$ are distinct Euclidean circles with intersection consisting of x and $\tau(x)$. Hence $f(x) = x$ or $f(x) = \bar{x}$. We may write $f(x) = \mathrm{Re}(x) + \varepsilon(x)\,\mathrm{Im}(x) \cdot i$, where $\varepsilon(x) = \pm 1$. As f is continuous, the map $\varepsilon : \mathbb{D} - \mathbb{R} \to \{+1, -1\}$ is continuous, hence constant on each half-disc. This constant is the same on both, for otherwise $f(x) = f(\bar{x})$ and f would not be injective. Therefore, f agrees with either $1_\mathbb{D}$ or τ on $\mathbb{D} - \mathbb{R}$, hence on $\mathbb{D}$ by continuity.

Proof of the Proposition Choose a point $c \in \mathbb{D} \cap \mathbb{R}_+^*$, set $a = f^{-1}(0)$ and $b = f^{-1}(c)$. There is an automorphism h de $\mathbb{D}$ such that $h(a) = 0$ and $h(b) \in \mathbb{R}_+$. As $d_\mu(0, h(b)) = d_\mu(a, b) = d_\mu(0, c)$, $h(b) = c$. Set $g = h \circ f^{-1}$. Then $g(0) = 0$, $g(c) = c$ and g is an isometry. By Lemma 6.8, $g = 1_\mathbb{D}$ or $g = \tau$, and so $f = h$ or $f = \tau \circ h$. $\square$

6.9.12

Proposition and Definition *Let Γ be a geodesic of $\mathbb{D}$. There is a unique isometry σ_Γ such that Γ is the set of points fixed by σ_Γ. The latter, called the geodesic **symmetry**, is antiholomorphic.*

Proof The case $\Gamma = \mathbb{R} \cap \mathbb{D}$ follows from 6.9.11, Lemma 6.8: then $\sigma_\Gamma = \tau$. The general case follows from 6.9.10, Corollary 6.16.

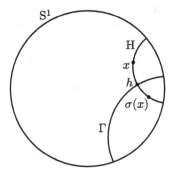

Remarks (1) The symmetry σ_Γ is induced by inversion with respect to the circle containing Γ.

(2) The map σ_Γ can be described as follows: let $x \in \mathbb{D}$; there is a unique geodesic H orthogonal to Γ and passing through x. It crosses Γ at a point h. if $x \neq h$, $y = \sigma_\Gamma(x)$ is the point of H distinct from x such that $d_\mu(h, y) = d_\mu(h, x)$.

6.9.13

Proposition and Definition *Let $a \in \mathbb{D}$, $\theta \in \mathbb{R}/2\pi\mathbb{Z}$; there is a unique isometry $r_{a,\theta}$ of $\mathbb{D}$ fixing a, whose tangent map at a is the rotation $t \mapsto e^{i\theta}t$. It is called the **Poincaré rotation** around the centre a with angle θ and is holomorphic. Any automorphism of $\mathbb{D}$ fixing a is of the form $r_{a,\theta}$.*

Proof The proposition is immediate if $a = 0$; the isometry $r_{a,\theta}$ is then the Euclidean rotation around O of angle θ.

The general case follows by transference using an automorphism h of $\mathbb{D}$ such that $h(a) = 0$, noting that

$$(h'(a))^{-1} \cdot 1 \circ e^{i\theta} \cdot 1 \circ h'(a) \cdot 1 = e^{i\theta} \cdot 1 \, ,$$

□

Corollary 6.19 *Let Γ and Γ' be geodesics of $\mathbb{D}$ intersecting at a point a with angle θ. Then $\sigma_{\Gamma'} \circ \sigma_\Gamma = r_{a,2\theta}$.*

Corollary 6.20 *Let $a \in \mathbb{D}$, $\theta \in \mathbb{R}/2\pi\mathbb{Z}$ and Γ a geodesic through a. Then $r_{a,\theta} \circ \sigma_\Gamma$, (resp. $\sigma_\Gamma \circ r_{a,\theta}$) is the geodesic symmetry*

$$r_{a,\frac{\theta}{2}}(\Gamma) \quad \left(resp. \ r_{a,-\frac{\theta}{2}}(\Gamma)\right).$$

6.9.14   Geodesically Convex Sets

Let a and b be points of $\mathbb{D}$ and γ the geodesic path from a to b.

The image de γ is called the *geodesic segment* with endpoints a and b, and is written $[a, b]$. Let α and β be such that $\gamma(\alpha) = a$ and $\gamma(\beta) = b$. For $u, v \in \mathbb{R}_+$ such that $u + v = 1$, we call the point $\gamma(u\alpha + v\beta)$ the *geodesic barycentre* of a and b with coefficients u and v, and write it $\mathrm{Bar}(a, u\,;\,b, v)$.

The map $(a, b, u) \mapsto \mathrm{Bar}(a, u\,;\,b, 1 - u)$ from $\mathbb{D} \times \mathbb{D} \times [0, 1]$ to $\mathbb{D}$ is continuous. Contrary to what happens in Euclidean geometry, if u, v and w are three numbers > 0 such that $u + v + w = 1$, then the points

$$\mathrm{Bar}(a, u\,;\,(b, v; c, w)) = \mathrm{Bar}\big(a, u;\, \mathrm{Bar}\big(b, \frac{v}{v + w}\,;\,c, \frac{w}{v + w}\big), v + w\big)$$

and

$$\mathrm{Bar}((a, u\,;\,b, v); c, w) = \mathrm{Bar}\big(\mathrm{Bar}\big(a, \frac{u}{u + v}\,;\,b, \frac{v}{u + v}\big), u + v\,;\,c, w\big)$$

are usually distinct.

A subset A of $\mathbb{D}$ is said to be *geodesically convex* if for all $a, b \in$ A, the geodesic segment $[a, b]$ is in A. All geodesically convex sets are connected.

Let Γ be a geodesic of $\mathbb{D}$. The complement of Γ has two connected components D_1 and D_2. The sets $H_1 = D_1 \cup \Gamma$ and $H_2 = D_2 \cup \Gamma$ are called the *half-disks* limited by Γ. They are closed and their boundary is Γ. They are geodesically convex: indeed, let $a, b \in H_1$ and $\gamma : [u, v] \to \mathbb{D}$ be the geodesic path from a to b. Suppose that for some $w \in [u, v], \gamma(w) \in D_2$. Then there exist $u_1 \in [u, w[$ and $v_1 \in]w, v]$ such that $\gamma(u_1) \in \Gamma$, and $\gamma(v_1) \in \Gamma$, which is impossible since the geodesic through a and b can only meet Γ at one point if it is distinct from Γ.

6.9.15   Geodesic Triangles

Let $a, b, c \in \mathbb{D}$ be points not on the same geodesic. Let $\Gamma_{b,c}$ denote the geodesic through b and c and H_a the half-disk limited by $\Gamma_{b,c}$ containing a. Define H_b and H_c likewise. The *geodesic triangle* with vertices a, b, c is $T = H_a \cap H_b \cap H_c$. The set T is geodesically convex, and hence is connected. As it is closed, the formula $\partial(A \cap B) = (\partial A \cap B) \cup (A \cap \partial B)$ which holds for any two arbitrary closed sets implies that its boundary is

$$\partial T = [a, b] \cup [b, c] \cup]c, a]\,.$$

The set T is bounded with respect to the Poincaré distance. Indeed, for all $x \in$ T, the Poincaré circle through x and with centre a contains points that are not in T since it contains points of each half-disk limited by $\Gamma_{a,b}$; hence it meets ∂T, and $d_\mu(a, x)$ is bounded above by

$$\sup_{y \in \partial T} d_\mu(a, y),$$

a number independent of x, which is finite since ∂T is compact. So the set T is compact (6.9.6, Corollary 6.14).

6.9.16 Sum of the Angles of a Geodesic Triangle

Let $a, b, c \in \mathbb{D}$, u (resp. v) the derivative at the origin of the geodesic path from a to b (resp. from a to c). There is a unique $\alpha \in]-\pi, +\pi[$ such that $v = e^{i\alpha} \cdot u$. We say that $|\alpha|$ is the (unorieunorientednted) angle at a of the geodesic triangle T with vertices a, b, c.

Theorem *(a) The sum of the three angles of a geodesic triangle is strictly less than π.*

(b) Given α, β, γ in $]0, \pi[$ such that $\alpha + \beta + \gamma < \pi$, there is a geodesic triangle in $\mathbb{D}$ with angles α, β and γ.

Proof (a) If f is an automorphism of $\mathbb{D}$, the angle of T at a equals the angle of f (T) at $f(a)$. Hence we may assume that $a = 0$. Then the geodesics $\Gamma_{a,b}$ and $\Gamma_{a,c}$ are Euclidean line segments, and a is in the region outside the circle containing $\Gamma_{b,c}$. Let ω be the Euclidean centre of this circle, α' and β' the angles at b and c of the Euclidean triangle with vertices a, b, c.

Then $\alpha = \alpha' - \frac{1}{2}\delta$ and $\beta = \beta' - \frac{1}{2}\delta$, where δ in the Euclidean angle $b\omega c$. Hence $\alpha + \beta + \gamma < \alpha + \beta' + \gamma' = \pi$.

(b) Set

$$\delta = \pi - (\alpha + \beta + \gamma), \quad \beta' = \beta + \frac{1}{2}\delta \text{ and } \gamma' = \gamma + \frac{1}{2}\delta.$$

We first construct an Euclidean triangle with vertices a, b, c having angles α, β', γ', then an isosceles Euclidean triangle with vertices ω, b, c, the angle at the vertex ω being δ, and such that the points a and ω are on opposite sides of the line bc. Let Ω be the circle through b and c centered at ω. As $\alpha = \pi - \delta - \beta - \gamma < \pi - \frac{\delta}{2}$, the point a is outside the region encircled by Ω. Let Θ_1 be the circle centered at a orthogonal to Ω. As $\frac{\delta}{2} < \beta'$, b is in the region inside Θ_1, and so is c. If necessary by applying a similitude, we may assume that $a = 0$ and $\Theta_1 = S^1$. The geodesic triangle with vertices a, b, c has the required property. □

6.9.17   Finiteness of the Automorphism Group ($g \geqslant 2$)

Theorem *Let* X *be a compact Riemann surface whose universal covering* $\tilde{X}$ *is isomorphic to* $\mathbb{D}$. *Then the automorphism group* G $= \mathrm{Aut}(X)$ *is finite.*

Proof Let $\pi : \mathbb{D} \to X$ be an analytic map transforming $\mathbb{D}$ into a universal covering of X. Then X can be identified with $\mathbb{D}/\Gamma$, where $\Gamma = \mathrm{Aut}_X(\mathbb{D})$ (4.5.1, Remark). Equip $\mathbb{D}$ with the Poincaré metric $d_{\mathbb{D}}$. The group Γ acts on $\mathbb{D}$ by isometries and equip X with the quotient metric

$$d_X(x, y) = \inf_{\tilde{y} \in \pi^{-1}(y)} d_{\mathbb{D}}(\tilde{x}, \tilde{y}) \,,$$

where $\tilde{x}$ is an arbitrary point of $\pi^{-1}(x)$.

Equip G with the metric of uniform convergence

$$d_G(g, g') = \sup_{x \in X} d_X(g(x), g'(x)) \,.$$

The group G acts on X by isometries and X is compact; so G is compact by the Ascoli theorem. We show that it is discrete.

Set $\Gamma^* = \Gamma - \{1_{\mathbb{D}}\}$, and for $u \in \mathbb{D}$, let $\tilde{\rho}(u) = \inf_{\gamma \in \Gamma^*} d_{\mathbb{D}}(u, \gamma(u))$. The map $\tilde{\rho} : \mathbb{D} \to \mathbb{R}_+^*$ is continuous, and $\tilde{\rho}(u)$ only depends on $x = \pi(u)$. Hence there is a continuous map $\rho : X \to \mathbb{R}_+^*$ such that $\tilde{\rho} = \rho \circ \pi$. Set $\delta = \inf_{x \in X} \rho(x)$. As X is compact, $\delta > 0$. $\qquad\square$

Lemma *Let* $g \in$ G. *If* $d_G(1_X, g) < \frac{\delta}{2}$, *then* $g = 1_{\mathbb{D}}$.

Proof of the Lemma Assume that $d_G(1_X, g) = \alpha < \frac{\delta}{2}$. For $u \in \mathbb{D}$, there is a unique $v \in \mathbb{D}$ such that $\pi(v) = g(\pi(u))$ and $d_{\mathbb{D}}(u, v) \leq \alpha$, and the map $\tilde{g} : u \mapsto v$ thus defined is continuous. Since $\pi \circ \tilde{g} = g \circ \pi$, $\tilde{g} : \mathbb{D} \to \mathbb{D}$ is holomorphic. With respect to the Poincaré metric, $d_{\mathbb{D}}(u, \tilde{g}(u)) \leq \frac{\delta}{2}$. With respect to the Euclidean metric, $|\tilde{g}(u) - u|$ tends to 0 as $|u|$ tends to 1 (6.9.8, Remark 2). But the function $u \mapsto \tilde{g}(u) - u$ is holomorphic, and hence by the maximum principle is null. So $\tilde{g} = 1_{\mathbb{D}}$ and $g = 1_X$. $\qquad\square$

End of the the proof of the theorem The lemma implies that the identity element 1_X in the group G is isolated. Hence G is discrete. As it is compact, it is finite. $\qquad\square$

Exercises 6.9. (Poincaré geometry)
1. *(The upper half-plane.)*— Let Π be the set of $z \in \mathbb{C}$ such that $\mathrm{Im}(z) > 0$.
   (a) Show that $z \mapsto \frac{z-i}{z+i}$ is an isomorphism from Π onto $\mathbb{D}$.
   (b) Show that the homographies preserving Π are those of the form

$$z \mapsto \frac{az + b}{cz + d} \text{ with } a, b, c, d \in \mathbb{R} \text{ and } ad - bc > 0 \,.$$

What are the automorphisms of Π (as a Riemann surface)?

(c) Show that there is a unique Riemannian metric on Π, invariant under the automorphisms, and define it.

2.—Prove the triangular inequality with respect to the distance defined by a Riemannian metric.

3. *(Isometries with respect to the Riemannian metric)* — Let U be an open subset of $\mathbb{C}$ and $\mu : U \to \mathbb{R}_+^*$ a continuous function.

(a) Let $a \in U$ and $r > 0$ be such that the Euclidean disk $D_{a,r}$ is contained in U. Let m and $M \in \mathbb{R}_+$ be such that

$$m \leqslant \mu(x) \leqslant M \text{ for all } x \in D_{a,r} .$$

Show that for all $x \in D_{a,r}$ and all differentiable path γ from a to x in U,

$$L_\mu(\gamma) \geqslant m|x - a| .$$

Deduce that

$$(\forall x \in D_{a,r}) \quad m|x - a| \leqslant d_\mu(a, x) \leqslant M|x - a| .$$

(b) Show that, if γ is a differentiable path in U with $\gamma(0) = a$, then the function $t \mapsto d_\mu(a, \gamma(t))$ is differentiable at 0 with derivative $\mu(a) \cdot |\gamma'(0)|$.

(c) Deduce that (iii) $\Rightarrow$ (i) in 6.9.4.

(d) With the notation of 6.9.4, let $g : U_1 \to U_2$ be a not necessarily holomorphic C^1 real map. Assume that g satisfies relation (iii).

Show that $\frac{\mu_2(g(x))}{\mu_1(x)} T_x g$ is an $\mathbb{R}$-linear isometry of $\mathbb{C}$. Deduce that g is either holomorphic or antiholomorphic.

(e) Does this result continue to hold if g is no longer assumed to be a C^1 real map?

4.—Let $a, b, c \in \mathbb{D}$ be points not on the same geodesic. Show that the map $\varphi :$ $(u, v, w) \mapsto \mathrm{Bar}(a, u; (b, v; c, w))$ from $T_0 = \{(u, v, w) \in \mathbb{R}_+^3 \mid u + v + w = 1\}$ to $\mathbb{D}$ is a homeomorphism from the Euclidean triangle T_0 onto the geodesic triangle T with vertices a, b, c.

5.—Let a and b be distinct points of $\mathbb{D}$. Show that the set of points of $\mathbb{D}$ equidistant from a and b with respect to the Poincaré distance is a geodesic, which will be called the bisector of (a, b).

Let $a, b, c \in \mathbb{D}$, M_1, M_2 and M_3 the bisectors of (a, b), (b, c) and (c, a). Show that, if M_1 meets M_2 at a point ω, then the geodesic M_3 passes through ω. Give an example where $M_1 \cap M_2 = \varnothing$. Show that, in all cases, the circles containing M_1, M_2 and M_3 are in the same linear sheaf (set of circles orthogonal to two given circles).

6.—Let $\Gamma = \gamma(\mathbb{R})$ be a geodesic with γ as in Definition 6.9.9, and $\lambda \in \mathbb{R}$.

(a) Show that there is a unique automorphism f of $\mathbb{D}$ such that $f(\gamma(t)) = \gamma(t + \lambda)$ for all $t \in \mathbb{R}$. Then f is said to be the translation of axis Γ oriented by γ and *unit* λ.

(b) Let $x \in \mathbb{D}$. Show that there is a unique geodesic H through x and orthogonal to Γ. Give an explicit description of $f(x)$.

(c) Let D_1 be one of the half-disks limited by Γ and $\rho > 0$. Show that the set of points of D_1 at a (Poincaré) distance ρ from Γ is the intersection of $\mathbb{D}$ with a circle of Σ which is not orthogonal to the limit circle S^1. Show that, if $\lambda \neq 0$, then Γ is the only geodesic preserved by f.

(d) Show that, for all $M \notin \Gamma$, $d_\mu(M, f(M)) > |\lambda|$.

7.—Let f be an antiholomorphic isometry of $\mathbb{D}$. Show that f extends to a homeomorphism (also written f) of $\overline{\mathbb{D}} = \mathbb{D} \cup S^1$. Show that $f|_{S^1}$ necessarily fixes two distinct points u and v.

Let Γ be the geodesic of $\mathbb{D}$ with endpoints u, v.

Show that f is the composition of a translation of axis Γ and of a symmetry with respect to Γ.

8. *(Automorphisms of $\mathbb{D}$)*—(a) Let φ be an isomorphism of the half-plane Π (6.9, Exercise 1) on $\mathbb{D}$ and $\lambda \in \mathbb{R}$, T_λ the translation $x \mapsto x + \lambda$ from Π to Π, and $a = \varphi(\infty) \in S^1$.

Show that $P = \varphi \circ T_\lambda \circ \varphi^{-1}$ is an automorphism of $\mathbb{D}$ whose extension to $\overline{\mathbb{D}}$ has a unique fixed point a. It is said to be a *parabolic transformation*. Show that all tangent circles to S^1 at a and in $\overline{\mathbb{D}}$ are preserved by P, and that if Γ is a geodesic with endpoint a, then so is $P(\Gamma)$.

(b) Show that an automorphism f of $\mathbb{D}$ is either a rotation, or a translation of axis a geodesic of $\mathbb{D}$, or a parabolic transformation (consider the fixed points of the homography extending f).

(c) Define a topology on the group $\text{Aut}(\mathbb{D})$ and give a description of the 1-parameter subgroups, i.e. of the continuous homomorphisms from $\mathbb{R}$ to $\text{Aut}(\mathbb{D})$.

9.—Let $a, b, c \in \mathbb{D}$ be points not on the same geodesic. Denote by M_a and M_a' the interior and exterior bisector geodesics of $([a, b], [a, c])$; similarly for M_b, etc.

Show that M_a, M_b and M_c are concurrent.

Show that if M_a and M_b' intersect at a point α, then the geodesic M_c' passes through α. Show that in general the circles containing M_a, M_b' and M_c' are in a same linear sheaf.

10.—(a) Show that in Poincaré geometry, the three heights of a triangle lie on three circles in the same linear sheaf.

(b) Does the same hold for the three bisectors?

11.—Let A, A', B, B' be four points of $\mathbb{D}$, I the midpoint of [A, A'] (i.e. $\text{Bar}\left(A, \frac{1}{2}; A', \frac{1}{2}\right)$) and J the midpoint of [B, B']). Show that

$$d_\mu(I, J) \leqslant \frac{1}{2}\left(d_\mu(A, B) + d_\mu(A', B')\right).$$

(Show that $C = r_{J,\pi}(A')$ can be obtained from A by applying a translation of axis the geodesic Δ passing through I and J and of unit $2d_\mu(I, J)$, then use 6.9, Exercise 6, (d).)

6.10 Tiling of the Disk

6.10.1 The Dihedral Group

Let X be a topological space. A *tiling* of X is a locally finite family $(A_i)_{i \in I}$ of closed subsets of X with disjoint interiors whose union is X.

Let n be an integer. Let Δ_1 and Δ_2 be two lines in $\mathbb{C}$, $\frac{\pi}{n}$ the angle between them, Π_1 and Π_2 two closed half-planes limited by Δ_1 and Δ_2 respectively, Π_2 being obtained from Π_1 by a rotation of angle $\frac{\pi}{n} + \pi$. Set $S = \Pi_1 \cap \Pi_2$, $\Delta_i^+ = \Delta_i \cap S$. Let u and v be the symmetries with respect to Δ_1 and Δ_2 respectively, G the isometry group of $\mathbb{C}$ (with respect to the Euclidean metric) generated by u and v.

Proposition and Definition *(a) The group G has 2n elements and is said to be the* **dihedral group** *of order 2n.*

(b) The family $(g(S))_{g \in G}$ is a tiling of $\mathbb{C}$.

(c) Let $g, g' \in G$ and $x, x' \in S$. If $g(x) = g'(x')$, then $x = x'$; and $g(x) = g'(x)$ if and only if

– $g' = g$ if $x \in \overset{\circ}{S}$;

– $g' = g$ or $g' = gw$ (resp. gv) if $x \in \Delta_1^+ - \{0\}$ (resp. $\Delta_2^+ - \{0\}$).

Lemma *Let G be a group generated by the two elements u and v such that*

$$u^2 = v^2 = (vu)^n = e$$

(identity element). Then any element of G can be written as $(vu)^k$ or $(vu)^k v$, with $0 \leqslant k \leqslant n - 1$. In particular, G has at most 2n elements.

Proof of the Lemma $(vu)^k$ can be defined for $k \in \mathbb{Z}/(n)$. Then,

$$u = u^{-1} = (vu)^{n-1} v, \quad v = v^{-1} = (vu)^0 v, \quad (vu)^k (vu)^{k'} = (vu)^{k+k'},$$

$$(vu)^k v.(vu)^{k'} = (vu)^k (uv)^{k'-1} u = (vu)^{n+k-k'} v,$$

$$(vu)^k v.(vu)^{k'} v = (vu)^{n+k-k'}.$$

The set H of elements of the above form is stable, and contains u, v and their inverses; so $H = G$. □

Proof of the Proposition We may assume that $\Delta_1 = \mathbb{R}$, $\Delta_2 = e^{\frac{\pi}{n} i} \mathbb{R}$, and that S is the set of $\rho.e^{i\theta}$ with $\rho \in \mathbb{R}_+$, $\theta \in \left[0, \frac{\pi}{n}\right]$. Then $v \circ u$ is the rotation of angle $\frac{2\pi}{n}$ and $(v \circ u)^n = 1$. If $x = \rho.e^{i\theta}$, then $(v \circ u)^k(x) = \rho.e^{i\theta'}$ with $\theta' = \frac{2k\pi}{n} + \theta$ and $(v \circ u)^k \circ v(x) = \rho.e^{i\theta'}$ with $\theta' = \frac{(2k+1)\pi}{n} + \frac{\pi}{n} - \theta$.

For all $y \in \mathbb{C}^*$, there exist $g \in G$ and $x \in S$ such that $y = g(x)$. Indeed, if $y = \rho.e^{i\theta'}$, then $\theta' = \frac{m}{n}\pi + \theta$ for some $m \in \mathbb{Z}/(2n)$ and $\theta \in \left[0, \frac{\pi}{n}\right]$. If θ' is not a multiple of

$\frac{\pi}{n}$, then the pair (m, θ) is unique. Now, $y = (v \circ u)^k(x)$ with $x = \rho.e^{i\theta}$ if $m = 2k$, and $y = (v \circ u)^k v(x)$ with $x = \rho.e^{i(\frac{\pi}{n} - \theta)}$ if $m = 2k + 1$, and y can be uniquely written as $g(x)$. This already shows that all $(v \circ u)^k(\overset{\circ}{S})$ and $(v \circ u)^k \circ v(\overset{\circ}{S})$ are disjoint; i.e. G has $2n$ elements and all $g(\overset{\circ}{S})$ are disjoint.

If $\theta' = \frac{2k\pi}{n}$ $\left(\text{resp } \frac{(2k+1)\pi}{n}\right)$, then y can be written as $g(x)$ in two different ways, namely $y = (v \circ u)^k(x)$ and $(v \circ u)^{k-1} \circ v(x)$ with $x = \rho$, (resp. $y = (v \circ u)^k(x)$ and $(v \circ u)^k \circ v(x)$ with $x = \rho.e^{i\frac{\pi}{n}}$). This proves (c) and completes the proof of (b).   □

6.10.2   A Tiling of $\mathbb{D}$

Let Ismet($\mathbb{D}$) denote the isometry group of the disk $\mathbb{D}$ with respect to the Poincaré metric, and $\sigma_{a,b}$ the Poincaré symmetry with respect to the geodesic through a and b. Let n, p and q be integers such that

$$\frac{1}{n} + \frac{1}{p} + \frac{1}{q} < 1,$$

and let T be a geodesic triangle in $\mathbb{D}$ with angles $\frac{\pi}{n}$, $\frac{\pi}{p}$ $\frac{\pi}{q}$, and vertices a, b, c (Theorem 6.9.16). We give a presentation by generators and relations of the subgroup of Ismet($\mathbb{D}$) generated by the symmetries $\sigma_{b,c}$, $\sigma_{c,a}$ and $\sigma_{a,b}$. Let $L = L(U, V, W)$ be the free group on three indeterminates. Define

$$\psi : L \to \text{Ismet}(\mathbb{D}) \text{ by } \psi(U) = \sigma_{b,c}, \ \psi(V) = \sigma_{c,a} \text{ and } \psi(W) = \sigma_{a,b}.$$

Then $\psi(U^2) = \psi(V^2) = \psi(W^2) = \psi((UV)^q) = \psi((VW)^n) = \psi((WU)^p) = 1$. Let N be the normal subgroup of L generated by U^2, V^2, W^2, $(UV)^q$, $(VW)^n$ and $(WU)^p$. Set $G = L/N$ and let $\chi : L \to G$ be the canonical map. Define $\varphi : G \to \text{Ismet}(\mathbb{D})$ by the commutative diagram:

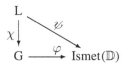

The group G acts on $\mathbb{D}$ by isometries via φ.

Let u, v, w be the images of U, V, W in G.

Theorem *The family $(g \cdot T)_{g \in G}$ of the transforms of* T *is a tiling of* $\mathbb{D}$. *In particular,* φ *is injective.*

The proof of this theorem is quite long. It will be completed in 6.10.9. We begin by constructing a space E containing T and acted on by G, for which it will be easy to see that for $g \in G$, $g \cdot T$ form a tiling of E. Define a map $\pi : E \to \mathbb{D}$ compatible with the actions of G. The difficulty lies in showing that π is a homeomorphism.

6.10.3 Construction of E

Let $\mathscr{R}$ be the equivalence relation on $G \times T$ identifying (g, x) with (gu, x) if $x \in [b, c]$, with (gv, x) if $x \in [a, c]$, with (gw, x) if $x \in [a, b]$. (In other words, $\mathscr{R}$ is the equivalence relation generated by the above pairs). Denote by E the quotient space $(G \times T)/\mathscr{R}$. The map $(g, x) \mapsto g \cdot x$ from $G \times T$ to $\mathbb{D}$ induces a continuous map $\pi : E \to \mathbb{D}$.

By passing to the quotient, the action $(h, (g, x)) \mapsto (hg, x)$ of G on $G \times T$ gives an action of G on E and π is compatible with the actions of G on E and $\mathbb{D}$.

Let G_a (resp. G_b, resp. G_c) be the subgroup of G generated by v and w (resp. w and u, resp. u and v). For $x \in\,]a, b[$, (resp. $]b, c[$, resp. $]c, a[$), set $G_x = \{e, w\}$ (resp. $\{e, u\}$, resp. $\{e, v\}$); for $x \in \overset{\circ}{T}$, set $G_x = \{e\}$. If $(g, x) \sim (g', x')$ with respect to $\mathscr{R}$, then $x = x'$. Indeed, the projection $\mathrm{pr}_2 : G \times T \to T$ is compatible with $\mathscr{R}$. Now, $(g, x) \sim (g', x)$ if and only if $g' \in g \cdot G_x$. Indeed, the relation thus described is an equivalence relation which implies $\mathscr{R}$ and is satisfied by all pairs generating $\mathscr{R}$.

Let χ be the canonical map from $G \times T$ to E. Then the map $\iota : x \mapsto \chi(e, x)$ from T to E is injective, and identifies T with its image in E.

6.10.4 The Star of a in E

For $x \in T$, let E_x be the image of $G_x \times T$ in E; this set is called the *star* of x. An *open star* of x is the image E'_x of $G_x \times T'_x$ in E, where T'_x is T without the sides that do not contain x (for example $T'_a = T - [b, c]$, $T'_x = T - ([b, c] \cup [c, a])$ if $x \in\,]a, b[$, $T'_x = \overset{\circ}{T}$ if $x \in \overset{\circ}{T}$).

For all x, the open start of x is open in E since it is the image of a saturated open subset of $G \times T$; the closed star of x is a compact neighbourhood of x.

Lemma 6.9 *The map π induces a homeomorphism from E_a onto a neighbourhood of a in $\mathbb{D}$.*

Proof (a) *Reduction to the case $a = 0$.* Let h be an automorphism of $\mathbb{D}$ such that $h(a) = 0$. Set $\widetilde{T} = h(T)$, and let $\widetilde{E}$ be the quotient of $G \times \widetilde{T}$ by the equivalence relation obtained by transferring $\mathscr{R}$. There is a commutative diagram

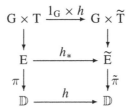

where the horizontal arrows are homeomorphisms. The space $\widetilde{E}$ is obtained from $\widetilde{T}$ in the same manner in which E is obtained from T. Hence we may assume that $a = 0$.

(b) *Proof of the lemma in this case.* The geodesics $\Gamma_{a,c}$ and $\Gamma_{a,b}$ are then part of Euclidean lines Δ_1 and Δ_2, the Poincaré symmetries $\sigma_{a,c}$ and $\sigma_{a,b}$ are induced by Euclidean symmetries with respect to Δ_1 and Δ_2, and T is a neighbourhood of 0 in a sector S of angle $\frac{\pi}{n}$.

The group G_a can be identified with the dihedral group of order $2n$. The equivalence relation induced on $G_a \times T$ by the map $(g, x) \mapsto g \cdot x$ is the one described in Proposition 6.10.1, (c); it implies the one induced by $\mathscr{R}$. Hence, $\pi|_{E_a}$ is injective. As T is a neighbourhood of a in S, the set $\pi(E_a) = \bigcup_{g \in G_a} g \cdot T$ is a neighbourhood of a in $\bigcup_{g \in G_a} g \cdot S = \mathbb{C}$.

As G_a is finite, $G_a \times T$ is compact, and so E_a is quasi-compact; and as $\mathbb{D}$ is Hausdorff, π induces a homeomorphism from E_a onto its image. $\qquad\square$

6.10.5

Lemma 6.10 *The map $\pi : E \to \mathbb{D}$ is etale.*

Proof The set $\overset{\circ}{T}$ in open in E, and π induces the identity on $\overset{\circ}{T}$. Hence π is etale on $\overset{\circ}{T}$.

For $x \in]a, b[$, the map $\pi|_{E_x}$ is injective since $T - [a, b]$ and its image under $\sigma_{a,b}$ are in disjoint open half-disks, and so are disjoint.

As T is a neighbourhood of x in the closed half-disk limited by $\Gamma_{a,b}$, the set $\pi(E_x) = T \cup \sigma_{a,b}(T)$ is a neighbourhood of x in $\mathbb{D}$. As E_x in quasi-compact, π induces a homeomorphism from E_x onto its image.

Hence, π is etale at x. Similarly, π is etale at all points either of $]b, c[$ or of $]c, a[$.

By Lemma 1 (6.10.4), π is etale at a. Similarly, π is etale at b and at c, hence on T.

Let $m = \chi(g, x) \in E$. There is a commutative diagram

$$
\begin{array}{ccc}
E & \xrightarrow{t \,\mapsto\, g \cdot t} & E \\
\pi \downarrow & & \downarrow \pi \\
\mathbb{D} & \xrightarrow{\quad g \quad} & \mathbb{D}
\end{array}
$$

where the horizontal arrows are homeomorphisms, and π is etale at x. Hence π is etale at $m = g \cdot x$. □

6.10.6

Lemma 6.11 *There exists $r > 0$ such that, with respect to the Poincaré metric, for all $t \in E$, there is a continuous section s of E defined on the ball of radius r centered at $\pi(t)$, with $s(\pi(t)) = t$.*

Proof For $t \in T$, let $R(t)$ be the upper bound in $\overline{\mathbb{R}}_+$ of all r for which there is a continuous section s of E defined on the ball $D_{t,r}$ of radius r centered at t relative to the Poincaré metric, with $s(t') = t'$ for all $t' \in D_{t,r} \cap T$. As E is etale, $R(t) > 0$ for all t. For $t' \in D_{t,r}$,

$$D_{t',r-d(t,t')} \subset D_{t,r} \ ;$$

Hence $R(t') \geqslant R(t) - d(t, t')$, and the function R is either constant, equal to $+\infty$, or Lipschitz with constant 1. In particular, it is continuous; as T is compact, it has a minimum $R_0 > 0$ and any $r < R_0$ has the required property for all $t \in T$.

Let $t = g \cdot t_0 \in E$ with $t_0 \in T$, and $s_0 : D_{t_0,r} \to E$ be a continuous section with $s_0(t_0) = t_0$. Then $s : x \mapsto g \cdot s_0(g^{-1}(x))$ is a continuous section of $D_{\pi(t),r}$ in E with $s(\pi(t)) = t$. Hence there is some $r < R_0$ has the required property for all $t \in E$. □

6.10.7

Lemma 6.12 *Let B be a metric space and E an etale Hausdorff space over B. Suppose there exists $r > 0$ such that, for all $t \in E$, there is a continuous section s of E defined on the ball of radius r centered at $\pi(t)$, with $s(\pi(t)) = t$. Suppose also that the intersection of two balls of B is connected. Then E is a covering of B.*

Proof Let $x \in B$ and U be the ball of radius r centered at x. Set $F = E(x)$. For all $t \in F$, there is a continuous section $s_t : U \to E$ such that $s_t(x) = t$. For $t \neq t'$, $s_t(y) \neq s_{t'}(y)$ for all $y \in U$. Indeed, U is connected and the set of y such that $s_t(y) = s_{t'}(y)$ is open since E is etale, closed since E is Hausdorff, and does not contain x; so is empty. The map $\varphi : (y, t) \mapsto s_t(y)$ from $U \times F$ to $E|_U$ is injective, a U-morphism, and is open. We show that it is surjective:

Let $t' \in E|_U$, $x' = \pi(t')$ and U' the ball of radius r centered at x'. There is a continuous section $s' : U' \to E$ with $s'(x') = t'$. Since $x' \in U, x \in U'$. Set $t = s'(x)$. The continuous sections s' and s_t agree at x; as E is etale and Hausdorff and $U \cap U'$ is connected, they agree on $U \cap U'$. In particular, $\varphi(x', t) = s_t(x') = t'$. Hence, φ is a homeomorphism and $E|_U$ is a trivial covering. □

6.10.8

Lemma 6.13 *The space* E *is Hausdorff and connected.*

Proof (a) E *is Hausdorff:* The projection $\text{pr}_2 : G \times T \to T$ is compatible with the equivalence relation $\mathscr{R}$, hence defines a continuous map $q : E \to T$. Therefore, $g \cdot t$ and $g' \cdot t'$ can be separated for $t \neq t'$. If $g \cdot t \neq g' \cdot t$, then the sets $g \cdot E'_t$ and $g' \cdot E'_t$ are open and disjoint.

(b) E *connectedness:* Let A be a clopen subset of E. For all g, the image of $\{g\} \times T$ is in A or in $E - A$. Let M be the set of $g \in G$ for which the image of $\{g\} \times T$ is in A. If $g \in M$, then $gu \in M$ since (gu, x) and (g, x) have the same image for $x \in [b, c]$. As $u^{-1} = u$ in G, $gu^{-1} \in M$. Similarly, gv, gv^{-1}, gw, gw^{-1} are in M. Since G is generated by u, v, w, $M = G$ or $M = \varnothing$, and so $A = E$ or $A = \varnothing$. $\square$

6.10.9   Proof of Theorem 6.10.2

In the space $\mathbb{D}$ equipped with the Poincaré distance, the balls are the Euclidean disks. Hence the intersection of two balls is connected. Thanks to lemmas 3 and 5, the assumptions of lemma 4 hold. Therefore E is a covering of $\mathbb{D}$. As $\mathbb{D}$ is simply connected, this covering is trivial. As E connected, it is of degree 1. In other words, π is a homeomorphism. $\square$

6.10.10   Crossed Products

Let A and B be two groups and $(b, a) \mapsto {}^b a$ an action of B on A by automorphisms (hence ${}^{bb'}a = {}^b({}^{b'}a)$ and ${}^b(aa') = {}^b a \cdot {}^b a'$ for $a, a' \in A$, $b, b' \in B$). For the sake of convenience, we assume that $A \cap B = \{e\}$, where e is the identity element common to A and B.

Proposition and Definition *There is a group* C *such that* A *is a normal subgroup of* C, B *a subgroup of* C, $b \cdot a \cdot b^{-1} = {}^b a$ *for* $a \in A$ *and* $b \in B$, *and such that canonical map* $\chi \circ \iota : B \to C/A$ *is an isomorphism. If* C *and* C' *satisfy these conditions, then there is a unique isomorphism from* C *onto* C' *inducing the identity on* A *and* B.

We say that C *is the* **crossed product** *(or* **semi-direct product***) of* A *by* B. *It is written* $C = A \rtimes B$.

Proof (a) *Uniqueness:* Let C be a group with the desired property. The map $(a, b) \mapsto a \cdot b$ from $A \times B$ to C is bijective. For $a, a' \in A$ and $b, b' \in B$, $a \cdot b \cdot a' \cdot b' = a \cdot {}^b a' \cdot b \cdot b'$. Hence C can be identified with the product set $A \times B$ equipped with the composition law

$$((a, b), (a', b')) \mapsto (a \, {}^b a', bb') \,.$$

(b) *Existence:* The composition law defined by

$$(a, b) \cdot (a', b') = (a\,{}^{b}a', bb')$$

on $A \times B$ is associative with

$$(a, b) \cdot (a', b') \cdot (a'', b'') = (a\;{}^{b}a'\;{}^{bb'}a'', bb'b''),$$

and each element has an inverse

$$(a, b)^{-1} = ({}^{b^{-1}}a^{-1}, b^{-1}).$$

The set $A \times B$ equipped with this law is a group C with the desired property.   □

6.10.11

Consider the group $G = L(U, V, W)/N$ defined in (6.10.2), and the homomorphism $\varepsilon : G \to \{+1, -1\}$ defined by $\varepsilon(u) = \varepsilon(v) = \varepsilon(w) = -1$.

Set $G_+ = \operatorname{Ker} \varepsilon = \varphi^{-1}(\operatorname{Aut} \mathbb{D})$. The group G_+ is a normal subgroup of index 2 in G, and G can be identified with the crossed product $G_+ \rtimes (u)$, where $(u) = \{e, u\}$.

Consider the elements $\alpha = vw$, $\beta = wu$ and $\gamma = uv$ in G_+. As $\alpha^n = \beta^p = \gamma^q = \alpha\beta\gamma = e$, the canonical map

$$\psi_1 : L(\alpha, \beta, \gamma) \to G_+$$

induces $\varphi_1 : \Lambda \to G_+$, where Λ is the quotient of $L(\alpha, \beta, \gamma)$ by the normal subgroup generated by $\alpha^n \beta^p$, γ^q and $\alpha\beta\gamma$.

Proposition *The homomorphism* $\varphi_1 : \Lambda \to G_+$ *is an isomorphism.*

Proof Define an action of (u) on Λ such that φ_1 is a (u)-homomorphism. The action of (u) on G_+ by inner automorphism in G satisfies

(∗)
$$\begin{cases} {}^{u}\alpha = \beta^{-1}\alpha^{-1}\beta, \\ {}^{u}\beta = \beta^{-1}, \\ {}^{u}\gamma = \gamma^{-1}. \end{cases}$$

Define an endomorphism $x \mapsto {}^{u}x$ of $L(\alpha, \beta, \gamma)$ by formulas (∗).

It can be checked to be involutive, i.e. ${}^{u}({}^{u}x) = x$. Hence there is an action of (u) on $L(\alpha, \beta, \gamma)$. Passing to the quotient, it can be checked to define an action of (u) on Λ by automorphisms. By construction, φ_1 is a (u)-homomorphism.

The homomorphism φ_1 defines a homomorphism

$$\overline{\varphi_1} : \Lambda \rtimes (u) \to G_+ \rtimes (u) = G.$$

We show that $\overline{\varphi_1}$ *is an isomorphism* by constructing an inverse. The homomorphism $U \mapsto u$, $V \mapsto u \cdot \gamma$, $W \mapsto \beta \cdot u$ from $L(U, V, W)$ to $\Lambda \rtimes (u)$ can be checked to pass to the quotient and to define a homomorphism $\sigma : G \to \Lambda \rtimes (u)$. As can be verified, this is an inverse of $\overline{\varphi_1}$.

In the commutative diagram

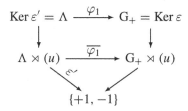

$\overline{\varphi_1}$ is an isomorphism, and so φ_1 is an isomorphism.   □

6.10.12

Proposition *Let* $g \in G_+$. *If* $\varphi(g) \in \operatorname{Aut} \mathbb{D}$ *has a fixed point in* $\mathbb{D}$, *then* g *is the conjugate in* G_+ *of a power of either* α, β *or* γ.

Proof We may assume that $g \neq e$. Let $x \in \mathbb{D}$ be such that $g \cdot x = x$. By Theorem 6.10.2, $x = h \cdot t$ for some $h \in G$, $t \in T$. Set $g' = h^{-1}gh$; then $g' \cdot t = t$. $t \notin \overset{\circ}{T}$ since the images of $\overset{\circ}{T}$ are disjoint. If $t \in]a, b[$, then $g' = w$, which is impossible since $g \in G_+$, which is normal. Similarly $t \notin]b, c[$ and $t \notin]c, a[$. Hence t is a vertex. It follows from the construction of E (6.10.3) that the stabilizer of a in G is G_a. Hence, if $t = a$, then $g' \in G_a \cap G_+ = \{1, \alpha, \dots, \alpha^{n-1}\}$. Similarly, if $t = b$ (resp. c), then g' is a power of β (resp. γ). This gives the result when $h \in G_+$. Otherwise $h = h'u$, and g is conjugate in G_+ to

$$u\alpha^k u^{-1} = \beta^{-1}\alpha^{n-k}\beta, \text{ or to } u\beta^k u^{-1} = \beta^{p-k}, \text{ or to } u\gamma^k u^{-1} = \gamma^{q-k}.$$

□

6.10.13

Proposition *Let* H *be a normal subgroup of* G_+. *Suppose that the images of* α, β *and* γ *in* G_+/H *are elements of orders* n, p *and* q *respectively.*

(a) The group H *acts freely on* $\mathbb{D}$ *and* $\mathbb{D}$ *is a covering of the quotient* $X = \mathbb{D}/H$. *The space* X *is Hausdorff.*

(b) There is a unique Riemann surface structure on X such that the canonical map
$\chi : \mathbb{D} \to X$ *is analytic.*

(c) The space X is compact if and only if H is of finite index in G.

Proof (a) Let $h \in H$ be such that $\varphi(h)$ has a fixed point in $\mathbb{D}$. By Proposition 6.10.12, h is conjugate to some α^k, β^k or γ^k. Its image $\chi(h) \in G/H$ is then conjugate to $\chi(\alpha)^k$ (resp. $\chi(\beta)^k$, resp. $\chi(\gamma)^k$); As it is zero, k is a multiple of n (resp. p, resp. q), and $h = 1$. Hence H acts freely on $\mathbb{D}$.

For $t \in T$, $g \in G$ and $h \neq e$ in H, $h \cdot g \cdot E'_t \cap g \cdot E'_t = \varnothing$ since $g^{-1}hg$ is a nonzero element of H, hence is not in G_t. Thus the assumptions of Theorem 4.3.13 hold, and $\mathbb{D}$ is a covering of X. As π identifies E with $\mathbb{D}$, the continuous map from E to T induced by $\mathrm{pr}_2 : G \times T \to T$ defines, by passing to the quotient, a continuous map $X \to T$. Thus the images of $g \cdot t$ and $g' \cdot t'$ can be separated for $t \neq t' \in T$. If $g \cdot t$ and $g' \cdot t$ have distinct images in X, then

$$g' \notin g \cdot H \cdot G_t = H \cdot g \cdot G_t \,,$$

and the open saturated subsets $H \cdot g \cdot E'_t$ and $H \cdot g' \cdot E'_t$ of $\mathbb{D}$ are disjoint. Hence X is Hausdorff.

(b) Every continuous section $\sigma : U \to \mathbb{D}$, where U is an open subset of X is a chart with domain U, and these charts form an atlas, the change of charts being induced by automorphisms of $\mathbb{D}$. The map χ is analytic with respect to the Riemann surface structure defined by this atlas, and it follows from 6.1.10 that this structure is the only one with the required property.

(c) Let $(g_i)_{i \in I}$ be a family of representatives for the elements of G/H. The image of $\bigcup_i g_i \cdot T$ is X. If G/H is finite, then so is I, and both $\bigcup_i g_i \cdot T$ and X are compact.

Let $t \in \overset{\circ}{T}$. The set $G \cdot t$ is discrete and closed in $\mathbb{D}$; hence so is its image in X. If X is compact, this image is finite; hence, as it is in bijective correspondence with G/H, G/H is finite if X is compact. □

6.10.14

We keep the assumptions of 6.10.13 and further assume that H is of finite index[7] in G. The group $\Gamma = G_+/H$ can be identified with an automorphism group of the Riemann surface X, and the space $Y = X/\Gamma$ can be identified with $\mathbb{D}/G_+$. Moreover, Y has a natural Riemann surface structure (6.7.1), and X is a ramified Galois covering of Y. Let π be the canonical map from X onto Y, and $\hat{a}$, $\hat{b}$ and $\hat{c}$ the images of a, b and c under $\pi \circ \chi$.

---

[7]In fact, this assumption can be shown to be superfluous for the results of this subsection and of 6.10.15.

Proposition *The ramification set of* X *over* Y *is* $\{\hat{a}, \hat{b}, \hat{c}\}$. *For* $\xi \in$ X *over* $\hat{a}$ *(resp.* $\hat{b}$, *resp.* $\hat{c}$*), the ramification index of* X *over* Y *at* ξ *is* n *(resp.* p, *resp.* q*).*

Proof For $\xi = \chi(x) \in$ X, the stabilizer Γ_ξ of ξ in Γ can be identified with the stabilizer $(G_+)_x$ of x in G_+. Indeed, if $\chi(g) \cdot \xi = \xi$, then $g \cdot x = h \cdot x$ for some $h \in$ H, and $h^{-1}g \in G_x$. Hence the canonical map $G_x \to \Gamma_\xi$ is surjective; its kernel is $\mathbb{G}_x \cap$ H and it is trivial since H acts freely.

Write x as $g \cdot t$ with $g \in$ G, $t \in$ T. The group $(G_+)_x$ is conjugate to $(G_+)_t$, hence nontrivial if and only if t is one of the 3 vertices of T. If t is a vertex, then $x = g' \cdot t$ with $g' \in G_+$. So $\xi = \chi(g') \cdot \chi(t)$, and $\pi(\xi) = \hat{a}$, $\hat{b}$ or $\hat{c}$. The ramification index of X at ξ is the cardinal of $(G_+)_t$, i.e. n if $\pi(\xi) = \hat{a}$, p if $\pi(\xi) = \hat{b}$ and q if $\pi(\xi) = \hat{c}$. $\square$

6.10.15

Proposition *Under the assumptions of 6.10.14, the Riemann surface* Y *has genus* 0.

Proof We construct a triangulation of Y. We keep the notation of this section as well as that of 6.5.1. Let $\theta : \mathbb{T} \to$ T be a C^1 homeomorphism on $\mathbb{T} - \{\alpha_0, \alpha_1, \alpha_2\}$ with Jacobian > 0, and such that $\theta \circ \iota_\nu$ is a geodesic path for $\nu = 1, 2$ or 3. Set

$$\theta_1(t) = u\big(\theta((1 + j)\bar{t})\big).$$

Set $f_0 = \pi \circ \chi \circ \theta$ and $f_1 = \pi \circ \chi \circ \theta_1$. Let a_1 (resp. a_2, resp. a_3) be the image under $\pi \circ \chi$ of the geodesic path from a to b (resp. from b to c, resp. from c to a), and $s_1 = \hat{a}, s_2 = \hat{b}, s_3 = \hat{c}$. This gives a triangulation τ of Y with 3 vertices, 3 edges and 2 faces. The Euler–Poincaré characteristic of Y (equipped with τ) is 2, and so its genus is 0. $\square$

Remark Using this triangulation, Y can be checked to be homeomorphic to the Riemann sphere.

6.10.16

Let d be the cardinal of Γ, i.e. the index of H in G_+.

Corollary *The genus* g *of* X *is* $1 + \frac{d}{2}\left(1 - \frac{1}{n} - \frac{1}{p} - \frac{1}{q}\right)$.

Proof There are $\frac{d}{n}$ ramification points of index n, $\frac{d}{p}$ of index p and $\frac{d}{q}$ of index q. The Riemann–Hurwitz formula 6.6.13 gives

$$g - 1 = -d + \frac{1}{2}\left(\frac{d}{n}(n - 1) + \frac{d}{p}(p - 1) + \frac{d}{q}(q - 1)\right).$$

□

6.10.17

Theorem *Let g, n, p and q be integers such that $g \geqslant 2$, $\frac{1}{n} + \frac{1}{p} + \frac{1}{q} < 1$, and $d = \frac{2g-2}{1-\frac{1}{n}-\frac{1}{p}-\frac{1}{q}}$ is integral. Assume that there is a group Γ of order d with three elements $\hat{\alpha}$, $\hat{\beta}$, $\hat{\gamma}$ of respective orders n, p and q satisfying $\hat{\alpha}\hat{\beta}\hat{\gamma} = e$. Then there is a Riemann surface X of genus g such that Γ is identified with an automorphism group of X.*

Proof By Proposition 6.10.11, Γ can be identified with a quotient of G_+ in such a way that $\hat{\alpha}$, $\hat{\beta}$ and $\hat{\gamma}$ are the respective images of α, β and γ. Then the assumptions 6.10.14 hold and the Riemann surface $X = \mathbb{D}/H$ has the required property. □

6.10.18

Corollary *There is a Riemann surface of genus 3 with an automorphism group of order 168.*

This corollary follows from Theorem 6.10.17 applied to $g = 3$, $n = 2$, $p = 3$, $q = 7$, which implies $d = 168$, and from the following lemma:

Lemma *Let $k = \mathbb{F}_2$ be the field $\mathbb{Z}/(2)$. The group $\Gamma = GL(k^3)$ of invertible 3×3 matrices with entries in k has 168 elements. There are three elements A, B, C in Γ of respective order 2, 3 and 7 such that $A \cdot B \cdot C = I$.*

Proof There is a bijection between the group Γ and the set of bases (V_1, V_2, V_3) for k^3. For V_1, there are 7 possible choices; V_1 being chosen, there are 6 ways of choosing V_2 and, V_1 and V_2 being chosen, 4 of choosing V_3. The order of G is therefore $7 \times 6 \times 4 = 168$.
Take

$$A = \begin{pmatrix} 1 & 0 & 1 \\ 0 & 1 & 0 \\ 0 & 0 & 1 \end{pmatrix} \quad B = \begin{pmatrix} 0 & 0 & 1 \\ 1 & 0 & 0 \\ 0 & 1 & 0 \end{pmatrix}$$

in Γ of respective orders 2 and 3. The "sagittal diagram" (Oh modern mathematics, what have you got into?)

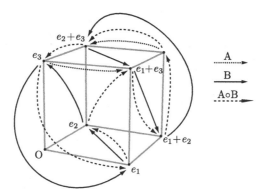

shows that A ∘ B has order 7. Then C = (A ∘ B)$^{-1}$ has the required property.   □

Exercises 6.10. (Disk tilings)
1.—Observe M. C. Escher's engraving. Ignoring the colours, it represents a tiling of the disk of the form $(gP)_{g \in G_+}$, where G_+ is defined as in 6.10.11 (the fundamental domain has the shape of a fish).

(a) What are the values of n, p, q?

(b) Let P be a fish. The elements $g \in G_+$ for which gP has the same colour as P form a subgroup of H. What is the index of H ?

Is the equivalence relation "gP and $g'P$ are of the same colour" between elements $g, g' \in G$, $g' \in gH$ or $g' \in Hg$?

Is the union of four fish of different colours necessarily a fundamental domain for the action of H?

(c) Give the permutations of $G/H = \{gH\}_{g \in G}$ defined by the action of G. How many are there?

(d) Consider the subgroup Γ of G_+ consisting of the elements preserving the colouring. Show that Γ is the largest normal subgroup of G_+ in H. Give its index in G_+.

(e) Show that Γ acts freely on $\mathbb{D}$ and that the Riemann surface $X = \mathbb{D}/\Gamma$ is compact.

(f) Give the equivalence relation identifying the space $B = \mathbb{D}/G_+$ with the quotient of P. Show that B is homeomorphic to the sphere S^2, and so is a Riemann surface of genus 0. What are the ramification indices of $\mathbb{D}$ over B? What is the genus of B/H? of B/Γ? (The Riemann–Hurwitz formula may be applied.)

(g) Are the spines of the fish geodesics?

The tiling of the Theorem 6.10.2 for $n = 2$, $p = 3$, $q = 7$

References

1. H. Cartan, *Théorie élémentaire des fonctions analytiques d'une ou plusieurs variables complexes* (Hermann, Paris, 1961)
2. J. Lelong-Ferrand, J.-M. Arnaudis, *Cours de mathématiques*, tome 1, *Algèbre* (Dunod, 1971)
3. R. Gunning, *Lectures on Riemann Surfaces*, 2nd edn. (Princeton University Press, Princeton, 1968)
4. H. Cartan, *Cours de calcul différentiel* (Hermann, Paris, 1977)
5. A. Weil, *Introduction à l'étude des variéétés kähalériennes* (Hermann, Paris, 1958)
6. N. Bourbaki, *Variétéés différentielles et analytiques, fascicule de réésultats, paragraphes 8 à 15* (Hermann, Paris, 1971)
7. M. Zisman, *Topologie algébrique éléémentaire* (Collection U, Armand Colin, 1972)
8. N. Bourbaki, *Algèbre, ch. 1 à 3* (Hermann, Paris, 1970); *Algèbre, ch. 10, algèbre homologique* (Masson, 1980)
9. D. Gilbarg, N.S. Trudinger, *Elliptic Partial Differential Equations of Second Order*, vol. 224, 2nd edn., Grundlehren der mathematischen Wissenschaften (Springer, Berlin, 1977, 2001)

Chapter 7
Dessins d'Enfants

Introduction

The Galois group $\mathbb{G} = \mathrm{Aut}_{\mathbb{Q}}(\overline{\mathbb{Q}})$, where $\overline{\mathbb{Q}}$ is an algebraic closure of $\mathbb{Q}$, let us say in $\mathbb{C}$, fascinates arithmeticians. This profinite group is hard to grasp. It certainly embeds in the product of $\mathfrak{S}(P^{-1}(0))$, as P runs through the set of irreducible polynomials of $\mathbb{Z}[X]$, but this set is itself not easy to understand if only because there is no natural way of numbering the roots of a polynomial.

The aim of Grothendieck's theory of dessins d'enfants is to make $\mathbb{G}$ act on finite sets defined combinatorially in order to gather information about it.

The "*arithmetic*" Riemann surfaces are those that can be defined over a "number field", i.e. a finite extension of $\mathbb{Q}$. In section 1, we explain what a Riemann surface defined over a subfield of $\mathbb{C}$ is. The theory begins with Belyi's theorem, according to which any arithmetic Riemann surface is a covering of the Riemann sphere ramified only over 0, 1 and ∞. The classification of these coverings becomes a combinatorial problem.

7.1 Definability

7.1.1 *Defining Polynomials*

The algebra of meromorphic functions for the Riemann sphere Σ $\mathscr{M}(\Sigma)$ can be identified with the field $\mathbb{C}(X)$ of rational fractions in one indeterminate X.

Let (S, π) be an analytic ramified covering of Σ, and consider $\mathscr{M}(S)$ as a $\mathbb{C}(X)$-algebra using $\pi^* : \mathbb{C}(X) \to \mathscr{M}(S)$. This is the situation in subsection 6.2.5, whose notation we keep (with Y instead of Z). Then $\mathscr{M}(S)$ is an etale algebra over $\mathbb{C}(X)$ (Theorem 6.2.3). If $\eta \in \mathscr{M}(S)$ is a primitive element with minimal polynomial

$$P(Y) = Y^d + a_{d-1}(X)Y^{d-1} + \cdots + a_0(X) \in \mathbb{C}(X)[Y],$$

then the algebra $\mathscr{M}(S)$ can be identified with $\mathbb{C}(X)[Y]/(P)$, and S is a compactification of

$$S' = V'_P = \{(x, y) \in \mathbb{C}^2 \mid x \in \mathbb{C} - \Delta_P , \ P(x, y) = 0\},$$

© Springer Nature Switzerland AG 2020
R. Douady and A. Douady, *Algebra and Galois Theories*,
https://doi.org/10.1007/978-3-030-32796-5_7

where $\Delta_P = \Delta_{\mathcal{M}(S),\eta}$ is the set of $x \in \mathbb{C}$ for which $P(x, Y) \in \mathbb{C}[Y]$ either has a multiple root or is not defined.

Each $a_i \in \mathbb{C}(X)$ can be uniquely written as $\frac{p_i}{q_i}$ with p_i and q_i coprime in $\mathbb{C}[X]$ and q_i monic. Let b_d be the (monic) l.c.m. of the polynomials q_i, $b_i = b_d a_i \in \mathbb{C}[X]$ and $Q = b_d Y^d + b_{d-1} Y^{d-1} + \cdots + b_0$. The polynomial $Q \in \mathbb{C}[X, Y]$ is called the polynomial obtained from $P \in \mathbb{C}(X)[Y]$ by *eliminating denominators*. It will be said to be a **defining polynomial** of (S, π).

The polynomial Q has the following property: *the decomposition of Q into prime factors has no repetition, nor any factor such as $X - a$.* A polynomial $Q \in \mathbb{C}[X, Y]$ with this property is called *regular* in Y. The set $\Delta_Q = \Delta_P$ is the set of $x \in \mathbb{C}$ for which $Q(x, Y) \in \mathbb{C}[Y]$ has either a multiple root or Y-degree strictly less than that of Q. The space S' can be identified with

$$V'_Q = V'_P = \{(x, y) \in \mathbb{C}^2 \mid x \notin \Delta_Q, \ Q(x, y) = 0\}.$$

Conversely, if $Q \in \mathbb{C}[X, Y]$ is a regular polynomial in Y, then the set

$$V'_Q = \{(x, y) \in (\mathbb{C} - \Delta_Q) \times \mathbb{C} \mid Q(x, y) = 0\},$$

together with $\pi : (x, y) \mapsto x$, is a finite covering of $\mathbb{C} - \Delta_Q$. By 6.1.10, this covering is induced by a finite analytic ramified covering $\widehat{V_Q}$ of Σ. $\widehat{V_Q}$ is said to be the analytic ramified covering defined by Q.

Proposition *The closure of S' in $\mathbb{C}^2$ is the algebraic set*

$$V_Q = \{(x, y) \in \mathbb{C}^2 \mid Q(x, y) = 0\}.$$

Proof The set V_Q is closed and contains V'_Q, hence so does its closure. We show the converse inclusion, which concerns the continuity of the roots of the polynomial (see Chap. 3, Exercise 8). Let (x_0, y_0) be a point such that $Q(x_0, y_0) = 0$, and (x_n) a sequence in $\mathbb{C} - \Delta_Q$ tending to x_0. Let ν be the multiplicity of y_0 as a root of $Q(x_0, Y)$. For sufficiently small $r > 0$, the point $Q(x_0, y)$ wraps around 0 ν times as y runs through the boundary of the disc $D_{y_0,r}$. Hence the same holds for $Q(x_n, y)$ for sufficiently large n. So for sufficiently large n, $Q(x_n, Y)$ has a root in $D_{y_0,r}$. Hence there is a sequence (y_n) tending to y_0 such that $(x_n, y_n) \in V'_Q$ for all sufficiently large n. $\qquad \square$

7.1.2 Definability of a Ramified Covering

Let K be a subfield of $\mathbb{C}$ and A an etale algebra over $K(X)$. Then the algebra $A_{\mathbb{C}} = \mathbb{C}(X) \otimes_{K(X)} A$ is an etale algebra over $\mathbb{C}(X)$. By (6.2.4), as a $\mathbb{C}(X)$-algebra, it is isomorphic to $\mathcal{M}(S)$, where (S, π) is an analytic ramified covering of Σ. A primitive element η of the K(X)-algebra A remains primitive in the $\mathbb{C}$(X)-algebra-$A_{\mathbb{C}}$, and S can be taken to be the Riemann surface $\widehat{\mathcal{S}}(A_{\mathbb{C}}, \eta)$.

Proposition and Definition *Let* (S, π) *be an analytic ramified covering of* Σ *and* K *a subfield of* $\mathbb{C}$. *The following conditions are equivalent:*

(i) there is a defining polynomial $Q \in K[X, Y]$ *of* (S, π);
(ii) there is a K(X)-subalgebra A in $\mathcal{M}(S)$ *such that the natural map* $\mathbb{C}(X) \otimes_{K(X)}$
      $A \to \mathcal{M}(S)$ *is an isomorphism.*

If these conditions hold, then (S, π) *is said to be* **definable** *over* K. *An algebra* A *satisfying condition (ii) is called a* K-**structure** *on* (S, π).

Proof (a) (i) $\Rightarrow$ (ii): Set $A = K(X)[Y]/(Q)$. Then $\mathbb{C}(X) \otimes_{K(X)} A = \mathbb{C}(X)[Y]/(Q)$ can be identified with $\mathcal{M}(S)$.

(b) (ii) $\Rightarrow$ (i): Let y be a primitive element of A over K(X), $P \in K(X)[Y]$ its minimal polynomial and $Q \in K[X, Y]$ obtained by eliminating the denominators. Then Q is a defining polynomial. $\square$

Remark $\mathbb{C} \otimes_K K[X] = \mathbb{C}[X]$, but in general $\mathbb{C} \otimes_K K(X) \neq \mathbb{C}(X)$. The algebra $\mathbb{C} \otimes_K K(X)$ is the ring of fractions with numerators in $\mathbb{C}[X]$ and denominators in K[X]; its an integral ring and $\mathbb{C}(X)$ can be identified with its field of fractions.

7.1.3 Semi-definability

Proposition and Definition *Let* (S, π) *be an analytic ramified covering of* Σ *and* K *a subfield of* $\mathbb{C}$. *The following conditions are equivalent:*

(i) (S, π) is definable over a finite extension L *of* K *in* $\mathbb{C}$;
(i') (S, π) is definable over the algebraic closure $\overline{K}$ *of* K *in* $\mathbb{C}$;
(ii) (S, π) is isomorphic to a ramified covering (union of connected components) of
       an analytic ramified covering definable over K.

If these conditions hold, (S, π) *will be said to be* **semi-definable** *over* K.

Proof The equivalence (i) $\Leftrightarrow$ (i') is immediate.

(a) (i) $\Rightarrow$ (ii): Let A be an etale algebra over L(X), where L is a finite extension of K in $\mathbb{C}$, such that $\mathcal{M}(S)$ can be identified with $\mathbb{C}(X) \otimes_{L(X)} A$. Let η be a primitive element of A and P its minimal polynomial, so that A can be identified with L(X)[Y]/(P). Let E be a Galois extension of K in $\mathbb{C}$ containing L, and G the Galois group of $Aut_K(E)$. As G acts on the coefficients of the polynomial of E(X, Y), it can be made to act on E(X, Y) by leaving fixed the indeterminates. Let g_*Q denote the image of Q under g; note that if Q is monic, then so is g_*Q.

Let $\widetilde{P}$ be the l.c.m. of g_*P: it is the product of polynomials g_*Q with $g \in G$ and Q an irreducible factor of P, taken without repetition.

Hence $\widetilde{P} = P \cdot P_1$ with P_1 and P coprime. Then $\widetilde{A}_E = E(X)[Y]/(\widetilde{P})$ is isomorphic to $A_E \times A'_E$, where $A_E = E(X)[Y]/(P) = E(X) \otimes_{L(X)} A$ and $A'_E = E(X)[Y]/(P_1)$. Set $\widetilde{A}_{\mathbb{C}} = \mathbb{C}(X) \otimes_{K(X)} \widetilde{A}_E$, $A'_{\mathbb{C}} = \mathbb{C}(X) \otimes_{K(X)} A'_{E}$, $\widetilde{S} = \widehat{\mathcal{S}}(\widetilde{A}_{\mathbb{C}}, \eta)$ 6.2.9, $S' = \widehat{\mathcal{S}}(A'_{\mathbb{C}}, \eta)$, with η denoting the image of Y. Then $\widetilde{A}_{\mathbb{C}}$ is isomorphic to $A_{\mathbb{C}} \times A'_{\mathbb{C}}$ and $\widetilde{S}$ is a sum, i.e. a disjoint union of S and S'.

Moreover $\widetilde{P} \in E(X)[Y]$ is left invariant by G; hence $\widetilde{P} \in K(X)[Y]$ and $\widetilde{S}$ together with its projection onto Σ is definable over K.

(b) (ii) $\Rightarrow$ (i): Assume that S is a clopen subset of $\widetilde{S}$ definable over K. We may assume that $\widetilde{S} = \widehat{\mathcal{S}}(A, \tilde{\eta})$, where $A = \mathbb{C}(X)[Y]/(\widetilde{P})$ and $\tilde{\eta}$ is the image of Y. Hence $\widetilde{S} = S \sqcup S_1$.

As in (6.2.5), let $\Delta_{\widetilde{P}}$ be the set of $x \in \mathbb{C}$ for which either one of the coefficients of $\widetilde{P}(x, Y)$ is not defined or this polynomial is a multiple root; it is a finite set. For all $x \in \mathbb{C} - \Delta_{\widetilde{P}}$,

$$\widetilde{P}(x, Y) = \prod_{y \in \widetilde{S}(x)} (Y - y) = \left(\prod_{y \in S(x)} (Y - y) \right) \cdot \left(\prod_{y \in S_1(x)} (Y - y) \right) = P(x) \cdot P_1(x),$$

where $S(x) = \tilde{\eta}(\pi_S^{-1}(x))$, etc.

The coefficients of $P(x)$ are the symmetric functions of z_i in $S(x)$; their dependence on x is holomorphic on $\mathbb{C} - \Delta_{\widetilde{P}}$, with growth of the order of $O\left(\frac{1}{|x-a|^k}\right)$ in the neighbourhood of $a \in \delta_{\widetilde{P}}$, and polynomial at infinity. Thanks to (6.2.2, Proposition), $P, P_1 \in \mathbb{C}(X)[Y]$. Then $S = \widehat{\mathcal{S}}(A, \eta)$, where $A = \mathbb{C}(X)[Y]/(P)$ and η is the image of Y.

We show that the coefficients of P are algebraic over K. Let Ω be an algebraic closure of $\mathbb{C}(X)$. Since $P = \prod_{i \in I}(Y - \eta_i)$ with algebraic η_i over $K(X)$ and $I \subset \widetilde{I}$, $\widetilde{P} = \prod_{i \in \widetilde{I}}(Y - \eta_i)$ in $\Omega[Y]$ Hence the coefficients of P as a polynomial in Y are algebraic elements of $\mathbb{C}(X)$ over $K(X)$. The proposition then follows from the next lemma:

Lemma *For all algebraic polynomials $f \in \mathbb{C}(X)$ over $K(X)$, there is a finite extension L of K such that $f \in L(X)$.*

Proof of the Lemma f can be uniquely written as $f = \frac{Q(X)}{Q_0(X)}$ with Q and Q_0 coprime in $\mathbb{C}[X]$ and Q_0 monic. Set $G = \mathrm{Aut}_K(\mathbb{C})$ and make G act on $\mathbb{C}(X)$ by leaving X fixed. The G-orbit of f is finite, and so the same holds for those of Q, Q_0 and of each of their coefficients. Hence, each coefficients of Q and Q_0 is algebraic over K, and L can be taken to be the extension of K in $\mathbb{C}$ generated by these coefficients. $\square$

7.1.4 Definability for Riemann Surfaces

Definition Let S be a compact Riemann surface and K a subfield of $\mathbb{C}$. S is said to be **definable** (resp. **semi-definable**) over K if there is a function $f \in \mathscr{M}^{**}(S)$ such that the ramified covering (S, f) is definable (resp. semi-definable) over K.

We remind the reader that $f \in \mathcal{M}^{**}(S)$ means that there is no connected component of S on which f is constant. A subalgebra A of $\mathcal{M}(S)$ is called a K-*structure* on S if there is a function $f \in \mathcal{M}^{**}(S)$ such that A is a K-structure on (S, f).

Proposition *Let* A *be a* K-*subalgebra of* $\mathcal{M}(S)$. *Then,* A *is a* K-*structure on* S *if and only if*

(i) A *is a finite product of extension fields of* K *with transcendence degree* 1;

(ii) $\lambda f \mapsto \lambda f$ *extends to an isomorphism from the total ring of fractions of* $\mathbb{C} \otimes_K$ A *onto* $\mathcal{M}(S)$.

Proof Suppose that A is a K-structure for (S, π), where $\pi = f$. Then A is an etale algebra over $K(X)$, hence a finite product of extensions of K with transcendence degree 1. The algebra $\mathcal{M}(S) = \mathbb{C}(X) \otimes_{K(X)} A$ is the algebra of fractions with numerators in A and denominators in $\mathbb{C}(X) - \{0\}$. Hence it is contained in the total ring of fractions of A, itself contained in $\mathcal{M}(S)$. Hence A satisfies (i) and (ii).

Conversely, suppose that A satisfies (i); let $f \in A$ be transcendental over K such that A is a finite algebra over $K(f) = K(X)$. Let g be a primitive element of $K(X)$-algebra A. Then A can be identified with $K(X)[Y]/(P)$, where P is the minimal polynomial of g. The algebra $\mathbb{C}(X) \otimes_{K(X)} A = \mathbb{C}(X)[Y]/(P)$ is the ring of fractions with numerators in $\mathbb{C}[X, Y]/(Q)$ (where Q is obtained from P by eliminating denominators) and denominators in $K[X] - \{0\}$. Hence it is contained in the total ring of fractions of $\mathbb{C} \otimes_K$ A. As it is an etale algebra over $K(X)$, it is a product of fields; it contains $\mathbb{C} \otimes_K$ A, hence also its total ring of fractions. Thus the total ring of fractions of $\mathbb{C} \otimes_K$ A is must be $\mathbb{C}(X) \otimes_{K(X)} A = \mathbb{C}(X)[Y]/(P) = \mathcal{M}(S)$, and A is a K-structure on (S, f). $\square$

Remarks (1) For any homographic function $f \in \mathcal{M}(\Sigma)$, i.e. of the form $\frac{aX+b}{cX+d}$ with $a, b, c, d \in \mathbb{C}$ and $ad - bc \neq 0$, the algebra $K(f)$ is a K-structure on Σ. Then, $K(f) = K(g)$ if and only if there exist $\alpha, \beta, \gamma, \delta \in K$ such that $\alpha\delta - \beta\gamma \neq 0$ and $g = \frac{\alpha f + \beta}{\gamma f + \delta}$. If K is countable (for example $\overline{\mathbb{Q}}$), this gives uncountably many distinct K-structures on Σ. These algebras are however isomorphic.

(2) If S is connected with genus ≥ 2 (or more generally if each connected component of S is of genus ≥ 2), and if K is algebraically closed, then there is at most one K-structure on S (Exercise 7.1.3).

Exercices 7.1 (Definability)
1.—Let S be a compact Riemann surface.

(a) Show that endowing S with an $\mathbb{R}$-structure is equivalent to defining an antiholomorphic homeomorphism $\tau : S \to S$ such that $\tau \circ \tau = I_S$.

(b) The real points of S are then the fixed points of τ. Give examples related to the same Riemann surface where the set $S(\mathbb{R})$ of real points in empty, and of others where it is not.

(c) Show that the number of $\mathbb{R}$-structures on S is bounded above by the cardinality of the automorphism group of S. Give examples with arbitrarily large numbers of $\mathbb{R}$-structures.

2.—Let S be a Riemann surface with a defining polynomial $Y^2 - f(X)$, where $f \in \mathbb{Q}[X]$ is non constant.

(a) Show that for $a \in \mathbb{Q}^*$, the $\mathbb{Q}$-subalgebra A_a generated by X and $\sqrt{a}.Y$ defines a $\mathbb{Q}$-structure on S.

(b) Show that this gives infinitely many distinct $\mathbb{Q}$-structures on S.

(c) Give examples where these distinct $\mathbb{Q}$-structures are isomorphic, and example where they are not.

3.—Let S be a compact Riemann surface all of whose connected components have genus $\geqslant 2$ and K an algebraically closed subfield of $\mathbb{C}$ (for example the algebraic closure $\overline{\mathbb{Q}}$ of $\mathbb{Q}$ in $\mathbb{C}$). The aim is to show that there is at most one K-structure on S.

Let A_1 and A_2 be K-structures on S, and $Q_1, Q_2 \in K[X, Y]$ defining polynomials of respective Y-degrees d_1 and d_2. The identity of $\mathcal{M}(S)$ gives an isomorphism ϕ from the total ring of fractions of $\mathbb{C} \otimes_K A_1$ onto that of $\mathbb{C} \otimes_K A_2$. The isomorphism ϕ is given by rational fractions $r_{i,\varepsilon}$ with $\varepsilon = 0, 1$, such that

$$\phi(X) = \sum_{i=0}^{d_2-1} r_{i,0}(X) \cdot Y^i, \quad \phi(Y) = \sum_{i=0}^{d_2-1} r_{i,1}(X) \cdot Y^i.$$

Consider the algebraic Galois group $G = \mathrm{Aut}_K(\mathbb{C})$. The action of G on the coefficients defines rational fractions ${}^g r_{i,\varepsilon} \in \mathbb{C}(X)$ for $g \in G$. Show that these ${}^g r_{i,\varepsilon}$ define an automorphism of S. Using (6.9.17), show that the G-orbit of the coefficients of $r_{i,\varepsilon}$ is finite. Deduce that these are elements of K and that $A_1 = A_2$.

7.2 Belyi's Theorem

7.2.1 Statement of the Theorem

For $f \in \mathcal{M}^{**}(S)$ (see 7.1.4), let $\Delta(f)$ be the set of critical values of f considered an analytic map from S to the Riemann sphere Σ. A polynomial $f \in \mathbb{C}[X]$ will be considered a map $\Sigma \to \Sigma$; in particular $\infty \in \Delta(f)$ if the degree of f is > 1.

The algebraic closure of a subfield K of $\mathbb{C}$ in $\mathbb{C}$ will be denoted $\overline{K}$.

Definition A compact Riemann surface is said to be **arithmetic** if it is semi-definable over $\mathbb{Q}$.

The aim of this section is to prove the following theorem:

Theorem *Let S be a compact Riemann surface. The following conditions are equivalent:*

(i) S is arithmetic;
(ii) there exists $f \in \mathcal{M}^{**}(S)$ such that $\Delta(f) \subset \overline{\mathbb{Q}} \cup \{\infty\}$;

(iii) there exists $f \in \mathcal{M}^{**}(S)$ such that $\Delta(f) \subset \{0, 1, \infty\}$.

The implication (iii) $\Rightarrow$ (ii) is trivial. The equivalence (i) $\Leftrightarrow$ (ii) is standard, but its detailed proof is long. The originality of the result due to Belyi lies in the implication (ii) $\Rightarrow$ (iii).

The equivalence (i) $\Leftrightarrow$ (ii) is a particular case of the more general statement:

Proposition *Let S be a compact Riemann surface and K a subfield of $\mathbb{C}$. The following conditions are equivalent:*

(i) S is semi-definable over K ;
(ii) there exists $f \in \mathcal{M}^{**}(S)$ such that $\Delta(f) \subset \overline{K} \cup \{\infty\}$.

In particular, if K is algebraically closed, for example if it is $\overline{\mathbb{Q}}$, then S is definable over K if and only if there exists $f \in \mathcal{M}^{**}(S)$ such that $\Delta(f) \subset K \cup \{\infty\}$.

Implication (i) $\Rightarrow$ (ii) will be proved in 7.2.2. Implication (ii) $\Rightarrow$ (i) will be proved in 7.2.5 using a density result stated in 7.2.3 as well as classic results of algebraic geometry. We will also give a method to obtain the number of distinct roots of a polynomial (7.2.4).

7.2.2 Implication (i) $\Rightarrow$ (ii)

Suppose that $S = \widehat{V_Q}$ (see 7.1.1), $Q \in L[X, Y]$, is regular of degree d in Y, where L is a finite extension of K. Then $\mathcal{M}(S) = \mathbb{C}(X)[Y]/(P)$, where P is the polynomial Q made monic in $\mathbb{C}(X)[Y]$. Set $A = L(X)[Y]/(P)$ and let f be an arbitrary element of $A^{**} = A \cap \mathcal{M}^{**}(S)$. We show that $\Delta(f) \subset \overline{L} \cup \{\infty\} = \overline{K} \cup \{\infty\}$.

Let $c \in S$ be a critical point of f and set $v = f(c)$. The function f is representable by a unique polynomial $F \in L(X)[Y]$ of degree $< d$:

$$F = f_0 + \cdots + f_{d-1}Y^{d-1}.$$

1st case: c is a point (x_c, y_c) of V'_Q and x_c is not a pole of any f_i.
The derivative $g = D_S f$ de f with respect to X along S is given by:

$$g = \frac{\partial F}{\partial X} - \frac{\partial Q/\partial X}{\partial Q/\partial Y} \cdot \frac{\partial F}{\partial Y}.$$

Hence it is in A.

The resultant $h = \text{Res}(g, P)$ in is $L(X)$. As $h(x) = \prod_{y \in S(x)} g(x, y)$ for $x \in \mathbb{C} - \Delta_Q$, and $h(x_c) = 0$, x_c is algebraic over L. Since $Q(x_c, y_c) = 0$, the ordinate y_c is also algebraic over L, and the same holds for $v = F(x_c)$.

2nd case: The image of c under the map $(X, Y) : \widehat{V_Q} \to \Sigma^2$ is a point (x_c, y_c) with $x_c \in \Delta_Q$, where x_c is a pole of some f_i.

In this case, x_c is obviously algebraic over L, and S can be analytically parametrized in the neighbourhood of c by T, where $T^k = X - x_c$. Then Y can be expressed as a series in T, with coefficients following from an algebraic calculation. Hence $Y \in \overline{L}[[T]]$. The function $f = F(X, Y)$ is represented by an element of $\overline{L}((T))$, i.e. by a fraction with numerator in $\overline{L}[[T]]$ and denominator a power of T. And $v = f(c)$, obtained by setting $T = 0$, is in $\overline{L} \cup \{\infty\}$.

This completes the proof of implication (i) $\Rightarrow$ (ii) in (7.2.1, proposition).

7.2.3 Density

Proposition *Let* K *be a subfield of* $\mathbb{C}$, *and* $\overline{K}$ *its algebraic closure in* $\mathbb{C}$. *Let* $\mathbf{f} = (f_1, ..., f_p) \in K[X_1, ..., X_n]^p$, *and* M *the set of points* $z = (z_1, ..., z_n) \in \mathbb{C}^n$ *such that* $f_1(z) = \cdots = f_p(z) = 0$. *Then* $M \cap \overline{K}^n$ *is dense in* M.

Proof The set M is algebraic in $\mathbb{C}^n$. The set M^* of regular points of M is open and dense in M [1, 2]. Let $x = (x_1, ..., x_n) \in M^*$, and d be the dimension of M^* at x. There is a partition of $\{1, ..., n\}$ in J' and J'' with Card $(J') = d$ such that, identifying $\mathbb{C}^n$ with $\mathbb{C}^{J'} \times \mathbb{C}^{J''}$, the vector space $T_x M^*$ is the graph of a linear map $\mathbb{C}^{J'} \to \mathbb{C}^{J''}$.

We then write $x = (x', x'')$. By the implicit function theorem, there are neighbourhoods U' of x' and U'' of x'' such that $M^* \cap (U' \times U'')$ is the graph of a $\mathbb{C}$-analytic map $h : U' \to U''$. This map is algebraic over $K[(X_i)_{i \in J'}]$, in the following sense: for all $j \in J''$, the map $h_j : U' \to \mathbb{C}$ satisfies a relation

$$a_{j,m_j} h_j^{m_j} + \cdots + a_{j,0} = 0,$$

where all $a_{j,k}$ are in $K[(X_i)_{i \in J'}]$ and $a_{j,m_j} \neq 0$. Hence $y' \in U' \cup \overline{K}^{J'}$ such that $a_{j,m_j}(y') \neq 0$ for all j, the coordinates of the point $y'' = h(y')$ are in $\overline{K}$, and $y = (y', y'') \in M^* \cap (U' \times U'') \cap \overline{K}^n$. Therefore, $M^* \cap (U' \times U'') \cap \overline{K}^n$ is dense in $M^* \cap (U' \times U'')$. It follows that $M^* \cap \overline{K}^n$ is dense in M^*, hence so in M. □

7.2.4 Polynomials with k Distinct Roots

Proposition *Let* $Q \in \mathbb{C}[X]$ *be a polynomial of degree* d, *with* k *distinct roots. Then the function* $h_Q : \mathbb{C} \to \mathbb{C}$ *assigning to* t *the discriminant* $\mathrm{discr}(Q - t)$ *has a zero of order* $d - k$ *at* 0.

Proof Let $\alpha_1, ..., \alpha_k$ be the distinct roots of Q with multiplicities $\mu_1, ..., \mu_k$. For t near 0, the polynomial $Q - t$ has μ_i roots $\xi_{i,1}, ..., \xi_{i,\mu_i}$ near α_i, with $\xi_{i,j} - \alpha_i$ infinitely small of the order of $|t|^{1/\mu_i}$ as t tends to 0, as well as $\xi_{i,j} - \xi_{i,j'}$ for $j \neq j'$. Indeed, by proposition 6.1.4, there is a chart φ de $\mathbb{C}$ in the neighbourhood of α_i such that $Q(z) = \varphi(z)^{\mu_i}$ for z in the domain of φ; then let $\xi_{i,j}$ be $\varphi^{-1}(s_j)$, where the s_j are the μ_ith roots of t. The product $\prod_{j \neq j'} (\xi_{i,j} - \xi_{i,j'})$ is of the order of $(|t|^{1/\mu_i})^{\mu_i^2 - \mu_i} = |t|^{\mu_i - 1}$;

and $\xi_{i,j} - \xi_{i',j'}$ is of the order of 1 for $i \neq i'$. Hence

$$\mathrm{discr}(Q - t) = \prod_{(i,j)\neq(i',j')} \left(\xi_{i,j} - \xi_{i',j'} \right)$$

is of the order of $|t|^{\nu}$, where $\nu = \sum \mu_i - 1 = d - k$. $\qquad\square$

Corollary *In the space of polynomials* $Q \in \mathbb{C}[X]$ *of degree* $\leqslant d$, *identifiable with* $\mathbb{C}^{d+1}$, *the set of polynomials of degree* $< d$ *or with at most k distinct roots is a* $\mathbb{Q}$-*algebraic set.*

Proof This is the set on which polynomial functions $Q \mapsto h_Q^{(r)}(0)$ with $0 \leqslant r < d - k$ over $\mathbb{Z}$ vanish. $\qquad\square$

7.2.5 Implication (ii) $\Rightarrow$ (i)

Proposition *Let* S *be a Riemann surface,* K *an algebraically closed subfield of* $\mathbb{C}$ *and* $f \in \mathcal{M}^{**}(S)$. *Suppose that* $\Delta(f) \subset K \cup \{\infty\}$. *Then* S *is definable over* K.

Proof Let $P(X, Y) = a_d(X)Y^d + \cdots + a_0(X)$ be a defining polynomial of (S, f). With the notation of 7.1.1,

$$\widehat{\Delta}_P = \Delta_P \cup \{\infty\} \supset \Delta(f).$$

Let $\xi_1,...,\xi_s$ be the points of $\widehat{\Delta}_P$, with $\Delta(f) = \{\xi_1,...,\xi_r\}$. For $1 \leqslant i \leqslant s$, let A_i be disjoint closed discs centered at ξ_i if $\xi_i \in \mathbb{C}$, of the form $\Sigma - \mathbb{D}_R$ if $\xi_i = \infty$. Set $A_i^* = A_i - \{\xi_i\}$ and let Γ_i be the boundary of A_i.

For $1 \leqslant i \leqslant r$, the ramified covering (S, f) induces a non trivial covering of Γ_i; with the notation of 6.2.5, for $r < i \leqslant s$, the following holds:
- if $\xi_i \in \Delta'$, then one of $\frac{a_i}{a_d}$ is not defined; so $a_d(\xi_i) = 0$, and the point $a_d(x) \in \mathbb{C}^*$ goes round $\nu_i \neq 0$ times as x runs through Γ_i;
- if $\xi_i = \infty$, one of $a_j(x)$ goes round a non-trivial number of times;
- if $\xi_i \in \Delta'' - \Delta'$, then the set $\pi^{-1}(A_i) \cap V_P$ is the union of graphs of d holomorphic functions $\eta_1,...,\eta_d : A_i \to \mathbb{C}$, and there are two indices $j \neq j'$ such that $\eta_{j'}(x) - \eta_j(x)$ goes round $\nu_i \neq 0$ times as x runs through Γ_i.

For $0 \leqslant i \leqslant d$, let e_i be the degree of $a_i \in \mathbb{C}[X]$, $e = e_d$, and E the set of polynomials $Q = b_d(X)Y^d + \cdots + b_0(X) \in \mathbb{C}[X, Y]$ such that b_i is of degree $\leqslant e_i$ for all i, and b_d is monic of degree e. This set can be identified with $\mathbb{C}^N$, where $N = \sum (e_i + 1) - 1$. Let E_K be the set of $Q \in E$ with coefficients in K.

Let M be the set of $Q \in E$ such that $\widehat{\Delta}_Q \supset \Delta(f)$ and $\mathrm{Card}\,(\Delta_Q) \leqslant \mathrm{Card}\,(\Delta_P)$.

Lemma 7.1 *For* $Q \in M$ *sufficiently near* P, *as a ramified covering of* Σ, *the Riemann surface* $\widehat{V}_Q$ *is isomorphic to* $\widehat{V}_P$, *hence to* S *equipped with* f.

Proof The set $B = \Sigma - \bigcup \overset{\circ}{A_i}$ is compact, and $X \mapsto \mathrm{discr}P(X)$ as well as a_d de not vanish on B. Let Ω be a connected relatively compact neighbourhood of P in E such that for $Q \in \Omega$, the functions $X \mapsto \mathrm{discr}Q(X)$ and b_d do not vanish on B. Then, by (4.3.12, Proposition),

$$Z = \{(Q, x, y) \in \Omega \times B \times \mathbb{C} \mid Q(x)(y) = 0\},$$

together with the projection $\pi : (Q, x, y) \mapsto (Q, x)$ is a covering of $\Omega \times B$. Indeed, π is etale by the implicit function theorem, and proper since the functions b_i are bounded and b_d is bounded below by a strictly positive constant. The injections $x \mapsto (P, x)$ and $x \mapsto (Q, x)$ are homotopic; so V_P and V_Q induce coverings isomorphic to B (4.3.17, Corollary 1).

In particular, for all i, the coverings of Γ_i induced by V_P and V_Q are isomorphic. Each set A_i has at least one point of $\widehat{\Delta}_Q$. Indeed, for $1 \leqslant i \leqslant r$, V_Q induces a non trivial covering of Γ_i; for i such that $\xi_i \in \Delta' \cup \{\infty\}$, $b_i(x)$ stays near $a_i(x)$ as x runs through Γ_i, and goes round 0 ν_i times; for i such that $\xi_i \in \Delta'' - \Delta'$, there are functions $\eta_1^Q,..., \eta_d^Q$ near $\eta_1,..., \eta_d$, where $\eta_j^Q(x) - \eta_j(x)$ goes round ν_i times as x runs through Γ_i.

As Card $(\widehat{\Delta}_Q) \leqslant$ Card $(\widehat{\Delta}_P)$, each set A_i has exactly one point ξ_i^Q of $\widehat{\Delta}_Q$, and the covering of $\overline{A}_i - \{\xi_i^Q\}$ induced by V_Q is determined by its restriction to Γ_i.

For $r < i \leqslant s$, this covering is trivial; Hence V_Q induces an (unramified) covering of $\Sigma - \bigcup_{i=1}^r A_i$, and an isomorphism between the coverings induced by $\widehat{V_P}$ and $\widehat{V_Q}$ over $\Sigma - \bigcup A_i$ extends to coverings induced over $\Sigma - \bigcup_{i=1}^r A_i$.

For $1 \leqslant i \leqslant r$, $\xi_i^Q = \xi_i$ and an isomorphism between the coverings induced by $\widehat{V_P}$ and $\widehat{V_Q}$ over $\Sigma - \bigcup_{i=1}^r A_i$ extends to an isomorphism of ramified coverings of Σ between $\widehat{V_P}$ and $\widehat{V_Q}$,   □

Lemma 7.2 *The set* $M \cap E_K$ *is dense in* M.

Proof The set M is K-algebraic. Indeed, the coefficients of the discriminant $\delta_Q = \mathrm{discr}_Y Q \in \mathbb{C}[X]$ are polynomials whose coefficients are integral with respect to the coefficients of Q. For each $\xi \in \Delta(f) - \{\infty\} \subset K$, the set of $Q = b_d Y^d + \cdots + b_0 \in$ E such that $(b_d \delta_Q)(\xi) = 0$ is K-algebraic. As a result of (7.2.4, Corollary) the set of $Q \in E$ such that Card $(\Delta_Q) \leqslant$ Card (Δ_P) is $\mathbb{Q}$-algebraic. The intersection M of all these sets is therefore a K-algebraic subset of E.

The lemma then follows from (7.2.3, Proposition).   □

End of the Proof of the Proposition With the above notation, S is isomorphic to $\widehat{V_P}$, hence also to $\widehat{V_Q}$ for $Q \in M$ sufficiently near P (Lemma 7.1). By Lemma 7.2, $Q \in M \cap E_K$ may be chosen sufficiently near P for this to hold. Then $\widehat{V_Q}$ is defined on K.   □

This completes the proof of (7.2.1, Proposition).

7.2.6 The Action $f_!$

Let S and S' be compact Riemann surfaces and $f : S \to S'$ an analytic map. For any finite subset A of S, set
$$f_!(A) = f(A) \cup \Delta(f)$$
where $\Delta(f)$ is the set of critical values of f. Then
$$f_!(\varnothing) = \Delta(f),$$
$$f_!(A \cup B) = f_!(A) \cup f_!(B) = f_!(A) \cup f(B),$$
$$(g \circ f)_!(A) = g_!(f_!(A)).$$

In particular,

$$\Delta(g \circ f) = g_!(\Delta(f)).$$

We will apply this to the case where $S = S' = \Sigma$ and f is a polynomial of degree $d \geqslant 2$. Then $\Delta(f) = \Delta^*(f) \cup \{\infty\}$, where $\Delta^*(f)$ is the set of critical values of $f : \mathbb{C} \to \mathbb{C}$.

7.2.7 Rational Critical Values

Proposition *For a compact Riemann surface of S, (ii) $\Rightarrow$ (ii'), where:*
(ii) there exists $f \in \mathcal{M}^{**}(S)$ such that $\Delta(f) \subset \overline{\mathbb{Q}} \cup \{\infty\}$,
(ii') there exists $f \in \mathcal{M}^{**}(S)$ such that $\Delta(f) \subset \mathbb{Q} \cup \{\infty\}$.

Lemma *Let $A \subset \mathbb{Q} \cup \{\infty\} \cup P^{-1}(0)$ be a finite set, with $P \in \mathbb{Q}[X]$ of degree d. Then there is a polynomial h such that $h_!(A) \subset \mathbb{Q} \cup \{\infty\} \cup P_1^{-1}(0)$ with $P_1 \in \mathbb{Q}[X]$ of degree $\leqslant d - 1$.*

Proof In fact, $h = P$ is suitable. Set $A = A' \cup A''$, where $A' \subset P^{-1}(0)$ and $A'' \subset \mathbb{Q} \cup \{\infty\}$. Then, $P(A') \subset \{0\}$ and $P(A'') \subset \mathbb{Q} \cup \{\infty\}$. The set $\Delta^*(P)$ of the finite critical values of P has at most $d - 1$ elements, and is invariant under $\mathrm{Aut}_{\mathbb{Q}}(\mathbb{C})$, and equals $P_1^{-1}(0)$, for some $P_1 \in \mathbb{Q}[X]$ of degree $\leqslant d - 1$. And
$$P_!(A) = P(A) \cup \Delta(P) \subset \mathbb{Q} \cup \{\infty\} \cup P_1^{-1}(0).$$ $\square$

A repeated application of this lemma gives:

Corollary *Let $A \subset \overline{\mathbb{Q}} \cup \{\infty\}$ be a finite set. Then there is a polynomial $h \in \mathbb{Q}[X]$ such that $h_!(A) \subset \mathbb{Q} \cup \{\infty\}$.*

Proof of the Proposition Let f be such that $\Delta(f) \subset \overline{\mathbb{Q}} \cup \{\infty\}$. By the above corollary, there exists $h \in \mathbb{Q}[X]$ such that $\Delta(h \circ f) = h_!(\Delta(f)) \subset \mathbb{Q} \cup \{\infty\}$. $\square$

7.2.8   Reduction of the Number of Critical Values

Proposition *Let* $A \subset \mathbb{Q} \cup \{\infty\}$ *be a finite set such that* $\infty \in A$ *and Card* $(A) > 3$. *There is a polynomial* $h \in \mathbb{Q}[X]$ *for which Card* $(h_!(A)) <$ *Card* (A).

Lemma *For* $f = f_{p,q} = X^p (1 - X)^q$ *and* $A = \left\{0, \frac{p}{p+q}, 1, \infty\right\}$, *Card* $(f_!(A)) = 3$.

Proof The critical points of f in $\mathbb{C}$ are 0, 1 and $\frac{p}{p+q}$; hence $\Delta^*(f) = \left\{0, f\left(\frac{p}{p+q}\right)\right\}$, and $f_!(A) = f(A) = \left\{0, f\left(\frac{p}{p+q}\right), \infty\right\}$.   □

Proof of the Proposition Let $A = \{a_1, a_2, ..., a_{n-1}, \infty\}$ with $a_1 < a_2 < a_3$. Let α be the affine map $\mathbb{C} \to \mathbb{C}$ such that $\alpha(a_1) = 0$ and $\alpha(a_3) = 1$. Then $\alpha(a_2) \in \mathbb{Q} \cap \ ]0, 1[$ is of the form $\frac{p}{p+q}$. Take $f = X^p (1 - X)^q$ and $h = f \circ \alpha$. Then $h_!(\{a_1, a_2, a_3, \infty\}) = f_!\left(\left\{0, \frac{p}{p+q}, \infty\right\}\right)$, and $h_!(A) = h_!(\{a_1, a_2, a_3, \infty\}) \cup h(a_4, ..., a_{n-1}))$ has at most $3 + n - 4 = n - 1$ elements.   □

Corollary 7.1 *For any finite set* $A \subset \mathbb{Q} \cup \{\infty\}$, *there is a polynomial* $h \in \mathbb{Q}[X]$ *such that* $h_!(A) \subset \{0, 1, \infty\}$.

Proof Repeated application of the proposition gives a polynomial $H \in \mathbb{Q}[X]$ such that Card $(H_!(A)) \leqslant 3$. This implies $\infty \in H_!(A)$, and so $H_!(A) \subset \{b_1, b_2, \infty\}$ with $b_1 < b_2$. Let β be affine and such that $\beta(b_1) = 0$ and $\beta(b_2) = 1$. For $h = \beta \circ H$, $h_!(A) \subset \{0, 1, \infty\}$.   □

Corollary 7.2 *For any finite set* $A \subset \overline{\mathbb{Q}} \cup \{\infty\}$, *there is a polynomial* $h \in \mathbb{Q}[X]$ *such that* $h_!(A) \subset \{0, 1, \infty\}$.

Proof The lemma follows from Corollary 7.1 and (7.2.7 Corollary).   □

End of the Proof of Belyi's Theorem We keep the notation of (7.2.1) and (7.2.7). It remains to prove that (ii') $\Rightarrow$ (iii). If $g \in \mathscr{M}^{**}(S)$ such that $\Delta(g) \subset \mathbb{Q} \cup \{\infty\}$, then by Corollary 7.1, there exists $h \in \mathbb{Q}[X]$ for which $h_!(\Delta(g)) \subset \{0, 1, \infty\}$. Then $\Delta(h \circ g) = h_!(\Delta(g)) \subset \{0, 1, \infty\}$.   □

7.3   Equivalence Between Various Categories

In a dark and deep unity,
Vast as the night and clarity
Ch. B. (F. d. M.)

In this section, we define equivalences between several categories seemingly from different domains: algebra, topology, combinatorics, complex analysis. This should establish *correspondences* between these different domains, thereby contributing to their enrichment.

7.3.1 Ramification Locus of an Algebra

Let K be a subfield of $\mathbb{C}$ and A an etale K(X)-algebra. Denote by $A_{\mathbb{C}}$ the $\mathbb{C}(X)$-algebra $\mathbb{C}(X) \otimes_{K(X)} A$. With the notation of (6.2.11), set $\Delta_A = \Delta_{A_{\mathbb{C}}}$: hence it is the ramification set of an analytic ramified covering (S, π) of Σ such that as a $\mathbb{C}(X)$-algebra, $\mathscr{M}(S)$ is isomorphic to $A_{\mathbb{C}}$.

7.3.2 Transfer of Δ

Let K be a field of characteristic 0 and A an etale K(X)-algebra; set α to be the ring homomorphism $K(X) \to A$ defining the algebra structure of A. Let ι_1 and ι_2 be two embeddings of K into $\mathbb{C}$. For $i = 1,\ 2$, set $K_i = \iota_i(K)$. Extend ι_i to an embedding $K(X) \to \mathbb{C}(X)$ by $\iota_i(X) = X$. Let A_i be the $K_i(X)$-algebra obtained by equipping the ring A of $\alpha_i = \alpha \circ \iota_i^{-1} : K_i \to A$, and Δ_i the subset Δ_{A_i} of $\mathbb{C}$.

Proposition Set $\sigma = \iota_2 \circ \iota_1^{-1} : K_1 \to K_2$, and let τ be an algebraic automorphism of $\mathbb{C}$ inducing σ. Then $\tau(\Delta_1) = \Delta_2$.

Proof Extend σ to $K_1(X)$ and τ to $\mathbb{C}(X)[Y]$ by $\sigma(X) = X, \tau(X) = X$ and $\tau(Y) = Y$. Then τ is a σ-morphism. Hence there is a morphism

$$\tau_* = \tau \otimes 1_A : (A_1)_{\mathbb{C}} = \mathbb{C}(X) \otimes_{K_1(X)} A_1 \to \mathbb{C}(X) \otimes_{K_2(X)} A_2 = (A_2)_{\mathbb{C}}.$$

If η_1 is a primitive element in $(A_1)_{\mathbb{C}}$, then $\eta_2 = \tau_*(\eta_1)$ is primitive in $(A_2)_{\mathbb{C}}$, and the automorphism τ_* of $\mathbb{C}(X)[Y]$ inducing τ on $\mathbb{C}$ and fixing X and Y maps the minimal polynomial P_1 of η_1 onto the minimal polynomial P_2 of η_2. The definition of $\Delta_{E,\eta}$ in (6.2.5) being of an algebraic nature,

$$\Delta_{(A_2)_{\mathbb{C}}, \eta_2} = \tau(\Delta_{(A_1)_{\mathbb{C}}, \eta_1}).$$

Thanks to (6.2.12, Proposition), taking the intersection over all primitive elements gives

$$\Delta_{(A_2)_{\mathbb{C}}} = \tau(\Delta_{(A_1)_{\mathbb{C}}}). \qquad\qquad \text{cqfd}$$

7.3.3

Corollary Let K be a subfield of $\mathbb{C}$ and A an etale K(X)-algebra. Then the set Δ_A is contained in $\overline{K}$ and generates a Galois extension of K.

Proof For any algebraic automorphism τ of $\mathbb{C}$ inducing the identity on K, $\tau(\Delta_A) = \Delta_A$. As Δ_A is finite, the elements of Δ_A only have finitely many conjugates. Hence they are algebraic over K. The extension generated by Δ_A is invariant under $\mathrm{Aut}_K(\overline{K})$, and so is Galois. □

7.3.4 The Category $\mathscr{A}$

Let $\mathscr{A}$ denote the category of etale algebras A over $\overline{\mathbb{Q}}(X)$ such that $\Delta_A \subset \{0, 1, \infty\}$. The morphisms are the homomorphisms of $\overline{\mathbb{Q}}(X)$-algebras.

The Galois group $\mathbb{G} = \mathrm{Aut}_{\mathbb{Q}}(\overline{\mathbb{Q}})$ acts on the right on $\mathscr{A}$ in the following sense: any element $g \in \mathbb{G}$ defines a (covariant) functor $g^* : \mathscr{A} \to \mathscr{A}$, $(g.h)^* = h^* \circ g^*$ and $(1_{\overline{\mathbb{Q}}})^* = 1_{\mathscr{A}}$. If A is an etale $\overline{\mathbb{Q}}(X)$-algebra, denoting by $\alpha : \overline{\mathbb{Q}}(X) \to A$ the ring homomorphism defining its structure, the algebra g^*A is the ring A equipped with $\alpha \circ g$. By (7.3.2, Proposition), $\Delta_{g^*A} = \Delta_A$. Hence g^*A is indeed an object of $\mathscr{A}$. If $\varphi : A \to B$ is a morphism of $\overline{\mathbb{Q}}(X)$-algebras, then the same map is a morphism $g^*A \to g^*B$. Hence, we may set $g^*\varphi = \varphi$.

Remark For each g the functor g^* is a covariant functor from $\mathscr{A}$ to itself, but g^* depends "contravariantly" on g.

7.3.5 Characterization of the Elements of an Etale Algebra

Proposition *Let* K *be an algebraically closed subfield of* $\mathbb{C}$ *and* A *an etale* $K(X)$-*algebra. Set* $E = \mathbb{C}(X) \otimes_{K(X)} A$; *identify* $K(X)$ *and* A *with subsets of* E *by* $f \mapsto f1$ *and* $a \mapsto 1a$. *Then* A *is the algebraic closure* $K(X)$ *in* E.

Proof The algebra A is finite over $K(X)$, and hence is in the algebraic closure of $K(X)$ in E. We show that no $f \in E - A$ is algebraic over $K(X)$.

Identify A with $K(X)[Y]/(P)$, where P is a polynomial of degree $d = [A : K(X)]$ (minimal polynomial of a primitive element). Then $(1, y, ..., y^{d-1})$ is a basis for A as a $K(X)$-vector space, and of E as a $\mathbb{C}(X)$-vector space. Any $f \in E$ can be uniquely written as $f_0 + f_1 y + \cdots + f_{d-1} y^{d-1}$. If $f \in E - A$, then for some i, f_i is in $\mathbb{C}(X) - K(X)$. This f_i can be uniquely written as $\frac{p}{q}$, where $p, q \in \mathbb{C}[X]$ such that $\mathrm{g.c.d.}(p, q) = 1$ and q monic: at least one of the coefficients of p or q is in $\mathbb{C} - K$, hence transcendental over K. The orbit of this coefficient under the action of $G_K = \mathrm{Aut}_K(\mathbb{C})$ is infinite. Hence so is the orbit of f under the action of G_K acting on $\mathbb{C}(X)[Y]/(P)$ by keeping fixed X and Y. As E is a finite product of extensions of $K(X)$, this shows that f is transcendental over $K(X)$. $\qquad\square$

7.3.6 The Categories $\mathscr{E}$ and $\mathscr{V}$

Let $\mathscr{E}$ be the category of etale $\mathbb{C}(X)$-algebras E such that $\Delta_E \subset \{0, 1, \infty\}$, the morphisms being the homomorphisms of $\mathbb{C}(X)$-algebras.

Let $\mathscr{V}$ be the category $\mathscr{V}_{\Sigma, \{0, 1, \infty\}}$ of analytic ramified coverings X of the Riemann sphere Σ such that $\Delta_X \subset \{0, 1, \infty\}$. The functor $\mathscr{M}$ which assigns to each

Riemann surface S the algebra $\mathcal{M}(S)$ of meromorphic functions on S defines an anti-equivalence of categories $\mathcal{M} : \mathcal{V} \to \mathcal{E}$ (6.2.4). (6.2.11).

As a result of (7.2.5), every algebra object E of $\mathcal{E}$ contains a $\overline{\mathbb{Q}}(X)$-subalgebra A such that E is identified with $A_{\mathbb{C}} = \mathbb{C}(X) \otimes_{\overline{\mathbb{Q}}(X)} A$, and by (7.3.5) this subalgebra is the algebraic closure of $\overline{\mathbb{Q}}(X)$ in E. Hence, the functor of scalar extension $A \mapsto A_{\mathbb{C}}$ from $\mathcal{A}$ to $\mathcal{E}$ and the functor $\alpha : \mathcal{E} \to \mathcal{A}$ which assigns to each $\mathbb{C}(X)$-algebra E the algebraic closure of $\overline{\mathbb{Q}}(X)$ in E are quasi-inverse equivalences of categories.

7.3.7 A Diagram of Categories

By (6.1.10, Corollary), the forgetful functor $\mathcal{V} \to \mathcal{RR}. = \mathcal{RR}_{\Sigma,\{0,1,\infty\}}$ (category of finite topological ramified coverings S of Σ such that $\Delta_S \subset \{0, 1, \infty\}$) is an equivalence of categories. By (6.1.11, Proposition) the restriction functor $S \mapsto S|_{\mathbb{C}-\{0,1,\infty\}}$ is an equivalence of categories from $\mathcal{RR}.$ onto the category $\mathcal{Covf}. = \mathcal{Covf}_{\mathbb{C}-\{0,1\}}$ of finite coverings of $\mathbb{C} - \{0, 1\}$.

Choose a basepoint z_0 in $\mathbb{C} - \{0, 1\}$, for example $z_0 = \frac{1}{2}$, and let Γ be the fundamental group $\pi_1(\mathbb{C} - \{0, 1\}, z_0)$. This is a (non commutative) free group on 2 generators γ_0 and γ_1, represented by loops around 0 and 1 respectively (6.4). Its profinite completion $\widehat{\Gamma} = \widehat{\Gamma}_T = \widehat{\pi}_1(\mathbb{C} - \{0, 1\}, z_0)$ is a profinite free group on 2 generators, which we also write γ_0 and γ_1. By (4.5.3, Theorem), the functor $S \mapsto S(z_0)$ defines an equivalence from $\mathcal{Covf}.$ onto the categories $\Gamma\text{-}\mathcal{Setf}$ (resp. $\widehat{\Gamma}\text{-}\mathcal{Setf}$) of finite sets on which Γ acts (resp. $\widehat{\Gamma}$ acts continuously), in other words of finite sets equipped with two permutations.

Let Ω be an algebraic closure of $\overline{\mathbb{Q}}(X)$, and $\Omega_{\{0,1,\infty\}}$ the (by (6.2.11) directed) union sub-extension in Ω of sub-extensions that are objets of $\mathcal{A}$. Let $\widehat{\Gamma}_{alg}$ be the Galois group of $\mathrm{Aut}_{\overline{\mathbb{Q}}(X)}\Omega_{\{0,1,\infty\}}$. By (5.9.4 Theorem), the functor $A \mapsto \mathrm{Hom}_{\overline{\mathbb{Q}}(X)\text{-}alg}$ $(A, \Omega_{\{0,1,\infty\}})$ defines an anti-equivalence of categories from $\mathcal{A}$ onto $\widehat{\Gamma}_{alg}\text{-}\mathcal{Setf}$.

This gives a diagram of functors all of which are equivalences or anti-equivalences of categories (the arrows representing contravariant functors are circled):

$$\widehat{\Gamma}_{alg}\text{-}\mathcal{Setf} \leftarrow\!\circ\!- \mathcal{A} \rightleftarrows \mathcal{E} \xleftarrow[]{\mathcal{M}}\!\circ\!- \mathcal{V} \to \mathcal{RR}. \to \mathcal{Covf}. \to \Gamma\text{-}\mathcal{Setf} \leftarrow \widehat{\Gamma}\text{-}\mathcal{Setf}$$

In the following, we expand this diagram.

Remarks (1) The categories $\widehat{\Gamma}_{alg}\text{-}\mathcal{Setf}$ and $\widehat{\Gamma}\text{-}\mathcal{Setf}$ being equivalent, the profinite groups $\widehat{\Gamma}$ and $\widehat{\Gamma}_{alg}$ are isomorphic (2.9.8, Corollary); in particular $\widehat{\Gamma}_{alg}$ is a profinite free group on 2 generators.

(2) The group $\mathbb{G} = \mathrm{Aut}_{\mathbb{Q}}(\overline{\mathbb{Q}})$ acts on $\mathcal{A}$. Hence it also acts on all categories in the above diagram. For example, to each $g \in \mathbb{G}$ can be assigned a functor g_* from the category $\mathcal{Covf}.$ to itself in such a way that the functors $(g \cdot h)_*$ and $g_* \circ h_*$ are isomorphic for g and h in $\mathbb{G}$, and $1_* \approx 1_{\mathcal{Covf}.}$. These are merely isomorphic functors following from the transference of the action of $\mathbb{G}$ on $\mathcal{A}$ using equivalences of categories without quasi-inverses. There are also anti-equivalences, but the functors

g_* are always covariant, and $(g.h)_* \approx g_* \circ h_*$ is always the case. In other words the group $\mathbb{G}$ always acts on the left.

Proposition *Let* E *and* E′ *be finite* $\widehat{\Gamma}$*-sets with* E′ $= g_*(\mathrm{E})$ *for* $g \in \mathbb{G}$. *Then there is an automorphism* ϕ *of* $\widehat{\Gamma}$ *such that* E′ $\approx \phi^* \mathrm{E}$. *In particular there is a bijection* $\beta : \mathrm{E} \to \mathrm{E}'$ *mapping* $\widehat{\Gamma}$ *(or* Γ*) onto the the group* $\Gamma_{\mathrm{E}'}$ *in* $\mathfrak{S}(\mathrm{E}')$, *which is the image of* Γ_{E} *(similarly defined) under* $f \mapsto \beta \circ f \circ \beta^{-1}$.

Proof The element $g \in \mathbb{G}$ defines an equivalence g_* from the category $\widehat{\Gamma}$-$\mathcal{Setf}$ to itself. By (2.9.8, Corollary) there is an automorphism ϕ of $\widehat{\Gamma}$ such that the functors g_* and ϕ^* are isomorphic. An isomorphism from ϕ^* onto g_* gives a bijection $\beta : \mathrm{E} \to \mathrm{E}'$ such that $\beta(\phi(g) \cdot x) = g \cdot \beta(x)$. Letting σ_g (resp. σ'_g) be the action of $g \in \Gamma$ on E (resp. on E′), $\beta \circ \sigma_{\phi(g)} = \sigma'_g \circ \beta$, i.e. $\sigma'_g = \beta \circ \sigma_{\phi(g)} \circ \beta^{-1}$.   □

7.3.8 Cyclic Order

Let O be a finite set of cardinality k. A **cyclic order** on O is an equivalence class consisting of bijections $\varphi : \mathrm{O} \to \mathbb{Z}/(k)$ for the equivalence relation identifying φ with $\varphi + a$, a being a constant. The bijections in this class are called the *cyclic charts* for the cyclically ordered set O.

The map $\sigma = \varphi \circ \mathrm{T} \circ \varphi^{-1} : \mathrm{O} \to \mathrm{O}$, where T is the translation $t \mapsto t + 1$ in $\mathbb{Z}/(k)$, is independent of the choice of the cyclic chart φ. Each x has a *next* element $\sigma(x)$. If E is a finite set and f a permutation of E, i.e. a bijection E → E, then there is a unique cyclic order on each orbit O of f such that the restriction of f to O is the map taking elements to the next one. Defining a cyclic order on O is thus equivalent to defining a permutation of O acting transitively (assuming that O $\neq \varnothing$: there is no cyclic order on $\varnothing$!)

Let O and O′ be two cyclically ordered sets of cardinalities k and k'. A map $h : \mathrm{O}' \to \mathrm{O}$ will be said to be **ramified** if $h \circ \sigma' = \sigma \circ h$, where σ and σ' are the maps assigning the next element in O and O′. As a result, k' is a multiple of k, and the expression $\varphi^{-1} \circ \varphi'$ in the cyclic charts is of the form $i \mapsto \chi(i) + a$, for some canonical map $\chi : \mathbb{Z}/(k') \to \mathbb{Z}/(k)$ and constant $a \in \mathbb{Z}/(k)$.

7.3.9 Graphs

Here we give a different definition of graphs from the one used in § 4.8.

A **finite graph** consists of:

(1) a finite set N whose elements are called *vertices*;
(2) a finite set A whose elements are called *directed edges*;
(3) two maps ε_0, $\varepsilon_1 : \mathrm{A} \to \mathrm{N}$ called *source* and *range*;
(4) a fixed point free *reversing* involution $\tau : \mathrm{A} \to \mathrm{A}$ such that $\varepsilon_0 \circ \tau = \varepsilon_1$ (whence $\varepsilon_1 \circ \tau = \varepsilon_0$).

In this chapter, for simplicity's sake a finite graph will be called a *graph*. The main differences with the definition given in (4.8.1) are:

(1) $(\varepsilon_0, \varepsilon_1) : A \to N^2$ is not assumed to be injective (two vertices may be connected by several edges), so A cannot be identified to a subset of $N \times N$,

(2) the image of $(\varepsilon_0, \varepsilon_1)$ is not assumed to avoid the diagonal: loops are allowed,

(3) N and A are assumed to be finite.

The *geometric realization* $|H|$ of a graph $H = (N, A, \varepsilon_0, \varepsilon_1)$ is the quotient of $N \sqcup (A \times [0, 1])$ by the equivalence relation generated by $(a, 0) \sim \varepsilon_0(a)$, $(a, 1) \sim \varepsilon_1(a)$ and $(a, t) \sim (\tau(a), 1 - t)$ for $a \in A$ and $t \in [0, 1]$.

The elements of A/τ are called (undirected) *edges*.

7.3.10 The Category $\mathscr{H}$

A **bicoloured graph** is a graph $H = (N, A, \varepsilon_0, \varepsilon_1, \tau)$ equipped with a map $\pi :$ $N \to \{0, 1\}$ such that $\pi(\varepsilon_1(a)) \neq \pi(\varepsilon_0(a))$ for $a \in A$. A vertex s is called *white* if $\pi(s) = 0$, *black* if $\pi(s) = 1$. The set A is the disjoint union of the set A^+ of *outgoing* directed edges (white source, black range) and of the set A^- of *incoming* directed edges (black source, white range).

The category of *bicoloured graphs cyclically ordered at each vertex.* will be denoted $\mathscr{H}$. Hence an object of $\mathscr{H}$ is a bicoloured graph $H = (N, A, \varepsilon_0, \varepsilon_1, \tau, \pi)$ equipped, for each vertex $s \in N$, with a cyclic order ω_s on the set A_s of directed edges with source s.

Defining cyclic orders ω_s is equivalent to defining the map $\sigma : A \to A$ assigning to each directed edge a, the next one with respect to the cyclic order $\omega_{\varepsilon_0(a)}$, and subject to the condition $\varepsilon_0 \circ \sigma = \varepsilon_0$ and to acting transitively on the fibres of ε_0.

A morphism $f : (H', (\omega'_{s'})) \to (H, (\omega_s))$ is given by two maps $f_N : N' \to N$ and $f_A : A' \to A$ such that $\varepsilon_0 \circ f_A = f_N \circ \varepsilon'_0$, $\varepsilon_1 \circ f_A = f_N \circ \varepsilon'_1$, $f_A \circ \tau' = \tau \circ f'_A$, $\pi \circ f_N = \pi'$, and such that f_A induces a branching map $f'_s : A'_{s'} \to A_{f(s')}$ for all vertices s' of H' (in other words, $f_A \circ \sigma' = \sigma \circ f_A$).

Remark A graph with a cyclic order at the vertices has no isolated vertex: the map $\varepsilon_0 : A \to N$ is surjective since the emptyset has no cyclic order, and so is ε_1 since $\varepsilon_1 = \varepsilon_0 \circ \tau$. The map f_N in a morphism of $\mathscr{H}$ is thus determined by f_A.

Let σ_0 (resp. σ_1) be the permutation of A^+ induced by σ (resp. by $\tau \circ \sigma \circ \tau$).

Proposition *The functor* $((H, (\omega_s)) \mapsto (A^+, \sigma_0, \sigma_1)$ *is an equivalence of categories from $\mathscr{H}$ onto the category $\check{\mathscr{H}}$ of finite sets equipped with two permutations σ_0, σ_1.*

Proof Define (A, σ) from (B, σ_0, σ_1), where σ_0 and σ_1 are permutations of B, by $A = B \times \{0, 1\}$ and $\sigma(b, i) = (\sigma_i(b), i)$. Then define the object $(H, (\omega_s))$ of $\mathscr{H}$ by taking N_0 to be the set B/σ_0 of orbits of σ_0, $N_1 = B/\sigma_1$, and $N = N_0 \sqcup N_1$. Set $\tau(b, i) = (b, 1 - i)$. Take $\varepsilon_i(b, j)$ to be the class of b in $N_{i+j \,(\mathrm{mod}\, 2)}$, $\pi = 0$ on N_0

and 1 on N_1, and ω_s to be the cyclic order with respect to which the next element is given by the restriction of σ.

This defines a functor $\check{\mathcal{H}} \to \mathcal{H}$. It is a quasi-inverse of the functor $\mathcal{H} \to \check{\mathcal{H}}$. $\square$

7.3.11 The Functor $\theta : \mathcal{RR}. \to \mathcal{H}$

If (S, π_S) is an object of $\mathcal{RR}. = \mathcal{RR}_{\Sigma, \{0, 1, \infty\}}$, then the space $\pi_S^{-1}([0, 1])$ is identified with the geometric realization of the bicoloured graph $H = (N, A, \varepsilon_0, \varepsilon_1, \tau, \pi)$ defined as follows:

- $N = \pi_S^{-1}(\{0, 1\})$; the map $\pi : N \to \{0, 1\}$ is the restriction of π_S ;
- A^+ (resp. A^-) is the set of continuous maps $a : [0, 1] \to S$ such that $\pi(a(t)) = t$ (resp. $= 1 - t$),
- $\tau(a)$ is the map $t \mapsto a(1 - t)$,
- $\varepsilon_0(a) = a(0)$, $\varepsilon_1(a) = a(1)$.

Indeed, the geometric realization $|H|$ of the graph H is a quotient of $A \times [0, 1]$, the map $(a, t) \mapsto a(t)$ of $A \times [0, 1]$ passes to the quotient and defines a continuous map $|H| \to \pi_S^{-1}([0, 1])$. It can be checked to be bijective; as $|H|$ is compact, it is a homeomorphism.

Define as follows a cyclic order at the vertices of H. If $s \in S(0)$ (resp. $S(1)$), there is (6.1.8) a chart φ for S in the neighbourhood of s such that $\pi(x) = (\varphi(x))^d$ (resp. $\pi(x) = (1 - \varphi(x))^d$), where d is the local degree of the projection $\pi : S \to \Sigma$ at s. The directed edges a_ℓ with source s are given by $\varphi(a_\ell(t)) = e^{2i\pi\ell/d}t$, for $\ell \in \mathbb{Z}/(d)$. The bijection $\ell \mapsto a_\ell$ defines a cyclic order ω_s on A_s, independent of the choice of φ. The bicoloured graph H, equipped with $(\omega_s)_{s \in N}$, is an object of $\mathcal{H}$ which will be written $\theta(S)$.

If $f : S' \to S$ is a morphism of $\mathcal{RR}.$, then the map f induces $f_N : N' \to N$ and $f_A : A' \to A$. These define a morphism f_* of bicoloured graphs. The map f turns S' into a ramified covering of S (this can be either directly seen or by using (6.1.10) and (6.1.5). Consequently, for $s' \in S'$, f_A induces a branching map $A'_{s'} \to A_{f(s')}$. Hence f_* is a morphism of $\mathcal{H}$, and θ becomes a functor $\mathcal{RR}. \to \mathcal{H}$.

We construct a quasi inverse functor η of a functor $\dot{\theta}$ related to θ, and then deduce that θ is an equivalence of categories.

7.3.12 The Functor η

Let $H = (N, A, \varepsilon_0, \varepsilon_1, \tau, \pi, (\omega_s))$ be an object of $\mathcal{H}$, i.e. a bicoloured graph cyclically ordered at each vertex. We construct a ramified covering $\dot{S}$ of $\mathbb{C}$ from $\mathcal{H}$, with $\Delta_{\dot{S}} \subset \{0, 1\}$.

Let $\overline{\mathbb{H}}$ be the closed upper half-plane: $\overline{\mathbb{H}} = \{x + iy \mid x \in \mathbb{R}, y \geqslant 0\}$, and consider $\Xi = A \times \overline{\mathbb{H}}$. Define $\pi_\Xi : \Xi \to \mathbb{C}$ by $\pi_\Xi(a, z) = z$ if a is an outgoing edge, $\pi_\Xi(a, z) = 1 - z$ if a is an incoming edge. Equip Ξ with the equivalence relation generated by:

- $(a, t) \sim (\tau(a), 1 - t)$ for $t \in [0, 1]$,
- $(a, t) \sim (\tau(\sigma(a)), 1 - t)$ for $t \leqslant 0$, where $\sigma(a)$ is the directed edge along a with respect to the cyclic order at the source vertex of a.

These conditions imply $(a, 0) \sim (a', 0)$ if a and a' have the same source, and $(a, 1) \sim (a', 1)$ if a and a' have the same range. Indeed, $(a, 0) \sim (\tau(\sigma(a)), 1) \sim (\sigma(a), 0)$, $(a, 1) \sim (\tau(a), 0) \sim (\sigma_1(a), 1)$ where $\sigma_1 = \tau \circ \sigma \circ \tau$, and σ (resp. σ_1) acts transitively on the set of directed edges having a given source (resp. range).

Let S be the quotient space of Ξ by the equivalence relation thus defined. Two equivalent points have the same image under the map π_Ξ. Hence there is a continuous map $\pi_S : S \to \mathbb{C}$. Let $\eta(H)$ be the space S equipped with π_S.

Proposition (a) *The space* S *equipped with* π_S *is a finite ramified covering of* $\mathbb{C}$, *with* $\Delta_S \subset \{0, 1\}$.

(b) *If* $f = (f_N, f_A) : H' \to H$ *is a morphism in* $\mathcal{H}$, *by passing to the quotient derived from* $f_A \times 1_{\overline{\mathbb{H}}} : A' \times \overline{\mathbb{H}} \to A \times \overline{\mathbb{H}}$ *gives a morphism* $f_* : S' \to S$ *in* $\mathcal{RR}_{\mathbb{C}, \{0, 1\}}$. *This defines a covariant functor* $\eta : \mathcal{H} \to \mathcal{RR}_{\mathbb{C}, \{0, 1\}}$.

(c) *The functor* θ *is of the form* $\dot{\theta} \circ \rho$, *where* $\rho : \mathcal{RR}. \to \mathcal{RR}_{\mathbb{C}, \{0, 1\}}$ *is the restriction functor, which is an equivalence of categories. The functor* η *is a quasi-inverse of* $\dot{\theta}$.

Proof (a) The restriction of S to $(\mathbb{C} - \mathbb{R}) \cup \,]0, 1[$ is a quotient of $A \times (\mathbb{H} \cup]0, 1[)$ by the equivalence relation identifying (a, t) with $(\tau(a), 1 - t)$ for $t \in \,]0, 1[$; it is a trivial covering of the open subset $(\mathbb{C} - \mathbb{R}) \cup \,]0, 1[$, with fibre A^+.

The restriction of S to $(\mathbb{C} - \mathbb{R}) \cup \mathbb{R}_*^*$ is the quotient of

$$(A^+ \times (\mathbb{H} \cup \mathbb{R}_-^*)) \cup (A^- \times (\mathbb{H} \cup]1, \infty[)$$

by the equivalence relation identifying (a, t) with $(\tau(\sigma(a)), 1 - t)$ for $a \in A^+, t < 0$; it is a trivial covering of the open subset $(\mathbb{C} - \mathbb{R}) \cup \mathbb{R}_*^*$, with fibre A^+.

The restriction of S to $(\mathbb{C} - \mathbb{R}) \cup \,]1, \infty[$ is the quotient of

$$(A^- \times (\mathbb{H} \cup \mathbb{R}_-^*)) \cup (A^+ \times (\mathbb{H} \cup]1, \infty[)$$

by the equivalence relation identifying (a, t) with $(\tau(\sigma(a)), 1 - t)$ for $a \in A^-, t < 0$; it is a trivial covering of the open subset $(\mathbb{C} - \mathbb{R}) \cup \,]1, \infty[$, with fibre A^+.

Hence, S induces a covering of $\mathbb{C} - \{0, 1\}$.

Let χ be the quotient map $\Xi \to S$, $a \in A$, and d the number of directed edges with source $s = \varepsilon_0(a)$. Define φ_a in the neighbourhood of $\dot{s} = \chi(a, 0)$ in S by $\varphi_a(\chi(\sigma^\ell(a)), z) = e^{2i\pi\ell/d} \cdot z^{1/d}$, and $\varphi_a(\chi(\tau\sigma^\ell(a)), z) = e^{2i\pi\ell/d} \cdot (1 - z)^{1/d}$, computed with $\text{Arg}(z) \in [0, \pi]$ and $\text{Arg}(1 - z) \in [-\pi, 0]$ for $z \in \overline{\mathbb{H}}$. Then φ_a is a chart for S in the neighbourhood of $\dot{s}$, and letting ψ denote the chart for $\mathbb{C}$ which is the identity if $a \in A^+$ and $z \mapsto 1 - z$ if $a \in A_-$, the expression of π_S in the charts φ_a and ψ is $z \mapsto z^d$.

The cyclic order defined by this chart on the set of directed edges originating at s agrees with ω_s.

(b) It suffices to check.

(c) It is a matter of defining for an object H of $\mathscr{H}$ an isomorphism $H \to H' = \dot{\theta}(\eta(H))$, and for a ramified covering S of $\mathbb{C}$ with $\Delta_S \subset \{0, 1\}$ an isomorphism $S \to S' = \eta(\dot{\theta}(S))$.

Let $H = (N, A, ...)$ be an object of $\mathscr{H}$; set $(S, \pi_S) = \eta(H)$ with $S = \Xi/\sim$ as above, and $H' = (N', A', ...) = \theta(S, \pi_S)$. The set $N'_0 = \pi_S^{-1}(0)$ is the quotient of A^+ by the equivalence relation identifying a with a' if $\varepsilon_0(a) = \varepsilon_0(a')$, hence can be identified with N_0. Likewise, N'_1 can be identified with N_1. Assigning to $a \in A$ the map $t \mapsto \chi(a, t)$ defines a bijection $A \to A'$ compatible with $\varepsilon_0, \varepsilon_1, \tau$ and π. Hence there is an isomorphism $H \to H'$ to $\mathscr{H}$.

Let (S, π_S) be a ramified covering of $\mathbb{C}$ with $\Delta_S \subset \{0, 1\}$, $H = (N, A, ...) = \dot{\theta}(S)$ and $(S', \pi_{S'}) = \eta(H)$. The space S' is a quotient of $\Xi' = A \times \overline{\mathbb{H}}$. For $a \in A$, the map $a : [0, 1] \to S$ is a unique continuous extension $\widehat{a}$ to $\overline{\mathbb{H}}$ such that $\pi_S \circ \widehat{a}(z) = z$ if $a \in A^+$, and $1 - z$ if $a \in A^-$. Hence there is a continuous map $\Xi' \to S$ compatible with the equivalence relation, and so giving a map $S' \to S$. It can be checked that the latter is a homeomorphism over $\mathbb{C}$, and that the entire construction depends functoriality on S. $\qquad\square$

Corollary *The functor $\theta : \mathscr{RR}. \to \mathscr{H}$ is an equivalence of categories.*

7.3.13 Action of $\mathbb{G}$ on $\mathscr{H}$

So there is a diagram of equivalences of categories

$$\widehat{\Gamma}\text{-}\mathcal{Setf}$$
$$\updownarrow$$

$$\widehat{\Gamma}_{alg}\text{-}\mathcal{Setf} \longleftrightarrow \mathscr{A} \longleftrightarrow \mathscr{E} \longleftrightarrow \mathscr{V} \longleftrightarrow \mathscr{RR}. \longleftrightarrow \mathscr{H}$$

The group $\mathbb{G} = \mathrm{Aut}_\mathbb{Q}(\overline{\mathbb{Q}})$ acts on $\mathscr{A}$, hence also on all the categories of the above diagram, in particular on $\mathscr{H}$. Here too it is an action in a weak sense, as in (7.3.7). However $\mathbb{G}$ acts on the set $\mathscr{X}$ of isomorphism classes of objects of $\mathscr{H}$.

The *valency* of a vertex s in a graph H which is an object of $\mathscr{H}$ is the number $v_H(s)$ of directed edges having source s. Letting X, E and A denote the objects of $\mathscr{RR}.$, $\mathscr{E}$, $\mathscr{A}$ corresponding to H, the valency $v_H(x)$ is the local degree at s_X corresponding to s under the projection $X \to \Sigma$. It can also be defined algebraically: it is the dimension of the vector space $A_s/\mathfrak{m}_{\pi(s)} \cdot A_s$ over $\overline{\mathbb{Q}}$, where A_s is the subalgebra of A consisting of functions defined at s, and $\mathfrak{m}_{\pi(s)}$ the ideal consisting of functions of $\overline{\mathbb{Q}}(X)$ defined and vanishing at $\pi(s)$.

A *valency function* ν_H is the map $\{0, 1\} \times \mathbb{N} \to \mathbb{N}$ assigning to (u, v) the number of vertices s of H such that $\pi(s) = u$ and $\nu_H(s) = v$. For $g \in \mathbb{G}$ and H an object of $\mathscr{H}$, the graphs H and $g^*(H)$ have the same valency function.

Several equivalence relations can be considered on the set $\mathscr{X}$: Given $x, x' \in \mathscr{X}$ represented by the objets H and H' of $\mathscr{H}$, we write

- $x \sim_{gb} x'$ or H $\sim_{gb}$ H', if H and H' are isomorphic as bicoloured graphs, irrespective of the cyclic order at the vertices;
- $x \sim_{\nu} x'$ or H $\sim_{\nu}$ H', if H and H' have the same valency function, i.e. $\nu_H = \nu_{H'}$.
- $x \sim_{\mathbb{G}} x'$ or H $\sim_{\mathbb{G}}$ H', if there exists $g \in \mathbb{G} = \mathrm{Aut}_{\mathbb{Q}}(\overline{\mathbb{Q}})$ mapping x onto x' (i.e. H into an object of $\mathscr{H}$ isomorphic to H').

The relation $\sim_{\nu}$ is weaker than the other two: $x \sim_{gb} x' \Rightarrow x \sim_{\nu} x'$ and $x \sim_{\mathbb{G}} x' \Rightarrow x \sim_{\nu} x'$ (this latter implication follows from the algebraic definition of valency).

Clearly, $x \sim_{\nu} x' \not\Rightarrow x \sim_{gb} x'$. In section (7.5) we will see that $x \sim_{gb} x' \not\Rightarrow x \sim_{\mathbb{G}} x'$, (and necessarily $x \sim_{\nu} x' \not\Rightarrow x \sim_{\mathbb{G}} x'$), and that $x \sim_{\mathbb{G}} x' \not\Rightarrow x \sim_{gb} x'$ (in particular the action of $\mathbb{G}$ is not trivial). In fact, already in section (7.4), we will see that the action of $\mathbb{G}$ on $\mathscr{X}$ is faithful.

All this continues to hold in the full subcategory $\mathscr{T}$ of $\mathscr{H}$ whose objets are the trees, and in the set $\mathscr{Y}$ of isomorphism classes of objects of $\mathscr{T}$.

7.3.14 A Numerical Equivalence Class

The graph along H_1 has

- 4 white vertices of valency 1;
- 2 white vertices of valency 3;
- 1 white vertices of valency 4;
- 7 black vertices of valency 2;

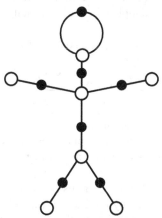

There are 31 isomorphism classes of objects of $\mathscr{H}$ in the same class of $\sim_{\nu}$ (same valency function).

A bicoloured graph all of whose black vertices have valency 2 can be uniquely obtained from an uncoloured graph—or, equivalently, monocoloured white—by subdividing each edge by a black vertex. Moreover, a cyclic order at the vertices of the monocoloured graph extends uniquely to the bicoloured graph. This gives a bijection between the set $\mathcal{X}_2$ of isomorphism classes of objects of $\mathcal{H}$ with bivalent black vertices, and the set $\mathcal{I}$ of isomorphism classes of monocoloured white graphs, cyclically ordered at each vertex.

The 31 points of $\mathcal{I}$ corresponding to those of the class of $\sim_\nu$ from H_1 to $\mathcal{X}_2$ are represented by the graphs of the page 429.

These objects are divided into 8 classes of $\sim_{gb}$:

$\{G_1, G_2, G_3\}, \{G_4, G_5\}, \{G_6, G_{11}\},$
$\{G_7, G_8, G_9, G_{10}, G_{12}, G_{13}\}, \{G_{14}, G_{15}\},$
$\{G_{16}, \ldots, G_{21}\}, \{G_{22}, \ldots, G_{28}\}, \{G_{29}, G_{30}, G_{31}\}.$

This does not tell us much on the action of the group $\mathbb{G}$. Proposition 7.3.7 gives a necessary condition for two dessins d'enfants to be in the same orbit of the action of $\mathbb{G}$. Two new invariants will give other necessary conditions: the *extended valency function* and the *automorphism group*.

7.3.15   The Extended Valency Function

To each object H of $\mathcal{H}$ corresponds an object X of the ramified covering $\mathcal{RR}$. of Σ, such that $\Delta_X \subset \{0, 1, \infty\}$. The integer $\nu_H(0, k)$ (resp. $\nu_H(1, k)$) is the number of points of X of local degree k over 0 (resp. 1). Similarly, let $\nu_H(\infty, k)$ be the number of points of X of local degree k over ∞. This defines an *extended valency function* $\widehat{\nu_H} : \{0, 1, \infty\} \times \mathbb{N} \to \mathbb{N}$.

If $H' = g^*H$ with $g \in \mathbb{G}$, then $\widehat{\nu_{H'}} = \widehat{\nu_H}$. This follows from the algebraic definition of the local degree. Define the *strengthened numerical equivalence* $\sim_{\widehat{\nu}}$ by setting $H \sim_{\widehat{\nu}} H'$ if $\widehat{\nu_{H'}} = \widehat{\nu_H}$. This equivalence relation is stronger that $\sim_\nu$ but weaker than $\sim_\mathbb{G}$. As will be seen in the example of 7.3.14, it is neither stronger nor weaker than $\sim_{gb}$.

Let $H = (N, A, \varepsilon_0, \varepsilon_1, \tau, (\omega_s)_{s\in N})$ be an object of $\mathcal{H}$ and X the object corresponding to $\mathcal{RR}$. The number of edges is

$$a = \sum_{k\in\mathbb{N}} k.\nu_H(0, k) = \sum_{k\in\mathbb{N}} k.\nu_H(1, k).$$

This is also the degree of X as a ramified covering of Σ. The number f of points of X over ∞ equals the number of connected components of $X - |H| = X - \pi^{-1}([0, 1])$; it is also the number of cycles of the permutation $\tau \circ \sigma$ of A, and $\nu_H(\infty, k)$ is the number of $2k$-cycles.

The space X is connected if and only if this is the case of H. If H is connected, then X is a surface of genus g given by the Euler formula $2 - 2g = n - a + f$, where $n = n_0 + n_1$ is the number of vertices, n_0 the number of white vertices and n_1 the number of black ones.

In the examples of 7.3.14, $n_0 = 7$, $n_1 = 7$, $a = 14$, $f = 2$, $g = 0$. The extended valency function is given by the local degrees of the points of X over ∞; it varies with the example but the sum total is always 14:

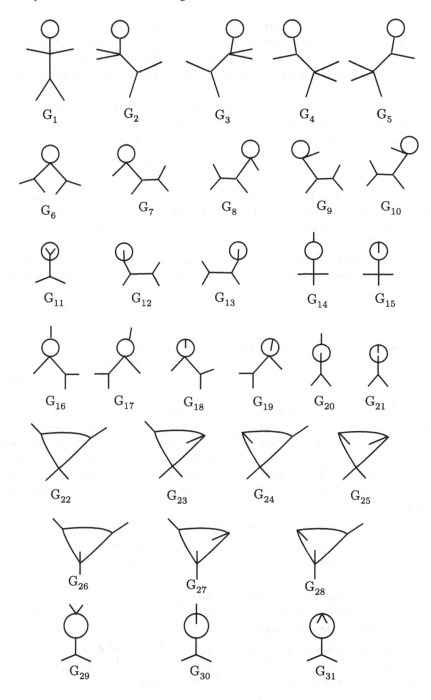

– in G_1 to G_{10}, there is a point of degree 1 and a point of degree 13;
– in G_{14}, G_{16}, G_{17} and G_{29}, there is a point of degree 2 and a point of degree 12;
– in G_{12}, G_{13} and G_{22}, there is a point of degree 3 and a point of degree 11;
– in G_{15}, G_{18}, G_{19}, G_{20} and G_{30}, there is a point of degree 4 and a point of degree 10;
– in G_{23}, G_{24} and G_{26}, there is a point of degree 5 and a point of degree 9;
– in G_{21} and G_{31}, there is a point of degree 6 and a point of degree 8;
– in G_{11}, G_{25}, G_{27} and G_{28}, there are two points of degree 7.

The class of $\sim_\nu$ considered decomposes into 7 classes of $\sim_{\widehat{\nu}}$. There are at least 7 classes of $\sim_{\mathbb{G}}$.

The only graphs having a non trivial automorphism (as object of $\mathcal{H}$) are G_{11}, G_{27} and G_{28}: for these three objets there is an automorphism of order 2 exchanging the two points over ∞. Hence G_{25} is not in the same class of $\sim_{\mathbb{G}}$ as these three, and the lower bound becomes 8.

All the graphs G_1 to G_{10} can be shown to be in the same class of $\sim_{\mathbb{G}}$. Complex conjugation induces an element of $\mathbb{G}$ exchanging G_{12} and G_{13}, G_{16} and G_{17}, G_{18} and G_{19}, G_{23} and G_{24}, G_{27} and G_{28}.

Hence, the number of classes of $\sim_{\mathbb{G}}$ in the given class of $\sim_\nu$ is bounded above by 17.

Exercise 7.3. (Equivalence between various categories)
1.—Let G be a group and $\mathscr{C}$ a category. We use the following language:

An *exact action* ρ of G on $\mathscr{C}$ consists of a functor $\rho(g) : \mathscr{C} \to \mathscr{C}$ for all $g \in G$ such that for $g_1, g_2 \in G$, $\rho(g_1.g_2) = \rho(g_1) \circ \rho(g_2)$.

A *weak action* ρ of G on $\mathscr{C}$ consists of of a functor $\rho(g) : \mathscr{C} \to \mathscr{C}$ for all $g \in G$ such that for $g_1, g_2 \in G$, the functors $\rho(g_1.g_2)$ and $\rho(g_1) \circ \rho(g_2)$ are isomorphic.

Two weak actions ρ and ρ' are *equivalent* if for all $g \in G$, the functors $\rho(g)$ and $\rho'(g)$ are isomorphic. A *strong action* is a weak action equivalent to an exact action.

(A) Suppose that $\mathscr{C}$ is a category with a single object ω such that $\mathrm{Hom}(\omega; \omega)$ is a group F. Set $H = \mathrm{Aut}(F)$, and let J be the subgroup of H consisting of the inner automorphisms of F.

(a) Show that J is a normal subgroup of H.

(b) Show that the exact actions of G on $\mathscr{C}$ correspond to homomorphisms $G \to H$. Show that the equivalence classes of weak actions correspond to homomorphisms $G \to H/J$.

(c) Give an example of a weak action which is not strong.
(B) Show that the actions of $\mathbb{G} = \mathrm{Aut}_{\mathbb{Q}}(\overline{\mathbb{Q}})$ on the categories described in Remark 2 of 7.3.7 are strong actions.

2.—Show that the composite functor $\breve{\theta} : \mathscr{R}\mathscr{R}. \to \breve{\mathscr{H}}$ of $\theta : \mathscr{R}\mathscr{R}. \to \mathscr{H}$ and of the functor $\mathscr{H} \to \breve{\mathscr{H}}$ defined in 7.3.10 assign to (S, f) the set $S^{-1}(\frac{1}{2})$ equipped with the permutations induced by the loops around 0 and 1 respectively.

7.4 Belyi Polynomials

7.4.1 Ramified Polynomial Coverings

Proposition and Definition *Let* (S, π) *be an analytic ramified covering of the Riemann sphere* Σ. *The following conditions are equivalent:*

(i) S is homeomorphic to the sphere S^2 *and the set* $\pi^{-1}(\infty)$ *is reduced to a point;*
(ii) there is an isomorphism $\phi : S \to \Sigma$ *and a polynomial* $f \in \mathbb{C}[X]$ *such that* $\pi = f \circ \phi$.

If these conditions hold, then (S, π) *is said to be a ramified* **polynomial** *covering.*

Proof The implication (ii) $\Rightarrow$ (i) is immediate. We show that (i) $\Rightarrow$ (ii). If the Riemann surface of S is homeomorphic to S^2, it is then isomorphic to Σ by the Riemann uniformization theorem. If moreover $\pi^{-1}(\infty)$ is reduced to a point e, then there is an isomorphism $\phi : S \to \Sigma$ such that $\phi(e) = \infty$. Then $f = \pi \circ \phi^{-1}$ is an analytic map $\Sigma \to \Sigma$ such that $f^{-1}(\infty) = \{\infty\}$, hence a polynomial. $\square$

Remark The isomorphism ϕ is uniquely defined up to left composition by an affine isomorphism of Σ, and the polynomial f up to right composition: if $\pi = f_1 \circ \phi_1 = f_2 \circ \phi_2$ as above, then there is an affine isomorphism $A : \Sigma \to \Sigma$ such that $f_2 = f_1 \circ A$ and $\phi_2 = A^{-1} \circ \phi_1$.

7.4.2

Let (S, π) be an analytic ramified covering of Σ such that $\Delta_\pi \subset \{0, 1, \infty\}$, H the graph $\pi^{-1}([0, 1])$, and set $S' = S - \pi^{-1}(\infty)$.

Proposition (S, π) *is a ramified polynomial covering if and only if the graph* H *is a tree (i.e. is simply connected).*

Lemma *The graph* H *is a deformation retract of* S'.

Proof Let $\overline{\mathbb{H}}$ (resp. $-\overline{\mathbb{H}}$) be the closed upper (resp. lower) half-plane in $\mathbb{C}$. Let $\rho : \mathbb{C} \to [0, 1]$ be a continuous retraction such that $\rho(\mathbb{R}_-) = \{0\}$ and $\rho([1, \infty[) = \{1\}$. Let $(t, z) \mapsto \rho^t(z)$ be a homotopy from $I_{\mathbb{C}}$ to ρ such that each ρ^t preserves $\mathbb{R}_-$ and $[0, \infty[$, and fixes each point of $[0, 1]$. Set $S^+ = \pi^{-1}(\overline{\mathbb{H}})$ and $S^- = \pi^{-1}(-\overline{\mathbb{H}})$. These may be written

$$S^+ = \bigcup_{i \in I^+} \sigma_i(\overline{\mathbb{H}}), \quad S^- = \bigcup_{i \in I^-} \sigma_i(-\overline{\mathbb{H}}),$$

where σ_i are the continuous sections. Set $\rho_i^t = \sigma_i \circ \rho^t \circ \pi$ onto the image of σ_i. Then, for each t, the maps ρ_i^t can be glued together to give a map $\tilde{\rho}^t : S' \to S'$, and this defines a homotopy from $I_{S'}$ to a retraction $\tilde{\rho}$ from S' onto H. $\square$

Proof of the Proposition If (S, π) is polynomial, then S' is isomorphic to $\mathbb{C}$, hence
is simply connected, and H is a tree. If H is a tree, then S' is simply connected, hence
isomorphic to $\mathbb{D}$, $\mathbb{C}$ or Σ by the Riemann uniformization theorem. The space S' is
a nonempty ramified covering of $\mathbb{C}$; so $\pi : S' \to \mathbb{C}$ is surjective, S' is not compact,
and Σ is excluded. Every bounded holomorphic function on S' is constant. Indeed,
if ζ is such a function, then the coefficients of the minimal polynomial of ζ over
$\mathbb{C}[X]$ are bounded holomorphic functions on $\mathbb{C}$, hence constant, and ζ can only take
finitely many values; as S' is connected, ζ is constant. Hence, $\mathbb{D}$ is excluded, and S'
is isomorphic to $\mathbb{C}$.

The set $S - S' = \pi^{-1}(\infty)$ has exactly one point. Indeed, there is an open neighbourhood U of ∞ in Σ such that each connected component of $\pi^{-1}(U)$ contains *a*
point of $\pi^{-1}(\infty)$ and a connected component of $\pi^{-1}(U) \cap S'$. Hence there is a bijection between $\pi^{-1}(\infty)$ and the set of relatively compact connected components of
$S' \cap \pi^{-1}(U) = S' - \pi^{-1}(\mathbb{C} - U)$ (see 6.1, Exercise 2, end compactification). Now,
$\mathbb{C} - U$ is compact; so is $\pi^{-1}(\mathbb{C} - U)$ since π is proper, and the complement of a
compact subset of $\mathbb{C}$ has exactly one relatively compact connected component; so
does S'. Hence Card $\pi^{-1}(\infty) = 1$.

Therefore, S is isomorphic to Σ. $\square$

7.4.3 Belyi Polynomials

If $\Delta(f) \subset \{0, 1, \infty\}$, $f \in \mathbb{C}[X]$ is called a **Belyi polynomial**. For any subfield K of
$\mathbb{C}$, let $\mathscr{Belpol}(K)$ denote the set of Belyi polynomials with coefficients in K, and
$\mathscr{Aff}(K)$ the group of affine transformations $z \mapsto az + b$ with $(a, b) \in K^* \times K$. It
acts on the right on $\mathscr{Belpol}(K)$ by $(A, f) \mapsto f \circ A$.

Proposition *If* $K \subset \mathbb{C}$ *is an algebraically closed subfield, then the set of isomorphism
classes of ramified polynomial coverings* (S, π) *of* Σ *such that* $\Delta_\pi \subset \{0, 1, \infty\}$ *can
be identified with the quotient* $\mathscr{Belpol}(K)/\mathscr{Aff}(K)$.

Proof Let (S, π) be a ramified polynomial d-fold covering of Σ such that $\Delta_\pi \subset
\{0, 1, \infty\}$. It is definable over $\overline{\mathbb{Q}}$ (7.2.5), and let $Q \in \overline{\mathbb{Q}}[X, Y]$ be a defining polyno-
mial. As (S, π) is polynomial, there is an isomorphism $\phi : S \to \Sigma$ mapping $\pi^{-1}(\infty)$
onto ∞. The inverse Ψ de ϕ induces a map $\psi : \Sigma' \to V_Q$, where Σ' is the complement of a finite set. The map ψ is of the form (f, g), where $f \in \mathbb{C}[T]$ is a Belyi
polynomial and $g \in \mathbb{C}(T)$.

The isomorphism ϕ is not uniquely defined, but its choice determines ψ, f and g.

Lemma ϕ *may be chosen so that* $f \in \overline{\mathbb{Q}}(T)$.

Proof Let (x_0, y_0) and (x_1, y_1) be distinct points of V_Q such that x_0 and x_1 are in
$\overline{\mathbb{Q}} - \Delta_Q$. Then $y_0, y_1 \in \overline{\mathbb{Q}}$ since $Q(x_i, y_i) = 0$. Choose ϕ so that $\psi(0) = (x_0, y_0)$
and $\psi(1) = (x_1, y_1)$. Then $f \in \mathscr{Belpol}(\mathbb{C})$ and $g \in \mathbb{C}(T)$ are characterized by
$Q(f(T), g(T)) = 0$ and $\deg(f) = d$. These are algebraic conditions defined on $\overline{\mathbb{Q}}$.

Hence the coefficients of f are left invariant by all algebraic automorphisms from $\mathbb{C}$ onto $\overline{\mathbb{Q}}$; hence they are in $\overline{\mathbb{Q}}$. □

End of the Proof of the Proposition We have shown that there exist $f_0 \in \mathscr{Belpol}(\overline{\mathbb{Q}}) \subset \mathscr{Belpol}(K)$ and an isomorphism $\phi_0 : S \to \Sigma$ such that $f_0 \circ \phi_0 = \pi$. If $f \in \mathscr{Belpol}(K)$ and $\phi : S \xrightarrow{\approx} \Sigma$ satisfy $f \circ \phi = \pi$, then there exists A of the form $t \mapsto at + b$ such that $\phi_0 = A \circ \phi$. So $\pi = f_0 \circ A \circ \phi = f \circ \phi$, and hence $f = f_0 \circ A$ since ϕ is surjective.

Comparing the leading coefficients gives $a^d \in K$; thus $a \in K$ since K is algebraically closed. Comparing the coefficients of the terms of degree $d - 1$ shows that $b \in K$. □

7.4.4 Statement of the Faithfulness Theorem

By 7.4.3, proposition and 7.3.13, the set of isomorphism classes of ramified polynomial coverings (S, π) of Σ such that $\Delta_\pi \subset \{0, 1, \infty\}$ can be naturally identified with the following sets:

- the quotient $\mathscr{Belpol}(\mathbb{C})/\mathscr{Aff}(\mathbb{C})$;
- the quotient $\mathscr{Belpol}(\overline{\mathbb{Q}})/\mathscr{Aff}(\overline{\mathbb{Q}})$;
- the set $\mathscr{T}$ of isomorphism classes of bicoloured trees cyclically ordered at each vertex.

The Galois group $\mathrm{Aut}_{\mathbb{Q}}(\overline{\mathbb{Q}})$ acts on $\overline{\mathbb{Q}}[X]$ by $(g, f) \mapsto g_*(f)$ (action on the coefficients with $g_*(X) = X$). The set $\mathscr{Belpol}(\overline{\mathbb{Q}})$ is invariant; so is $\mathscr{Aff}(\overline{\mathbb{Q}})$; Hence $\mathrm{Aut}_{\mathbb{Q}}(\overline{\mathbb{Q}})$ acts on the quotient $\mathscr{Belpol}(\overline{\mathbb{Q}})/\mathscr{Aff}(\overline{\mathbb{Q}})$ and on $\mathscr{T}$.

The following theorem states that this action is faithful.

Theorem (Lenstra–Schneps) *For all nontrivial $g \in \mathrm{Aut}_{\mathbb{Q}}(\overline{\mathbb{Q}})$, there is a Belyi polynomial $f \in \mathscr{Belpol}(\overline{\mathbb{Q}})$ such that $g_*(f)$ cannot be written as $f \circ A$, where $A \in \mathscr{Aff}(\overline{\mathbb{Q}})$.*

The proof will be given in (7.4.6). It is taken from [3] and rests on a uniqueness property up to action of $\mathscr{Aff}$ of the factorization of polynomials by composition. It will be given in (7.4.5).

7.4.5 Comparing Factorizations

Proposition *Let K be a subfield of $\mathbb{C}$, f_1, f_2, g_1 and g_2 polynomials of degrees > 0 such that $g_1 \circ f_1 = g_2 \circ f_2$. Suppose that f_1 and f_2 have the same degree d with coefficients in K, and that g_1 and g_2 have the same degree d'. Then there is an automorphism $A \in \mathscr{Aff}(K)$ such that $f_2 = A \circ f_1$ and $g_2 = g_1 \circ A^{-1}$.*

Remark Set $h = g_1 \circ f_1 = g_2 \circ f_2$. As $\deg(h) = \deg(f_1) \cdot \deg(g_1) = \deg(f_2) \cdot \deg(g_2)$, each of the assumptions $\deg(f_1) = \deg(f_2)$ and $\deg(g_1) = \deg(g_2)$ implies the other.

For any polynomial $f \in \mathbb{C}[X]$, define the equivalence relation $\sim_f$ on $\mathbb{C}$ by $x \sim_f y \Leftrightarrow f(x) = f(y)$. We will say it is defined by f.

Lemma *Under the assumptions of the proposition, f_1 and f_2 define the same relation on $\mathbb{C}$.*

Proof Set $h = g_1 \circ f_1 = g_2 \circ f_2$; let L_1, L_2 and M be the graphs of the equivalence relations defined respectively by f_1, f_2 and h: $M = \{(x, y) \in \mathbb{C}^2 \mid h(x) = h(y)\}$ and $L_j = \{(x, y) \in \mathbb{C}^2 \mid f_j(x) = f_j(y)\}$. Then $L_j \subset M$.

Let π be the projection $(x, y) \mapsto x$ and Δ_M the set of $x \in \mathbb{C}$ such that the polynomial $h(Y) - h(x) \in \mathbb{C}[Y]$ has a multiple root. As $\Delta_M = h^{-1}(\Delta_h)$, where Δ_h is the set of critical values of h, the set Δ_M is finite. The space M equipped with π induces a (non ramified) covering M' of $\mathbb{C} - \Delta_M$ of degree $d \cdot d'$. Defining Δ_{L_j} likewise, $\Delta_{L_j} \subset \Delta_M$. Set $L_j' = L_j - \pi^{-1}(\Delta_M)$. The spaces L_1' and L_2' are subcoverings of M' over $\mathbb{C} - \Delta_M$.

In the neighbourhood of infinity, the covering M' is trivial: for sufficiently large R, there are $d \cdot d'$ analytic sections $\tau_k : \mathbb{C} - \mathbb{D}_R \to M'$, for $0 \leqslant k \leqslant d \cdot d' - 1$, such that the restriction of M' to $\mathbb{C} - \mathbb{D}_R$ is the union of their images, $\frac{\tau_k(x)}{x}$ tending to $e^{2i\pi \frac{k}{d \cdot d'}}$ as x tends to ∞. This follows from the fact that the part of highest degree in $h(Y) - h(X)$ is of the form $c \cdot (Y^{dd'} - X^{dd'})$.

Similarly the restriction of L_j' to $\mathbb{C} - \mathbb{D}_R$ is the union of the images of d sections σ_k^j for $1 \leqslant k \leqslant d - 1$, with $\frac{\sigma_k^j(x)}{x}$ tending to $e^{2i\pi \frac{k}{d}}$ as x tends to ∞. However L_j' is a subcovering of M'. This forces $\sigma_k^1 = \sigma_k^2 = \tau_{k.d'}$. So L_1' and L_2' agree over $\mathbb{C} - \mathbb{D}_R$. As these are subcoverings of M' over $\mathbb{C} - \Delta_M$, the latter being connected, $L_1' = L_2'$.

By (7.1.1, Proposition), L_j is the closure of L_j'. Thus $L_1 = L_2$. $\qquad\square$

Proof of the Proposition A non constant polynomial $f : \mathbb{C} \to \mathbb{C}$ is an open map, and so defines a homeomorphism $f_* : \mathbb{C}/\sim_f \to \mathbb{C}$. The maps $(f_1)_*$ and $(f_2)_*$ are homeomorphisms of $\mathbb{C}/\sim_{f_1} = \mathbb{C}/\sim_{f_2}$ over $\mathbb{C}$, and $A = (f_2)_* \circ (f_1)_*^{-1}$ is a homeomorphism $\mathbb{C} \to \mathbb{C}$ satisfying $f_2 = A \circ f_1$.

In the neighbourhood of a point z which is not a critical value of f_1, $A = f_2 \circ \psi$, where ψ is a branch of f_1^{-1}. Hence A is holomorphic. The map A, being continuous and holomorphic on the complement of a finite set, is holomorphic. Similarly for A^{-1}, and so A is a $\mathbb{C}$-analytic automorphism of $\mathbb{C}$; therefore $A \in \mathscr{Aff}(\mathbb{C})$.

Setting $A(z) = a.z + b$, $f_2 = a \cdot f_1 + b$. Comparing the leading terms, $a \in K$, and $b = f_2 - a \cdot f_1 \in K$. $\qquad\square$

7.4.6 Proof of the Faithfulness Theorem

Let $\gamma \in \mathrm{Aut}_{\mathbb{Q}}(\overline{\mathbb{Q}})$ be a nontrivial element. Then there exists $\alpha_1 \in \overline{\mathbb{Q}}$ such that $\alpha_2 = \gamma(\alpha_1) \neq \alpha_1$.

Let $f_1 \in \overline{\mathbb{Q}}[X]$ be a polynomial whose derivative f_1' has 0, 1 and α_1 as roots with distinct multiplicities m_0, m_1, m_α. The polynomial $f_2 = \gamma_*(f_1)$ has a derivative f_2' whose roots are 0, 1 and α_2 with the same multiplicities m_0, m_1, m_α.

Now, $\Delta(f_1) \subset \overline{\mathbb{Q}} \cup \{\infty\}$. By (7.2.8, Corollary), there exists $g \in \mathbb{Q}[X]$ such that $\Delta(g \circ f_1) = g_!(\Delta(f_1)) \subset \{0, 1, \infty\}$. Then $h_1 = g \circ f_1 \in \mathcal{B}\text{\it{el}}\text{\it{pol}}(\overline{\mathbb{Q}})$. So $\Delta(f_2) = \gamma(\Delta(f_1))$, and $h_2 = g \circ f_2 \in \mathcal{B}\text{\it{el}}\text{\it{pol}}(\overline{\mathbb{Q}})$.

The faithfulness theorem then follows from the next lemma:

Lemma *The polynomial h_2 cannot be written as $h_1 \circ A$ with $A \in \mathcal{A}\text{\it{ff}}(\mathbb{C})$.*

Proof Suppose that $h_2 = h_1 \circ A$ with $A \in \mathcal{A}\text{\it{ff}}(\mathbb{C})$, i.e. $g \circ f_2 = g \circ f_1 \circ A$. By (7.4.5, Proposition), there exists $B \in \mathcal{A}\text{\it{ff}}(\mathbb{C})$ such that $f_2 = B \circ f_1 \circ A$ and $g = g \circ B^{-1}$.

Write $A(z) = a \cdot z + a_0$ and $B(z) = b \cdot z + b_0$. Then $f_2 = B \circ f_1 \circ A$ has derivative $f_2' = ab(f_1' \circ A)$. The roots of f_1' are the images under A of the roots of f_2' with their multiplicities preserved. Hence, $A(0) = 0$, $A(1) = 1$, $A(\alpha_2) = \alpha_1$. The first two equalities imply that A is the identity, and then $\alpha_1 = \alpha_2$ by the last one, giving a contradiction. □

This completes the Proof of Theorem 7.4.4.

7.5 Two Examples

In this section we give two examples:

   • an example of two Belyi polynomials giving non-isomorphic trees, regardless of the colouring and of the cyclic order at vertices, but which are conjugate under the action of the group $\mathbb{G} = \mathrm{Aut}_{\mathbb{Q}}(\overline{\mathbb{Q}})$;

   • an example, taken from Zapponi [4], of two Belyi polynomials giving isomorphic bicoloured trees with distinct cyclic order at each vertex, and which are not conjugate under the action of $\mathbb{G}$.

7.5.1 First Example: Setting of the Problem

Consider the following bicoloured trees

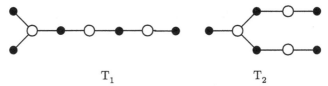

$$T_1 \qquad\qquad\qquad\qquad T_2$$

They have the same valency function: they both have two white vertices of valency 2, a white vertex of valency 3, three black vertices of valency 1 and two black vertices of valency 2. Up to isomorphism, these are the only bicoloured trees with this valency function, and this remains true if the cyclic order at each vertex is taken account of.

Are they in the same orbit under the action of $\mathbb{G} = \mathrm{Aut}_{\mathbb{Q}}(\overline{\mathbb{Q}})$? In other words, is there some $g \in \mathbb{G}$ such that $g^*(T_1) = T_2$? Proposition 7.5.4 will provide the answer.

7.5.2 The Polynomial Q_c

The bicoloured trees T_1 and T_2 correspond to the Belyi polynomials P_1 and P_2 of degree 7, with one triple root and two double ones, unique up to pre-composition with an affine map. Thus the origin can be taken to be the triple root, and then such a polynomial can be written as $P = az^3(z^2 - 2bz + c)^2$.

For P_1 and P_2, $b \neq 0$. Otherwise P would be an odd polynomial and 1 would be a critical value only if it so was -1. Hence we may assume that $b = 1$ since the action of a homothety on the variable can be expressed as $b \mapsto \lambda b$, $c \mapsto \lambda^2 c$.

Let Q_c be the polynomial $z^3(z^2 - 2z + c)^2$. Let us find the values of c for which Q_c has two nonzero equal critical values, given by two distinct critical points. For such a value v of c, $P = \frac{1}{v} Q_c$ with equal to one of P_1, P_2; and P_1 and P_2 can be obtained in this manner.

7.5.3 Study of $Q = Q_c$

Derivative. $Q'(z) = z^2(z^2 - 2z + c)G(z)$, where $G(z) = 7z^2 - 10z + 3c$.

Critical points (other than the roots). They are given by the equation $G(\alpha) = 0$. The reduced discriminant is $\Delta = 25 - 21c$. Set $\delta = \sqrt{\Delta}$. Then $c = \frac{1}{21}(25 - \delta^2)$; the critical points other than the roots are $\alpha = \frac{1}{7}(5 + \delta)$ and $\alpha' = \frac{1}{7}(5 - \delta)$.

We express the critical value N corresponding to α in terms of δ, that corresponding to α' will follow by replacing δ by $-\delta$:
$$\alpha^2 = \tfrac{1}{49}(25 + 10\delta + \delta^2);$$
$$\alpha^3 = \tfrac{1}{343}(125 + 75\delta + 15\delta^2 + \delta^3);$$
$$\alpha^2 - 2\alpha + c = \tfrac{4}{147}(10 - 3\delta - \delta^2);$$
$$(\alpha^2 - 2\alpha + c)^2 = \tfrac{16}{9 \cdot 7^4}(100 - 60\delta - 11\delta^2 + 6\delta^3 + \delta^4).$$
Non-zero critical values.

$$Q(\alpha) = \alpha^3(\alpha^2 - 2\alpha + c)^2$$
$$= \tfrac{16}{9.7^7}(12500 - 4375\delta^2 - 875\delta^3 + 350\delta^4 - 154\delta^5 + 21\delta^6 + \delta^7).$$

Difference of these critical values.

$$Q(\alpha) - Q(\alpha') = \tfrac{32}{9.7^7}\delta^3(-875 - 154\Delta + \Delta^2).$$

The values of c for which Q_c has two nonzero equal critical values correspond to the values of δ for which this expression vanishes.

7.5.4 Interpretation of These Results

The value $\delta = 0$ gives $c = \tfrac{25}{21}$. For this value, Q_c has a double critical point (hence of local degree 3) with nonzero critical value. The Belyi polynomial $P = \tfrac{1}{12500}Q_c$ gives the tree

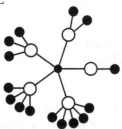

It is the only bicoloured tree with this valency function.

The polynomial $R(\Delta) = \Delta^2 - 154\Delta - 875$ has two real roots $\Delta^+ > 0$ and $\Delta^- < 0$. In both cases, $c = \tfrac{1}{21}(25 - \Delta) \in \mathbb{R}$, the coefficients of Q_c and P are real, and the tree realized in $\mathbb{C}$ is symmetric with respect to $\mathbb{R}$. The root Δ^+ gives two nonzero real values for δ, hence two distinct real values for the critical point α, i.e. two white points on the real axis; it corresponds to the tree T_1. The root Δ^- gives two distinct complex conjugate values for δ, hence also for α; it corresponds to the tree T_2.

The reduced discriminant of R is $77^2 + 875 = 6804 = 18^2 \cdot 21$. It follows that

Proposition *An element $g \in G = \mathrm{Aut}_{\mathbb{Q}}(\overline{\mathbb{Q}})$ exchanges the trees T_1 and T_2 if and only if it exchanges the two square roots of 21.*

7.5.5 Second Example: "Leila's bouquet"

Consider the bicoloured tree L

with 1 black pentavalent vertex O, 15 monovalent black vertices, and 5 white vertices $S_1, \ldots, S_5$, S_i being of valency $d_i = i + 1$. Up to isomorphism, L is the only bicoloured tree with this valency function. Let a_i denote the edge from O to S_i. For any cyclic order ω on $\{1, ..., 5\}$, let L_ω be an objet of $\mathcal{H}$ obtained by equipping L with the order ω on the edges a_i (the cyclic order at white vertices is unimportant). This gives, up to isomorphism, all the objects of $\mathcal{H}$ with this valency function. A cyclic order will be defined by a bijection $\varphi : \{1, ..., 5\} \to \{1, ..., 5\}$.

Are all the objects L_ω in the same orbit under the action of $\mathbb{G} = \mathrm{Aut}_{\mathbb{Q}}(\overline{\mathbb{Q}})$? The corollary of the above Theorem 7.5.8 shows that this is not the case. The Proof of Theorem 7.5.8 will be completed in 7.5.16; to state this theorem, we need to define two rationals $\delta(P) = \delta_{alg}(P)$ and $\delta_{top}(P)$, and these definitions rest on the calculation given in 7.5.6.

The figure below shows two examples of objects L_ω.

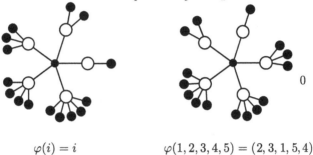

$$\varphi(i) = i \qquad\qquad\qquad \varphi(1, 2, 3, 4, 5) = (2, 3, 1, 5, 4)$$

7.5.6 Computing Discriminants

Let P be a degree 20 Belyi polynomial such that $P^{-1}([0, 1])$ is a bicoloured tree isomorphic to L, in other words such that $\theta(P)$ is an object of $\mathcal{H}$ isomorphic to one of L_ω. For each ω, there is a unique such polynomial P up to pre-composition with an affine transformation. Hence P may be chosen to be monic with the black pentavalent vertex O at 0.

Letting x_i be the root of P corresponding to the white vertex S_i,

$$P(Z) = \prod_{i=1}^{5}(Z - x_i)^{d_i} \quad \text{where} \quad d_i = i + 1 .$$

Here we use the "*algebraic*" numbering of x_i given by their multiplicity as roots. This numbering provides a cyclic order which may be different from the "*topologi-*

cal" cyclic order induced by the extension of the tree in ℤ ℂ.

Let Q be the monic polynomial having x_i as simple roots:

$$Q(Z) = \prod_{i=1}^{5} (Z - x_i).$$

Let Δ be the discriminant of Q:

$$\Delta = \prod_{i,j \mid i \neq j} (x_i - x_j) = \prod_{i,j \mid i < j} (x_i - x_j)^2$$

(note that there are 10 pairs (i, j) such that $i < j$ and that 10 is even. This is why the latter equality has a $+$ sign).

Proposition

$$\Delta = \left(\frac{200}{3}\right)^2 \cdot \prod_i x_i^4.$$

Lemma

$$\frac{P'(Z)}{P(Z)} = \sum_i \frac{d_i}{Z - x_i} = 20\frac{Z^4}{Q(Z)}.$$

Proof (a) As $P(Z) = \prod(Z - x_i)^{d_i}$, $\frac{P'}{P} = \log'(P) = \sum \frac{d_i}{Z - x_i}$.

(b) The only zero of the rational fraction $\frac{P'}{P}$ in $\mathbb{C}$ is one of multiplicity 4 at the point 0, and the only pole is a simple at each x_i; hence $\frac{P'}{P}$ is of the form $k \cdot \frac{Z^4}{Q(Z)}$. As Z tends to infinity, $\frac{P'}{P}$ is equivalent to $\frac{20}{Z}$, and soù $k = 20$. $\square$

Proof of the Proposition

$$20Z^4 = Q \cdot \frac{P'}{P} = \sum_i d_i \cdot \frac{Q}{Z - x_i} = \sum_i d_i \prod_{j \mid j \neq i} (Z - x_j).$$

All terms of the sum vanish at $Z = x_{i_0}$ except the one corresponding to $i = i_0$; so

$$20x_i^4 = d_i \prod_{j \mid j \neq i} (x_i - x_j).$$

Taking the product over the different values of i gives:

$$20^5 \cdot \prod_i x_i^4 = \left(\prod_i d_i\right) \cdot \Delta.$$

Hence, as $\prod d_i = 6! = 720 = 2^4 \cdot 3^2 \cdot 5$,

$$\frac{\Delta}{\prod x_i^4} = \frac{2^{10}5^5}{2^43^25} = \frac{2^65^4}{3^2} = \left(\frac{2^35^2}{3}\right)^2 = \left(\frac{200}{3}\right)^2. \qquad \square$$

7.5.7 The Invariant $\delta_{a\ell g}$

If P is a Belyi polynomial satisfying the conditions of 7.5.6, define $\delta(P) = \delta_{a\ell g}(P)$ by

$$\prod_{i,j\,|\,i<j} (x_i - x_j) = \delta(P) \cdot \prod_i x_i^2 .$$

It follows from Proposition 7.5.6. that $\delta_{a\ell g}(P) = \pm\frac{200}{3}$. In particular $\delta_{a\ell g}(P)$ is rational.

Proposition *The rational $\delta_{a\ell g}(P)$ only depends on the orbit of the isomorphism class of the object $\theta(P)$ of $\mathcal{H}$ under the action of $\mathbb{G} = \operatorname{Aut}_{\mathbb{Q}}(\overline{\mathbb{Q}})$.*

Proof Let P_1 and P_2 be two polynomials satisfying the conditions of (7.5.6). Let $x_i^{(1)}$ and $x_i^{(2)}$ be the respective roots of P_1 and P_2 of multiplicity $i + 1$.

(a) If the objects $\theta(P_1)$ and $\theta(P_2)$ of $\mathcal{H}$ are isomorphic, then there exists λ such that $P_2(Z) = P_1(\lambda Z)$. So $x_i^{(1)} = \lambda x_i^{(2)}$, and hence

$$\prod_{i<j}(x_i^{(1)} - x_j^{(2)}) = \lambda^{10} \prod_{i<j}(x_i^{(2)} - x_j^{(2)})$$
$$\prod(x_i^{(1)})^2 = \lambda^{10} \prod(x_i^{(2)})^2 .$$

Finally, $\delta_{a\ell g}(P_1) = \delta_{a\ell g}(P_2)$.

(b) If there is an automorphism $g \in \mathbb{G}$ such that $P_2 = g_*(P_1)$, then $x_i^{(2)} = g(x_i^{(1)})$. So $\delta(P_2) = g(\delta(P_1))$. As $\delta(P_1) \in \mathbb{Q}$, this implies $\delta_{a\ell g}(P_2) = \delta_{a\ell g}(P_1)$. □

7.5.8 Definition of $\delta_{top}(P)$

Under the assumptions of (7.5.6), for a permutation φ of $\{1, ..., 5\}$, define $\delta_\varphi(P)$ by

$$\prod_{i,j\,|\,\varphi(i)<\varphi(j)} (x_i - x_j) = \delta_\varphi(P) \cdot \prod_i x_i^2 .$$

Then $\delta_\varphi(P) = \varepsilon(\varphi) \cdot \delta(P)$, where $\varepsilon(\varphi)$ is the signature of φ. In particular, $\delta_\varphi(P)$ only depends on the cyclic order defined by φ on the set of x_i (note that 5 is odd!). We write $\delta_\xi(P) = \delta_\varphi(P)$ if ξ is the cyclic order defined by φ.

Set $\delta_{top}(P) = \delta_\omega(P)$, where ω is the cyclic order on all a_i such that $\theta(P)$ is isomorphic to L_ω, i.e. the cyclic order of the graph $H = P^{-1}([0, 1])$ at 0 induced by its inclusion in $\mathbb{C}$.

Theorem (Zapponi) *For any Belyi polynomial P satisfying the conditions of (7.5.6),*

$$\delta_{top}(P) = -\frac{200}{3}.$$

Corollary *Let ω_1 and ω_2 be two cyclic orders on $\{1, ..., 5\}$, and P_1 and P_2 the Belyi polynomials satisfying conditions of 7.5.6 and realizing L_{ω_1} and L_{ω_2} respectively. Let φ_1 and φ_2 be bijections from $\{1, ..., 5\}$ onto $\mathbb{Z}/(5)$ defining ω_1 and ω_2. Set $\sigma = \varphi_2 \circ \varphi_1^{-1}$ (so that σ transforms ω_1 into ω_2).*

If $\varepsilon(\sigma) = -1$, then $\delta_{alg}(P_2) = -\delta_{alg}(P_1)$, and there is no $g \in \mathbb{G}$ transforming P_1 into P_2.

The rest of this section is dedicated to the proof of the theorem. It will be completed in (7.5.16). If A is an affine transformation, then $\delta_{top}(P) = \delta_{top}(P \circ A)$. The focus will be on Q (equipped with a numbering of its roots) and not on P. We construct a connected space $\mathscr{B}^+$ such that, for all P, there is an affine A such that the polynomial Q corresponding to P $\circ$ A is in $\mathscr{B}^+$; we also construct $\theta_{top} : \mathscr{B}^+ \to \mathbb{R}^*$ such that $\delta_{top}(P) = \theta_{top}(Q)$. The function θ_{top}, being continuous, its sign remains invariant; this is why the sign of $\delta_{top}(P)$ does not depend on P.

If P is a Belyi polynomial satisfying the conditions of (7.5.6), then

$$\frac{1}{20}\frac{P'(Z)}{P(Z)} = \frac{Z^4}{Q(Z)} = \sum \frac{r_i}{Z - x_i}$$

with $r_i = \frac{1}{20} \cdot d_i$. We investigate the most general situation where there is a monic polynomial Q of degree 5, such that $f_Q(Z) = \frac{Z^4}{Q(Z)}$ can be written as $\sum \frac{r_i}{Z-x_i}$, with $r_i \in \mathbb{R}_+^*$.

7.5.9 Notation

Let $\mathscr{A}$ be the set of monic polynomials Q of degree 5 with 5 distinct roots, and $\widetilde{\mathscr{A}}$ the set of $\widetilde{Q} = (Q, x_1, ..., x_5)$ such that $Q \in \mathscr{A}$ and $Q^{-1}(0) = \{x_1, ..., x_5\}$. The sets $\mathscr{A}$ and $\widetilde{\mathscr{A}}$ are $\mathbb{C}$-analytic submanifolds of dimension 5 in $\mathbb{C}^6$ and $\mathbb{C}^{11}$ respectively, and the projection $\varpi : (Q, x_1, ..., x_5) \mapsto Q$ makes $\widetilde{\mathscr{A}}$ a covering of degree $5! = 120$ of $\mathscr{A}$ (there are 5! ways to number the roots).

For $Q \in \mathscr{A}$, set $f_Q(Z) = \frac{Z^4}{Q(Z)}$. The rational fraction f_Q has simple poles at the zeros x_i of Q, and it is of the form $\frac{1}{Z} + O(\frac{1}{Z^2})$ as $Z \to \infty$. Hence

$$f_Q(Z) = \sum \frac{r_i}{Z - x_i},$$

where r_i are given by

$$r_i = r_i(\widetilde{Q}) = \frac{x_i^4}{Q'(x_i)} = \frac{x_i^4}{\prod_{j|j \neq i}(x_i - x_j)};$$

They are nonzero and $\sum r_i = 1$. This defines a map

$$\rho : \tilde{Q} = (Q, x_1, ..., x_5) \mapsto (r_1, ..., r_5)$$

from $\tilde{\mathscr{A}}$ to the subset $\mathscr{W}$ of $\mathbb{C}^5$ consisting of $r = (r_i)$ such that $(\forall i) \; r_i \neq 0$ and $\sum r_i = 1$. The set $\mathscr{W}$ is a submanifold of dimension 4 in $\mathbb{C}^5$, and the map ρ is analytic.

Let $\mathscr{Z}$ be the set of $r = (r_i) \in \mathscr{W}$ such that there is a nonempty subset J of $\{1, ..., 5\}$ with $\sum_{i \in J} r_i = 0$, and $\mathscr{W}^* = \mathscr{W} - \mathscr{Z}$. Set $\tilde{\mathscr{A}}^* = \rho^{-1}(\tilde{\mathscr{W}}^*) \subset \tilde{\mathscr{A}}$, $\mathscr{A}^* = \varpi(\tilde{\mathscr{A}}^*)$, and $\mathscr{W}^+ = \{r = (r_i) \in \mathscr{W} \mid (\forall i) \; r_i \in \mathbb{R}_+^*\}$. Let $\mathscr{W}^+ \subset \mathscr{W}^*$. Set $\tilde{\mathscr{A}}^+ = \rho^{-1}(\mathscr{W}^+)$ and $\mathscr{A}^+ = \varpi(\tilde{\mathscr{A}}^+)$.

The polynomial Q is determined by the x_i. Multiplying the x_i by a same factor $\lambda \neq 0$, changes the polynomial Q to $\lambda^5 Q(\frac{z}{\lambda})$, the fraction f_Q to $\frac{1}{\lambda} f_Q(\frac{z}{\lambda})$, and leaves the r_i invariant.

Let $\mathscr{B}$ be the set of $Q \in \mathscr{A}$ such that $Q(0) = -1$, i.e. $\prod x_i = 1$. Set $\tilde{\mathscr{B}} = \varpi^{-1}(\mathscr{B})$, $\tilde{\mathscr{B}}^* = \tilde{\mathscr{A}}^* \cap \tilde{\mathscr{B}}$, $\tilde{\mathscr{B}}^+ = \tilde{\mathscr{A}}^+ \cap \tilde{\mathscr{B}}$, $\mathscr{B}^* = \mathscr{A}^* \cap \mathscr{B}$, $\mathscr{B}^+ = \mathscr{A}^+ \cap \mathscr{B}$.

The map $(\lambda, Q, (x_i)) \mapsto (\lambda^5 Q(\frac{z}{\lambda}), (\lambda x_i))$ makes $\mathbb{C}^* \times \tilde{\mathscr{B}}$ a 5-fold covering of $\tilde{\mathscr{A}}$.

7.5.10   A Covering Property

Proposition *The map ρ makes $\tilde{\mathscr{B}}^*$ into a finite covering of $\mathscr{W}^*$.*

Corollary *The map ρ makes $\tilde{\mathscr{B}}^+$ a trivial finite covering of $\mathscr{W}^+$.*

Note that $\mathscr{W}^+ = \{r \in \,]0, 1[^5 \mid \sum r_i = 1\}$ is contractible.

The proposition then follows by (4.3.12, Proposition) and by Lemmas (7.5.11) and (7.5.12) given below.

7.5.11   The Tangent Linear Map ρ

Lemma *For all $\tilde{Q} = (Q, x_1, ..., x_5) \in \tilde{\mathscr{B}}$, the tangent linear map $T_{\tilde{Q}}\rho$ induces an isomorphism $T_{\tilde{Q}}\tilde{\mathscr{B}} \to T_r\mathscr{W}$.*

Proof (a) *Differentiation of* $Q \mapsto f_Q$ *on* $\mathscr{A}$.

Let $Q \in \mathscr{A}$ and consider a family $(Q_\varepsilon)_{\varepsilon \in V}$, where $Q_\varepsilon = Q + \varepsilon H$, with H a polynomial of degree $\leqslant 4$, and V a neighbourhood of 0 in $\mathbb{C}$, which will be assumed to be connected and which we will allow ourselves to contract.

For $z \in \mathbb{C}$ such that $Q(z) \neq 0$, there is a finite expansion

$$f_{Q_\varepsilon}(z) = f_Q(z) - \frac{z^4 H(z)}{Q(z)^2}\varepsilon + O(\varepsilon^2) \, .$$

It holds uniformly as z runs through a compact subset of $\mathbb{C} - Q^{-1}(0)$.

(b) *Differentiation of* ρ *on* $\tilde{\mathscr{A}}$.

Let $\widetilde{Q} = (Q, x_1, ..., x_5) \in \widetilde{\mathscr{A}}$. By the implicit functions theorem, there are holomorphic functions ξ_i on V such that $\xi_i(0) = x_i$ and $Q_\varepsilon(\xi_i(\varepsilon)) = 0$. Set $\widetilde{Q}_\varepsilon = (Q_\varepsilon, \xi_1(\varepsilon), ..., \xi_5(\varepsilon))$. Since

$$\rho_i(\widetilde{Q}_\varepsilon) = \frac{1}{2i\pi} \int_{\gamma_i} Q_\varepsilon(z)dz ,$$

where γ_i is a loop only going round the roots x_i, and doing so only once. It follows that

$$\rho_i(\widetilde{Q}_\varepsilon) = \rho_i(Q_\varepsilon) - \text{Residue}_{x_i} \left(\frac{Z^4H}{Q^2} \right) \cdot \varepsilon + O(\varepsilon^2) .$$

Therefore the tangent linear map to ρ is

$$T_{\widetilde{Q}}\rho : H \mapsto \left(- \text{Residue}_{x_i} \left(\frac{Z^4H}{Q^2} \right) \right)_{i \in \{1, ..., 5\}}.$$

(c) *Injectivity of* $T_{\widetilde{Q}}\rho$ *on* $T_{\widetilde{Q}}\widetilde{\mathscr{B}}$.

The space $T_{\widetilde{Q}}\widetilde{\mathscr{B}} = T_Q\mathscr{B}$ consists of all $H \in \mathbb{C}[Z]$ of degree $\leqslant 4$ such that $H(0) = 0$. Let $H \in T_{\widetilde{Q}}\widetilde{\mathscr{B}}$ and suppose that

$$(\forall i) \quad \text{Residue}_{x_i} \left(\frac{Z^4H}{Q^2} \right) = 0 .$$

Then $\frac{Z^4H}{Q^2}$ is of the form $\sum \frac{a_i}{(Z-x_i)^2} = R'$, where $R = \sum \frac{-a_i}{Z-x_i}$.
The rational fraction $R_0 = R - R(0)$ is bounded at infinity and has simple poles only at x_i; hence it is of the form $\frac{G}{Q}$, where G is a polynomial of degree $\leqslant 5$. As $R(0) = 0$ and $R' = \frac{Z^4H}{Q^2}$ has a zero of order $\geqslant 4$ at 0, R_0 has a zero of order $\geqslant 5$ at 0. Therefore R_0 is of the form $\lambda \frac{Z^5}{Q(Z)}$, and

$$\frac{Z^4H}{Q^2} = R_0' = \lambda \frac{5Z^4Q - Z^5Q'}{Q^2} ;$$

so $H = \lambda(5Q - ZQ')$ and $H(0) = -5\lambda$. Hence as $H(0) = 0$, $\lambda = 0$ and $H = 0$.

Therefore the map $T_{\widetilde{Q}}\widetilde{\mathscr{B}} \to T_r\mathscr{W}$ induced by $T_{\widetilde{Q}}\rho$ is injective. The initial and end spaces having dimension 4, it is an isomorphism. □

7.5.12 A Properness Property

Lemma *The map* $\rho : \widetilde{\mathscr{B}}^* \to \mathscr{W}^*$ *is proper.*

Proof For $\tilde{Q} = (Q, x_1, ..., x_5) \in \tilde{\mathscr{A}}$ set $R(Q) = \sup(|x_i|)$ and $N(\tilde{Q}) = (N(Q), Y_1, ..., Y_5)$, where $Y_i = \frac{x_i}{R(Q)}$ and $N(Q) = \prod(Z - Y_i)$. Then $r_i = \text{Residue}_{x_i} f_Q = \text{Residue}_{Y_i} f_{N(Q)}$, where $f_{N(Q)} = \frac{Z^4}{N(Q)}$.

Let $(\tilde{Q}_n) = ((Q_n, x_1^{(n)}, ..., x_5^{(n)}))$ be a sequence in $\tilde{\mathscr{B}}$ such that $\rho(\tilde{Q}_n)$ tends to a point $r = (r_i) \in \mathscr{W}^*$. If necessary extracting a subsequence, $x_i^{(n)}$ may be supposed to converge to a limit $x_i \in \Sigma = \mathbb{C} \cup \{\infty\}$. We need to show that all x_i are in $\mathbb{C}^*$ and are distinct.

If all x_i are finite, then $\prod x_i = 1$, and so $x_i \in \mathbb{C}^*$ for all i, and

$$(\forall i) \quad \prod_{j \,|\, j \neq i} (x_i - x_j) = \frac{x_i^4}{r_i} \neq 0 \,;$$

hence the x_i are distinct.

Suppose that some x_i equals ∞. Then $R_n = R(Q_n) \to \infty$, and $Y_i^{(n)} = \frac{x_i^{(n)}}{R_n}$ satisfies $|Y_i^{(n)}| \leqslant 1$. If necessary extracting a subsequence, $Y_i^{(n)}$ may be supposed to converge to a limit $Y_i \in \mathbb{C}$. As $\prod Y_i = \lim R_n^{-5} = 0$, the set J of i such that $Y_i = 0$ is nonempty, and its cardinal k satisfies $0 < k < 5$.

The polynomial $N(Q_n)$ tends to $\prod(Z - Y_i) = Z^k \cdot \prod_{i \notin J}(Z - Y_i)$, and the rational fraction $f_{N(Q)}$ tends to

$$g = \frac{Z^{4-k}}{\prod_{i \notin J}(Z - Y_i)}$$

which has no pole at 0. Hence $\sum_{i \in J} r_i^{(n)} \to 0$, and $\sum_{i \in J} r_i = 0$, giving a contradiction. ☐

This completes the Proof of Proposition 7.5.10.

7.5.13 A Uniqueness Property

Proposition Let $\tilde{Q} = (Q, x_1, ..., x_5) \in \tilde{\mathscr{B}}$ and $r = (r_i) = \rho(\tilde{Q})$. If $r_i = \frac{1}{5}$ for all i, then $Q = Z^5 - 1$.

Proof The function $5 f_Q = \frac{5Z^4}{Q(Z)} = \sum \frac{1}{Z - x_i}$ is the logarithmic derivative of $Q = \prod(Z - x_i)$. So $Q' = 5Z^4$, and Q is of the form $Z^5 - c$. As $c = \prod x_i = 1$, where $Q(Z) = Z^5 - 1$,

Corollary The covering map $\rho : \tilde{\mathscr{B}}^* \to \mathscr{W}^*$ is of degree $5! = 120$.

7.5.14 Topological Numbering

For $Q \in \mathscr{B}^+$, there is a natural topological way to number the roots of Q.

First let $Q \in \mathscr{B}$ be arbitrary. On a sufficiently small connected neighbourhood V of 0, there is a unique holomorphic function F such that $F' = f_Q = \frac{Z^4}{Q}$ and $F(0) = 0$. As $f_Q(z) = -z^4 + O(z^5)$, $F(z) = -\frac{1}{5}z^5 + O(z^6)$. If necessary contracting V, we find a holomorphic map $\phi : V \to \mathbb{C}$ such that $F(z) = -\frac{1}{5}(\phi(z))^5$, $\phi(0) = 0$ and $\phi'(0) = 1$, and ϕ may be assumed to be an isomorphism from V onto a disc $\mathbb{D}_\ell$.

Now suppose that $Q \in \mathscr{B}^+$. Then there are $r_i > 0$ such that $f_Q(Z) = \sum \frac{r_i}{Z - x_i}$; set $h(z) = \sum r_i \log |z - x_i|$ and $U = \{z \mid h(z) < 0\}$. On V, $h = \Re(F)$.

The set U is an open relatively compact subset in $\mathbb{C}$ since as z tends to infinity, $h(z) \sim \log |z|$ tends to infinity. The open $V \cap U = \{z \mid \Re((\phi(z))^5 > 0\}$ is the union of 5 sectors $S_1, ..., S_5$, where

$$S_k = \left\{ z \in V - \{0\} \mid \left| \arg(\phi(z)) - \frac{2\pi k}{5} \right| < \frac{\pi}{10} \right\}$$

(here, for $\theta \in \mathbb{T} = \mathbb{R}/\mathbb{Z}$, set $|\theta| = \inf |t|$, where t is a representative of θ).

Proposition *(a) For $Q \in \mathscr{B}^+$, the open subset U has five connected components $(U_i)_{i \in \mathbb{Z}/(5)}$ with $U_i \cap V = S_i$;*

(b) Each connected component U_i contains a unique root x_i^{top} of Q.

(c) As Q varies in $\mathscr{B}^+$, for each $i \in \mathbb{Z}/(5)$ the root $x_i^{top}(Q)$ depends continuously on Q.

Proof (a) and (b) Let U_i be the connected component of U containing S_i. The U_i are distinct, for otherwise there would be a Jordan curve in $\overline{U}$ around some connected component of $V - \overline{U}$, contradicting the maximum principle.

Thanks to this principle (or rather to the minimum one), each connected component of U contains a root of Q; so U at most 5 connected components.

Hence, U has exactly 5 connected components, each meets V along a unique S_i and contains a unique root of Q.

(c) We make Q vary in $\mathscr{B}^+$; We will write $S_i(Q)$, $x_i^{top}(Q)$, etc. Let $Q_0 \in \mathscr{A}^+$ and $i \in \mathbb{Z}/(5)$. We choose a point $a \in S_i(Q_0)$ and a path γ from a to $x_i^{top}(Q_0)$ in U_{Q_0}. The implicit function theorem enables us to find a continuous function $Q \mapsto \xi(Q)$ defined in the neighbourhood of Q_0, such that $\xi(Q)$ is a root of Q and $\xi(Q_0) = x_i^{top}(Q_0)$, and a path γ_Q from a to $\xi(Q)$ in $\mathbb{C}$ continuously dependent on Q. For Q sufficiently near Q_0, $a \in S_i(Q)$, and the path γ_Q is contained in U_Q. Then $\xi(Q)$ is in the same connected component as $a \in S_i(Q)$; thus $\xi(Q) = x_i^{top}(Q)$, which implies that $x_i^{top}(Q)$ is continuously dependent on Q. $\qquad\square$

Corollary 7.3 *The space $\mathscr{B}^+$ is homeomorphic to $\mathscr{W}^+$. In particular, it is connected.*

Proof Let E be a connected component of $\mathscr{B}^+$. The map $Q \mapsto (Q, x_0^{top}(Q), ..., x_4^{top}(Q))$ is a continuous section $\mathscr{B}^+ \to \widetilde{\mathscr{B}}^+$; it induces a homeomorphism from E onto a connected component of $\widetilde{\mathscr{B}}^+$, and ρ induces a homeomorphism from this component onto $\mathscr{W}^+$ since the latter is connected. This gives a homeomorphism $\psi : E \to \mathscr{W}^+$.

In particular there exists $Q_1 \in \mathcal{B}^+$ such that $\psi(Q_1) = \rho(\widetilde{Q}_1) = (\frac{1}{5}, ..., \frac{1}{5})$ for some $\widetilde{Q}_1$ over Q_1. However this implies that $Q_1 = Z^5 - 1$. Hence $\mathcal{B}^+$ only has one connected component, namely $Z^5 - 1$. ☐

Corollary 7.4 *The space $\mathcal{A}^+$ is connected.*

Proof The map $(\lambda, Q) \mapsto \lambda^5 Q(\frac{Z}{\lambda})$ from $\mathbb{C}^* \times \mathcal{B}^+$ to $\mathcal{A}^+$ is continuous and surjective, since there is a lifting $(\lambda, \widetilde{Q}, (x_i)) \mapsto (\lambda^5 Q(\frac{Z}{\lambda}), (\lambda x_i))$, which is a covering. Hence $\mathcal{A}^+$ is connected. ☐

7.5.15 The Invariant $\theta_{top}(Q)$

Let $\widetilde{Q} = (Q, x_1, ..., x_5) \in \widetilde{\mathcal{A}}$ and $r = (r_i) = \rho(\widetilde{Q})$. As $f_Q(Z) = \frac{Z^4}{Q(Z)} = \sum \frac{r_i}{Z - x_i}$ and $r_i^{-1} = x_i^{-4} \prod_{j | j \neq i} (x_i - x_j)$, $\prod r_i^{-1} = \prod_{i, j | j \neq i} (x_i - x_j)$.
   Set

$$\theta(\widetilde{Q}) = \prod x_i^{-2} \cdot \prod_{i, j | i < j} (x_i - x_j).$$

The number of factors in this product is 10, which is even. Thus $(\theta(\widetilde{Q}))^2 = \prod r_i^{-1}$. If σ is a permutation of $\{1, ..., 5\}$, then

$$\theta(Q, x_{\sigma(1)}, ..., x_{\sigma(5)}) = \varepsilon(\sigma) \cdot \theta(Q, x_1, ..., x_5).$$

As 5 is odd, any cyclic permutation is even, and $\theta(\widetilde{Q})$ only depends on the chosen cyclic order on the roots of Q.
   So $\theta(\lambda^5 Q(\frac{Z}{\lambda}), \lambda x_1, ..., \lambda x_5) = \theta(Q, x_1, ..., x_5)$.
   If $\widetilde{Q} \in \widetilde{\mathcal{A}}^+$, then $(\theta(\widetilde{Q}))^2 = \prod r_i^{-1} \in \mathbb{R}_+^*$, and so $\theta(\widetilde{Q}) \in \mathbb{R}^*$.
   For $Q \in \mathcal{B}^+$, set $\theta_{top}(Q) = \theta(Q, (x_i^{top}(Q))_{i \in \mathbb{Z}/(5)})$. For $Q \in \mathcal{A}^+$, there exists λ such that $Q_1 = \lambda^5 Q(\frac{Z}{\lambda}) \in \mathcal{B}^+$. Then set $\theta_{top}(Q) = \theta_{top}(Q_1)$; there are 5 possible values for λ, but they all give the same cyclic order on the set of roots of Q, so that $\theta_{top}(Q)$ is well defined.

Proposition *For all $Q \in \mathcal{A}^+$, $\theta_{top}(Q) < 0$.*

Proof The map $\theta_{top} : \mathcal{A}^+ \to \mathbb{R}^*$ is continuous by (7.5.14, Proposition), (c); hence its sign remains constant. It suffices to check that $\theta_{top}(Z^5 - 1) < 0$.
   Setting $x_k = e^{2i\pi k/5}$, for $k = -2, ..., 2$,

$$\theta_{top}(Z^5 - 1) = ((x_{-2} - x_{-1})(x_1 - x_2)) \cdot ((x_{-2} - x_0)(x_0 - x_2)) \cdot ((x_{-1} - x_0)$$
$$(x_0 - x_1)) \cdot ((x_{-1} - x_2)(x_2 - x_1)) \cdot ((x_{-2} - x_2)(x_{-1} - x_1)).$$

In this product of 5 factors, the first 4 are < 0 as they are the opposite of the product of two conjugates, and the last one is a product of two purely imaginary numbers of the same sign. Hence $\theta_{top}(Z^5 - 1) < 0$. ☐

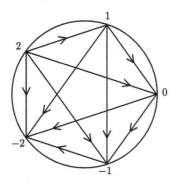

7.5.16   Relation Between $\delta_{top}(P)$ and $\theta_{top}(Q)$

Let P be a Belyi polynomial satisfying the conditions of (7.5.6). Then $Q(Z) = \prod(Z - x_i) \in \mathscr{A}$, with $r_i = \frac{d_i}{20}$. The function $h(Z) = \sum r_i \log |Z - x_i|$ equals $\frac{1}{20} \sum d_i \log |Z - x_i| = \frac{1}{20} \log |P(Z)|$. Hence the edges connected to 0 in the graph $P^{-1}([0, 1])$ are each contained in a connected component of the open subset $U = \{Z \mid h(Z) < 0\}$, and the cyclic order defined on the set of x_i by the topological numbering of (7.5.14) agrees with that defined by the branches of the graph $P^{-1}(0)$.

Therefore

$$\theta_{top}(Q) = \delta_{top}(P) .$$

By Proposition 7.5.15, $\delta_{top}(P) = \theta_{top}(Q) < 0$, proving Theorem 7.5.8.

References

1. R. Hartshorne, *Algebraic Geometry*, Graduate Texts in Mmathematics (Springer, Berlin, 1977)
2. D. Mumford, *Complex Projective Varieties* (Springer, Berlin, 1976)
3. L. Schneps, Dessins d'enfants, *The Grothendieck Theory of Dessins d'Enfants*, vol. 200, London Mathematical Society Lecture Notes Series (Cambridge University Press, Cambridge, 1994), pp. 25–46. (Luminy 1993)
4. L. Zapponi, *Dessins d'enfants et Action Galoisienne*, Ph.D. thesis. University of Franche-Comté (Besan on), février (1998)

Index of Notation

Classical notation

Card X:	cardinal of X.
A − B:	$A \cap \complement B$.
$\mathfrak{P}(X)$:	set of all subsets of X.
$x \sim_\mu y$:	x and y are equivalent with respect to relation μ.
$\exists!$:	there exists a unique
inf, sup:	infimum, supremum.
l.c.m.:	least common multiple.
g.c.d.:	greatest common divisor.
L(E, F):	space of linear maps from E to F.
ℓ^2:	space of square integrable sequences.
$\langle x, y \rangle$:	scalar product.
$\widehat{E}$:	completion of E.
$A[[X_1, ..., X_n]]$:	ring of formal series with coefficients in A.
$\mathbb{C}\{X_1, ..., X_n\}$:	convergent power series in the neighbourhood of O.
$\aleph_0$:	countably infinite cardinal.
$P^n\mathbb{R}, P^n\mathbb{C}$:	n-dimensional real projective space, resp. complex.
S^n:	unit sphere in $\mathbb{R}^{n+1}$.
X/G:	quotient of X by the equivalence relation whose classes are the G-orbits.
O_n (resp. U_n):	group of linear isometries of $\mathbb{R}^n$ (resp. $\mathbb{C}^n$).
$\mathfrak{S}(E)$:	permutation group of E.
$\mathfrak{A}(E)$:	alternating group.
SO_n (resp. SU_n):	subgroup of O_n (resp. U_n) of matrices with determinant $+1$.
$\mathfrak{P}(E)$:	set of all subsets of E.

© Springer Nature Switzerland AG 2020
R. Douady and A. Douady, *Algebra and Galois Theories*,
https://doi.org/10.1007/978-3-030-32796-5

Bibliography

1. G. Belyĭ, Galois extensions of a maximal cyclotomic field. Izv. Akad. Nauk SSSR, Ser. Mat. **43**, 267–276 (in Russian); English Translation. In: Math. USSR Izv. **14**(1979), 247–256 (1979)
2. H. Farkas, I. Kra, *Riemann Surfaces* (Springer, Berlin, 1980)
3. O. Forster, *Lectures on Riemann Surfaces* (Springer, Berlin, 1981), (corrected 2nd printing, 1991)
4. R. Godement, *Cours d'algèbre*, 2nd edn. (Hermann, Paris, 1997)
5. R. Godement, *Analyse mathématique*, vol 3. (Fonctions analytiques, différentielles et variétés, surfaces de Riemann) (Springer, Berlin, 2002)
6. A. Grothendieck, *Esquisse d'un programme, Grundlehren der mathematischen Wissenschaften 290* (Springer, Berlin, 1988)
7. J. Oesterlé, Dessins d'enfants, *Seminaire Bourbaki*, exposé 907, juin 2002
8. K. M. Pilgrim, Dessins d'enfants and Hubbard trees. Ann. Sci. École Norm. Sup. **33**(4), 671–693 (2000)
9. *Revêtements étales et groupe fondamental*, Séminaire de Géométrie Algébrique du Bois-Marie 1960/1961 (SGA1), dirigé par A. Grothendieck, Lecture Notes in Mathematics, vol. 224 (Springer, Berlin, 1970)
10. L. Zapponi, Fleurs, arbres et cellules. Compositio Mathematica **122**(2), 113–133 (2000)

© Springer Nature Switzerland AG 2020
R. Douady and A. Douady, *Algebra and Galois Theories*,
https://doi.org/10.1007/978-3-030-32796-5

Notation Defined in Text

© Springer Nature Switzerland AG 2020

R. Douady and A. Douady, *Algebra and Galois Theories*,

https://doi.org/10.1007/978-3-030-32796-5

Index

© Springer Nature Switzerland AG 2020
R. Douady and A. Douady, *Algebra and Galois Theories*,
https://doi.org/10.1007/978-3-030-32796-5

Printed in the United States
by Baker & Taylor Publisher Services